Surface Modification to Improve Properties of Materials

Surface Modification to Improve Properties of Materials

Special Issue Editor

Miran Mozetič

MDPI • Basel • Beijing • Wuhan • Barcelona • Belgrade

MDPI

Special Issue Editor
Miran Mozetič
Department of Surface Engineering and Optoelectronics
Jozef Stefan Institute
Slovenia

Editorial Office
MDPI
St. Alban-Anlage 66
4052 Basel, Switzerland

This is a reprint of articles from the Special Issue published online in the open access journal *Materials* (ISSN 1996-1944) from 2018 to 2019 (available at: https://www.mdpi.com/journal/materials/special_issues/Surface_Modification)

For citation purposes, cite each article independently as indicated on the article page online and as indicated below:

LastName, A.A.; LastName, B.B.; LastName, C.C. Article Title. *Journal Name* **Year**, *Article Number*, Page Range.

ISBN 978-3-03897-796-4 (Pbk)
ISBN 978-3-03897-797-1 (PDF)

Cover image courtesy of Miran Mozetič.

Contents

About the Special Issue Editor

Miran Mozetič, Professor. After completing his doctoral degree in Electronic Vacuum Technologies from the University of Maribor, Slovenia, professor Mozetič developed numerous technologies for surface modification to improve properties of materials. Currently, he serves as the Head of Department of Surface Engineering at Jozef Stefan Institute, Ljubljana, Slovenia. His group is involved in surface engineering of solid materials by non-equilibrium gaseous plasma treatments. His current research is focused on R&D niches where scientific knowledge on the generation of plasma with specific properties is beneficial to obtain the desired surface finish of solid materials. He has specialized in the generation of non-equilibrium plasma of extensive radiation in the hard UV range and cold plasma of extremely high dissociation fraction of gaseous molecules.

materials

MDPI

Editorial
Surface Modification to Improve Properties of Materials

Miran Mozetič

Department of Surface Engineering, Jozef Stefan Institute, Jamova cesta 39, 1000 Ljubljana, Slovenia; miran.mozetic@guest.arnes.si

Received: 29 January 2019; Accepted: 30 January 2019; Published: 31 January 2019

Abstract: Surface properties of modern materials are usually inadequate in terms of wettability, adhesion properties, biocompatibility etc., so they should be modified prior to application or any further processing such as coating with functional materials. Both the morphological properties and chemical structure/composition should be modified in order to obtain a desired surface finish. Various treatment procedures have been employed, and many are based on the application of non-equilibrium gaseous media, especially gaseous plasma. Although such treatments have been studied extensively in past decades and actually commercialized, the exact mechanisms of interaction between reactive gaseous species and solid materials is still inadequately understood. This special issue provides recent trends in nanostructuring and functionalization of solid materials with the goal of improving their functional properties.

Keywords: surface properties; nanostructuring; functionalization; grafting

---

1. Introduction

An important property of solid materials is surface morphology. The morphology governs the effective surface area which is always larger than the macroscopic geometrical area. Most techniques employed in mass production of materials, such as casting, injection molding, extrusion and rolling, enable limited abilities for increasing the surface area well above the geometric one. Solid materials are often prepared from liquids whose surface energy facilitates smooth surfaces. Even upon solidification, such a smooth surface is often preserved. In numerous applications, however, a solid material of a large surface area (much larger than the geometric one) is preferred. A rough surface often enables better adhesion of a coating and is essential in numerous applications. The surface area is usually increased well over the geometric one either by etching or deposition. Etching is usually performed using liquid or gaseous media. Etching by liquid media is often a fast process and thus beneficial in industrial applications, but it has drawbacks, such as inadequate morphology and ecological concerns due to the production of large quantities of waste chemicals. The dry gaseous etching is often much slower but is usually ecologically sustainable and provides almost arbitrary surface morphology. Etching of heterogeneous materials containing different components results in increased surface area due to preferential removal of one component. Homogeneous materials, however, assume rich morphology upon etching due to various mechanisms, including preferential removal of a material at grain boundaries, electrochemical effects, kinetic effects due to bombardment with ions, and self-assembly. The latter is particularly beneficial but limited to specific materials. A trivial method for increasing surface morphology in a highly controlled manner is the deposition of a mask followed by a suitable etching procedure. The etching using a mask is often anisotropic, in particular when the technology of reactive ion etching is applied. An alternative to etching is deposition of an arbitrary material. This technique enables almost arbitrary surface finish.

Apart from morphology, the functional properties of materials are governed by the composition and structure of the surface film. These are always different from the bulk counterparts because of the simple reason that unlike the bulk, where the atoms are surrounded by other atoms in all directions, the surface atoms are bonded to other atoms only from one side. The surface atoms therefore feel the attractive force of other atoms only from one side resulting in the surface energy. Furthermore, the gas-phase atoms or molecules feel the attractive force of surface atoms; therefore, most materials are covered with foreign molecules in practical cases. The surface of most metals and alloys, for example, are oxidized and covered with a thin film of organic molecules that come to the surface from the surrounding air. The composition of the surface is thus usually different from the composition of the bulk material. The ability for bonding foreign atoms and molecules is beneficial in cases where the surface functional properties are inadequate. For example, the surface properties of polymers are tailored rather arbitrarily by functionalization with specific functional groups. These functional groups allow for almost arbitrary wettability of the solid materials. A widely used technique for surface functionalization is a brief treatment with non-equilibrium gaseous plasma.

2. This Special Issue

Scientific papers selected for this issue deal with modification of surface properties of various materials by advanced techniques. Asadollahi et al. [1] describe a method for synthesis of stable super-hydrophobic surfaces through an economical and practical process by development of an organosilicon-based coating using an atmospheric-pressure plasma jet technique. Such a surface finish emerged from the investigation of natural surfaces with a high contact angle often known as the lotus leaf effect [2]. The superhydrophobic characteristics of the micro-nanostructured and wax-coated surface of the lotus leaf was first studied by Dettre and Johnson in 1964 [3]. Since then, numerous authors studied methods for creating such a surface finish. Asadollahi et al. [1] developed a technique using atmospheric pressure plasma polymerization of hexamethyl disiloxane in the jet of nitrogen plasma produced by a rotating arc discharge. The plasma jet was modified by mounting a quartz tube on the jet head, thus confining the plasma jet in a smaller volume, which was found to be beneficial for deposition of high-quality coatings.

Porous silicon has been found useful in a broad range of industries [4]. Numerous methods for synthesizing such materials have been reported, and the review paper by Lee et al. summarizes breakthroughs and recent trends [5]. Despite extensive work, there is still a great demand for further development of advanced surface modification methods for this material. Recently developed techniques include hydrolytic condensation, ring-opening click chemistry, and synthesis of calcium or magnesium silicates. The next-generation of surface modification methods forecasted in this review [5] will focus on reproducibility suitable for mass application in industry, improved bio-applicability and low toxicity, methods for preventing pore collapse, and multi-functional surface finish.

An assembly of organosilanes on porous silicon using visible light is reported by Rodriguez et al. [6]. The functionalization of semiconductor nanostructures with organic monolayers is regarded as essential for tailoring the surface chemistry for bioconjugation [7]. Porous silicon is often a matrix of silicon quantum dots immersed into amorphous network containing silica and silicon. It is different from classical nanodots in amorphous silicon networks because of extremely high surface area [8]. Rodriguez et al. employed a condensation process to synthesize organosilane functionalized porous silicon films using aminopropyl-triethoxy-silane and perfluorodecyl-triethoxy-silane at low concentrations. Visible light activation promoted surface oxidation of porous silicon, thus stimulating reaction with organosilanes. The process enabled a rather homogeneous surface with no traces of silane-derived colloidal structures, thus making the process useful as a model for the analysis of interfaces between organosilanes and porous silicon.

Aluminum alloys should be subjected to various postproduction treatments in order to assure required functional properties [9]. An optimal surface finish is particularly important for alloys subjected to severe weather conditions, such as for aircraft [10]. The surface modifications should inhibit oxidation, promote adhesion of further coatings, or reduce staining. Wet chemical etching enables production of porous textured surfaces of aluminum-alloys [11]. Not only an appropriate combination of chemicals but the order of their applications is crucial to prepare porous surfaces, irrespective from the original roughness.

An alternative method for improving surface properties of alloys is the ultrasonic nanocrystal surface modification process [12]. The technique was used for modification of nickel-rich super-alloy Inconel, which is used in aerospace and nuclear industries [13] due to its excellent properties, excepting fretting wear resistance [14]. A high surface compressive residual stress was obtained by ultrasonification together with increased roughness and hardness. Furthermore, the fretting wear resistance was improved, so the technique may be useful for extending the lifespan of aircraft and nuclear components [12]. The technique was also used for the treatment of tantalum of a rather high purity [15]. The treatment caused both an increase in hardness and induced compressive residual stress. The effect was particularly pronounced at high material temperature upon sonification. A plastically deformed layer with the refined nano-grains was observed and correlated to enhanced wear resistance [15].

Yet another technique suitable for surface modification of alloys is shot-peening. This technique affects a thin surface layer and causes tensile plastic strain, resulting in favorable compressive residual stresses. The technique was used for treatment of aluminum alloy Al 6061-T6 welds [16]. This material is also known for its superior properties in airspace industry. The shot-peening was performed by bombarding the surface with glass shot beads according to aerospace recommended levels. The surface morphology was modified upon the treatment, and lower hardness was obtained as compared to the base metal alloy in the heat-affected zone even after short shot-peening treatments [16].

A review of methods for surface texturing of the most commonly used titanium alloy (Ti_6Al_4V) was also published in this issue [17]. The techniques for achieving appropriate surface morphology of this material include mechanical, electrochemical, and localized heat treatments, in particular laser shock peening, electro spark surface texturing, electrical discharge machining, reactive ion etching, lithography, abrasive jet machining, and anodization. The surface texturing is beneficial in different applications, from self-lubrication of special tools [18] to vascular stents [19].

Surface properties of metals and alloys can be also tailored by the deposition of different coatings. Of particular importance are hard coatings, as demonstrated in Reference [20]. The wear performance increased significantly after depositing nanocomposite coatings with different structures. The films were deposited on aluminum die casting mold tool substrates using gaseous plasma sustained by a pulsed arc discharge. The best results in terms of hardness, soldering behavior, stress, and oxidation resistance were achieved for the $AlCrN/Si_3N_4$ nanocomposite coatings [20]. Such a coating of appropriate thickness enabled improved die-mold service life.

Cemented carbide has been used widely in engineering applications because of a high surface hardness, good thermal stability, outstanding chemical properties, and excellent wear resistance [21]. Calcium fluoride is a widely utilized solid lubricant at high temperatures. The friction coefficient of this lubricant decreases gradually with increasing temperature and exhibits an excellent lubricating effect even at 1000 °C [22]. The performances of microhole-textured carbide tools filled with CaF_2 was elaborated by Song et al. [23]. They found such materials suitable for promoting machining performance. The textured carbide improved the tribological performance compared to the non-textured material at high machining speed.

An appropriate surface finish is also important for the development of paper with specific functional properties. In many applications, the paper should exhibit antibacterial properties [24]; therefore, it should be coated with antibacterial nanoparticles. Such nanoparticles do not attach to the cellulose surface unless an appropriate plasma treatment is employed [25]. Schlemmer et al. [26] report a reliable, fast, and eco-friendly method to fabricate paper fines sheets impregnated with silver nanoparticles, which are known for their antimicrobial activity. The standard route for paper sheet formation was modified with an additional step using colloidal nanoparticles. The key observation reported in this paper is that a highly stable nanoparticle solution is essential for good dispersion of nanoparticles into a paper solution, and thus reasonably homogeneous distribution in the final product.

Surface properties of organic materials are also important in the development of alternative drugs, in particular for curing cancer diseases. Promising materials include lectins, which are important for cell communication and signaling in many physiologic as well as pathophysiologic processes [27]. Recently, it was reported that lectin interactions with tumor-specific glycan epitopes promote tumor growth and immune modulation [28]. A critical issue is adsorption of lectins onto suitable substrates, as reported by Niegelhell et al. [29]. They found the largest adsorbed amounts and the fastest adsorption kinetics on the surface of polystyrene. The adsorption kinetics of the examined lectins were found comparable to bovine serum albumin. The polysaccharide layers are also found to be prone to swelling.

The modification of polymer materials by non-equilibrium gaseous plasma has attracted enormous attention in the past decades due to industrial demands. Recent advances in the complex phenomena occurring upon interaction of reactive gaseous species with polymer surfaces have been summarized in an extensive review [30]. Particularly important are fluorinated polymers such as Teflon due to broad applications from kitchenware to medicine. Surface functionalization of this material is challenging, especially when exotic functional groups should be grafted to mimic biomaterials [31]. The paper by Lopez-Garcia et al. [32] reports morphological and structural modifications of Teflon surfaces upon treatment with non-equilibrium plasma sustained by different discharges. The treatment resulted in the modification of surface morphology, whereas functionalization with polar groups was found to be moderate even though a variety of treatment parameters were tested in this paper.

A broad application of polymers occurs in food packaging. The demand for packed food is increasing rapidly, but a major obstacle is the life span of fresh products. The polymer foils suitable for mass application in food packaging lack antimicrobial properties and are rather permeable by oxygen; therefore, they should be coated with suitable coatings. Due to the hydrophobic character of standard foils, the adhesion of any coating should be improved, and a natural choice is the application of gaseous plasma [33]. The plasma parameters, however, should be chosen carefully to prevent any damage to other foil properties which might result from VUV radiation from gaseous plasma [34]. An alternative technique which prevents such effects is the application of extremely short treatments in weak plasma where radiation is almost absent, but the concentration of neutral reactive species are still comparable to that in ordinary plasma. The efficiency of such an approach was demonstrated in paper [35].

Polymer composites are among the most widely used materials. Their properties depend on the type of polymer blends and fillers, as well as the distribution and even orientation of fillers in the polymer matrix [36]. The fillers tend to agglomerate in the polymer matrix, which is particularly important for two-dimensional carbon materials such as graphene [37]. Researchers worldwide are developing novel methods for the modification of fillers' surface properties in order to obtain the desired quality of products. Of particular important are biofillers extracted from natural sources. Shah et al. [38] elaborated the modification of egg shell particles using stearic acid and their reinforcement in the epoxy-polymer matrix. They report excellent toughness, elongation increase, and reduced brittleness of such composites.

Antimicrobial properties, biodegradability, and biocompatibility are important properties for polymer materials used in food packaging, medical, and pharmaceutical applications, as well as in cosmetics. Such properties of polymers could be achieved by appropriate immobilization of active coatings onto a polymer surface, but a serious obstacle is often poor adhesion due to the inadequate wettability, which should be modified [39]. Antimicrobial agents can be incorporated directly into polymers, or they are attached via the side chains. Polyvinyl alcohol (PVA) substrates were treated with plasma created by coplanar surface barrier atmospheric pressure discharge and cross-linked with glutaric acid. Such a pretreatment allowed for optimal nisin adhesion [40].

While polymer foils or plastic components are rather quickly activated using appropriate plasma parameters, the technology is more demanding in cases where small granules should be treated. This topic was addressed by Šourkova et al. [41]. The authors used a low pressure plasma reactor powered with a pulsed microwave discharge with a stirring devise inserted into the plasma reactor. Such an experimental configuration allowed for a low density of charged particles in gaseous plasma, but the density of neutral oxygen atoms was as high as 2×10^{21} m^{-3} in a large volume at a relatively low discharge power density. As a result, the polyethylene granules were effectively functionalized with polar groups without measurable etching of the polymer material. Such rapid functionalization is a consequence of a very large affinity of the polymer surface to atomic oxygen [42].

While functionalization with polar groups is beneficial for adhesion of various coatings on polymer materials, there are applications where the surface finish should prevent sticking of unwanted liquids to the polymer surface. In such cases, the polymer should be functionalized with non-polar groups, and the best are fluorine-containing groups. Several fluorine-containing gases dissociate to F atoms under plasma conditions, but the surface finish depends on the type of the precursor. Resnik et al. [43] compared plasma treatment of polyethylene terephthalate using two gases and reveal effects that have not been observed before. The treatment of this polymer with plasma sustained both in tetrafluoromethane and sulfur hexafluoride revealed a high concentration of fluorine in the surface layer probed by X-ray photoelectron spectroscopy (XPS) only up to a certain pressure. Thereafter, both the fluorine content and the water contact angle decreased significantly, which was explained by the lack of F atoms at elevated pressure using low power density plasma.

Biomimetics is a hot topic in interdisciplinary materials science. A review paper has been published on various examples of successful synthesis and application of materials mimicking nature [44]. The paper describes wetting properties of man-made materials mimicking both super-hydrophobic and super-hydrophilic surfaces of plants and animals, with a particular emphasis of combining both effects on a scale measured in micrometers. Such materials have extremely high potential for application for body implants because the adsorption and conformation of proteins depends enormously on the surface wettability. The activation of blood platelets on such surfaces is reduced significantly; therefore, such a surface finish represents an alternative to biomaterials of laterally uniform surface finish [45,46].

Hemo-compatibility of body implants made from polymeric materials has attracted enormous attention in the past decades due to medical applications. Currently, cardiovascular diseases represent the main causes of mortality in the modern world. Treatments include implanting stents, artificial heart valves, and vascular grafts. An ideal surface finish of such implants should not only prevent activation of blood platelets, but also inhibit scar formation and facilitate rapid endothelization. These requirements are contradictory; therefore, researchers worldwide are investigating methods for producing such a surface finish that would meet all requirements. A review paper on recent advances in biocompatibility of plasma-treated polymeric implants was prepared by Recek [47]. An extensive literature review led to the conclusion that there are actually no available standardized methods for testing the hemocompatibility of biomaterials. In this review paper, the most promising methods to gain biocompatibility of synthetic materials are reported, and several hypotheses to explain the improvement in hemocompatibility of plasma treated polymer materials are offered.

Mushrooms represent an important nutritional component of the human diet and are valuable in many countries; therefore, the demand often exceeds the supply. A method for increased growth employs treatment of the substrates by pulsed electrical fields [48]. Several power supplies have been tested for the promotion of mushroom growth by electrical stimulation reaching voltages on the order of 100 kV [49]. The efficiency of such treatments depended enormously on the type of mushroom; therefore, an appropriate voltage has to be adopted according to particular conditions [39]. Unlike conventional methods, the innovative power supply reported in this issue by Takahashi et al. [48] enabled application also in hilly and mountainous areas and resulted in a significant improvement in mushroom production.

Acknowledgments: We would like to acknowledge Maryam Tabrizian, Editor-in-Chief, Chelsea Mu, Assistant Editor, and all the staff of the Materials Editorial Office for their great support during the preparation of this Special Issue. We would also like to thank all the authors for their great contributions, and the reviewers for the time they dedicated to reviewing the manuscripts. The Editor of this Special Issue acknowledges the financial support from the Slovenian Research Agency (research core funding No. P2-0082).

Conflicts of Interest: The authors of individual papers declare no conflict of interest.

References

1. Asadollahi, S.; Profili, J.; Farzaneh, M.; Stafford, L. Development of organosilicon-based superhydrophobic coatings through atmospheric pressure plasma polymerization of HMDSO in nitrogen plasma. *Materials* **2019**, *12*, 219. [CrossRef] [PubMed]
2. Guo, Z.; Liu, W.; Su, B.-L. Superhydrophobic surfaces: From natural to biomimetic to functional. *J. Colloid Interface Sci.* **2011**, *353*, 335–355. [CrossRef] [PubMed]
3. Johnson, R.E.; Dettre, R.H. Contact angle hysteresis. Iii. Study of an idealized heterogeneous surface. *J. Phys. Chem.* **1964**, *68*, 1744–1750. [CrossRef]
4. Sailor, M.J. *Porous Silicon in Practice: Preparation, Characterization and Applications*; John Wiley & Sons: Hoboken, NJ, USA, 2012.
5. Lee, S.H.; Kang, J.S.; Kim, D. A mini review: Recent advances in surface modification of porous silicon. *Materials* **2018**, *11*, 2557. [CrossRef] [PubMed]
6. Rodriguez, C.; Muñoz Noval, A.; Torres-Costa, V.; Ceccone, G.; Manso Silván, M. Visible light assisted organosilane assembly on mesoporous silicon films and particles. *Materials* **2019**, *12*, 131. [CrossRef] [PubMed]
7. Sacarescu, L.; Roman, G.; Sacarescu, G.; Simionescu, M. Fluorescence detection system based on silicon quantum dots-polysilane nanocomposites. *Express Polym. Lett.* **2016**, *10*, 990–1002. [CrossRef]
8. Mäkilä, E.; Bimbo, L.M.; Kaasalainen, M.; Herranz, B.; Airaksinen, A.J.; Heinonen, M.; Kukk, E.; Hirvonen, J.; Santos, H.A.; Salonen, J. Amine modification of thermally carbonized porous silicon with silane coupling chemistry. *Langmuir* **2012**, *28*, 14045–14054. [CrossRef] [PubMed]
9. Sheasby, P.G.; Pinner, R.; Wernick, S. *The Surface Treatment and Finishing of Aluminium and Its Alloys*; ASM International, Finishing Publications: Materials Park, OH, USA, 2001.
10. Heinz, A.; Haszler, A.; Keidel, C.; Moldenhauer, S.; Benedictus, R.; Miller, W.S. Recent development in aluminium alloys for aerospace applications. *Mater. Sci. Eng. A* **2000**, *280*, 102–107. [CrossRef]
11. Kadlečková, M.; Minařík, A.; Smolka, P.; Mráček, A.; Wrzecionko, E.; Novák, L.; Musilová, L.; Gajdošík, R. Preparation of textured surfaces on aluminum-alloy substrates. *Materials* **2018**, *12*, 109. [CrossRef] [PubMed]
12. Amanov, A.; Umarov, R.; Amanov, T. Increase in strength and fretting resistance of alloy 718 using the surface modification process. *Materials* **2018**, *11*, 1366. [CrossRef] [PubMed]
13. Vesel, A.; Drenik, A.; Elersic, K.; Mozetic, M.; Kovac, J.; Gyergyek, T.; Stockel, J.; Varju, J.; Panek, R.; Balat-Pichelin, M. Oxidation of Inconel 625 superalloy upon treatment with oxygen or hydrogen plasma at high temperature. *Appl. Surf. Sci.* **2014**, *305*, 674–682. [CrossRef]
14. Ott, E.A.; Groh, J.R.; Banik, A.; Dempster, I.; Gabb, T.P.; Helmink, R.; Liu, X.; Mitchell, A.; Sjoberg, G.P.; Wusatowska-Sarnek, A. *Superalloy 718 and Derivatives*; John Wiley & Sons: New York, NY, USA, 2012.
15. Chae, J.-M.; Lee, K.-O.; Amanov, A. Gradient nanostructured tantalum by thermal-mechanical ultrasonic impact energy. *Materials* **2018**, *11*, 452. [CrossRef] [PubMed]
16. Atieh, A.M.; Rawashdeh, N.A.; AlHazaa, A.N. Evaluation of surface roughness by image processing of a shot-peened, TIG-welded aluminum 6061-T6 alloy: An experimental case study. *Materials* **2018**, *11*, 771. [CrossRef]
17. Lin, N.; Li, D.; Zou, J.; Xie, R.; Wang, Z.; Tang, B. Surface texture-based surface treatments on Ti_6Al_4V titanium alloys for tribological and biological applications: A mini review. *Materials* **2018**, *11*, 487. [CrossRef] [PubMed]
18. Tang, W.; Zhou, Y.K.; Zhu, H.; Yang, H.F. The effect of surface texturing on reducing the friction and wear of steel under lubricated sliding contact. *Appl. Surf. Sci.* **2013**, *273*, 199–204. [CrossRef]
19. Flasker, A.; Kulkarni, M.; Mrak-Poljsak, K.; Junkar, I.; Cucnik, S.; Zigon, P.; Mazare, A.; Schmuki, P.; Iglic, A.; Sodin-Semrl, S. Binding of human coronary artery endothelial cells to plasma-treated titanium dioxide nanotubes of different diameters. *J. Biomed. Mater. Res. A* **2016**, *104*, 1113–1120. [CrossRef] [PubMed]
20. Paiva, J.M.; Fox-Rabinovich, G.; Locks Junior, E.; Stolf, P.; Seid Ahmed, Y.; Matos Martins, M.; Bork, C.; Veldhuis, S. Tribological and wear performance of nanocomposite PVD hard coatings deposited on aluminum die casting tool. *Materials* **2018**, *11*, 358. [CrossRef]
21. Haubner, R.; Lessiak, M.; Pitonak, R.; Kopf, A.; Weissenbacher, R. Evolution of conventional hard coatings for its use on cutting tools. *Int. J. Refract. Met. Hard Mater.* **2017**, *62*, 210–218. [CrossRef]
22. Deng, J.X.; Cao, T.K.; Ding, Z.L.; Liu, J.H.; Sun, J.L.; Zhao, J.L. Tribological behaviors of hot-pressed Al_2O_3/TiC ceramic composites with the additions of CaF_2 solid lubricants. *J. Eur. Ceram. Soc.* **2006**, *26*, 1317–1323.

23. Song, W.; Wang, S.; Lu, Y.; Xia, Z. Tribological performance of microhole-textured carbide tool filled with CaF$_2$. *Materials* **2018**, *11*, 1643. [CrossRef]

24. Breitwieser, D.; Spirk, S.; Fasl, H.; Ehmann, H.M.A.; Chemelli, A.; Reichel, V.E.; Gspan, C.; Stana-Kleinschek, K.; Ribitsch, V. Design of simultaneous antimicrobial and anticoagulant surfaces based on nanoparticles and polysaccharides. *J. Mater. Chem. B* **2013**, *1*, 2022–2030. [CrossRef]

25. Gorjanc, M.; Mozetic, M.; Vesel, A.; Zaplotnik, R. Natural dyeing and UV protection of plasma treated cotton. *Eur. Phys. J. D* **2018**, *72*, 41. [CrossRef]

26. Schlemmer, W.; Fischer, W.; Zankel, A.; Vukušić, T.; Filipič, G.; Jurov, A.; Blažeka, D.; Goessler, W.; Bauer, W.; Spirk, S.; et al. Green procedure to manufacture nanoparticle-decorated paper substrates. *Materials* **2018**, *11*, 2412. [CrossRef] [PubMed]

27. Sze, K.L.; Tzi, B.N. Lectins: Production and practical applications. *Appl. Microbiol. Biotechnol.* **2011**, *89*, 45–55.

28. Taniguchi, N.; Kizuka, Y. Glycans and Cancer: Role of N-Glycans in Cancer Biomarker, Progression and Metastasis, and Therapeutics. In *Advances in Cancer Research*; Drake, R.R., Ball, L.E., Eds.; Academic Publisher: Waltham, MA, USA, 2015.

29. Niegelhell, K.; Ganner, T.; Plank, H.; Jantscher-Krenn, E.; Spirk, S. Lectins at interfaces—An atomic force microscopy and multi-parameter-surface plasmon resonance study. *Materials* **2018**, *11*, 2348. [CrossRef] [PubMed]

30. Alenka, V.; Miran, M. New developments in surface functionalization of polymers using controlled plasma treatments. *J. Phys. D Appl. Phys.* **2017**, *50*, 293001.

31. Vesel, A.; Kovac, J.; Zaplotnik, R.; Modic, M.; Mozetic, M. Modification of polytetrafluoroethylene surfaces using H$_2$S plasma treatment. *Appl. Surf. Sci.* **2015**, *357*, 1325–1332. [CrossRef]

32. López-García, J.; Cupessala, F.; Humpolíček, P.; Lehocký, M. Physical and morphological changes of poly(tetrafluoroethylene) after using non-thermal plasma-treatments. *Materials* **2018**, *11*, 2013. [CrossRef]

33. Pankaj, S.K.; Bueno-Ferrer, C.; Misra, N.N.; Milosavljevic, V.; O'Donnell, C.P.; Bourke, P.; Keener, K.M.; Cullen, P.J. Applications of cold plasma technology in food packaging. *Trends Food Sci. Technol.* **2014**, *35*, 5–17. [CrossRef]

34. Zhang, Y.; Ishikawa, K.; Mozetič, M.; Tsutsumi, T.; Kondo, H.; Sekine, M.; Hori, M. Polyethylene terephthalate (PET) surface modification by VUV and neutral active species in remote oxygen or hydrogen plasmas. *Plasma Process. Polym.* **2019**. [CrossRef]

35. Vukušić, T.; Vesel, A.; Holc, M.; Ščetar, M.; Jambrak, A.R.; Mozetič, M. Modification of physico-chemical properties of acryl-coated polypropylene foils for food packaging by reactive particles from oxygen plasma. *Materials* **2018**, *11*, 372. [CrossRef] [PubMed]

36. Sunny, A.T.; Mozetic, M.; Primc, G.; Mathew, S.; Thomas, S. Tunable morphology and hydrophilicity to epoxy resin from copper oxide nanoparticles. *Compos. Sci. Technol.* **2017**, *146*, 34–41. [CrossRef]

37. Huskic, M.; Bolka, S.; Vesel, A.; Mozetic, M.; Anzlovar, A.; Vizintin, A.; Zagar, E. One-step surface modification of graphene oxide and influence of its particle size on the properties of graphene oxide/epoxy resin nanocomposites. *Eur. Polym. J.* **2018**, *101*, 211–217. [CrossRef]

38. Shah, A.H.; Zhang, Y.; Xu, X.; Dayo, A.Q.; Li, X.; Wang, S.; Liu, W. Reinforcement of stearic acid treated egg shell particles in epoxy thermosets: Structural, thermal, and mechanical characterization. *Materials* **2018**, *11*, 1872. [CrossRef]

39. Asadinezhad, A.; Novák, I.; Lehocký, M.; Sedlařík, V.; Vesel, A.; Junkar, I.; Sáha, P.; Chodák, I. A physicochemical approach to render antibacterial surfaces on plasma-treated medical-grade PVC: Irgasan coating. *Plasma Process. Polym.* **2010**, *7*, 504–514. [CrossRef]

40. Kolarova Raskova, Z.; Stahel, P.; Sedlarikova, J.; Musilova, L.; Stupavska, M.; Lehocky, M. The effect of plasma pretreatment and cross-linking degree on the physical and antimicrobial properties of nisin-coated PVA films. *Materials* **2018**, *11*, 1451. [CrossRef]

41. Šourková, H.; Primc, G.; Špatenka, P. Surface functionalization of polyethylene granules by treatment with low-pressure air plasma. *Materials* **2018**, *11*, 885. [CrossRef]

42. Vesel, A.; Zaplotnik, R.; Kovac, J.; Mozetic, M. Initial stages in functionalization of polystyrene upon treatment with oxygen plasma late flowing afterglow. *Plasma Sources Sci. Trans.* **2018**, *27*, 094005. [CrossRef]

43. Resnik, M.; Zaplotnik, R.; Mozetic, M.; Vesel, A. Comparison of SF$_6$ and CF$_4$ plasma treatment for surface hydrophobization of pet polymer. *Materials* **2018**, *11*, 311. [CrossRef]

44. Avrămescu, R.-E.; Ghica, M.V.; Dinu-Pîrvu, C.; Prisada, R.; Popa, L. Superhydrophobic natural and artificial surfaces—A structural approach. *Materials* **2018**, *11*, 866. [CrossRef]

45. Sopotnik, M.; Leonardi, A.; Krizaj, I.; Dusak, P.; Makovec, D.; Mesaric, T.; Ulrih, N.P.; Junkar, I.; Sepcic, K.; Drobne, D. Comparative study of serum protein binding to three different carbon-based nanomaterials. *Carbon* **2015**, *95*, 560–572. [CrossRef]

46. Humpolicek, P.; Kucekova, Z.; Kasparkova, V.; Pelkova, J.; Modic, M.; Junkar, I.; Trchova, M.; Bober, P.; Stejskal, J.; Lehocky, M. Blood coagulation and platelet adhesion on polyaniline films. *Colloid. Surf. B* **2015**, *133*, 278–285. [CrossRef] [PubMed]

47. Recek, N. Biocompatibility of plasma-treated polymeric implants. *Materials* **2019**, *12*, 240. [CrossRef] [PubMed]

48. Takahashi, K.; Miyamoto, K.; Takaki, K.; Takahashi, K. Development of compact high-voltage power supply for stimulation to promote fruiting body formation in mushroom cultivation. *Materials* **2018**, *11*, 2471. [CrossRef] [PubMed]

49. Takaki, K.; Yoshida, K.; Saito, T.; Kusaka, T.; Yamaguchi, R.; Takahashi, K.; Sakamoto, Y. Effect of electrical stimulation on fruit body formation in cultivating mushrooms. *Microorganisms* **2014**, *2*, 58. [CrossRef] [PubMed]

materials   MDPI

Article

Development of Organosilicon-Based Superhydrophobic Coatings through Atmospheric Pressure Plasma Polymerization of HMDSO in Nitrogen Plasma

Siavash Asadollahi [1,2]**, Jacopo Profili** [2]**, Masoud Farzaneh** [1] **and Luc Stafford** [2,*]

[1] Canada Research Chair on Engineering of Power Network Atmospheric Icing (INGIVRE),
    Université du Québec à Chicoutimi, Saguenay, QC G7H 2B1, Canada;
    siavash.asadollahi@gmail.com (S.A.); MasoudFarzaneh@uqac.ca (M.F.)
[2] Département de Physique, Université de Montréal, Montréal, QC H3T 1J4, Canada;
    profili.jacopo@gmail.com
*   Correspondence: luc.stafford@umontreal.ca; Tel.: +1-514-343-6542

Received: 30 November 2018; Accepted: 28 December 2018; Published: 10 January 2019

Abstract: Water-repellent surfaces, often referred to as superhydrophobic surfaces, have found numerous potential applications in several industries. However, the synthesis of stable superhydrophobic surfaces through economical and practical processes remains a challenge. In the present work, we report on the development of an organosilicon-based superhydrophobic coating using an atmospheric-pressure plasma jet with an emphasis on precursor fragmentation dynamics as a function of power and precursor flow rate. The plasma jet is initially modified with a quartz tube to limit the diffusion of oxygen from the ambient air into the discharge zone. Then, superhydrophobic coatings are developed on a pre-treated microporous aluminum-6061 substrate through plasma polymerization of HMDSO in the confined atmospheric pressure plasma jet operating in nitrogen plasma. All surfaces presented here are superhydrophobic with a static contact angle higher than 150° and contact angle hysteresis lower than 6°. It is shown that increasing the plasma power leads to a higher oxide content in the coating, which can be correlated to higher precursor fragmentation, thus reducing the hydrophobic behavior of the surface. Furthermore, increasing the precursor flow rate led to higher deposition and lower precursor fragmentation, leading to a more organic coating compared to other cases.

Keywords: atmospheric pressure plasma jets; plasma polymerization; superhydrophobicity; wetting

1. Introduction

A superhydrophobic surface is defined as a surface for which the equilibrium water contact angle (WCA) is higher than 150° [1] and contact angle hysteresis is lower than 10° [2]. The concept of superhydrophobicity initially emerged from the investigation of natural surfaces with high contact angle and low contact angle hysteresis, notably the lotus leaf (Nelumbo) surface [3]. The superhydrophobic characteristics of the micro-nanostructured and wax coated surface of the lotus leaf was first studied by Dettre and Johnson in 1963 [4]. Since then, several other examples of natural superhydrophobic surfaces have been identified [5]. During the past few decades, many studies have been done trying to mimic some of the structures observed on natural superhydrophobic leaves to develop artificial superhydrophobic surfaces [3,6]. Such surfaces may find a wide range of applications from textile industry to power network design and maintenance [7]. Many studies have been done on water-repellent self-cleaning fabrics [8,9]. Superhydrophobic surfaces may be used in biomedical applications, vessel replacements or wound management [2,10,11]. Since icephobicity (i.e., low adhesion force between ice and the substrate) shows a correlation with superhydrophobicity [12], superhydrophobic coatings can be considered as suitable candidates to reduce the ice accumulation on various structures, notably power network equipment [13–15]. Construction industry can benefit from the development of superhydrophobic surfaces for manufacturing self-cleaning windshields and windows [16–18]. In marine industry, superhydrophobic coatings can be used to develop anti-fouling surfaces [19] or to assist in oil-water separation [20]. Superamphiphobic surfaces in particular (surfaces with both superhydrophobic and superoleophobic characteristics) may be used for such applications [21]. Furthermore, due to their potential in minimizing the liquid/surface contact area, hydrophobic and superhydrophobic surfaces can be used in anticorrosion application [6,22–24].

One of the most promising approaches to surface modification and coating deposition are plasma-based surface treatment methods, due to their high controllability, relatively low cost, low pollution levels, and short treatment times. Such treatments typically involve removing molecules from the surface (plasma etching/sputtering) and/or depositing a different material on the surface (plasma polymerization) using high-energy plasma-generated-species [25]. During the past few decades, plasma-based processes have been used for a wide range of applications, such as deposition of various functional coatings [10,25,26], modifying surface topography and micro/nano texturing [27], treatment of tumors and infectious wounds [28–30], or even treatment of food products [31,32]. Depending on the gas pressure in which the plasma is generated, plasma treatment may be carried out in low-pressure or atmospheric-pressure. Low-pressure plasma treatment typically results in more uniform films and may be used for 2D or 3D treatment due to the spatial homogeneity of the reactive species. Atmospheric-pressure plasmas, on the other hand, are easier to generate and maintain since they do not require cost-intensive vacuum pumps and chambers [33]. However, due to the open-air configuration in atmospheric-pressure treatment, plasma treatment of oxidation-sensitive materials becomes limited.

In this study, the development of an organosilicon-based superhydrophobic surface through atmospheric-pressure plasma deposition of hexamethyldisiloxane (HMDSO) is reported. In this specific paper, emphasis is placed on the precursor fragmentation dynamics and the effects of 'available energy per precursor molecule' on coating properties. Molecular fragmentation of HMDSO in plasma leads to the deposition of low-surface-energy methyl groups on the substrate, thus reducing the wettability of the surface. It is often argued that the wetting characteristics is directly linked to the degree of precursor fragmentation since the energy required to break Si-C bond (318 kJ/mol) is less than the energy required to break Si-O bond (452 kJ/mol). Therefore, higher plasma energies typically lead to more polar oxide functions on the surface, which will in turn decrease the surface hydrophobicity [25,34,35]. Using a dielectric barrier discharge operating at atmospheric-pressure, Siliprandi et al. have shown that for low HMDSO concentrations (less than 0.3% in their study), the deposition process strongly depends on precursor presence in the plasma. However, beyond this threshold value the deposition process is mostly controlled by the plasma generation power, indicating a power-deficient regime [34]. In other words, fragmentation can be essentially controlled by adjusting the available energy per precursor molecule; this a well-established concept in low-pressure plasma deposition. Gas phase fragmentation and recombination reactions are also dependent on the residence time of plasma-generated-species. Longer residence times can be linked to higher precursor fragmentation, which correlated to higher oxide content in the case of HMDSO deposition. This has been confirmed by investigating the effects of plasma gas velocity and precursor injection position on surface chemical composition [36]. For more information on the applications of plasma technology in development of superhydrophobic surfaces, see Reference [25,37].

More specifically, this work reports the development of superhydrophobic coatings on pre-treated alumina-based substrates through atmospheric-pressure plasma deposition of HMDSO in the jet of an open-to-air nitrogen plasma produced by rotating arc discharges. The effects of precursor flow rate and generation power on precursor fragmentation in the discharge and thus surface chemical composition is demonstrated. Furthermore, the wetting behaviors of all coatings are studied through static and dynamic water contact angle measurement, and the results are correlated with the precursor fragmentation dynamics, surface chemical composition, and surface morphology.

2. Experimental Procedure

30 mm × 50 mm × 1.8 mm samples are cut from Al-6061 sheets provided by ALCAN. For all plasma treatment procedures, a commercial OpenAir AS400 atmospheric pressure plasma jet manufactured by PlasmaTreat® (PlasmaTreat GmbH, Steinhagen, Germany) with a PFW10 nozzle is used. The schematics of the jet are presented in Figure 1. The process is open to ambient air and is carried out at room temperature and in uncontrolled humidity conditions. At first, the samples are exposed to multiple passes of air plasma treatment at very short jet-to-substrate distances to generate an alumina-based micro-roughened porous surface structure on the aluminum substrate (Figure 2). The details of this process, which is driven by the formation of electric discharges between the rotating arc inside the jet body and the substrate, are previously studied by the authors [38]. These "pre-treated" samples are then cleaned in an ultrasonic bath of ethanol and de-ionized water (15 min each at room temperature) to remove any surface contamination prior to coating deposition. For the deposition step, the plasma jet is slightly modified with a quartz tube mounted on the jet-head. This will be discussed in more detail later. HMDSO (Sigma-Aldrich®, St. Louis, MO, USA, >98%) was used as the growth precursor. Nominally pure nitrogen gas is purchased from Praxair and is used as received. All samples studied here are prepared by the injection of HMDSO vapor (vaporized at 125 °C) mixed with a carrier gas (nitrogen) into the flowing afterglow region of a nitrogen plasma on pre-treated aluminum surfaces. In total, 3 samples are studied. Two samples are prepared with two different monomer flow rates: 3 g/h and 5 g/h, referred to as PT3 and PT5 samples, respectively. A third sample, referred to as PT5P75, was prepared with the same conditions as PT5 except for the plasma power, which is increased from 500 W to 750 W. It should be noted that plasma deposition is performed under pulsed plasma conditions (duty cycle of 50%) to prevent excessive precursor fragmentation and to limit powder formation. Plasma parameters used for deposition process are also presented in Table 1.

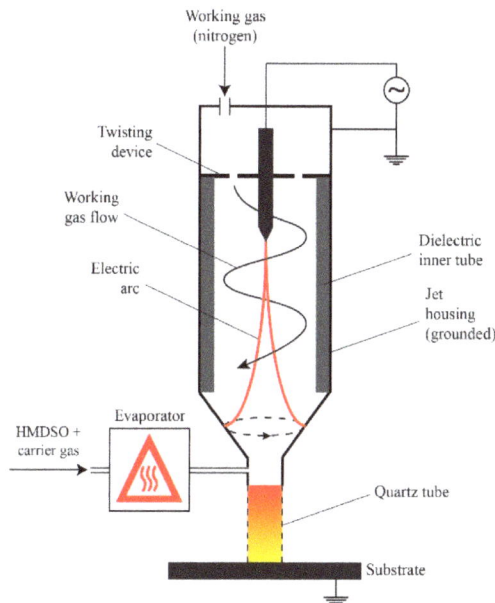

Figure 1. Schematics of the modified APPJ used in this study.

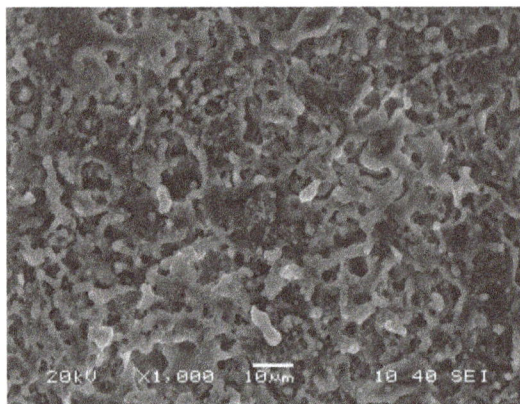

Figure 2. Surface structure of the aluminum surface after pre-treatment [38]. This structure is used as the substrate for deposition of organosilicon based coatings.

Table 1. Plasma conditions used for preparation of samples.

Sample Name	Monomer Flow Rate	Plasma Power	Plasma Duty Cycle	Jet Speed	Jet-Substrate Distance	Ionization Gas Flow Rate	Carrier Gas Flow Rate
PT3	3 g/h	500 W	50%	1 m/min	30 mm	500 L/h	400 L/h
PT5	5 g/h	500 W	50%	1 m/min	30 mm	500 L/h	400 L/h
PT5P75	5 g/h	750 W	50%	1 m/min	30 mm	500 L/h	400 L/h

An LR2-T optical emission spectrometer (Toshiba®, Tokyo, Japan) with a spectral resolution of 2 nm is used to analyze plasma properties. This spectrometer can acquire optical emission spectra in the range of 200–1200 nm, and the detector is thermoelectrically cooled to optimize the signal-to-noise ratio. All optical emission spectra presented in this paper are recorded with an exposure time of 200 ms. The fiber was placed close to the sample surface (<1 mm above the surface) at 3 cm from the jet axis. A JSM-7600F scanning electron microscope (SEM) manufactured by JEOL® (Tokyo, Japan) is used to acquire images of coating's surface structure. To improve image clarity, samples are coated with a nanometric layer of carbon prior to SEM analysis. Presence of various chemical functions on the surface was studied through Fourier transform infrared spectroscopy (FTIR). A Bruker Vertex 70 FTIR system is used in ATR mode with a resolution of 2 cm^{-1}. Each one of the FTIR spectra presented in this paper is averaged over 32 consecutive scans on the sample. Further chemical analysis on the chemical composition of the surface and the chemical state of silicon is performed using an X-ray photoelectron spectrometer (XPS) with a non-monochromatic Al (max energy 1486.6 eV) anode manufactured by Staib Instruments® (Langenbach, Germany). Scanning parameters used for the acquisition of survey and high-resolution spectra are presented in Table 2. All analysis on the spectra is performed using CasaXPS software (developed by Casa Software Ltd., Telgnmouth, UK). Charge compensation was performed by fitting a synthetic peak in C 1s envelope and then calibrating the peak position it to 284.5 eV.

Table 2. Acquisition parameters for survey and high-resolution X-ray photoelectron spectrometer (XPS) spectra.

Scan Type	Start Energy	End Energy	Step Width	dE	Dwell Time	# of Scans	Beam Power
Survey	1300 eV	0 eV	1 eV	4 eV	100 ms	5	150 W
High resolution	A window of ~20 eV width around the peak.		0.1 eV	0.3 eV	300 ms	10	300 W

All water contact angle measurements are performed using a DSA100 goniometer manufactured by Kruss® (Kruss GmbH, Hamburg, Germany). Static contact angle is measured by placing a 4 µL droplet on the surface and calculating the angle at the 3-phase interface based on Young-Laplace approximation. Static contact angle values presented here are the average of at least 15 measurements on different locations and/or samples. To evaluate the dynamic wetting behavior, at first the needle is placed close to the surface and a 4 µL droplet is deposited on the surface. Then, the volume of this droplet is steadily increased to 13 µL with a rate of 3 µL/min. The contact angle on each side of the droplet is measured 5 times per second using the tangent method. Advancing contact angle can then be defined based on the *contact angle* vs. *time* curve. Similarly, the receding contact angle is defined by reducing the droplet volume back to its initial value and constantly measuring the contact angle of the moving interfaces. For more information on this procedure, see [39].

As mentioned before, in the deposition of hydrophobic or superhydrophobic coatings, oxygen incorporation in the coating needs to be limited, particularly in open-to-air atmospheric-pressure plasma conditions. In this context, the plasma jet described above was slightly modified to confine the flowing afterglow region and limit the diffusion of oxygen into the reactive plasma zone. This was achieved by mounting a quartz tube (ID 25 mm, OD 27 mm, and length of 30) on the jet nozzle using a handmade copper clamp (Figure 3). It was immediately observed that confining the plasma jet allows for ignition and maintenance of much weaker plasmas by reducing the loss of ionized species into the ambient atmosphere. This is of significant interest in this work, since lower energy leads to lower precursor fragmentation and thus higher retention of organic functions in the coating structure.

Figure 3. Plasma jet after modification with the quartz tube.

Plasma composition in the presence of the quartz tube was studied using OES and the results are presented in Figure 4. It should be noted that the optical spectra presented here were acquired using a different quartz tube (not pictured here) with an opening at the lowest part. The fiber is placed in front of this opening at a suitable distance (3 cm) so that the optical emission is gathered directly from the discharge zone without interference from the quartz or any possible deposition on the optical fiber. The spectra presented in Figure 4 are acquired with and without the quartz tube surrounding the jet and under PT5 conditions. Only a limited wavelength range (150–450 nm) is presented here, since the emission outside this range was very weak. The spectra are dominated by the nitrogen second positive system, which is typically observed in atmospheric-pressure nitrogen plasmas open to ambient air. Emission intensity is significantly stronger in the spectra acquired with the quartz tube, and since the plasma gas flow rate and the precursor flow rate are identical in both cases, higher intensity is suggestive of a modification of the plasma density and/or the plasma temperatures, at least at the position of OES measurements (<1 mm from sample surface). It is worth mentioning that the presence of Si lines is suggestive of significant precursor fragmentation: atomic Si lines are rarely observed during plasma polymerization in cold, atmospheric-pressure plasmas; this can most likely be attributed to the much higher electron densities and temperatures in the arc-based discharges and flowing afterglow regions than in other systems such as homogeneous dielectric barrier discharges [40]. In fact, Pulpytel et al. have observed atomic Si lines in the optical emission spectrum of air plasma with HMDSO as the precursor in the same reactor operated in comparable experimental conditions [41].

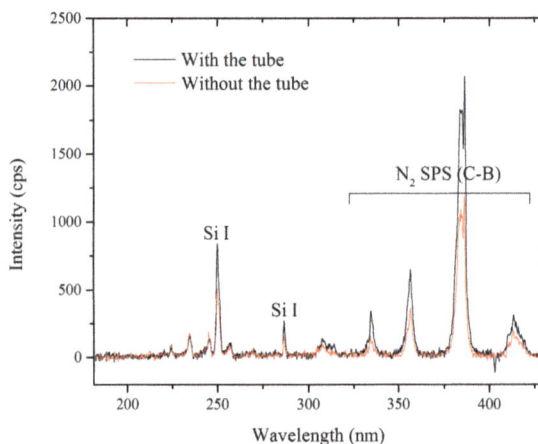

Figure 4. OES spectra acquired from the plasma with and without the quartz tube surrounding the jet-head.

Chemical composition of the samples acquired from nitrogen plasma deposition on flat aluminum with and without the quartz tube surrounding the jet was studied using FTIR spectroscopy, and the results are presented in Figure 5. The spectral features observed in Figure 5 are similar to those observed in typical organosilicon-based depositions. Comparing the intensity of Si-O-Si band in both cases suggests a lower amount of deposition in the sample prepared without the quartz tube. This can be due to the loss of reactive species due to the jet movement or turbulence in the ambient atmosphere when the jet is not confined. To compare the chemical composition of the coating regardless of the deposition thickness, the fingerprint region of the spectra is also normalized according to the Si-O-Si band and the result is presented as an inset in Figure 5.

The intense band between 1000 cm^{-1} and 1200 cm^{-1} is generally assigned to Si-O-Si and Si-O-C asymmetric stretching modes [42,43]. This band is usually assumed to be the sum of three Gaussian components which correspond to different bond angles in Si-O-Si: TO$_2$ mode at 1120 cm^{-1} (170°–180°

bond angle), TO$_1$ mode at 1070 cm^{-1} (140° bond angle), and TO$_3$ mode at 1030 cm^{-1} (120° bond angle) [43]. TO$_2$ mode is often associated with fragments of Si-Si chains, but in organic films this wavenumber is also populated by Si-O-C stretching mode. TO$_1$ is assigned to the in-phase asymmetric stretching vibrational mode of the neighboring SiO$_2$ moieties (−O−Si−O−) in a quartz-like structure while TO$_3$ mode in SiO$_x$C$_y$H$_z$ films is often observed because of the methyl environment of the bond [44]. The position of Si-O-Si band in the sample prepared with the quartz tube seems to be between the theoretical position of the TO$_1$ and TO$_3$ mode (max intensity at 1056 cm^{-1}).

In comparison, in the absence of the quartz tube, Si-O-Si band is shifted to higher wave numbers (maximum intensity at 1105 cm^{-1}), suggesting that without the tube this region is dominated by either Si-O-C bonds or Si-O-Si bond in TO$_2$ mode, which may be associated with the fragments of Si-O-Si chains. This, along with the lower deposition in the case without the quartz tube, suggests that the addition of the quartz tube to the plasma jet leads to higher deposition and longer chains of SiO$_x$C$_y$H$_z$ structures.

In both spectra, Si-(CH$_3$)$_n$ group is easily recognizable by a strong, sharp band at around 1275 cm^{-1} (CH symmetric deformation in Si-CH$_3$) together with three bands in the 865–750 cm^{-1} range [45]. This suggests that, even if with the quartz tube the coating seem more inorganic, the Si-O-Si network is at least partially functionalized with methyl groups.

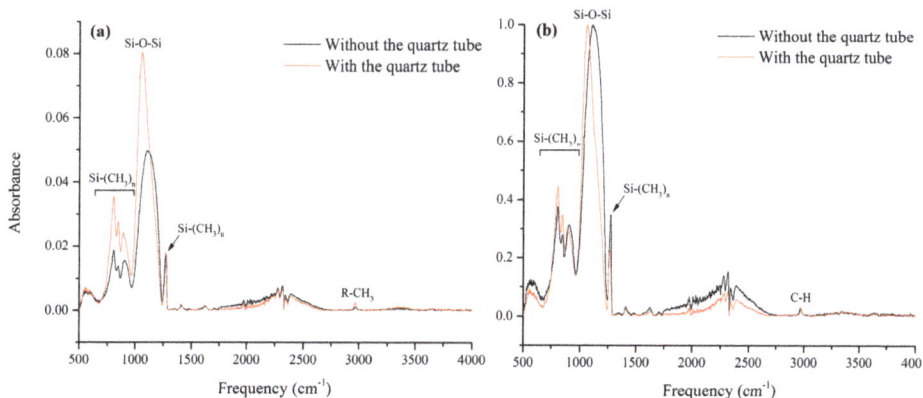

Figure 5. (**a**) Fourier transform infrared spectroscopy (FTIR) spectra of samples prepared with PT5 conditions, with and without the quartz tube surrounding the jet. (**b**) shows the same spectra normalized according to Si-O-Si band.

To study the effect of this modification on surface morphology, two samples were prepared under PT5 conditions on the pre-treated aluminum with and without the quartz tube surrounding the jet. Figure 6 compares the surface structure of the coating with and without the quartz tube. Without the quartz tube, the deposition consists of spherical structures covering the micro features obtained after the pre-treatment step (Figure 6a,b). These particles seem to be distributed uniformly on the surface and their diameter range from a few to hundreds of nanometers. On the other hand, surface structure in the presence of the quartz tube (Figure 6c,d) is consisted of dendrite-like structures with multiple levels of roughness, ranging from only a few nanometers to tens of micrometers. Such a hierarchical structure is shown to be ideal for hydrophobic and superhydrophobic applications [46]. Similar features have been reported by Kilicaslan et al. during the deposition of HMDSO-based nanomaterials in atmospheric-pressure plasmas sustained by microwave electromagnetic fields [47]. While spherical silica nanoparticles were obtained in the presence of O$_2$ in Ar/HMDSO plasma, dendrite- like SiOCH nanostructures were formed in the absence of O$_2$.

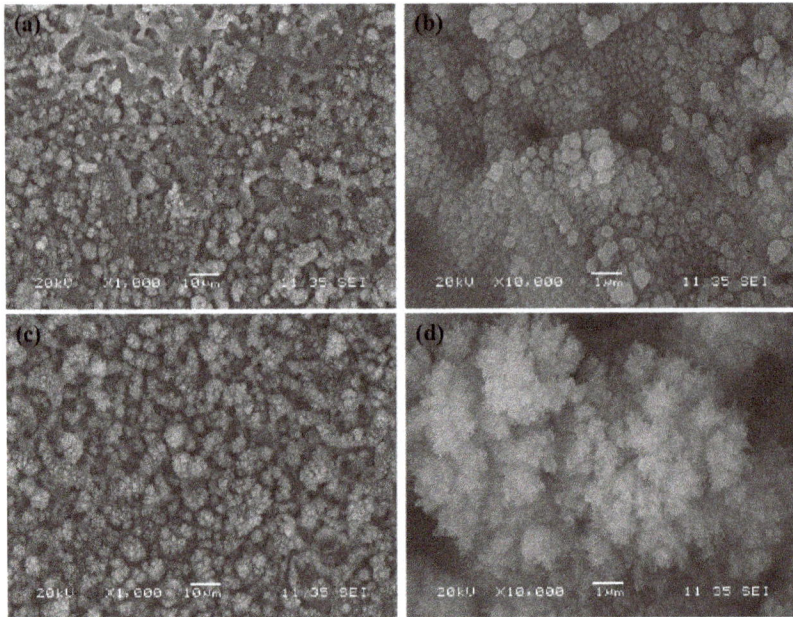

Figure 6. Surface structure of the coatings deposited without a quartz tube surrounding the jet (**a,b**) and with the quartz tube (**c,d**).

In conclusion, it is shown that the addition of a quartz tube around the atmospheric-pressure plasma jet can affect the treatment and deposition dynamics in several ways:

- Facilitates the ignition and maintenance of weaker plasmas, which leads to lower monomer fragmentation and is generally favorable in hydrophobic applications;
- Limits the diffusion of oxygen from the ambient air into the discharge zone;
- Increases the coating thickness by reducing the loss of reactive species into the ambient atmosphere;
- Increases the cross-linking of the silica-like network by increasing the Si-O-Si chain lengths.

All the results presented in the following sections are acquired from the modified jet (i.e., with the quartz tube surrounding the flowing afterglow region).

3. Results and Discussion

3.1. Optical Emission Spectroscopy

To characterize the discharge zone, optical emission spectra were acquired from all conditions, and the results are presented in Figure 7. In all cases, spectra is dominated by the nitrogen second positive system located between 325 nm and 425 nm (Figure 7b) [41]. Signal intensity is significantly stronger in PT5P75 in all regions. This is due to the higher plasma power, which leads to a more emitting plasma (note the different scales on Y axis in Figure 7). In the near UV region (Figure 7a), NOγ system is clearly observed for PT3 and PT5 through four bands located at 234, 245, 257, and 269 nm. As the plasma power increases, the emission from NO disappeared, which may be explained by a higher presence of quenching species in the discharge zone. On the other hand, atomic O lines may be observed between 700 nm and 850 nm in the case of PT5P75 (Figure 7c), which is consistent with higher precursor fragmentation in the case of higher plasma power.

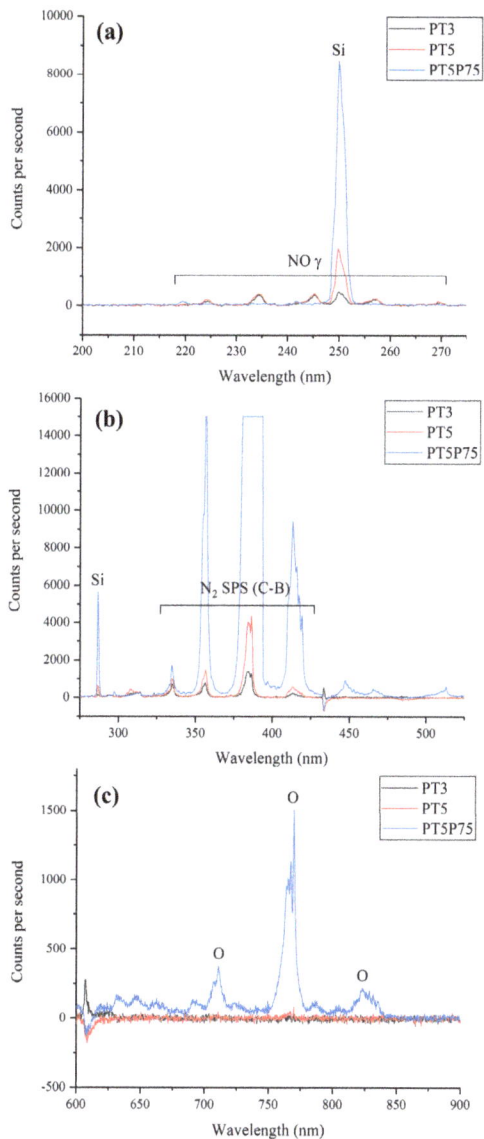

Figure 7. Optical emission spectra from PT3, PT5, and PT5P75 conditions. Three ranges of wavelength are chosen to highlight some of the differences. (**a**) 200–275 nm, (**b**) 275–525 nm, and (**c**) 600–900 nm.

3.2. Surface Morphology

Surface morphology for different conditions was studied through scanning electron microscopy, and the results are presented in Figure 8. Comparing different surface structures, it is observed that in PT3 (Figure 8a), the precursor flow rate is not high enough for complete coverage of the pre-treated aluminum surface. On the other hand, in PT5 (Figure 8b), the substrate is fully covered by the deposition material, while the morphological features originated from the pre-treatment procedure are retained. In the case of PT5P75 (Figure 8c), micro-features from the pre-treated substrate are completely buried under the deposited material, resulting in the loss of an important roughness level. This may be due to the higher precursor fragmentation, leading to an increased presence of oxygen in the discharge zone, promoting the formation of silica powders and increasing the size of the deposited particles [48]. The presence of silica powder in the coating structure (such as the larger deposited agglomerates in Figure 8c) has an adverse effect of the mechanical stability of the coating, since larger particles become increasingly unstable and easier to remove under external forces.

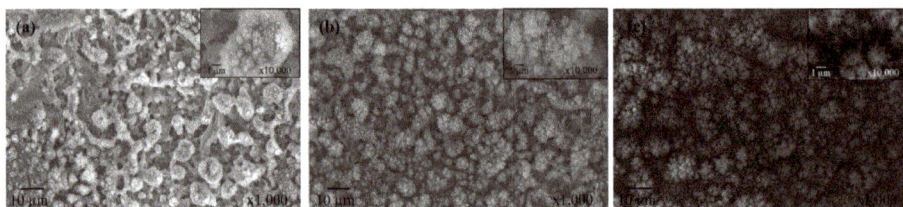

Figure 8. Scanning electron microscopy (SEM) images of (**a**) PT3, (**b**) PT5, and (**c**) PT5P75.

3.3. Chemical Composition

FTIR spectroscopy and XPS were used to study the chemical composition of different samples. Figure 9 shows the full range of FTIR spectra acquired from uncoated pre-treated aluminum, PT3, PT5, and PT5P75. Comparing the spectra related to the pre-treated substrate with those from the coatings, one can readily conclude that most features observed in the spectra are originated from the deposition.

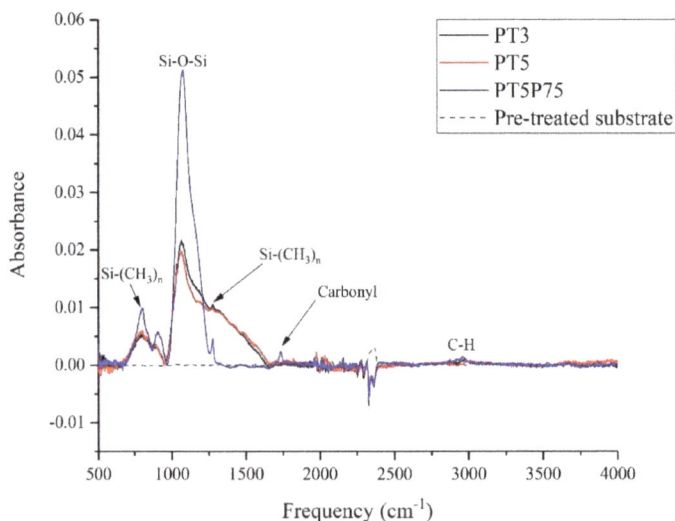

Figure 9. Full range of Fourier transform infrared spectroscopy (FTIR) spectra acquired from PT3, PT5, and PT5P75.

In all cases, the fingerprint region of the spectrum is dominated by features common to siloxane-based coatings. The presence of bands related to Si-$(CH_3)_n$ at around 800 and 1280 cm^{-1}, along with the bands related to C-H groups at around 3000 cm^{-1}, confirms the deposition of organic groups through plasma polymerization. Since Si-O-Si band intensity is strongest in the case of PT5P75, the deposition thickness is significantly larger for PT5P75 compared to PT5 and PT3. The presence of carbonyl groups (C=O), evident by the sharp peak at around 1750 cm^{-1}, is suggestive of higher precursor fragmentation with higher generation power. This increases the presence of oxygen in the discharge zone, which in turn enhances the formation of larger silica-based particles [48]. This is in agreement with larger deposited features observed in SEM images (Figure 8).

In the IR spectra of siloxane-based surfaces, the 500–1700 cm^{-1} range can provide valuable information regarding the Si-O network [43]. However, since this region is populated with several organic and organosilicon-based species, the spectra should be deconvoluted into its components to reliably quantify the peak intensity and surface area. In this context, the fingerprint region of the spectra presented in Figure 9 was deconvoluted to distinguish and quantify some of the organosilicon-based species present on the surface. Figure 10 shows the results of this deconvolution and shows the locations of the synthetic curves along with their assigned vibrations.

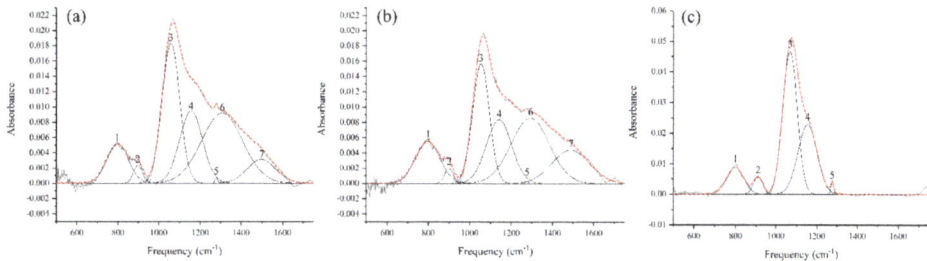

Figure 10. Synthetic curve models developed on the Fourier transform infrared spectroscopy (FTIR) spectra acquired for (**a**) PT3, (**b**) PT5, and (**c**) PT5P75.

For a brief description of what numbered components represent, see Table 3.

Table 3. Locations and assigned chemical groups for the synthetic curves presented in Figure 10.

Peak Index	Approximate Position (cm^{-1})	Assigned to
1	800	Si-C rocking vibration in Si-(CH$_3$)$_n$ [36,49,50]
2	900	Si-OH bending [50]
3	1060	Si-O-Si asymmetric stretching (TO$_1$)
4	1150	Si-O-Si asymmetric stretching (TO$_2$)
5	1275	C-H symmetric deformation in Si-(CH$_3$)$_n$
6	1300–1500	
7		

A notable feature of the FTIR spectra presented in Figure 10 is the continuously descending signal from 1300 cm^{-1} to 1500 cm^{-1} in the case of PT3 and PT5. Typically, this range is populated by many C-H bending vibrations in methyl (-CH$_3$), methylene (=CH$_2$) and methyne (=CH) [51]. However, C-H bending vibrations are usually narrower and therefore more distinct than peak indexes 6 and 7. Furthermore, this signal is not observed in the case of PT5P75, where C-H bending vibrations are also expected to occur. Therefore, it is highly unlikely that these peaks are due to C-H bending vibrations. A few other studies have observed several bands in the 1300–1500 cm^{-1} range in the case of highly porous structures with micrometric pore sizes [52,53]. These bands are not specifically identified in these studies but are only reported for highly porous surfaces. Therefore, an alternative, and more likely, explanation for this descending signal would be through the effects of surface porosity on the spectral background of the FTIR data. In PT3 and PT5, it is likely that the substrate's pores interfere with the IR absorbance, and therefore the effects of porosity are more pronounced. In PT5P75, deposited material covers a major part of substrate porosity (see Figure 8), leading to the absence of any signal in 1300–1500 cm^{-1} range. In any case, this signal introduces a significant amount of uncertainty to any deconvolution procedure performed on this region. The main characteristic peak of Si-$(CH_3)_n$ is located at the lower boundary of this range (1275 cm^{-1}), and therefore its shape and intensity is heavily affected by peak indexes 6 and 7. In this study, to avoid the uncertainty originated from the overlapping components, a different peak located at around 800 cm^{-1} (peak index 1) is used as a representative of the organic deposition. It should be noted that Peak 1 may be further deconvoluted into three separate components based on the value of n in Si-$(CH_3)_n$ [36,49,50]. In this study however, the resolution of the FTIR measurement is not small enough to distinguish between these components, and therefore the total surface area under this peak is used as a representative of the organic content. These various states of organic silicon will be investigated later while discussing the results of high resolution Si 2p core peak XPS spectra. Another consideration is the presence of Si-O-Si bending vibration at around 780–800 cm^{-1} [47,54], which may overlap Peak 1 in the above calculations. However, in organosilicon coatings, the intensity of Si-O-Si bending vibration is typically negligible compared to the Si-C rocking [48,55,56].

In order to determine the amount of organic functions in the Si-O-Si network, the ratio between the surface area under the Si-$(CH_3)_n$ component (peak index 1) and the surface area under quartz-like Si-O-Si component (peak index 3) is calculated. Furthermore, to determine the structural integrity of the Si-O-Si network, the ratio between the surface area under the Si-O-Si TO_2 component (peak index 4) and the surface area under the Si-O-Si TO_1 component (peak index 3) is calculated. Since TO_1 is correlated with a quartz-like structure while TO_2 is correlated with fragments of siloxane chains, a higher TO_2/TO_1 ratio corresponds to shorter siloxane chains and more disorder in the silica network. These ratios were calculated for PT3, PT5, and PT5P75 and are presented in Table 4, where A_n denotes the surface area under the nth peak index.

Table 4. Ratios between surface areas under some of the components in the Fourier transform infrared spectroscopy (FTIR) curve-fitting model (Figure 10).

	A_1/A_3 (Si-$(CH_3)_n$/Si-O-Si)	A_4/A_3 (TO_2/TO_1)
PT3	0.35	0.67
PT5	0.52	0.64
PT5P75	0.22	0.81

Since the plasma power is identical for PT3 and PT5, the amount of available energy per HMDSO molecule is higher in the case of PT3. Therefore, higher monomer fragmentation is expected in lower precursor flow rates, which leads to lower Si-$(CH_3)_n$/Si-O-Si ratio. This is manifested as a lower A_1/A_3 ratio for PT3 than PT5. Similarly, since the precursor flow rate is identical for PT5 and PT5P75, higher power leads to higher energy per precursor molecule, which in turn increases the fragmentation and leads to lower Si-$(CH_3)_n$/Si-O-Si ratio in the case of PT5P75.

To study the structural integrity of the siloxane network, A_4/A_3 ratio is investigated. For PT3 and PT5, this ratio is almost identical, which is suggestive of similar siloxane structures in both cases. However, as discussed before, the main difference between PT3 and PT5 is in coating thickness and the amount of deposition (Figure 8). In the case of PT5P75, it is shown that increasing the generation power has a significant effect on the Si-O-Si network, leading to shorter siloxane chains (higher TO_2 intensity). This is consistent with higher precursor fragmentation with higher plasma energy.

Complementary quantification of the surface chemical composition was acquired through X-ray photoelectron spectroscopy. Figure 11 shows the atomic percentages of Si, C, and O based on the survey spectra. In PT5, lower precursor fragmentation (due to the lower energy per precursor molecule) leads to the highest organic content on the surface, which is manifested as higher C percentage and lower O percentage. PT5P75 has a higher content of silicon and oxygen and a lower content of carbon, which can be linked to a decreased carbon content due to the higher precursor fragmentation. This is also in agreement with the FTIR data, where it was shown that the organic content is reduced with increasing the power. Finally, comparing PT3 with PT5, it is observed that the increased amount of energy available per precursor molecule in the case of PT3 leads to smaller amounts of deposition (lower silicon percentage) with less organic content (lower carbon percentage).

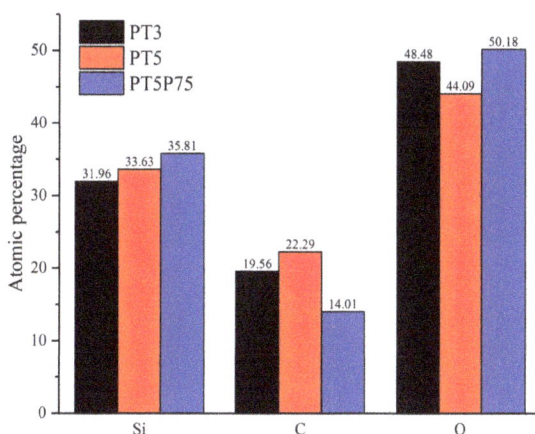

Figure 11. Atomic percentages of silicon, carbon and oxygen acquired from X-ray photoelectron spectrometer (XPS) for PT3, PT5, and PT5P75.

High resolution spectra of Si 2p core peak were used to determine the chemical state of silicon atoms in the siloxane structure. Si 2p core peak analysis is done based on the method developed by O'Hare et al. [57,58] for the analysis of siloxane-based coatings using X-ray high resolution spectra of Si 2p core peak. In this method, four different chemical states of silicon atoms are identified based on the number of bonds with oxygen: Q, T, D, and M which are in contact with 4, 3, 2, and 1 oxygen atoms, respectively. These chemical states are shown schematically in Figure 12 and some further details are provided in Table 5.

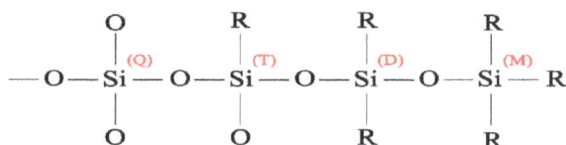

Figure 12. Silicon chemical states in a typical siloxane-based molecular structure.

Table 5. Chemical states of silicon in siloxane coatings [58].

	Binding Energy	Energy Shift	Function
$Q\ [SiO_{4/2}]$	103.69 eV	0 eV	cross-linking
$T\ [CH_3SiO_{3/2}]$	102.89 eV	0.80 eV	cross-linking
$D\ [(CH_3)_2SiO_{2/2}]$	102.21 eV	0.68 eV	propagation
$M\ [(CH_3)_3SiO_{1/2}]$	101.85 eV	0.36 eV	termination

Based on these chemical states and their respective binding energy in X-ray photoelectron spectra, synthetic curve-fitting models (not presented here) were developed on Si 2p core peak to determine the amount of deposited organosilicon species. These models where developed by (1) restricting the position of (Q) component to 103.69 ± 1 eV based on the calibration experiments on pure quartz samples, (2) restricting the position of (T), (D), and (M) based on their respective shifts from the position of (Q), and (3) forcing equal peak widths for all components. The results from component quantification are presented in Figure 13.

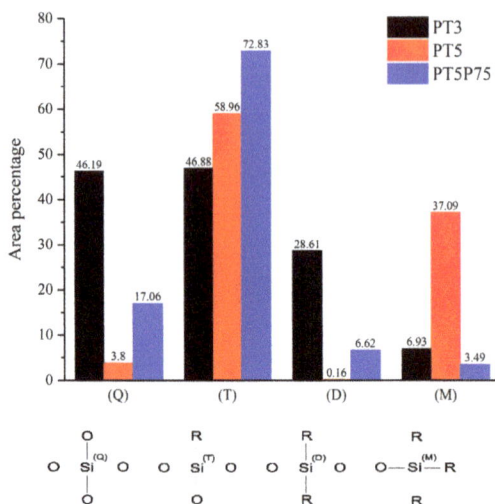

Figure 13. Component quantification of siloxane-based coatings determined through high resolution spectroscopy of Si 2p core peak.

When HMDSO molecules are vaporized and injected into the flowing afterglow region of the atmospheric-pressure plasma jet, they may go through several stages of fragmentation. Si-O bonds, which are weaker than Si-C bonds, break first, generating (M) silicon states which may be deposited on the surface. Further fragmentation results in broken Si-C bonds, removing methyl groups from silicon and replacing them with oxygen atoms in the open-to-air plasma. Therefore, the presence of various silicon chemical states may be interpreted as various degrees of fragmentation. In the case presented here, higher presence of (M) groups and lower presence of (Q) groups in the topmost layer of PT5 is suggestive of lower precursor fragmentation. Even before curve-fitting, higher organic content in PT5 is evident by a 0.5 eV shift observed in Si 2p core peak position towards lower energies. This is consistent with the FTIR data, where it was shown that PT5 exhibits higher $Si-(CH_3)_n/Si-O-Si$ ratio with a more organized structure. For PT3 and PT5P75, more than 90% of silicon atoms are in contact with 4 or 3 oxygen atoms (Q and T respectively), which is suggestive of high precursor fragmentation. However, in the case of PT3, (Q) and (T) are represented almost equally on the topmost layer of the coating, while in PT5P75 silicon is mostly at (T) state. Based on what has been discussed so far, it is evident that both coatings are the outcome of heavy precursor fragmentation. However, higher presence of precursor molecules in the case of PT5P75 results in the higher implementation of organic functions in the siloxane network, leading to a high concentration of (T) state in PT5P75. Similarly, lower presence of precursor molecules in PT3 reduces the functionalization of siloxane network with organic functions, leading to higher concentration of fully oxidized silicon (Q).

At this point, it should be noted that in a typical curve fitting process, the results may be examined by developing a curve model on a different peak related to a different element and confirming the agreement between the results acquired from both models. In the case presented here, this validation can potentially be done by creating synthetic models on C 1s peak based on (T), (D), and (M) functions along with other possible carbon states (such as C-C, C-H, and C=O). However, the binding energies for C 1s$^{(T)}$, C 1s$^{(D)}$, and C 1s$^{(M)}$ are in a 0.5 eV range (284.7 eV, 284.5 eV, and 284.2 eV, respectively) [57]. Furthermore, the binding energy of adventitious carbon (carbon originated from exposure to air or other sources of contamination), which is typically the most prominent source of carbon in XPS data, is also in the same range. Due to the relatively large FWHM of the high-resolution spectra (~2 eV), distinguishing the peaks in such a short range is practically impossible. However, the models presented here were developed with as many physically and chemically relevant restrictions as possible (on peak position, peak width, peak shape, etc.) to ensure mathematical, physical, and chemical accuracy of component quantification. Furthermore, the results from this quantification are in complete agreement with FTIR results and the expectations based on the literature. Therefore, despite some potential limitations, we feel confident that these models can accurately represent the chemical composition of the surface.

3.4. Wetting Behavior

The wetting behavior of PT3, PT5, and PT5P75 is studied through static and dynamic water contact angle measurements. Numerous studies and reviews have emphasized the importance of dynamic wetting studies. It is often suggested that if a surface shows hysteresis, i.e., a difference between advancing and receding angles, static contact angle becomes an arbitrary value between advancing and receding angles [59–61] and tends to miss a significant amount of information regarding the wetting characteristics of a surface. In fact, it has been suggested that since the advancing angle is more sensitive to low-energy components of the surface while the receding angle is more sensitive to high-energy components [62], one can study them individually to gain a deeper insight on surface functionalities.

In this work, dynamic wetting is studied through placing a 4 µL droplet on the surface while keeping the needle in contact with the resting droplet and increasing or decreasing the droplet volume while measuring contact angle at the three-phase interfacial point five times per second (Figure 14). These measurements are then plotted against time and the curve is smoothed using the LOWESS method to account for minor variations (±1°) while retaining the general trend of the curves. More details on this procedure is provided in [63]. These curves are presented in Figure 15 for PT3, PT5, and PT5P75.

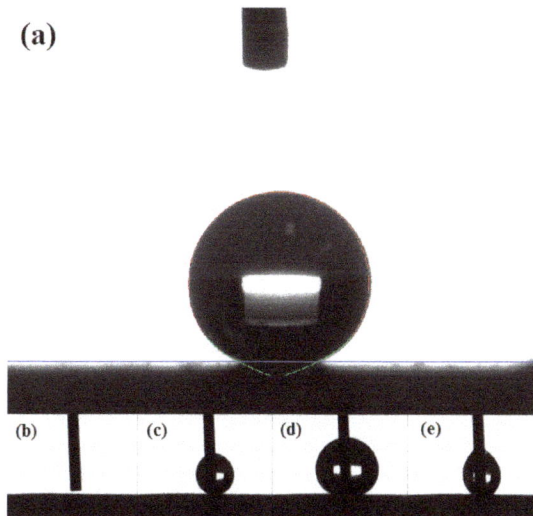

Figure 14. (**a**) a demonstration of the Tangent approximation method and (**b–e**) the procedure for determination of advancing and receding angles.

As the droplet volume slowly increases, the baseline initially remains the same, resulting in an increase in the measured contact angles. Eventually, the weight of the droplet becomes large enough to expand the baseline, which may be shown in the contact angle vs. time curves as either a drop in the measured values followed by a further gradual increase (sudden baseline expansion), or as a constant measured value with volume after the initial increase (continuous baseline expansion). Advancing contact angle is then determined as the measured value just before baseline expansion [39,60]. Similarly, reducing the droplet volume leads to an initial decrease with a sudden increase (sudden baseline shrinkage) or a constant value (continuous baseline shrinkage). In this case, receding angle is determined as the measured value just before the baseline shrinkage (see Figure 15). The fundamental physics behind this procedure have inspired several groups to discuss and compare their knowledge of surface chemistry, surface topography, and interfacial science [39,59,60,62,63]. However, the details of advancing and receding measurements are rarely discussed in detail in the literature, and therefore we are not certain whether other groups have observed the same behavior or their interpretation of how the baseline reacts to increasing/decreasing volume is consistent with ours.

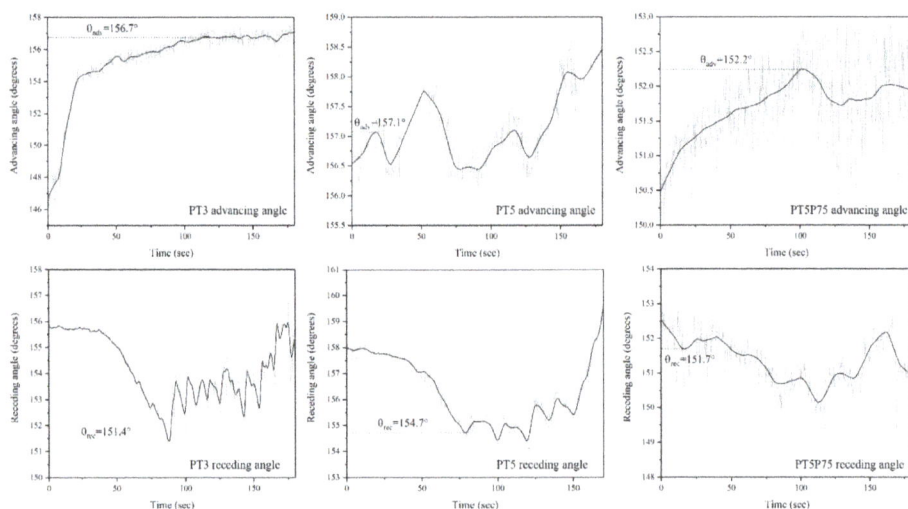

Figure 15. Advancing and receding angle curves for PT3, PT5 and PT5P75.

Based on the advancing and receding angle values presented in Figure 15, contact angle hysteresis was calculated for all coatings, and the results are presented in Figure 16 along with the static contact angle values. It should be noted that theoretically, the static contact angle should be between advancing and receding values. However, in the results presented here, some discrepancies may be observed due to different measurement methods: Static contact angle measurements were carried out using the Young–Laplace model, since it approximates the entire shape of the droplet, thus considering the effects of droplet weight or any other distortions in the droplet shape. However, during dynamic measurements, the needle is in contact with the droplet at all times, and therefore Young–Laplace approximation cannot identify the circular shape of the droplet. Therefore, a different approximation method (namely, the tangent method) was used, which only considers the three phase interfacial points and attempts to draw a line at this point tangent to the droplet shape to determine the contact angle.

All surfaces are shown to be superhydrophobic (WCA > 150°) with statistically insignificant variations in static contact angle due to the different deposition conditions. However, a significant difference in contact angle hysteresis (CAH) is observed for different samples, which is directly related to surface roughness. Traditionally, it is argued that hysteresis originates from defects, and thus rougher surfaces have higher CAH [59,64]. However, several studies have shown that the effects of roughness on CAH depends on the wetting regime. When water is in contact with the whole surface profile (Wenzel wetting regime), surface features act as obstacles to water motion, reducing the droplet mobility and increasing CAH. On the other hand, when roughness features are small enough so that the capillary effect prevents liquid penetration into the surface asperities (Cassie-Baxter wetting regime), water is only in contact with a small area fraction of surface features, and therefore its motion is not hindered by surface structures [62,65], increasing the droplet mobility and decreasing CAH.

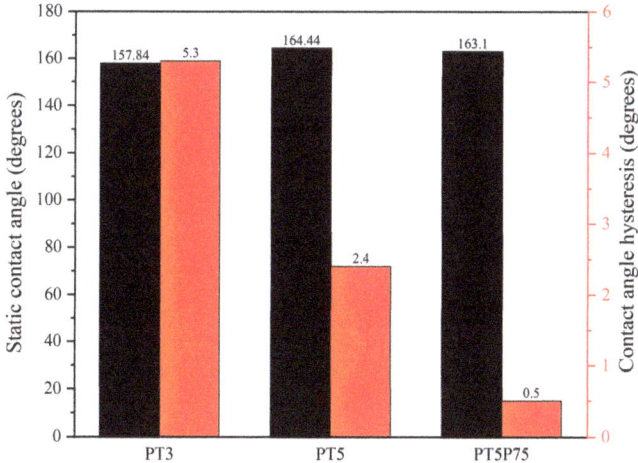

Figure 16. Static and dynamic contact angle for PT3, PT5, and PT5P75.

In the present work, the lowest contact angle hysteresis is observed in the case of PT5P75. While presenting the chemical characteristics of the developed coatings, it was shown that precursor fragmentation in PT5P75 is higher, leading to more polar oxide content on the surface, which is expected to render the surface less hydrophobic. However, since PT5P75 exhibits higher surface roughness due to larger silica-based particles with surface features as small as only a few nanometers (see Figure 8), contact angle hysteresis is very small. In the case of PT3, surface roughness is lower due to the lower amount of deposition and large micrometric surface features originated from the pre-treatment. Finally, PT5 exhibits a CAH higher than PT3 due to the full coverage of the surface with multi-leveled structures, and lower than PT5P75 due to the smaller deposited structures.

4. Conclusions

In this study, superhydrophobic coatings are developed through atmospheric pressure plasma polymerization of HMDSO in the jet of nitrogen plasma produced by rotating arc discharges. The details regarding the plasma deposition process and precursor fragmentation dynamics are discussed. The plasma jet is modified by mounting a quartz tube on the jet head, thus confining the plasma jet in a smaller volume. It is shown that this modification leads to structures that are similar to what is observed in atmospheric pressure plasma polymerization of HMDSO in the absence of O_2. After jet modification, the effects of precursor flow rate and plasma power on surface structure, wetting behavior and surface chemistry is studied. It is shown that increasing the flow rate while keeping the plasma power constant increases precursor fragmentation, leading to higher oxide deposition. On the other hand, increasing the plasma power while keeping the precursor flow rate constant results in faster deposition rates and subsequently thicker coatings, but with a higher oxide content. This demonstrates the significance of "available energy per precursor molecule" parameter, which can significantly affect the precursor fragmentation in the discharge. All conditions studied here lead to superhydrophobic surfaces with static contact angles higher than 150° and contact angle hysteresis lower than 6°.

Author Contributions: Conceptualization: S.A.; Methodology: S.A., L.S.; Software: S.A.; Validation: S.A., J.P., M.F., L.S.; Formal Analysis: S.A., J.P.; Investigation: S.A.; Resources: M.F., L.S.; Data Curation: S.A.; Writing—Original Draft: S.A.; Writing—Review and Editing: S.A., J.P.; Visualization: S.A.; Supervision: M.F., L.S.; Project Administration: M.F., L.S.; Funding Acquisition: M.F., L.S.

Funding: This work was carried out within the framework of the NSERC/Hydro-Quebec/UQAC Industrial Chair on Atmospheric Icing of Power Network Equipment (CIGELE), the Canada Research Chair on Engineering of Power Network Atmospheric Icing (INGIVRE) at Universite du Quebec a Chicoutimi, and the Chaire de Recherche du Canada en Science et Applications des Plasmas Froids Hautement Reactifs at Université de Montréal. The authors would like to thank the CIGELE partners (Hydro-Quebec, Hydro One, Reseau Transport d'Electricite (RTE), General Cable, K-Line Insulators, Dual-ADE, and FUQAC) whose financial support made this research possible.

Conflicts of Interest: The authors declare no conflict of interest.

References

1. Saraf, R.; Lee, H.J.; Michielsen, S.; Owens, J.; Willis, C.; Stone, C.; Wilusz, E. Comparison of three methods for generating superhydrophobic, superoleophobic nylon nonwoven surfaces. *J. Mater. Sci.* **2011**, *46*, 5751–5760. [CrossRef]

2. Ma, M.; Hill, R.M. Superhydrophobic surfaces. *Curr. Opin. Colloid Interface Sci.* **2006**, *11*, 193–202. [CrossRef]

3. Guo, Z.; Liu, W.; Su, B.L. Superhydrophobic surfaces: From natural to biomimetic to functional. *J. Colloid Interface Sci.* **2011**, *353*, 335–355. [CrossRef] [PubMed]

4. Johnson, R.E.; Dettre, R.H. Contact Angle Hysteresis. III. Study of an Idealized Heterogeneous Surface. *J. Phys. Chem.* **1964**, *68*, 1744–1750. [CrossRef]

5. Guo, Z.; Liu, W. Biomimic from the superhydrophobic plant leaves in nature: Binary structure and unitary structure. *Plant Sci.* **2007**, *172*, 1103–1112. [CrossRef]

6. Yan, Y.Y.; Gao, N.; Barthlott, W. Mimicking natural superhydrophobic surfaces and grasping the wetting process: A review on recent progress in preparing superhydrophobic surfaces. *Adv. Colloid Interface Sci.* **2011**, *169*, 80–105. [CrossRef]

7. Li, X.-M.; Reinhoudt, D.; Crego-Calama, M. What do we need for a superhydrophobic surface? A review on the recent progress in the preparation of superhydrophobic surfaces. *Chem. Soc. Rev.* **2007**, *36*, 1350–1368. [CrossRef]

8. Kale, K.H.; Palaskar, S. Atmospheric pressure plasma polymerization of hexamethyldisiloxane for imparting water repellency to cotton fabric. *Text. Res. J.* **2010**, *81*, 608–620. [CrossRef]

9. Cheng, Y.; Zhu, T.; Li, S.; Huang, J.; Mao, J.; Yang, H.; Gao, S.; Chen, Z.; Lai, Y. A novel strategy for fabricating robust superhydrophobic fabrics by environmentally-friendly enzyme etching. *Chem. Eng. J.* **2019**, *355*, 290–298. [CrossRef]

10. Favia, P.; D'Agostino, R. Plasma treatments and plasma deposition of polymers for biomedical applications. *Surf. Coat. Technol.* **1998**, *98*, 1102–1106. [CrossRef]

11. Ellinas, K.; Kefallinou, D.; Stamatakis, K.; Gogolides, E.; Tserepi, A. Is There a Threshold in the Antibacterial Action of Superhydrophobic Surfaces? *ACS Appl. Mater. Interfaces* **2017**, *9*, 39781–39789. [CrossRef] [PubMed]

12. Yao, X.; Song, Y.; Jiang, L. Applications of bio-inspired special wettable surfaces. *Adv. Mater.* **2011**, *23*, 719–734. [CrossRef]

13. Farzaneh, M. Atmospheric icing of power networks. In *Atmospheric Icing of Power Networks*; Springer: Berlin, Germany, 2008; pp. 1–381, ISBN 978-1-40-208530-7.

14. Alizadeh, A.; Yamada, M.; Li, R.; Shang, W.; Otta, S.; Zhong, S.; Ge, L.; Dhinojwala, A.; Conway, K.R.; Bahadur, V.; et al. Dynamics of ice nucleation on water repellent surfaces. *Langmuir* **2012**, *28*, 3180–3186. [CrossRef] [PubMed]

15. Farzaneh, M. Ice accretions on high–voltage conductors and insulators and related phenomena. *Philos. Trans. R. Soc. Lond. A Math. Phys. Eng. Sci.* **2000**, *358*, 2971–3005. [CrossRef]

16. Farzaneh, M.; Sarkar, D.K. Nanostructured Superhydrophobic Coatings. *J. CPRI* **2008**, *4*, 135–147.

17. Marmur, A. Wetting on hydrophobic rough surfaces: To be heterogeneous or not to be? *Langmuir* **2003**, *19*, 8343–8348. [CrossRef]

18. Li, S.; Page, K.; Sathasivam, S.; Heale, F.; He, G.; Lu, Y.; Lai, Y.; Chen, G.; Carmalt, C.J.; Parkin, I.P. Efficiently texturing hierarchical superhydrophobic fluoride-free translucent films by AACVD with excellent durability and self-cleaning ability. *J. Mater. Chem. A* **2018**, *6*, 17633–17641. [CrossRef]

19. Genzer, J.; Efimenko, K. Recent developments in superhydrophobic surfaces and their relevance to marine fouling: A review. *Biofouling* **2006**, *22*, 339–360. [CrossRef]

20. Ge, M.; Cao, C.; Huang, J.; Zhang, X.; Tang, Y.; Zhou, X.; Zhang, K.; Chen, Z.; Lai, Y. Rational design of materials interface at nanoscale towards intelligent oil-water separation. *Nanoscale Horizons* **2018**, *3*, 235–260. [CrossRef]

21. Liu, H.; Wang, Y.; Huang, J.; Chen, Z.; Chen, G.; Lai, Y. Bioinspired Surfaces with Superamphiphobic Properties: Concepts, Synthesis, and Applications. *Adv. Funct. Mater.* **2018**, *28*, 1707415. [CrossRef]

22. Wang, S.; Feng, L.; Jiang, L. One-step solution-immersion process for the fabrication of stable bionic superhydrophobic surfaces. *Adv. Mater.* **2006**, *18*, 767–770. [CrossRef]

23. Hermelin, E.; Petitjean, J.; Lacroix, J.C.; Chane-Ching, K.I.; Tanguy, J.; Lacaze, P.C. Ultrafast electrosynthesis of high hydrophobic polypyrrole coatings on a zinc electrode: Applications to the protection against corrosion. *Chem. Mater.* **2008**, *20*, 4447–4456. [CrossRef]

24. Zhang, F.; Zhao, L.; Chen, H.; Xu, S.; Evans, D.G.; Duan, X. Corrosion resistance of superhydrophobic layered double hydroxide films on aluminum. *Angew. Chem. Int. Ed.* **2008**, *47*, 2466–2469. [CrossRef] [PubMed]

25. Jafari, R.; Asadollahi, S.; Farzaneh, M. Applications of plasma technology in development of superhydrophobic surfaces. *Plasma Chem. Plasma Process.* **2013**, *33*, 177–200. [CrossRef]

26. Martinu, L.; Poitras, D. Plasma deposition of optical films and coatings: A review. *Vac. Sci. Technol. A* **2000**, *18*, 2619–2645. [CrossRef]

27. Gogolides, E.; Constantoudis, V.; Kokkoris, G.; Kontziampasis, D.; Tsougeni, K.; Boulousis, G.; Vlachopoulou, M.; Tserepi, A. Controlling roughness: From etching to nanotexturing and plasma-directed organization on organic and inorganic materials. *J. Phys. D Appl. Phys.* **2011**, *44*, 174021. [CrossRef]

28. Saccaro, S.; Fonseca, R.; Veillon, D.M.; Cotelingam, J.; Nordberg, M.L.; Bredeson, C.; Glass, J.; Munker, R. Primary plasma cell leukemia: Report of 17 new cases treated with autologous or allogeneic stem-cell transplantation and review of the literature. *Am. J. Hematol.* **2005**, *78*, 288–294. [CrossRef]

29. Lloyd, G.; Friedman, G.; Jafri, S.; Schultz, G.; Fridman, A.; Harding, K. Gas plasma: Medical uses and developments in wound care. *Plasma Process. Polym.* **2010**, *7*, 194–211. [CrossRef]

30. Kong, M.G.; Kroesen, G.; Morfill, G.; Nosenko, T.; Shimizu, T.; Van Dijk, J.; Zimmermann, J.L. Plasma medicine: An introductory review. *New J. Phys.* **2009**, *11*, 115012. [CrossRef]

31. Hertwig, C.; Reineke, K.; Ehlbeck, J.; Knorr, D.; Schlüter, O. Decontamination of whole black pepper using different cold atmospheric pressure plasma applications. *Food Control* **2015**, *55*, 221–229. [CrossRef]

32. Pasquali, F.; Stratakos, A.C.; Koidis, A.; Berardinelli, A.; Cevoli, C.; Ragni, L.; Mancusi, R.; Manfreda, G.; Trevisani, M. Atmospheric cold plasma process for vegetable leaf decontamination: A feasibility study on radicchio (red chicory, *Cichorium intybus* L.). *Food Control* **2016**, *60*, 552–559. [CrossRef]

33. Merche, D.; Vandencasteele, N.; Reniers, F. Atmospheric plasmas for thin film deposition: A critical review. *Thin Solid Films* **2012**, *520*, 4219–4236. [CrossRef]

34. Siliprandi, R.A.; Zanini, S.; Grimoldi, E.; Fumagalli, F.S.; Barni, R.; Riccardi, C. Atmospheric pressure plasma discharge for polysiloxane thin films deposition and comparison with low pressure process. *Plasma Chem. Plasma Process.* **2011**, *31*, 353–372. [CrossRef]

35. Foroughi Mobarakeh, L.; Jafari, R.; Farzaneh, M. Superhydrophobic Surface Elaboration Using Plasma Polymerization of Hexamethyldisiloxane (HMDSO). *Adv. Mater. Res.* **2011**, *409*, 783–787. [CrossRef]

36. Lommatzsch, U.; Ihde, J. Plasma polymerization of HMDSO with an atmospheric pressure plasma jet for corrosion protection of aluminum and low-adhesion surfaces. *Plasma Process. Polym.* **2009**, *6*, 642–648. [CrossRef]

37. Gogolides, E.; Ellinas, K.; Tserepi, A. Hierarchical micro and nano structured, hydrophilic, superhydrophobic and superoleophobic surfaces incorporated in microfluidics, microarrays and lab on chip microsystems. *Microelectron. Eng.* **2015**, *132*, 135–155. [CrossRef]

38. Asadollahi, S.; Farzaneh, M.; Stafford, L. Highly porous micro-roughened structures developed on aluminum surface using the jet of rotating arc discharges at atmospheric pressure. *J. Appl. Phys.* **2018**, *123*, 073302. [CrossRef]

39. Kietzig, A.M. Comments on "an essay on contact angle measurements"—An illustration of the respective influence of droplet deposition and measurement parameters. *Plasma Process. Polym.* **2011**, *8*, 1003–1009. [CrossRef]

40. Müller, M.; Oehr, C. Comments on "an essay on contact angle measurements" by Strobel and Lyons. *Plasma Process. Polym.* **2011**, *8*, 19–24. [CrossRef]

41. Levasseur, O.; Stafford, L.; Gherardi, N.; Naudé, N.; Beche, E.; Esvan, J.; Blanchet, P.; Riedl, B.; Sarkissian, A. Role of substrate outgassing on the formation dynamics of either hydrophilic or hydrophobic wood surfaces in atmospheric-pressure, organosilicon plasmas. *Surf. Coat. Technol.* **2013**, *234*, 42–47. [CrossRef]

42. Pulpytel, J.; Kumar, V.; Peng, P.; Micheli, V.; Laidani, N.; Arefi-Khonsari, F. Deposition of organosilicon coatings by a non-equilibrium atmospheric pressure plasma jet: Design, analysis and macroscopic scaling law of the process. *Plasma Process. Polym.* **2011**, *8*, 664–675. [CrossRef]

43. Gunde, M.K. Vibrational modes in amorphous silicon dioxide. *Phys. B Condens. Matter* **2000**, *292*, 286–295. [CrossRef]

44. Kirk, C.T. Quantitative analysis of the effect of disorder-induced mode coupling on infrared absorption in silica. *Phys. Rev. B* **1988**, *38*, 1255. [CrossRef]

45. Petersen, J.; Bardon, J.; Dinia, A.; Ruch, D.; Gherardi, N. Organosilicon Coatings Deposited in Atmospheric Pressure Townsend Discharge for Gas Barrier Purpose: Effect of Substrate Temperature on Structure and Properties. *ACS Appl. Mater. Interfaces* **2012**, *4*, 5872–5882. [CrossRef] [PubMed]

46. Launer, P.J. *Infrared Analysis of Organosilicon Compounds: Spectra-Structure Correlations*; Burnt Hills: New York, NY, USA, 1987.

47. Barthlott, W.; Neinhuis, C. Purity of the sacred lotus, or escape from contamination in biological surfaces. *Planta* **1997**, *202*, 1–8. [CrossRef]

48. Kilicaslan, A.; Levasseur, O.; Roy-Garofano, V.; Profili, J.; Moisan, M.; Cote, C.; Sarkissian, A.; Stafford, L. Optical emission spectroscopy of microwave-plasmas at atmospheric pressure applied to the growth of organosilicon and organotitanium nanopowders. *J. Appl. Phys.* **2014**, *115*, 113301. [CrossRef]

49. Ricci, M.; Dorier, J.L.; Hollenstein, C.; Fayet, P. Influence of argon and nitrogen admixture in HMDSO/O2 plasmas onto powder formation. *Plasma Process. Polym.* **2011**, *8*, 108–117. [CrossRef]

50. Park, E.S.; Ro, H.W.; Nguyen, C.V.; Jaffe, R.L.; Yoon, D.Y. Infrared Spectroscopy Study of Microstructures of Poly (silsesquioxane)s. *Chem. Mater.* **2008**, *20*, 1548–1554. [CrossRef]

51. Morent, R.; De Geyter, N.; Van Vlierberghe, S.; Dubruel, P.; Leys, C.; Gengembre, L.; Schacht, E.; Payen, E. Deposition of HMDSO-based coatings on PET substrates using an atmospheric pressure dielectric barrier discharge. *Prog. Org. Coat.* **2009**, *64*, 304–310. [CrossRef]

52. Coates, J. Interpretation of Infrared Spectra, A Practical Approach. In *Encyclopedia of Analytical Chemistry*; John Wiley & Sons, Ltd.: Hoboken, NJ, USA, 2000; pp. 10815–10837, ISBN 978-0-47-002731-8.

53. Gurav, A.B.; Latthe, S.S.; Kappenstein, C.; Mukherjee, S.K.; Rao, A.V.; Vhatkar, R.S. Porous water repellent silica coatings on glass by sol-gel method. *J. Porous Mater.* **2011**, *18*, 361–367. [CrossRef]

54. Zhu, Q.; Chu, Y.; Wang, Z.; Chen, N.; Lin, L.; Liu, F.; Pan, Q. Robust superhydrophobic polyurethane sponge as a highly reusable oil-absorption material. *J. Mater. Chem. A* **2013**, *1*, 5386–5393. [CrossRef]

55. Laroche, G.; Fitremann, J.; Gherardi, N. FTIR-ATR spectroscopy in thin film studies: The importance of sampling depth and deposition substrate. *Appl. Surf. Sci.* **2013**, *273*, 632–637. [CrossRef]

56. Raynaud, P.; Despax, B.; Segui, Y.; Caquineau, H. FTIR plasma phase analysis of hexamethyldisiloxane discharge in microwave multipolar plasma at different electrical powers. *Plasma Process. Polym.* **2005**, *2*, 45–52. [CrossRef]

57. Maurau, R.; Boscher, N.D.; Guillot, J.; Choquet, P. Nitrogen introduction in pp-HMDSO thin films deposited by atmospheric pressure dielectric barrier discharge: An XPS study. *Plasma Process. Polym.* **2012**, *9*, 316–323. [CrossRef]

58. O'Hare, L.A.; Hynes, A.; Alexander, M.R. A methodology for curve-fitting of the XPS Si 2p core level from thin siloxane coatings. *Surf. Interface Anal.* **2007**, *39*, 926–936. [CrossRef]

59. O'Hare, L.A.; Parbhoo, B.; Leadley, S.R. Development of a methodology for XPS curve-fitting of the Si 2p core level of siloxane materials. *Surf. Interface Anal.* **2004**, *36*, 1427–1434. [CrossRef]

60. Strobel, M.; Lyons, C.S. An essay on contact angle measurements. *Plasma Process. Polym.* **2011**, *8*, 8–13. [CrossRef]

61. Montes Ruiz-Cabello, F.J.; Rodríguez-Valverde, M.A.; Cabrerizo-Vílchez, M.A. Additional comments on "an essay on contact angle measurements" by M. Strobel and C. S. Lyons. *Plasma Process. Polym.* **2011**, *8*, 363–366. [CrossRef]

62. Di Mundo, R.; Palumbo, F. Comments regarding "an essay on contact angle measurements". *Plasma Process. Polym.* **2011**, *8*, 14–18. [CrossRef]

63. Gao, L.; McCarthy, T.J. Wetting 101°. *Langmuir* **2009**, *25*, 14105–14115. [CrossRef]

64. Morra, M.; Occhiello, E.; Garbassi, F. Knowledge about polymer surfaces from contact angle measurements. *Adv. Colloid Interface Sci.* **1990**, *32*, 79–116. [CrossRef]

65. Di Mundo, R.; Palumbo, F.; D'Agostino, R. Nanotexturing of polystyrene surface in fluorocarbon plasmas: From sticky to slippery superhydrophobicity. *Langmuir* **2008**, *24*, 5044–5051. [CrossRef] [PubMed]

materials

MDPI

Article

Visible Light Assisted Organosilane Assembly on Mesoporous Silicon Films and Particles

Chloé Rodriguez [1], Alvaro Muñoz Noval [1], Vicente Torres-Costa [1,2], Giacomo Ceccone [3] and Miguel Manso Silván [1,*]

[1] Departamento de Física Aplicada and Instituto de Ciencia de Materiales Nicolás Cabrera, Universidad Autónoma de Madrid, 28049 Madrid, Spain; chloe.rodriguez@uam.es (C.R.); alvaro.betelgeuse@gmail.com (A.M.N.); vicente.torres@uam.es (V.T.-C.)
[2] Centro de Microanálisis de Materiales, Universidad Autónoma de Madrid, 28049 Madrid, Spain
[3] European Commission, Joint Research Center, 21020 Ispra (Va), Italy; giacomo.ceccone@ec.europa.eu
* Correspondence: miguel.manso@uam.es; Tel.: +34-914974918

Received: 25 November 2018; Accepted: 25 December 2018; Published: 3 January 2019

Abstract: Porous silicon (PSi) is a versatile matrix with tailorable surface reactivity, which allows the processing of a range of multifunctional films and particles. The biomedical applications of PSi often require a surface capping with organic functionalities. This work shows that visible light can be used to catalyze the assembly of organosilanes on the PSi, as demonstrated with two organosilanes: aminopropyl-triethoxy-silane and perfluorodecyl-triethoxy-silane. We studied the process related to PSi films (PSiFs), which were characterized by X-ray photoelectron spectroscopy (XPS), time of flight secondary ion mass spectroscopy (ToF-SIMS) and field emission scanning electron microscopy (FESEM) before and after a plasma patterning process. The analyses confirmed the surface oxidation and the anchorage of the organosilane backbone. We further highlighted the surface analytical potential of ^{13}C, ^{19}F and ^{29}Si solid-state NMR (SS-NMR) as compared to Fourier transformed infrared spectroscopy (FTIR) in the characterization of functionalized PSi particles (PSiPs). The reduced invasiveness of the organosilanization regarding the PSiPs morphology was confirmed using transmission electron microscopy (TEM) and FESEM. Relevantly, the results obtained on PSiPs complemented those obtained on PSiFs. SS-NMR suggests a number of siloxane bonds between the organosilane and the PSiPs, which does not reach levels of maximum heterogeneous condensation, while ToF-SIMS suggested a certain degree of organosilane polymerization. Additionally, differences among the carbons in the organic (non-hydrolyzable) functionalizing groups are identified, especially in the case of the perfluorodecyl group. The spectroscopic characterization was used to propose a mechanism for the visible light activation of the organosilane assembly, which is based on the initial photoactivated oxidation of the PSi matrix.

Keywords: porous silicon; visible light assisted organosilanization; solid state NMR; XPS; ToF-SIMS

1. Introduction

The functionalization of semiconductor nanostructures with organic monolayers is a requirement for tailoring their surface chemistry for ulterior bioconjugation [1,2]. Porous Silicon (PSi) can be described as a matrix of silicon quantum dots (Si QDs) immersed in an amorphous Si/silica network. It differs from Si QDs in amorphous Si networks in the high specific surface areas (hundreds of m^2/g) [3], tailorable within a wide range of pore structures. Although both micro- and mesopores can be used to achieve high porosity, mesoporous structures are generally preferred because they exhibit better mechanical stability in comparison to the microporous ones [4]. With respect to porous silica networks, they differ mainly in their optoelectronic properties (photo and electroluminescence). PSi is generally engineered as a film supported on a Si wafer (PSiF), but has been described also in the form of colloidal particles (PSiPs) for a wide range of applications.

Engineered PSi is especially attractive in the biomedical field due to its high biocompatibility, biodegradability [5] and role as multifunctional actuator [6]. By means of functionalization, PSi had gained a great deal of attention in applications such as magnetic focusing [7] or drug delivery [8], targeted contrast agents in biomedical imaging [9], tissue engineering through polymer [10] or hydroxyapatite loads [11], photothermal cancer treatment through plasmonic composites [12], biocatalytic substrate upon functionalization with enzymes [13] or biosensing through interferometric effects [14]. In order to avoid an unspecific reactivity of PSi, a conjugation with organic molecules/biomolecules is essential.

Surface modification of PSi through organosilane assemblies allows for specific functionalities and biomolecular selectivity [15]. Relevantly, to enable the organosilane assembly, a prior oxidation of the PSi surface is required, which has been traditionally carried out by thermal or chemical oxidation [16]. Such oxidation of PSi has previously been induced by visible light activation, but aimed at a functionalization with metal nanoparticles. Light assisted redox reactions with residual water are in fact responsible for a mild surface oxidation [17], which we have previously proposed for the activation of the assembly of glycydyloxy-trimethyl-silane [18].

In this work, we aim to propose a generalization for the process of visible light assisted organosilanization (VLAO) with two additional organosilanes, as well as a reaction mechanism for the process. We therefore rely on an advanced characterization of the modified surfaces. We outline the analytical potential of solid-state nuclear magnetic resonance (SS-NMR) at the surface level and complement with traditional surface spectroscopic techniques, such as time of flight secondary ion mass spectroscopy (ToF-SIMS), and X-ray photoelectron spectroscopy (XPS), in order to study the properties of organosilane functionalized PSi. The surface sensitivity of SS-NMR has already been highlighted in porous silica based systems as a technique to decipher the influence of surfactants in the structure of molecular sieves [19]. Within the more restricted area of PSi, SS-NMR has been used for the characterization of the chemistry upon formation [20] and the correlation of structure with luminescence [21].

Contrary to bulk Si, where the surface is inaccessible to NMR, PSi provides a sufficient number of surface nuclei to overcome NMR's inherently low surface sensitivity and achieve a satisfactory signal-to-noise ratio [22]. Relevantly, although the sensitivity of SS-NMR for amine surface capping has already been outlined for Si nanoparticles [23], a prospective study on the analytical potential of SS-NMR on organosilane functionalized PSiPs could provide further insight into the chemical structure of the formed hybrid materials. A surface/interface sensitive approach of this sort has already been used to characterize the organosilane adsorption on highly porous clay materials [24]. Transmission electron microscopy (TEM), scanning electron microscopy (SEM), Fourier transformed infrared spectroscopy (FTIR), ToF-SIMS and XPS are used on convenient PSiF and PSiP models to characterize the organosilane assemblies and correlate with the SS-NMR information.

2. Experimental

2.1. Preparation of PSiFs and PSiPs

The back side of the p-type boron-doped (resistivity 0.05–0.1 Ω·cm) (100) oriented Si wafers was first coated with an aluminum layer to provide low resistance ohmic electrical contacts. Si was then cut and mounted into a Teflon electrochemical cell to form PSiFs by anodic etching of the silicon wafers in an aqueous electrolyte composed of a mixture of hydrofluoric acid (HF) (40%) and absolute ethanol (98%) (volume ratio 1:1). The current density was fixed at 78 mA/cm^2 and the anodization time at 20 s, leading to a 1 μm thick PSiF. For the formation of PSiPs, the current was periodically (every 50 s) pulsed for 1 s to a higher value (104 mA/cm^2). The waveform was repeated for 30 cycles, producing highly porous and mechanically fragile layers spaced at predetermined points in the porous film [25].

The thickness-induced instability in the PS/Si interface caused a fragmentation of the layer, which could be easily scrapped from the surface. Through these artificial cleavage planes in the PSi film, we favored the extraction of homogeneous PSiPs by sonication in ethanol for 3 h. The resulting PSiPs were rinsed with ethanol and used without any filtration.

2.2. Organosilane Assembly on PSi

Two different chemical functionalities were considered using 3-aminopropyl-triethoxy-silane (APTS) and 1*H*,1*H*,2*H*,2*H*-perfluorodecyl-triethoxy-silane (PFDS) (both from Sigma Aldrich, St. Louis, MO, USA) (Figure 1). The organosilane-based solutions were prepared by diluting the aminosilane (0.2% *v*/*v*) in EtOH (absolute, 99.8%, Sigma Aldrich). The VLAO process took place by illuminating the concerned PSi objects through the silane dilutions so that both the PSi structures and the organosilanes were exposed to the visible light simultaneously. The visible light source consisted of a 150 W halogen lamp with a maximum of 10% power emission at 700 nm and a power emission below 2% for radiation under 360 nm. The VLAO process on the PSiPs/PSiFs took place for 10 min (APTS) or 30 min (PFDS). Temperature rises (T < 40 °C) were limited with a heat dissipation bath [18]. The silane concentration used for the functionalization was low in order to remain within the sub-monolayer regime and minimize undesired reactions with solvent moisture. Residual water was however not extracted from the solvent due to its principal role in the photo-assisted oxidation of the PSi surface [26]. The samples were then cleaned by rinsing in the respective solvent used for the reaction and dried under an N_2 flow. The whole process was carried out in a glove box.

(a)   (b)

Figure 1. Molecular structure of (a) 3-aminopropyl-triethoxy-silane (APTS) and (b) 1*H*,1*H*,2*H*,2*H*-perfluorodecyl-triethoxy-silane (PFDS).

The surface characterization of PSiFs was performed on samples bearing an internal negative control. After visible light activated organosilane assembly, PSiFs were patterned using a Si mask in an etching process carried out in a capacitive plasma reactor, for 5 min, with a 50 W RF (13.56 MHz) discharge, 30 sccm Ar and a working pressure of 65 Pa.

2.3. Characterization

Morphological characterization of the PSiPs was carried out using a 2100F TEM (JEOL, Akishima, Tokyo, Japan) operated at 200 kV after dispersing the PSiPs from an acetone solutions onto carbon coated Cu grids. Field emission SEM (FESEM) images from PSiPs and PSiFs were acquired in a XL30S microscope (FEI/Philips, Hillsboro, OR, United States) operated operated at 10 keV. The characterization of the functionalized PSiPs was performed using a Vector 22 FTIR spectrometer (Brucker, Billerica, MA, US), resolution 8 cm^{-1}, 4000–400 cm^{-1}, 32 scans at 10 kHz) in the transmission configuration after the preparation of KBr disks. NMR spectra were obtained at room temperature in a AV-400-WB (Bruker, Billerica, MA, US) incorporating a 4 mm triple probe channel using ZrO rotors with KeI-F stopper. The rotor speed was 10 kHz. ^{13}C spectra were obtained in a cross-polarization (CP-MAS) between ^{1}H and ^{13}C nuclear spins with dipolar decoupling of ^{1}H at 80 kHz. The working frequencies were 400.13 MHz for ^{1}H and 100.61 MHz for ^{13}C. The ^{1}H excitation pulse was 2.75 μs, the spectral width was 35 kHz, the contact time was 3 ms, the relaxation time was 4 s, and 1 k scans were accumulated. TMS and Adamantano (CH$_2$ 29.5 ppm) were the primary and secondary references, respectively. ^{19}F spectra were acquired with a double probe channel of 2.5 mm and Vespel stopper rotors using 2 μs pulses of 90° and a spectral width of 30 kHz. The recovery time was 120 s due to ^{19}F nuclei having large spin-lattice relaxation times [27]. The rotation speed was 20 kHz and the spectra were acquired overnight. CFCl$_3$ and Na$_2$SiF$_6$ (−152.46 ppm) were used as the primary and secondary references, respectively. For the ^{29}Si cross polarization magic angle spinning (CP/MAS) experiment, an excitation pulse of 3 μs was used for ^{1}H, corresponding to an angle of $\pi/2$. The relaxation time was 5 s and the spectral width 40 kHz. TMS and Caolin (−91.2 ppm) were used as the primary and secondary references, respectively. The 'magic angle' of 54.74° relative to the direction of the static magnetic field allow the elimination of broadening.

XPS data were obtained with a with a PHOIBOS 150 9MCD energy analyzer (SPECS, Berlin, Germany). XPS spectra of the functionalized PSiFs were acquired with electron emission angles of 0° and source-to-analyzer angles of 90° using non-monochromatic Mg K$_\alpha$ excitation with pass energies of 75 eV for the survey and 25 eV for the core-level spectra, respectively. The data were processed using CasaXPS v16R1 (Casa Software, UK) taking the aliphatic contribution to the C1s core level at 285.0 eV as a binding energy (BE) reference. The PFDS micropatterned PSiFs were characterized with ToF-SIMS. The analysis was conducted using a reflector type SIMS IV spectrometer incorporated with Surface Lab software v6.4 (ION-TOF GmbH, Münster, Germany) and a 25 keV liquid metal ion gun (LMIG) operating with bismuth primary ions. Spectra were acquired in static mode (primary ion fluence <10^{12} ions·cm^{-2}). Charge compensation was applied to the analysis using low-energy (~20 eV) electron flooding. Mass calibration was obtained using the peaks C- (12 m/z), C2- (24 m/z), C3- (36 m/z), C4- (48 m/z), and C5- (60 m/z) for negative ion and H+, C+ (12 m/z), CH+ (13 m/z), CH2+ (14 m/z) and CH3+ (15 m/z) for positive ion spectra. Analyses were done using a squared area of 250 × 250 μm^2 in the high mass resolution burst mode (resolution M/ΔM > 6000). Values of m/z are given as dimensionless in keeping with international union of pure and applied chemistry (IUPAC) recommendations, even though m represents the unified atomic mass unit in u.

3. Results and Discussion

3.1. Characterization of PSiFs

The effects of the VLAO process on the PSiFs was initially evaluated using FESEM. The results illustrate that the process respects the initial topography of a PSiF control (Figure 2a), where the different magnifications of the process performed using both APTS (Figure 2b) and PFDS (Figure 2c) denote no pore occlusion or drastic adsorption of heterogeneously induced colloidal structures [28], which are often the by-product of silanization processes. Although the porosity remains within the same order of magnitude for both samples, the resulting mean pore size is slightly bigger for APTS functionalization (roughly 40 nm) as compared to the PFDS functionalization (approximately 25 nm). The origin of this difference resides in the different porosities observed on batch-to-batch observation due to slight differences in the radial distribution of the electric field on the active Si electrode and the loss of chemical potential of the electrolyte during the progression of one series of samples.

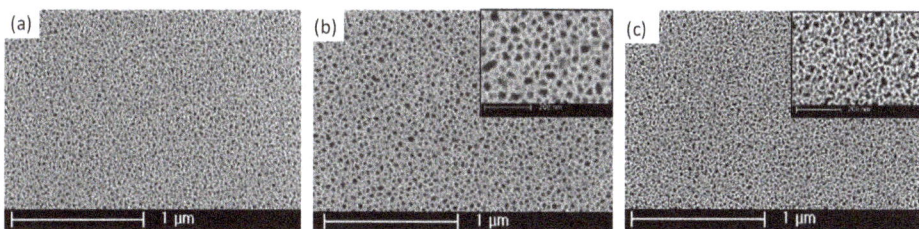

Figure 2. Field emission scanning electron microscopy (FESEM) images of the surface of PSiFs before (**a**) and after assembly of APTS (**b**) and PFDS (**c**).

To further explore the properties of VLAO processed PSiFs, we performed a surface analysis of the samples functionalized with the two organosilanes before and after an Ar plasma etching process. The APTS functionalized PSiFs were characterized by XPS as shown in Figure 3, by acquiring spectra from macro-patterned areas (i.e., a 1 × 1 cm² Si mask-exposed half of the APTS-PSiF to the plasma protecting the rest of the sample). The protected area (see spectra on the top of Figure 3) showed a considerably high content of C (22.0 at.%) and N (4.4 at.%), sustaining the integration of the aminopropyl group on the surface. O (41.2 at.%) and Si (32.4 at.%) however, were the dominant elements on the surface as shown in the top, widescan spectrum image in Figure 3. The analysis of the most relevant core level spectra confirmed on the one hand, the variety of the surface organic species, and on the second hand, the predominance of the oxidized state of the Si species. In fact, the Si2p core level spectrum (center) shows a dominant component with a BE at 103.0 eV, compatible with a fully oxidized Si, which can arise from both the oxidized PSiF [29] and the siloxane bonds from the adsorbed organosilane [30]. Relevantly, the C1s core level is characterized by a strong asymmetry, compatible with four different contributions [31,32]: The aliphatic C–C at BE 285.0 eV, the C–N at BE 286.0 eV, the C–O at BE 287. 2 eV and the C=O at BE 288.5 eV. The first two contributions of the C1s spectrum (right) confirm the high level of integration of the aminopropyl group, while the N1s spectrum in the inset shows that the primary amines are present in neutral and charged forms [33,34] for the low and high BE contributions, respectively.

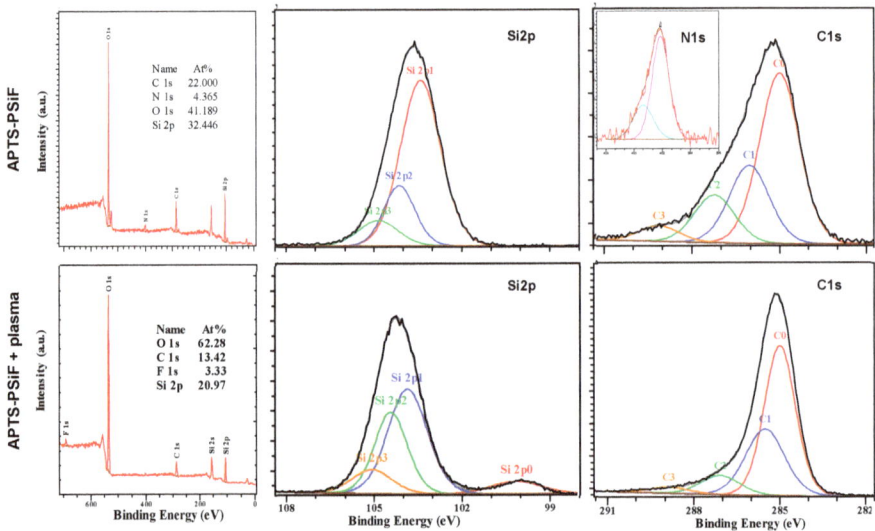

Figure 3. X-ray photoelectron spectroscopy (XPS) analysis of an APTS functionalized PSiF before (**top**) and after (**bottom**) an Ar plasma etching process. From left to right: widescan, Si2p and C1s spectra. Note: N1s inset to the fresh APTS-PSiF sample.

After the selective Ar plasma etching process, the surface composition of the APTS functionalized PSiFs changed considerably as shown in the set of spectra in the bottom part of Figure 3. The etching treatment had a drastic effect on the organic content of the PSiFs with a reduction from 22 to 13 at.% in C and an almost complete deletion of surface N (only traces below 1% could be detected, see corresponding widescan spectrum). Further, the Si content was drastically affected with a reduction from 35 to 21 at.%.

The surface suffers a drastic C depletion after the etching process with a resulting increase of O content from approximately 41 up to 62 at.%. Additionally, the presence of F produced as a by-product during the PSiF synthesis was confirmed after etching of the adsorbed APTS (up to 3.3 at.%). The preferential effect on deletion of the adsorbed APTS was confirmed from the Si2p spectrum. After etching, a secondary wide minor contribution at the low BE side was observed, denoting the presence of Si–Si clusters near the surface, which are compatible with the structure of non-fully oxidized PSiFs [29]. With regard to the C1s spectrum, the relative intensity of the contributions at a high BE was drastically reduced, which denotes the drastic reduction in variety of organic structures on the PSiF surface.

The PSiFs functionalized with PFDS were characterized using ToF-SIMS, as illustrated in Figure 4 for two different regions (positive and negative on the left and right, respectively). During the Ar plasma etching process, a micro-mask was used to exploit the imaging mode of the equipment used. In the positive ions spectra, the mask protected areas denoted the presence of $C_2H_3O^+$ and $SiHO^+$ ions, which are characteristic derivative fragments of the core of perfluorinated silanes [35]. On the other hand, the plasma exposed areas showed contributions from $KNaF^+$ and K_2F^+, which are characteristic of F by-products of the PSiF formation process [36]. The same kind of relationships were found in a second region used for the analysis of negative ions. Characteristic ions from masked protected areas, such as $Si_3HO_7^-$ or $Si_3C_3HO^-$, can be ascribed to a certain level of 'clusterization' of siloxane bonds from PFDS, suggesting that the VLAO process leads to some homogeneous condensation of PFDS. For the exposed areas, the SiF_5^- and F^- ion maps illustrate once again, that the Ar plasma process induces an efficient surface cleaning, exposing by-products of the PSiF synthesis.

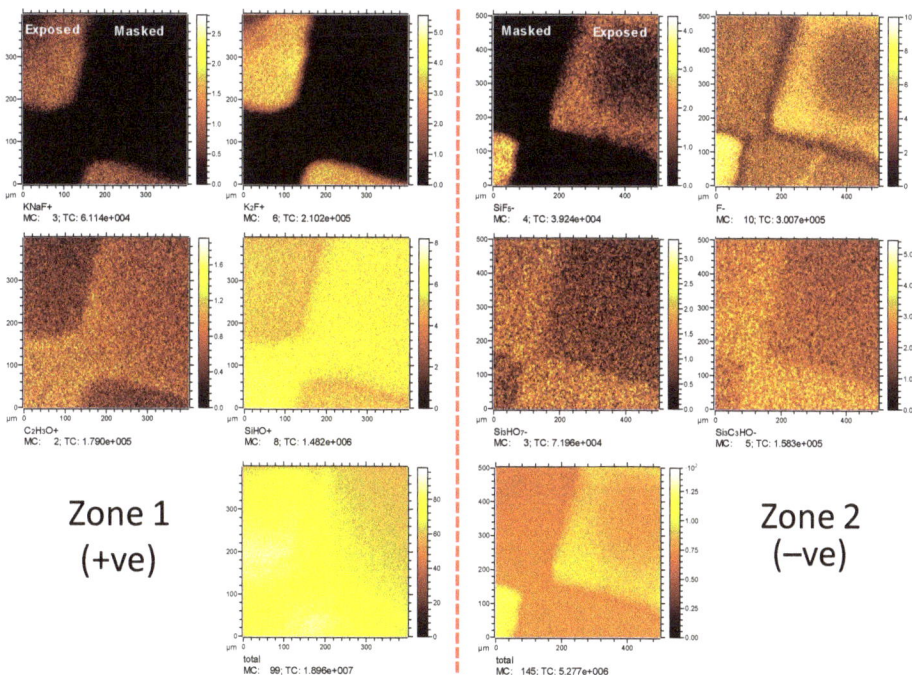

Figure 4. Time of flight secondary ion mass spectroscopy (ToF-SIMS) chemical maps from a micropatterned PSiF surface after PFDS functionalization. **Left**: $KNaF^+$, K_2F^+, $C_2H_2O^+$, $SiOH^+$ and total positive ions (from left to right and top to bottom). **Right**: SiF_5^-, F^-, $Si_3HO_7^-$, $Si_3C_3HO^-$ and total negative ions (from left to right and top to bottom).

Overall it can be concluded that APTS- and PFDS-rich PSiFs can be formed by a VLAO process and that the activation of a certain degree of poly-silane bonds on the surface is strongly suggested.

A selective Ar plasma etching process can reverse the presence of the organosilanes locally, re-exposing the underlying PSiF to the surface.

3.2. Characterization of PSiPs

The internal microstructure of PSiPs was studied using TEM before functionalization. Figure 5a,b shows the internal morphology of the PSiPs as prepared, which showed, due to the modified synthesis parameters, higher particle size and dispersion (1.5 ± 0.5 μm) with respect to previously synthesized particles [25]. These are intrinsically anisotropic with well-defined columnar pores, which present an open structure as determined from the top view image (Figure 5b), in which electron beam and PSiPs pores presented identical directions. The images from single pores (Figure 5c) established a difference between the wall edge and the corrugated internal pore. Higher magnification images (see Figure 5d) allowed the identification of Si nanocrystals in the internal pore wall by slightly tilting the sample until Si(100) planes were resolved.

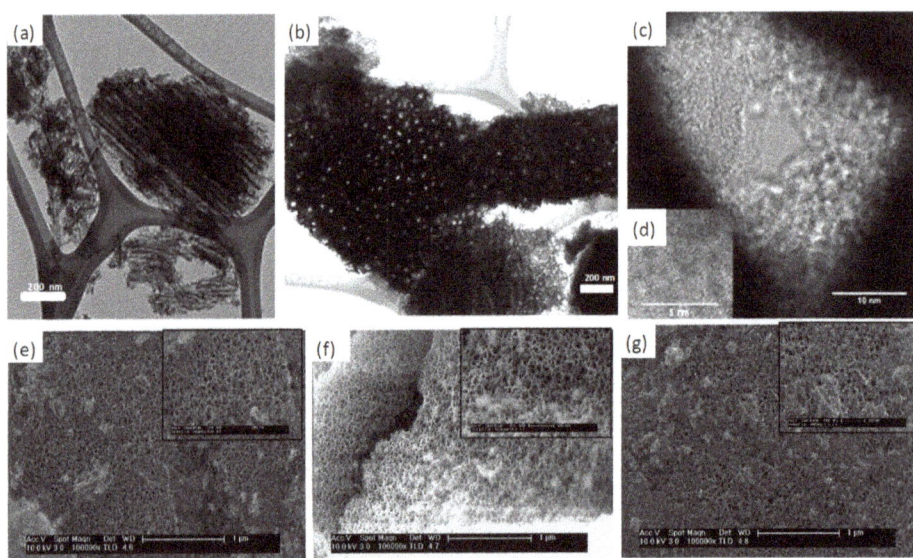

Figure 5. Transmission electron microscopy (TEM) lateral (**a**) and top (**b**) images of columnar PSiPs and (**c**) TEM image of a pore and (**d**) detail of a Si nanocrystal revealing (100) planes. FESEM images of (**e**) the surface of PSiPs, and PSiPs functionalized with (**f**) APTS and (**g**) PFDS.

The SEM images shown in the bottom part of Figure 5 highlight the surface morphology of the functionalized PSiPs in comparison with the morphology of the original PSiPs. Larger particles were chosen to highlight the homogeneity of the surface termination. The latter shows a spongy surface with an average pore aperture of 40 nm. Wide fields were selected to identify the homogeneity of the surface at the scales where the pores start to be observable, while insets show a higher magnification view of the pore structure and distribution.

We note the similarity between the pristine PSiPs (Figure 5e) and the functionalized structures (Figure 5f,g). In particular, there is no film or colloidal structure associated with the organosilane modification obstructing the PSiPs pores. These results support the view that the modification takes place through a mild assembly of the organosilane. In other words, the assembly is heterogeneous and does not consist of adsorbed structures stemming from a previous (in liquid) homogeneous nucleation.

In conclusion, this morphological study confirms that the activated assembly process gives rise to a structure with no remarkable topographic addition.

It is widely accepted that as prepared PSiPs exhibit Si–H bonds on the surface [37]. The FTIR spectrum of freshly-etched PSiPs (Figure 6), reflects the presence of these species. We can clearly distinguish the scissoring modes at 624, 662 and 901 cm^{-1}, and on the other hand, the stretching modes at 2085 and 2108 cm^{-1} [38]. After VLAO process, the peaks corresponding to SiH$_x$ modes tend to disappear, confirming the chemical modification of the pore surface. We confirmed the mildly activated oxidation of PSiPs through bands at 940–977 cm^{-1} (δ(-O$_y$Si-H$_x$)) and 2195–2249 cm^{-1} (ν(-O$_y$Si-H$_x$)). Additionally, we observed in APTS and PFDS silanized samples, the presence of new peaks due to the aliphatic carbon chains at 575, 1318, 1382, 1472, 2849 and 2917 cm^{-1} corresponding to surface integrated CH$_2$–C bonds.

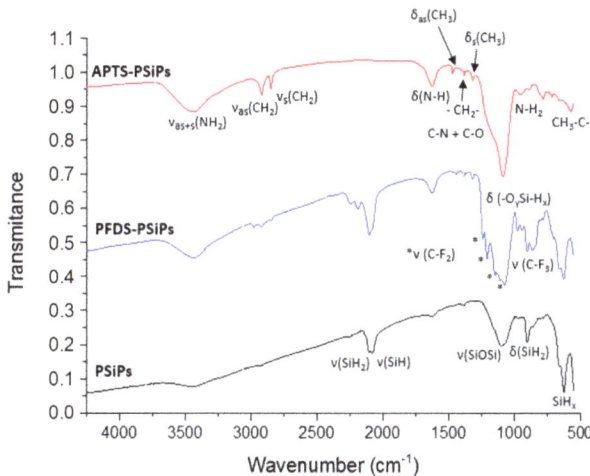

Figure 6. Fourier transformed infrared spectroscopy (FTIR) spectra of APTS and PFDS functionalized PSiPs compared with the spectrum corresponding to freshly formed PSiPs.

Oxidation and aliphatic presence are especially evident in APTS-PSiPs. The peaks at 780, 1088, 1627 and 3435 cm^{-1} can be attributed to N–H$_2$, C–N, δ(N–H) and ν_{AS+S} (NH$_2$), respectively. However, in the case of the two latter modes, their identification is compromised due to an overlapping with the broad water absorption bands caused by the moisture absorbed on the KBr disk. However, in the case of the band at 3435 cm^{-1} we can see that it increases once the PSiPs surface has been functionalized with APTS. This is due to water adsorption, which is consistent with an increasingly polar character induced by the amino groups.

In the spectra corresponding to samples with PFDS, the SiH$_x$ modes partially remain after VLAO. Additionally, the CF$_3$ group presents a sharp characteristic band at the low wavenumber edge (858 cm^{-1}). The CF$_2$ group, owing to its four characteristic bands at 1133, 1149, 1208 and 1238 cm^{-1}, also appears to be an ideal group for tracing the functionalization [39]. Overall, the FTIR results demonstrate that oxidation processes of PSiPs take place during VLAO. Furthermore, an organosilane assembly is induced on the PSiPs surface.

The overlapping of some characteristic bands in the FTIR spectra of PSiPs makes the univocal identification of molecular groups difficult. In order to overcome this issue, we used SS-NMR spectroscopy, taking advantage of its chemical sensitivity. Figure 7a shows the ^{29}Si NMR spectra of the PSiPs with different organosilanes compared with freshly formed PSiPs. In the spectra of the functionalized PSiPs, the chemical shift of the dominant signal is centered at −96 ppm, with a line width (full width at half maximum, FWHM) of 20 ppm, which can be assigned to the SiH or SiH_2 structural elements [37]. No signal corresponding to amorphous silicon (*a*-Si) (a very broad signal centered at −40 ppm) was observed.

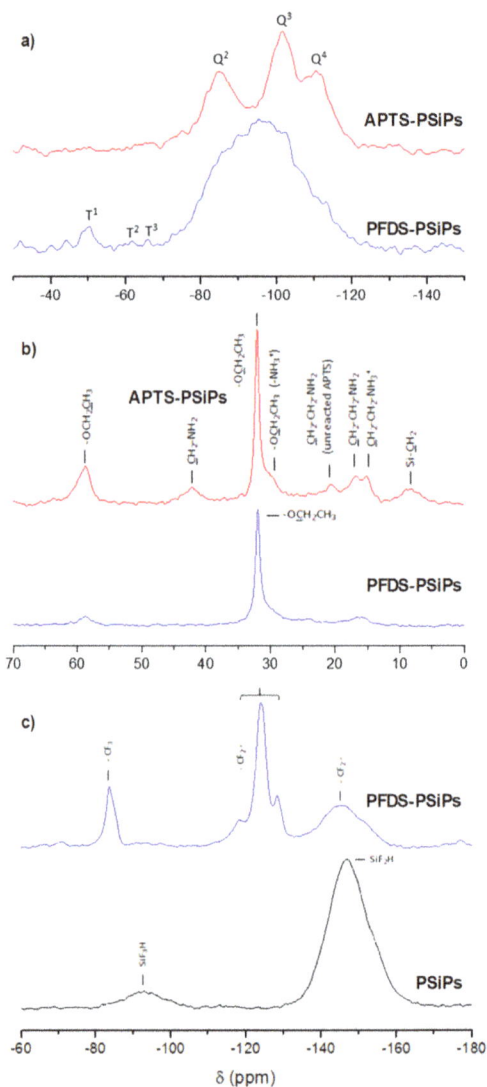

Figure 7. ^{29}Si (**a**) and ^{13}C (**b**) nuclear magnetic resonance (NMR) spectra of the APTS and PFDS functionalized PSiPs. ^{19}F NMR spectra of PFDS PSiPs and fresh PSiPs (**c**).

After functionalization, several peaks appeared, indicating the presence of different environments, which complement the signal of the underlying PSiPs. Relevantly and with respect to the effect of the VLAO on PSiPs oxidation, alterations in Q^n contributions were identified—the peaks at −85, −101 and −109 ppm corresponding to germinal hydroxyl silanol sites [$(O)_2Si(OH)_2$, Q^2], hydroxyl containing silicon sites [$(O)_3SiOH$, Q^3] and cross-linked $Si[(O)_4Si$, Q^4], respectively [40]. The high intensity of these peaks indicates that a significant part of the PSiPs surface is oxidized/hydrolyzed to form Q^n

structures. In the case of PSiPs functionalized with PFDS, these signals appear as shoulders of a main large peak at −95 ppm corresponding to the unreacted PSi. This means that the fluorosilane is effectively protecting the Si surface against hydration and hydroxylation.

The spectra provide further details of the presence of Si–C bonds and exhibit low intensity T^n peaks at −50, −63 and −67 ppm assigned to the Si atoms covalently bonded to organic groups R (T^1 [(SiO)Si(OEt)$_2$(R) and R = –CH$_2$CH$_2$CH$_2$NH$_2$ and –CH$_2$CH$_2$(CF$_2$)$_7$CF$_3$ for APTS and PFDS respectively], T^2 [(SiO)$_2$Si(OEt)(R) and identical meaning for R] and T^3 [(SiO)$_3$Si(R), idem for R]). These peaks confirm the reaction of PSiPs with the two silanes.

As the electronic shielding of the central Si increases, the chemical shift becomes increasingly negative with each additional Si–O–Si linkage [41,42]. It is remarkable that the dominant intensity is observed at T1, which would indicate that the organosilanes are not fully condensed, especially in the case of PFDS (most intense T1 signal). This type of information on the reactivity at the surface is extremely relevant and cannot be extracted by using traditional spectroscopic techniques used for surface characterization of organosilanes such as XPS.

Figure 7b presents the ^{13}C spectra of the organosilane functionalized PSiPs. In the spectra corresponding to APTS-PSiPs and PFDS-PSiPs, we observed peaks at 32 and 58 ppm due to the CH$_3$ and CH$_2$ from unreacted ethoxy groups (OCH$_2$CH$_3$) [43]. Moreover, in the case of the APTS-PSiPs, the expected three characteristic signals of the aminosilane are present at 8.3 ppm (Si–CH$_2$), 17 ppm (CH$_2$–CH$_2$–NH$_2$) and 42.2 ppm (CH$_2$–NH$_2$). The unfolding and the widening of the latter two signals, as well as the shoulder of the CH$_3$ peak at 32 ppm, indicate the presence of a small portion of protonated amine. On the other hand, the signal at 21 ppm corresponds to the intermediate CH$_2$ of the aliphatic chain of amine [44].

In the spectrum corresponding to the PFDS-PSiPs, we observed small peaks at 16 ppm (Si–CH$_2$–) and 24 ppm (Si–CH$_2$–CH$_2$–) corresponding to the aliphatic chain. They are both shifted down-field with respect to the equivalent carbons on the aminopropyl chain because of the adjacent CF$_2$ groups. The small intensity of the signals confirms the data obtained in the ^{29}Si spectrum, indicating that the organosilane condensation is not complete.

Because ^{19}F is one of the most useful NMR active nuclei, the ^{19}F NMR spectrum was obtained from the PFDS-PSiPs. Indeed, the ^{19}F NMR spectra of PSiPs and PFDS-PSiPs in Figure 7c show no similarities. The spectrum corresponding to the PSiPs reference presents two peaks at −93 and −147 ppm corresponding to SiF$_3$H and SiF$_2$H species, respectively. In the spectrum of PFDS-PSiPs, we can clearly identify the peak corresponding to the CF$_3$ group at 83.8 ppm. The CF$_2$ groups give rise to peaks at 124.1 and 145 ppm [27].

In addition to the structural information provided by SS-NMR, quantitative studies can be carried out owing to the direct proportionality between the signal intensity and the number of contributing nuclei. In the case of the functionalization with PFDS, we obtained a relation between the peaks corresponding to the CF$_2$ and CF$_3$ bonds in agreement with the molecular structure of the monolayer precursor (~7).

Combining the results obtained from FTIR spectra with those from NMR spectra, the overall information is consistent with the formation of the organosilane functionalized PSiPs. It appears however, that the ratio of organosilane molecules with three siloxane bonds saturated to the PSiPs surface is low.

3.3. Reaction Mechanism

The results from the characterization of PSiFs and PSiPs subject to the VLAO process have shown evidence of both PSi surface and organosilane transformation. Relevantly, the different analytical techniques reinforced the idea that an activation of a surface oxidation of PSi is induced and an instability in hydrolysable siloxane bonds of the organosilane is activated, which favors the formation of Si–O–Si bonds. Figure 8 shows two proposed parallel reactions capable of explaining the observed changes on the PSi surfaces. From the side of the fresh PSi surface, contrary to ultra-violet (UV) light, visible light at room temperature does not provide photons with sufficient energy to promote the homolytic cleavage of the Si–H bonds. However, excitons (electron-hole pairs) are generated during the illumination of PSi at the H–Si surface due to photo-excitation [45]. This facilitates the nucleophilic attack of trace water found in absolute ethanol, leading to a transformation of the original Si–H to a Si–OH bond (Figure 8a). Thus, a hydroxylated PSi surface is formed avoiding the thermal [46] or chemical [16] oxidation processes previously proposed. From the side of the organosilane, it is widely accepted that their tendency towards hydrolysis and condensation of alkoxysilanes, such as APTS and PFDS, is lower than that of transition metal alkoxydes. However, it is known that surface hydroxyl groups are the driving force for the heterogeneous condensation of these alkoxydes (Figure 8b). It is likely that during the progress of the heterogeneous condensation, the dangling bonds from adsorbed APTS or PDFS bind a neighboring molecule, which would be illustrated by the molecular structures identified on PSiFs and PSiPs compatible with slightly oligomerized organosilanes.

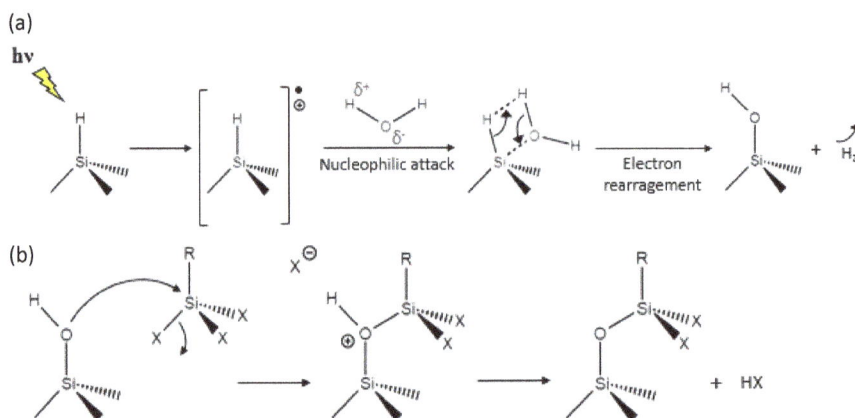

Figure 8. Proposed mechanism for the (**a**) visible light activation of PSi and (**b**) further condensation of the organosilanes on the hydroxylated PSi surface. X = OEt. R = –CH$_2$CH$_2$(CF$_2$)$_7$CF$_3$ and –CH$_2$CH$_2$CH$_2$NH$_2$ for PFDS and APTS, respectively.

Because the proposed VLAO conditions imply a simultaneous exposure of the PSi structures and organosilanes to visible light, alternative reactions could be activated at the organosilanes. At this point, it is relevant to maintain the low concentration of the organosilanes (0.2%), which ensures that the light absorbed, is well below the light absorbed in PSi. Under these conditions, the rates of production of homogeneous silane–silane reactions (whatever their nature can be) must remain well below the production of heterogeneous PSi-silane reactions. This gives further support to the proposed chemical route for the VLAO process on PSi structures.

4. Conclusions

A condensation process has been applied to the synthesis of organosilane functionalized PSiFs and PSiPs using APTS and PFDS at low concentrations of 0.2%. Visible light activation was shown to promote PSi surface oxidation making the reaction with organosilanes viable. The process leads to homogeneous surfaces with no traces of silane derived colloidal structures. This suggests that organosilane coverage is below one monolayer and makes the process useful as a model for the analysis of organosilane-PSi interfaces. In particular, SS-NMR and FTIR proved to be powerful techniques to obtain spectroscopic features regarding the organosilane assemblies on PSiPs. Analogue information is obtained by using XPS and ToF-SIMS on PSiFs. Indeed, while the FTIR and XPS analyses strongly supported organic surface bound moieties resulting from the functionalization reactions, SS-NMR and

ToF-SIMS provided more detailed information from the interface species, in particular, the surface oxidation induced during VLAO process and the presence of slightly oligomerized organosilane molecules. Using ^{13}C SS-NMR, we differentiated aliphatic carbons within the same chain (i.e., as in the perfluorodecyl group) and with respect to the other silanes (i.e., the perfluorodecyl, with respect to the aminopropyl). This is especially relevant as there is no surface spectroscopic technique with the requisite chemical sensitivity to reveal these features.

The combined use of ^{29}Si and ^{19}F SS-NMR was found to be extremely useful in the determination of silane binding efficiency. PFDS showed a lower degree of condensation with PSiPs (dominant T1 in ^{29}Si spectra), suggesting that it provides the highest hydroxylation protection (reduced formation of Q^n structures). This difference has its origin in the influence of the non-hydrolysable radical of the organosilane in the distribution of charge in the molecule. This is known to slightly influence the susceptibility of the alkoxy groups to homogeneous and heterogeneous condensation. In fact, the hydrophobic nature of the perfluorodecyl group may have a direct impact in the retardation of formation of the condensation by-product molecules, constituting a steric hindrance effect.

As a prospective remark, for the sake of functionalization efficiency, a study of the optimization of organosilane concentration towards increased coverage should be considered. Since PSi can potentially be used to allocate nanomaterials and biomolecules to provide additional functionalities, this study highlights the interest on the multi-technique surface analysis of any nano or biomolecular complexes conjugated with PSi.

Author Contributions: C.R. and M.M.S. designed the experiments and suggested the different organosilanes. C.R. and V.T.-C. carried out PSi fabrication and visible-light activated organosilane assembly. A.M.N. performed electron microscopy characterization. G.C. carried out photoelectron and mass spectroscopy analyses. C.R. carried out infrared and nuclear magnetic resonance characterization. C.R. and M.M.S. produced the first draft and the remaining authors contributed until final version.

Funding: We acknowledge MSC funding provided by the European Commission through FP7 grant THINFACE (ITN GA 607232) and by Ministerio de Economía y Competitividad through grant NANOPROST (RTC-2016-4776-1).

Acknowledgments: The authors thank Luis García Pelayo from DFA-UAM, María José de la Mata Segarra from SIdI-UAM and Cristina García from ICP-CSIC for their technical assistance during materials processing and data acquisition.

Conflicts of Interest: The authors declare no conflicts of interest.

References

1. Badr, Y.A.; El-Kader, K.M.A.; Khafagy, R.M. Raman spectroscopic study of CdS, PVA composite films. *J. Appl. Polym. Sci.* **2004**, *92*, 1984–1992. [CrossRef]
2. Sacarescu, L.; Roman, G.; Sacarescu, G.; Simionescu, M. Fluorescence detection system based on silicon quantum dots-polysilane nanocomposites. *Express Polym. Lett.* **2016**, *10*, 990–1002. [CrossRef]
3. Makila, E.; Bimbo, L.M.; Kaasalainen, M.; Herranz, B.; Airaksinen, A.J.; Heinonen, M.; Kukk, E.; Hirvonen, J.; Santos, H.A.; Salonen, J. Amine Modification of Thermally Carbonized Porous Silicon with Silane Coupling Chemistry. *Langmuir* **2012**, *28*, 14045–14054. [CrossRef]
4. Kao, H.M.; Liao, C.H.; Hung, T.T.; Pan, Y.C.; Chiang, A.S.T. Direct synthesis and solid-state NMR characterization of cubic mesoporous silica SBA-1 functionalized with phenyl groups. *Chem. Mater.* **2008**, *20*, 2412–2422. [CrossRef]
5. Tanaka, T.; Godin, B.; Bhavane, R.; Nieves-Alicea, R.; Gu, J.; Liu, X.; Chiappini, C.; Fakhoury, J.R.; Amra, S.; Ewing, A.; et al. In vivo evaluation of safety of nanoporous silicon carriers following single and multiple dose intravenous administrations in mice. *Int. J. Pharm.* **2010**, *402*, 190–197. [CrossRef] [PubMed]
6. Rodriguez, C.; Torres-Costa, V.; Ahumada, O.; Cebrian, V.; Gomez-Abad, C.; Diaz, A.; Silvan, M.M. Gold nanoparticle triggered dual optoplasmonic-impedimetric sensing of prostate-specific antigen on interdigitated porous silicon platforms. *Sens. Actuators B Chem.* **2018**, *267*, 559–564. [CrossRef]

7. Dominguez, R.B.; Alonso, G.A.; Munoz, R.; Hayat, A.; Marty, J.L. Design of a novel magnetic particles based electrochemical biosensor for organophosphate insecticide detection in flow injection analysis. *Sens. Actuators B Chem.* **2015**, *208*, 491–496. [CrossRef]

8. Dorvee, J.R.; Sailor, M.J.; Miskelly, G.M. Digital microfluidics and delivery of molecular payloads with magnetic porous silicon chaperones. *Dalton Trans.* **2008**, 721–730. [CrossRef] [PubMed]

9. Godin, B.; Tasciotti, E.; Liu, X.W.; Serda, R.E.; Ferrari, M. Multistage Nanovectors: From Concept to Novel Imaging Contrast Agents and Therapeutics. *Acc. Chem. Res.* **2011**, *44*, 979–989. [CrossRef]

10. Fan, D.M.; De Rosa, E.; Murphy, M.B.; Peng, Y.; Smid, C.A.; Chiappini, C.; Liu, X.W.; Simmons, P.; Weiner, B.K.; Ferrari, M.; et al. Mesoporous Silicon-PLGA Composite Microspheres for the Double Controlled Release of Biomolecules for Orthopedic Tissue Engineering. *Adv. Funct. Mater.* **2012**, *22*, 282–293. [CrossRef]

11. Chadwick, E.G.; Clarkin, O.M.; Tanner, D.A. Hydroxyapatite formation on metallurgical grade nanoporous silicon particles. *J. Mater. Sci.* **2010**, *45*, 6562–6568. [CrossRef]

12. Zhu, G.X.; Liu, J.T.; Wang, Y.Z.; Zhang, D.C.; Guo, Y.; Tasciotti, E.; Hu, Z.B.; Liu, X.W. In Situ Reductive Synthesis of Structural Supported Gold Nanorods in Porous Silicon Particles for Multifunctional Nanovectors. *ACS Appl. Mater. Interfaces* **2016**, *8*, 11881–11891. [CrossRef] [PubMed]

13. DeLouise, L.A.; Miller, B.L. Quantatitive assessment of enzyme immobilization capacity in porous silicon. *Anal. Chem.* **2004**, *76*, 6915–6920. [CrossRef] [PubMed]

14. Kim, S.G.; Kim, S.; Ko, Y.C.; Cho, S.; Sohn, H. DBR-structured smart particles for sensing applications. *Colloids Surf. A Physicochem. Eng. Asp.* **2008**, *313*, 398–401. [CrossRef]

15. Dong, J.P.; Wang, A.F.; Ng, K.Y.S.; Mao, G.Z. Self-assembly of octadecyltrichlorosilane monolayers on silicon-based substrates by chemical vapor deposition. *Thin Solid Films* **2006**, *515*, 2116–2122. [CrossRef]

16. Naveas, N.; Torres Costa, V.; Gallach, D.; Hernandez-Montelongo, J.; Martin Palma, R.J.; Predenstinacion Garcia-Ruiz, J.; Manso-Silvan, M. Chemical stabilization of porous silicon for enhanced biofunctionalization with immunoglobulin. *Sci. Technol. Adv. Mater.* **2012**, *13*. [CrossRef] [PubMed]

17. Munoz-Noval, A.; Gallach, D.; Angel Garcia, M.; Ferro-Llanos, V.; Herrero, P.; Fukami, K.; Ogata, Y.H.; Torres-Costa, V.; Martin-Palma, R.J.; Ciment-Font, A.; et al. Characterization of hybrid cobalt-porous silicon systems: Protective effect of the Matrix in the metal oxidation. *Nanoscale Res. Lett.* **2012**, *7*. [CrossRef]

18. Rodriguez, C.; Ahumada, O.; Cebrian, V.; Costa, V.T.; Silvan, M.M. Biofunctional porous silicon micropatterns engineered through visible light activated epoxy capping and selective plasma etching. *Vacuum* **2018**, *150*, 232–238. [CrossRef]

19. Blasco, T.; PerezPariente, J.; Kolodziejski, W. A solid-state NMR study of the molecular sieve VPI-5 synthesized in the presence of a CTABr surfactant. *Solid State Nucl. Magn. Reson.* **1997**, *8*, 185–194. [CrossRef]

20. Petit, D.; Chazalviel, J.N.; Ozanam, F.; Devreux, F. Porous silicon structure studied by nuclear magnetic resonance. *Appl. Phys. Lett.* **1997**, *70*, 191–193. [CrossRef]

21. Brandt, M.S.; Ready, S.E.; Boyce, J.B. Si-29 nuclear magnetic resonance of luminescent silicon. *Appl. Phys. Lett.* **1997**, *70*, 188–190. [CrossRef]

22. Faulkner, R.A.; DiVerdi, J.A.; Yang, Y.; Kobayashi, T.; Maciel, G.E. The Surface of Nanoparticle Silicon as Studied by Solid-State NMR. *Materials* **2013**, *6*, 18–46. [CrossRef] [PubMed]

23. Giuliani, J.R.; Harley, S.J.; Carter, R.S.; Power, P.P.; Augustine, M.P. Using liquid and solid state NMR and photoluminescence to study the synthesis and solubility properties of amine capped silicon nanoparticles. *Solid State Nucl. Magn. Reson.* **2007**, *32*, 1–10. [CrossRef] [PubMed]

24. Shanmugharaj, A.M.; Rhee, K.Y.; Ryu, S.H. Influence of dispersing medium on grafting of aminopropyltriethoxysilane in swelling clay materials. *J. Colloid Interface Sci.* **2006**, *298*, 854–859. [CrossRef] [PubMed]

25. Qin, Z.T.; Joo, J.; Gu, L.; Sailor, M.J. Size Control of Porous Silicon Nanoparticles by Electrochemical Perforation Etching. *Part. Part. Syst. Charact.* **2014**, *31*, 252–256. [CrossRef]

26. Munoz-Noval, A.; Torres-Costa, V.; Martin-Palma, R.J.; Herrero-Fernandez, P.; Angel Garcia, M.; Fukami, K.; Ogata, Y.H.; Manso Silvan, M. Electroless nanoworm Au films on columnar porous silicon layers. *Mater. Chem. Phys.* **2012**, *134*, 664–669. [CrossRef]

27. Tacke, R.; Becht, J.; Lopezmras, A.; Sheldrick, W.S.; Sebald, A. Syntheses, x-ray crystal-structure analyses, and solid-state NMS-studies of some zwitterionic organofluorosilicates. *Inorg. Chem.* **1993**, *32*, 2761–2766. [CrossRef]

28. Leichle, T.; Silvan, M.M.; Belaubre, P.; Valsesia, A.; Ceccone, G.; Rossi, F.; Saya, D.; Pourciel, J.B.; Nicu, L.; Bergaud, C. Nanostructuring surfaces with conjugated silica colloids deposited using silicon-based microcantilevers. *Nanotechnology* **2005**, *16*, 525–531. [CrossRef]

29. Munoz Noval, A.; Sanchez Vaquero, V.; Punzon Quijorna, E.; Torres Costa, V.; Gallach Perez, D.; Gonzalez Mendez, L.; Montero, I.; Martin Palma, R.J.; Climent Font, A.; Garcia Ruiz, J.P.; et al. Aging of porous silicon in physiological conditions: Cell adhesion modes on scaled 1D micropatterns. *J. Biomed. Mater. Res. Part A* **2012**, *100A*, 1615–1622. [CrossRef]

30. Gallach, D.; Sanchez, G.R.; Noval, A.M.; Silvan, M.M.; Ceccone, G.; Palma, R.J.M.; Costa, V.T.; Duart, J.M.M. Functionality of porous silicon particles: Surface modification for biomedical applications. *Mater. Sci. Eng. B-Adv. Funct. Solid-State Mater.* **2010**, *169*, 123–127. [CrossRef]

31. Beamson, G.; Briggs, D. *High Resolution XPS of Organic Polymers—The Scienta ESCA300 Database*; Wiley Interscience: New York, NY, USA, 1992.

32. Manso-Silvan, M.; Valsesia, A.; Hasiwa, M.; Rodriguez-Navas, C.; Gilliland, D.; Ceccone, G.; Garcia Ruiz, J.P.; Rossi, F. Micro-spot, UV and wetting patterning pathways for applications of biofunctional aminosilane-titanate coatings. *Biomed. Microdevices* **2007**, *9*, 287–294. [CrossRef] [PubMed]

33. Vanea, E.; Simon, V. XPS study of protein adsorption onto nanocrystalline aluminosilicate microparticles. *Appl. Surf. Sci.* **2011**, *257*, 2346–2352. [CrossRef]

34. Naveas, N.; Hernandez-Montelongo, J.; Pulido, R.; Torres-Costa, V.; Villanueva-Guerrero, R.; Ruiz, J.P.G.; Manso-Silvan, M. Fabrication and characterization of a chemically oxidized-nanostructured porous silicon based biosensor implementing orienting protein A. *Colloids Surf. B Biointerfaces* **2014**, *115*, 310–316. [CrossRef] [PubMed]

35. Ohlhausen, J.A.; Zavadil, K.R. Time-of-flight secondary ion mass spectrometry measurements of a fluorocarbon-based self-assembled monoloyer on Si. *J. Vac. Sci. Technol. A* **2006**, *24*, 1172–1178. [CrossRef]

36. Kempson, I.M.; Barnes, T.J.; Prestidge, C.A. Use of TOF-SIMS to Study Adsorption and Loading Behavior of Methylene Blue and Papain in a Nano-Porous Silicon Layer. *J. Am. Soc. Mass Spectrom.* **2010**, *21*, 254–260. [CrossRef] [PubMed]

37. Pietrass, T.; Bifone, A.; Roth, R.D.; Koch, V.P.; Alivisatos, A.P.; Pines, A. Si-29 high resolution solid state nuclear magnetic resonance spectroscopy of porous silicon. *J. Non-Cryst. Solids* **1996**, *202*, 68–76. [CrossRef]

38. Tsuboi, T.; Sakka, T.; Ogata, Y.H. Structural study of porous silicon and its oxidized states by solid-state high-resolution Si-29 NMR spectroscopy. *Phys. Rev. B* **1998**, *58*, 13863–13869. [CrossRef]

39. Innocenzi, P.; Brusatin, G.; Guglielmi, M.; Bertani, R. New synthetic route to (3-glycidoxypropyl)trimethoxysilane-base hybrid organic-inorganic materials. *Chem. Mater.* **1999**, *11*, 1672–1679. [CrossRef]

40. Chang, W.K.; Liao, M.Y.; Gleason, K.K. Characterization of porous silicon by solid-state nuclear magnetic resonance. *J. Phys. Chem.* **1996**, *100*, 19653–19658. [CrossRef]

41. Silvan, M.M.; Messina, G.M.L.; Montero, I.; Satriano, C.; Ruiz, J.P.G.; Marletta, G. Aminofunctionalization and sub-micrometer patterning on silicon through silane doped agarose hydrogels. *J. Mater. Chem.* **2009**, *19*, 5226–5233. [CrossRef]

42. Vasil'ev, S.G.; Volkov, V.I.; Tatarinova, E.A.; Muzafarov, A.M. A Solid-State NMR Investigation of MQ Silicone Copolymers. *Appl. Magn. Reson.* **2013**, *44*, 1015–1025. [CrossRef] [PubMed]

43. Al-Sabagh, A.M.; Harding, D.R.K.; Kandile, N.G.; Badawi, A.M.; El-Tabey, A.E. Synthesis of Some Novel Nonionic Ethoxylated Surfactants Based on alpha-Amino Acids and Investigation of Their Surface Active Properties. *J. Dispers. Sci. Technol.* **2009**, *30*, 427–438. [CrossRef]

44. Lu, H.T. Synthesis and characterization of amino-functionalized silica nanoparticles. *Colloid J.* **2013**, *75*, 311–318. [CrossRef]

45. Sano, H.; Maeda, H.; Matsuoka, S.; Lee, K.H.; Murase, K.; Sugimura, H. Self-assembled monolayers directly attached to silicon substrates formed from 1-hexadecene by thermal, ultraviolet, and visible light activation methods. *Jpn. J. Appl. Phys.* **2008**, *47*, 5659–5664. [CrossRef]
46. Meskini, O.; Abdelghani, A.; Tlili, A.; Mgaieth, R.; Jaffrezic-Renault, N.; Martelet, C. Porous silicon as functionalized material for immunosensor application. *Talanta* **2007**, *71*, 1430–1433. [CrossRef] [PubMed]

materials

MDPI

Article

Preparation of Textured Surfaces on Aluminum-Alloy Substrates

Markéta Kadlečková [1,2], Antonín Minařík [1,2,*], Petr Smolka [1,2], Aleš Mráček [1,2], Erik Wrzecionko [1,2], Libor Novák [1], Lenka Musilová [1,2] and Radek Gajdošík [1]

[1] Department of Physics and Materials Engineering, Faculty of Technology, Tomas Bata University in Zlín, Vavrečkova 275, 760 01 Zlín, Czech Republic; m1_kadleckova@utb.cz (M.K.); smolka@utb.cz (P.S.); mracek@utb.cz (A.M.); wrzecionko@utb.cz (E.W.); novak.libor7@seznam.cz (L.N.); lmusilova@utb.cz (L.M.); radek.gajda@seznam.cz (R.G.)

[2] Centre of Polymer Systems, Tomas Bata University in Zlín, Třída Tomáše Bati 5678, 76001 Zlín, Czech Republic

* Correspondence: minarik@utb.cz; Tel.: +420-57-603-5086

Received: 30 November 2018; Accepted: 25 December 2018; Published: 31 December 2018

Abstract: The ways of producing porous-like textured surfaces with chemical etching on aluminum-alloy substrates were studied. The most appropriate etchants, their combination, temperature, and etching time period were explored. The influence of a specifically textured surface on adhesive joints' strength or superhydrophobic properties was evaluated. The samples were examined with scanning electron microscopy, profilometry, atomic force microscopy, goniometry, and tensile testing. It was found that, with the multistep etching process, the substrate can be effectively modified and textured to the same morphology, regardless of the initial surface roughness. By selecting proper etchants and their sequence one can prepare new types of highly adhesive or even superhydrophobic surfaces.

Keywords: aluminum; alloy; duralumin; etching; surface texture; porous-like; adhesive bonding; superhydrophobic

1. Introduction

Aluminum alloys have long been used in many industrial applications, especially for their anticorrosion properties and low specific weight [1]. They have been utilized in construction, structural, cover, and front parts in the automotive, aviation, and astronautics industries, for the production of molds, etc. [2]. Especially for front and functional surfaces, which come into contact with other materials and weather conditions, aluminum alloys are essential postproduction processes for surface treatment [2–4]. These modifications should improve the appearance of the final product and inhibit surface oxidation [5–7], promote adhesion [8–11], or reduce staining thanks to their self-cleaning properties [4,12].

The composition of etchants for aluminum-alloy etching has been intensively studied [3,13–21]. The following chemicals are among those used: a hot sodium hydroxide solution with subsequent nitric acid application [3], phosphoric acid [13], acids with the addition of surfactants [19], salts [20], ferrous ions [14], mixtures of phosphoric and nitric acid with copper or ammonia ions [15], a mixture of hydrochloric acid with sulfuric acid or ethylene glycol [16], and hydrochloric acid alone [12,21].

Alloy composition and the surface-machining method also affect the course of the etching process and its result [17]. The etching rate can be controlled by temperature [3], and current density in the case of electrochemical etching [16,22,23].

The selected processes result in either smooth [13,23] or textured surfaces [4,12,18,21,22], which can exhibit hydrophobic properties when subsequent surface modification is applied [12,18,21,22].

Chemicals that are used for these modifications include fluorocarbons [12], stearic acid [5,6,18,24], polypropylene [25], and silanes [7,26].

In this work, the ways of producing porous-like textured surfaces on aluminum-alloy substrates with multistep etching were studied. The aim was to provide a method that allows preparing comparable final surfaces regardless of initial surface roughness and machining, with minimum material loss, and, from a practical point of view, to show how these new specifically textured surfaces influence adhesive bonding or wetting properties.

2. Materials and Methods

2.1. Materials

The studied material was an aluminum alloy (designed as duralumin, outlined further in the text) with a composition of 96.8 ± 0.1% Al, 2.6 ± 0.1% Mg, 0.5 ± 0.1% Fe, and others. The composition was determined with a delta element X-ray fluorescence spectrometer. Standard chemicals were purchased from Sigma-Aldrich (St. Louis, MO, USA) in p.a. purity. Ultrapure water with a resistivity of 18.2 MΩ·cm was used (Direct-Q ®3UV, Merck, NJ, USA). Flexible polyurethane resin U4291 (ABchemie, Corbelin, France) was used for the preparation of adhesive bonds.

2.2. Preparation of Duralumin Samples

Duralumin sheets of 1 mm thickness were used. Samples were cut into 2 cm × 6 cm and 2 cm × 1 cm pieces. The surface of the rolled sheets with an initial roughness of Ra ≤ 0.5 μm was modified by grinding or corundum blasting to a final roughness of Ra ≤ 3 μm and Ra > 6 μm, respectively. The sandpaper used for grinding was of a 180 grade, and the corundum particles had a mean size of about 90 μm. The mechanically treated samples were rinsed with acetone, water, and ethanol, and dried at 23 °C for 20 min prior to each experimental step. Prior to etching, the samples were conditioned to etching-bath temperature.

2.3. Etchant Composition

Etchants contained a base (NaOH) or a mixture of acids (37% HCl, 65% HNO$_3$, 85% H$_3$PO$_4$, 96% H$_2$SO$_4$) with methanol (CH$_3$OH) and sodium nitrite (NaNO$_2$), and sodium nitrate (NaNO$_3$). Etchants were prepared in glass beakers and conditioned to the desired temperature prior and during etching.

2.4. Etching Process

The etching process was performed in glass beakers and the etchants were conditioned to a temperature in the range of 25 to 100 °C. Etching time was 1 to 10 min. The samples were rinsed with water and ethanol and allowed to dry at 23 °C between the individual etching-process steps. Afterward, the samples were kept in LDPE (low density polyethylene) bags in a desiccator.

2.5. Preparation of Superhydrophobized Surfaces

The selected duralumin samples were also hydrophobized with stearic acid similarly to the literature [5,6,18,24]. First, the stearic acid was dissolved by stirring in an ethanol:water 1:1 weight fraction solution at 60 °C for 20 min. Then, the duralumin samples were introduced into the solution and left there for 29 h at 60 °C. Finally, the samples were removed and rinsed with ethanol and water.

2.6. Preparation of Samples for Adhesion Testing

The 6 cm × 2 cm duralumin samples were joined with the U4291 polyurethane. The initial viscosity of the resin was 0.6 Pa·s for the material to reproduce the surface texture of the duralumin samples well. The samples were joined over the 2 cm × 2 cm area (Figure 1). Curing time was at least 72 h at 23 °C.

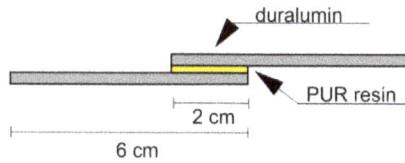

Figure 1. Sample for adhesion tests.

2.7. Scanning Electron Microscopy

Changes in surface appearance were analyzed by a Phenom Pro (Phenom-World BV, Eindhoven, The Netherlands) scanning electron microscope (SEM). The samples were observed at an acceleration voltage of 10 kV in backscattered electron mode.

2.8. Atomic Force Microscopy

Changes in the surface topography of the selected samples were characterized using a Ntegra-Prima (NT-MDT Spectrum Instruments, Moscow, Russia) atomic force microscope (AFM). Measurements were performed at a scan speed of 0.5 Hz with a resolution of 512×512 pixels, in tapping mode, at room temperature, in air atmosphere. A silicone-nitride probe with a resonant frequency of 150 ± 50 kHz and a stiffness constant of 5.5 N/m (NSG01, AppNano, Applied NanoStructures, Inc., Mountain View, CA, USA) was used. The data from the AFM measurement were processed in Gwyddion 2.5 software (Czech Metrology Institute, Jihlava, Czech Republic).

2.9. Profilometry

Changes in surface roughness (Ra) were characterized by a DiaVite DH-8 contact profilometer (Bülach, Switzerland). A diamond tip with a curvature radius of 2 microns was used. The evaluation of the surface roughness was performed according to the ASME B46.1 standard. Mean Ra values were determined from 15 individual measurements at various locations on three samples. Sample thickness was measured with a Mitutoyo 543-561D digital indicator (Kawasaki, Japan).

2.10. Goniometry

The apparent and sliding contact angles of water on the stearic acid-modified duralumin surface were characterized with a drop shape analyzer DSA30, Krüss (Hamburg, Germany). Measurement was performed at 23 °C temperature. A drop with a volume of 3 μL (for the apparent contact angle) or 10 μL (for the sliding contact angle) was deposited on the measured surface. Ultrapure water with a resistance of 18.2 MΩ·cm was used for the measurement. All measurements were repeated 10 times; the mean values and standard deviations are presented in the results.

2.11. Adhesion Testing

The samples were prepared according to Section 2.6. Adhesion joint-strength evaluation was performed with an Instron 3345 universal testing machine (Norwood, MA, USA) with a 5 kN force sensor. Travel speed was 2 mm/min. The tests were performed in quintuplicate.

3. Results and Discussion

The first section of the results deals with the effect of etchant composition, temperature, and etching time on the duralumin surface topography. Based on these data, four etchants were chosen in order to prepare either a specifically structured surface or smooth surfaces. The last part is devoted to the effect of a porous-like surface structure on adhesive joint strength or wetting properties.

3.1. Etchant Composition

Etchant composition is vital for an effective etching process [3,13–21]. Various etchants were tested and optimized. Figure 2a–o shows the effect of etchant composition on surface relief and the roughness of the sandblasted duralumin. Etchant compositions, along with the achieved Ra values, are shown in Table 1. These data show that the most significant roughness reduction was achieved with the HNO_3 + HCl or H_3PO_4 + HNO_3 + HCl mixture, namely, from Ra 6.6 to 2.6 or 2.7 µm, respectively. In other cases, the effect was not so significant.

Table 1. Etchant composition and resultant surface roughness. Sandblasted duralumin etched at 70 °C for 4 min.

Label	Etching Mixture	Mixture Ration	Ra (µm)
(a)	None (sandblasted surface)	-	6.6 ± 0.4
(b)	H_3PO_4	10 mL	6.7 ± 0.7
(c)	HNO_3	10 mL	5.2 ± 0.6
(d)	H_3PO_4 + HCl	5:5 mL	4.7 ± 0.9
(e)	HNO_3 + HCl	5:5 mL	2.6 ± 0.8
(f)	H_3PO_4 + HNO_3	5:5 mL	7.5 ± 2.2
(g)	H_3PO_4 + HNO_3 + HCl	3.5:3.5:3.5 mL	2.7 ± 0.5
(h)	H_3PO_4 + HNO_3 + HCl + H_2SO_4	2.5:2.5:2.5:1 mL	5.3 ± 0.8
(i)	H_3PO_4 + HNO_3 + HCl + H_2SO_4 + H_2O	2.5:2.5:2.5:1:1 mL	5.6 ± 0.7
(j)	H_2O + NaOH	10 mL:0.4 g	6.0 ± 2.4
(k)	H_2O + NaOH + $NaNO_2$	10 mL:0.4 g:8 g	4.9 ± 0.8
(l)	H_2O + NaOH + $NaNO_3$	10 mL:0.4 g:2 g	7.5 ± 0.7
(m)	H_2O + NaOH + $NaNO_2$ + $NaNO_3$	10 mL:0.4 g:8 g:2 g	6.8 ± 1.5
(n)	H_2O + HNO_3 + H_3PO_4 + H_2SO_4 + $NaNO_3$	10 mL:9.8 mL:7.8 mL:6 mL:4 g	6.0 ± 0.4
(o)	HNO_3 + H_3PO_4 + H_2SO_4	3.5:3.5:3.5 mL	5.2 ± 0.2

Based on this preliminary testing and optimization, the following four etchants were chosen for further experiments:

- Etch Mix I (21 mL H_2O + 9 g NaOH)
- Etch Mix II (21 mL H_3PO_4 + 3 mL HNO_3 + 6 mL H_2SO_4)
- Etch Mix III (10 mL CH_3OH + 10 mL HCl + 10 mL HNO_3)
- Etch Mix IV (20 mL H_2O + 9.8 mL HNO_3+ 7.8 mL H_3PO_4 + 6 mL H_2SO_4 + 4 g $NaNO_3$)

Etch Mix I for pre-etching and edge-chamfering, Etch Mix II for smooth surface, Etch Mix III for the generation of a specific texture, and Etch Mix IV for a porous-like pattern.

The choice of etchants was based on the comparison of data in Table 1 and Figure 2. Etch Mix I was chosen for its elimination of sharp edges in the sandblasted surface (Figure 2). The smoothing of surfaces can be observed in Figure 2o, Etch Mix II. The evolution of the specific surface relief and surface smoothing can be observed in Figure 2e, Etch Mix III. A porous-like pattern at a minimal change of the initial surface can be observed in Figure 2n, Etch Mix IV. The concentration of the components in individual etchants was modified so that the desired effect was achieved at the lowest possible temperature and time. In the case of the Etch Mix III, the experiments have shown that methanol can accelerate etching at a lower temperature.

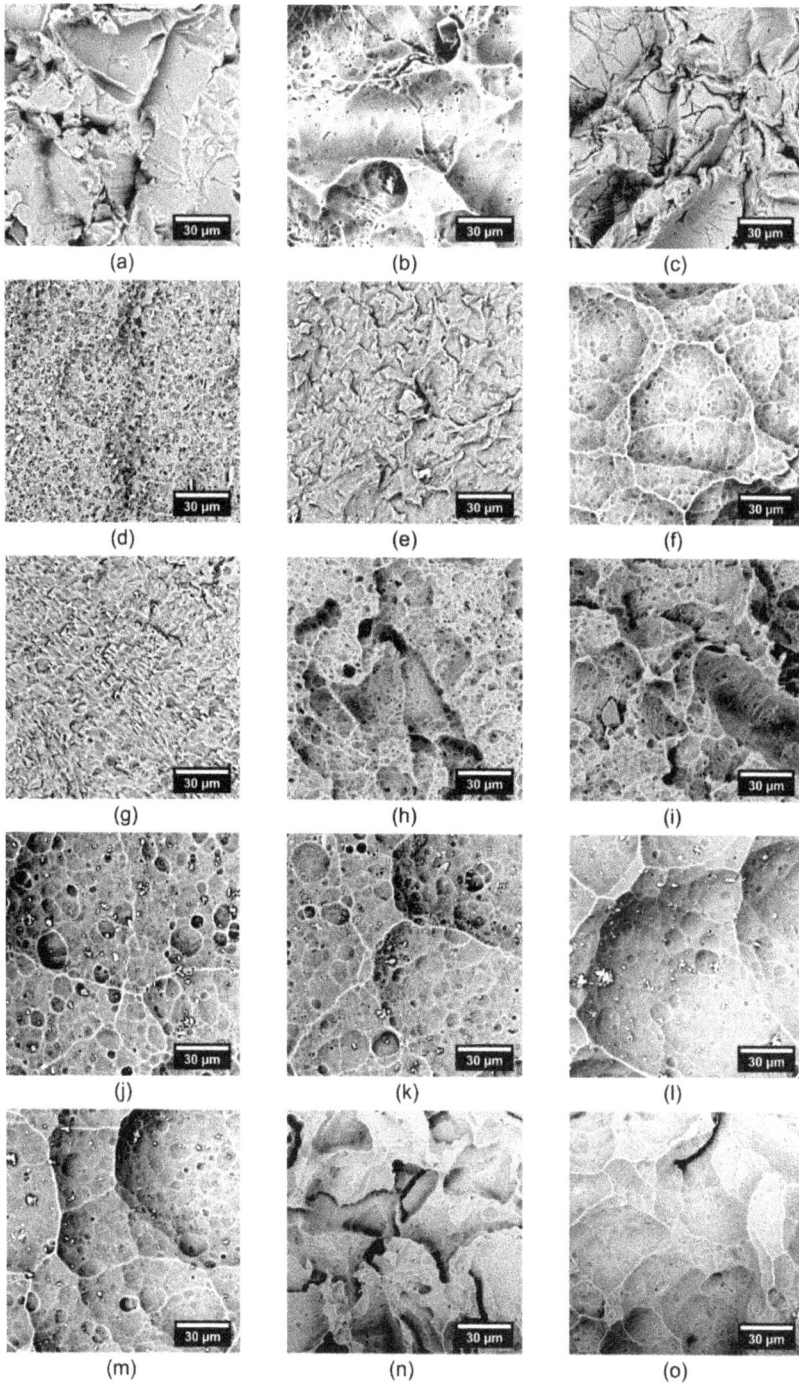

Figure 2. Scanning electron microscope (SEM) micrographs of etched duralumin surfaces. Designation corresponds with Table 1. Etching time 4 min, temperature 70 °C.

3.2. Temperature and Time Period

Important parameters affecting the etching process are the etchant temperature and etching time. A two-step process was utilized to demonstrate these effects (Figure 3). In the first step, the sample was etched for 5 min at 23 °C with Etch Mix I, and subsequently Etch Mix II for 5 min, either at various temperatures, or at one constant temperature and varying etching time. At 5 min etched time, the Ra value decreased with temperature, from 6.6 to 0.6 μm (Figure 3a). A similar trend can be observed in Figure 3b, where a given temperature decreased sample thickness with etching time. These observations correspond with the literature data for various etchants [3] and are essential from a practical point of view. Lowering the temperature from 100 °C to 90 °C results in a lower etching rate to such an extent that Ra reduction was no longer possible with some of the etchants. This fact demonstrates the dominating role of temperature in the etching process.

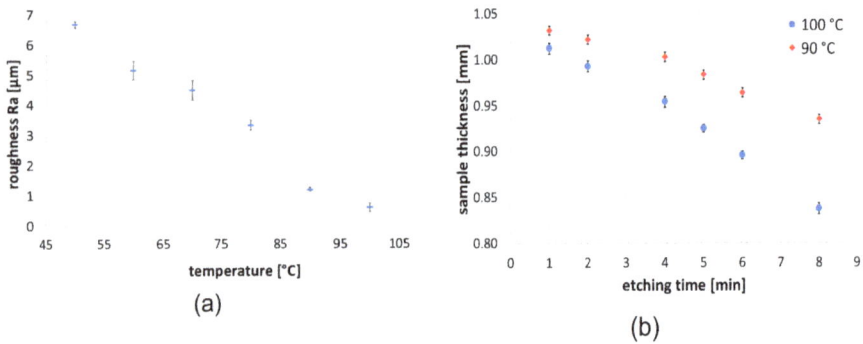

Figure 3. (a) Effect of temperature on surface roughness and (b) effect of etching time on sample-thickness reduction on sandblasted duralumin with Etch Mix I and Etch Mix II. The etching process with Etch Mix I was the same for both cases (5 min at 23 °C), the second step with Etch Mix II proceeded at either (a) a fixed time of 5 min and varying temperature or (b) a fixed temperature and varying etching time.

3.3. Initial Surface Roughness

It was found that the etching rate on sandblasted surfaces is much higher than on rolled or grinded (Table 2). This phenomenon was connected with a larger surface area and accelerated propagation of the etching process in the surface dents. The Ra value on grinded samples decreased from 2.6 ± 0.1 μm to 1.1 ± 0.1 μm, and thickness was reduced by 9%. In contrast, on the sandblasted samples the Ra value went from 6.6 ± 0.4 μm to 1.9 ± 0.2 μm with 25% thickness reduction. One can thus conclude that Etch Mix III was more aggressive on a surface with higher roughness.

Table 2. Roughness and thickness reduction in samples with various initial surface characteristics. Etched with Etch Mix III for 3 min at 23 °C.

Label	Sample	Thickness (mm)	Ra (μm)
(a)	Rolled surface	1.08 ± 0.01	0.3 ± 0.1
(b)	Etched rolled surface	1.05 ± 0.01	0.7 ± 0.1
(c)	Grinded surface	1.09 ± 0.01	2.6 ± 0.1
(d)	Etched grinded surface	1.00 ± 0.01	1.1 ± 0.1
(e)	Sandblasted surface	1.09 ± 0.01	6.6 ± 0.4
(f)	Etched sandblasted surface	0.82 ± 0.01	1.9 ± 0.2

Figure 4 deals with the effect of the initial sample surface on the etching process with Etch Mix III at 23 °C for 3 min. Ra values and thickness data are shown in Table 2. The experiments revealed that a very similar surface can be prepared relatively rapidly at mild temperatures, regardless of texture of the initial surface. For comparison, see Figure 4b,d,f. Only on the smooth surface is the Ra value higher after etching compared to the initial surface. In the case of grinded and sandblasted surfaces, the trends are opposite and the Ra values decrease. In either case, all etched surfaces contain some kind of flakelike features, where the characteristic dimensions of the flakes are related to the texture of the initial surface, i.e., larger flakes were observed in the sample with a higher initial surface roughness, namely, Ra 0.7 ± 0.1, 1.1 ± 0.1, and 1.9 ± 0.2 µm in the treated rolled, ground, and sandblasted samples, respectively. For complete data, please refer to Table 2.

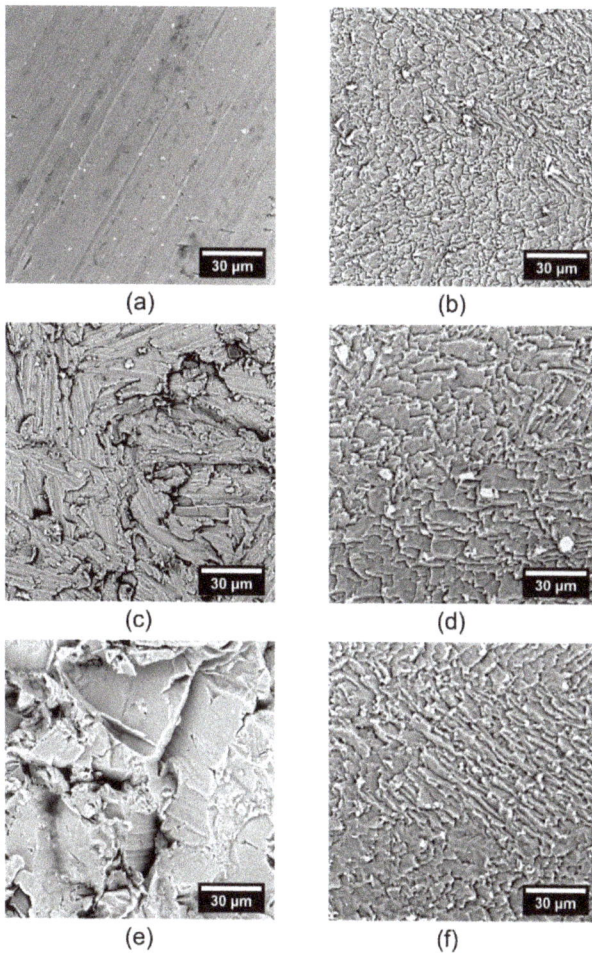

Figure 4. Effect of initial sample surface on the etching process with Etch Mix III at 23 °C for 3 min. Designation corresponds with the data in Table 2. (**a**) rolled surface; (**b**) etched rolled surface; (**c**) grinded surface; (**d**) etched grinded surface; (**e**) sandblasted surface; (**f**) etched sandblasted surface.

3.4. Preparation of Porous-Like Textured Surfaces

To prepare porous-like surfaces, the samples in Figure 4b,d,f were exposed to further etching with Etch Mix IV at 80 °C for 3 min (see Figure 5). The Ra value increased at the rolled surface to 1.0 ± 0.1 µm and at the grinded surface to 0.9 ± 0.1 µm, and decreased at the sandblasted surface to 1.6 ± 0.2 µm. The surfaces thus had a similar appearance and roughness. The sandblasted samples were also analyzed with AFM (Figure 5d) and the area roughness parameter was determined, Sa = 0.4 µm.

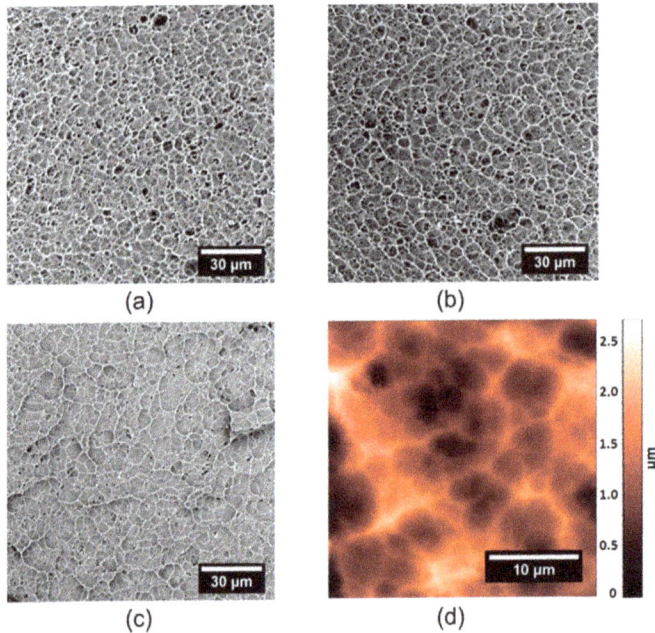

Figure 5. Porous-like surfaces prepared at (**a**) rolled, (**b**) grinded and (**c,d**) sandblasted duralumin. Surfaces etched with Etch Mix III and Etch Mix IV. (**a–c**) SEM and (**d**) atomic force microscope (AFM) micrographs.

When Etch Mix III and Etch Mix IV were applied in reverse order, i.e., first Etch Mix IV and then Etch Mix III, the surfaces were similar to those in Figure 4b,d,f. This means that the order of etchant application cannot be changed if we want to prepare porous-like surfaces.

3.5. Preparation of Smooth Surface

The literature describes many ways for the preparation of a smooth duralumin surface [3]. Our experiments, presented in Figure 2 and Table 1, revealed that one-step etching at a mild temperature does not lead to surfaces with Ra under 1 µm. Thus, a multistep approach was developed, consisting of pre-etching with Etch Mix I at 23 °C for 5 min, with a subsequent application of Etch Mix II at 100 °C for 5 min (Figure 6). In the first step, the edges or surface microstructures are chamfered, and the Ra slightly decreases from 6.6 ± 0.4 µm (Figure 6a) to 6.2 ± 0.1 µm (Figure 6b). In the second step. the Ra value drops dramatically to 0.6 ± 0.1 µm (Figure 6c). The final surface was also analyzed with AFM, and the area roughness parameter was Sa = 0.08 µm.

Further text demonstrates that the prepared surfaces can be used for many distinct applications.

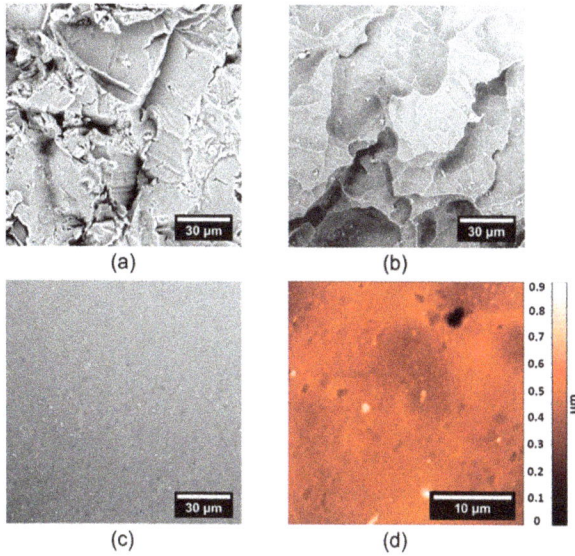

Figure 6. Flat surface prepared from sandblasted duralumin. (**a**) Initial surface, (**b**) etching with Etch-Mix-I, (**c,d**) etching with Etch-Mix-I and Etch-Mix-II. (**a–c**) SEM and (**d**) AFM micrographs.

3.6. Adhesive Bonding of Textured Surfaces

The strength of the adhesive joints is determined by chemical composition of substrates and their surface texture. It is known that textured surfaces influence material-utility properties, increase adhesion [8–11], and implicate the development of cell systems [27,28].

Prepared porous-like (Figure 5) and smooth (Figure 6) surfaces were tested by means of adhesive joint strength (Figure 7a). The overlapping 2 cm × 2 cm area was strained according to Figure 7b. The following forces at break were recorded: 1188 N for smooth duralumin, 1390 N for sandblasted duralumin, and 1528 N for porous-like duralumin (Figure 7a). These data indicate that periodic surface texture can be more beneficial in increasing adhesion strength than a high Ra value—porous-like surface vs. sandblasted surfaces. There is a limit to this statement, as seen in the adhesion data for smooth surface with an Ra equal 0.6 μm, which had the worst adhesion strength.

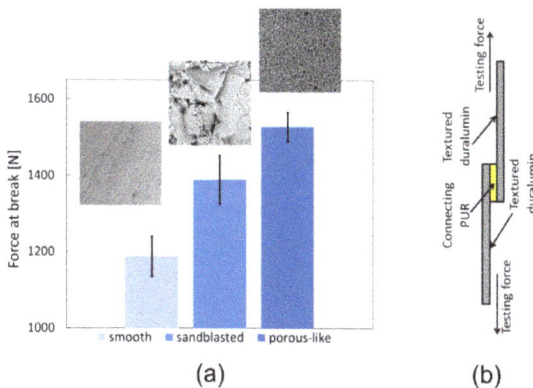

Figure 7. (**a**) Adhesive joint strength according to substrate surface texture. (**b**) Experimental setup.

3.7. Wetting Properties

The literature suggests that it is possible to increase duralumin's hydrophobic properties with stearic acid coating [6,18,24]. However, none of the described approaches was as efficient as our procedure. It was found that the most important step is the exposition of duralumin to chrome sulfuric acid for 2 min at 23 °C to remove surface oxides and etching-process residues. The process is followed by boiling the samples in water for 5 min. Then, the samples were exposed to stearic acid, as described in Section 2.5. With this approach, surfaces with an apparent water contact angle of almost 170° and a sliding angle of about 4° can be achieved, which is the highest reported value for stearic acid modification [5,6,18]. As shown in Figure 8b,c, stearic acid forms a nanotexture on the porous-like surface, thus contributing to the hydrophobic properties. Superhydrophobic surfaces feature a combination of a specific chemical composition along with specific surface micro- and nanotexture [4,12,18,21,22]. Without the chrome sulfuric acid treatment, only apparent water contact angles of about 150° can be achieved. The apparent water contact angles of stearic acid-treated surfaces are presented in Table 3.

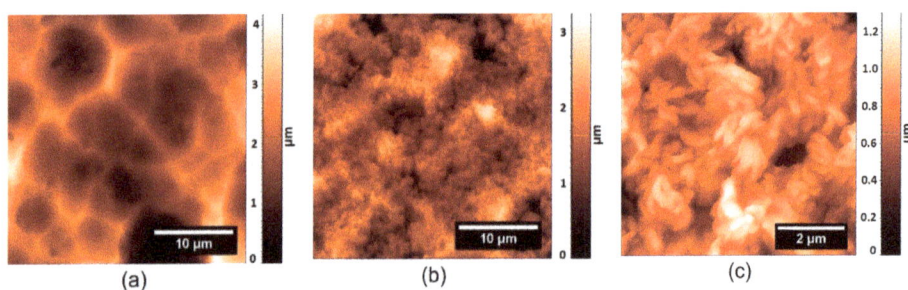

Figure 8. AFM micrographs of a porous-like duralumin surface (**a**) after chrome sulfuric acid application, and (**b**,**c**) stearic acid application.

Table 3. Water contact angles of duralumin surfaces with applied stearic acid

Surface	Water Contact Angle (°)		Ra before Stearic Acid (μm)
	Apparent	Sliding	
Rolled	155 ± 1	15 ± 3	0.3 ± 0.1
Grinded	157 ± 2	12 ± 3	2.6 ± 0.1
Sandblasted	165 ± 2	10 ± 2	6.6 ± 0.4
Porous-like	169 ± 1	4 ± 1	1.6 ± 0.2

4. Conclusions

This work deals with the combination of different etching mixtures, their composition, temperature, and etching time on the modification of aluminum-alloy substrates in order to prepare either porous-like structured or smooth surfaces. It was found that the appropriate combination of etchants applied in a precise order can be used to prepare porous-like textured surfaces with an Ra of 1.6 μm or smooth substrates with an Ra below 0.8 μm. Such modifications are possible regardless of initial surface roughness or surface machining.

In order to prepare a specific surface microtexture, it is convenient to especially combine the mixture of nitric and hydrochloric acids with methanol. The methanol addition significantly promotes the etching process at rough surfaces and allows temperature reduction from 80 °C to 23 °C.

From a practical point of view, it was demonstrated that the porous-like surfaces can either promote adhesive bonding or allow the effective preparation of superhydrophobic surfaces featuring a self-cleaning effect due to high apparent water contact angles.

Author Contributions: Conceptualization, A.M. (Antonín Minařík); Methodology, M.K. and A.M. (Antonín Minařík); Validation, A.M. (Antonín Minařík), M.K., and L.M.; Formal analysis, M.K., R.G., E.W., L.M., P.S., and A.M. (Antonín Minařík); Investigation, M.K., A.M. (Antonín Minařík),; Resources, M.K., L.N., and R.G.; Data curation, M.K., P.S., L.N, A.M. (Antonín Minařík), and A.M. (Aleš Mráček); Writing—original draft preparation, A.M. (Antonín Minařík), M.K., E.W. P.S., and A.M. (Aleš Mráček); Visualization, M.K., A.M. (Antonín Minařík), and E.W.; Supervision, A.M. (Antonín Minařík) and A.M. (Aleš Mráček).

Funding: This work was supported by the Ministry of Education, Youth, and Sports of the Czech Republic—Program NPU I (LO1504), and by the European Regional Development Fund (Grant No. CZ.1.05/2.1.00/19.0409), as well as by TBU grant nos. IGA/FT/2017/011 and IGA/FT/2018/011 funded from the resources of specific university research.

Conflicts of Interest: The authors declare no conflict of interest.

References

1. Heinz, A.; Haszler, A.; Keidel, C.; Moldenhauer, S.; Benedictus, R.; Miller, W.S. Recent development in aluminium alloys for aerospace applications. *Mater. Sci. Eng. A* **2000**, *280*, 102–107. [CrossRef]
2. Du, Y.J.; Damron, M.; Tang, G.; Zheng, H.; Chu, C.J.; Osborne, J.H. Inorganic/organic hybrid coatings for aircraft aluminum alloy substrates. *Prog. Org. Coat.* **2001**, *41*, 226–232.
3. Sheasby, P.G.; Pinner, R.; Wernick, S. *The Surface Treatment and Finishing of Aluminium and Its Alloys*; ASM International, Finishing Publications: Materials Park, OH, USA, 2001.
4. He, M.; Zhou, X.; Zeng, X.; Cui, D.; Zhang, Q.; Chen, J.; Li, H.; Wang, J.; Cao, Z.; Song, Y.; et al. Hierarchically structured porous aluminum surfaces for high-efficient removal of condensed water. *Soft Matter* **2012**, *8*, 6680–6683. [CrossRef]
5. Huang, Y.; Sarkar, D.K.; Chen, X.G. Fabrication of corrosion resistance micro-nanostructured superhydrophobic anodized aluminum in a one-step electrodeposition process. *Metals* **2016**, *6*, 47. [CrossRef]
6. Huang, Y.; Sarkar, D.K.; Grant Chen, X. Superhydrophobic aluminum alloy surfaces prepared by chemical etching process and their corrosion resistance properties. *Appl. Surf. Sci.* **2015**, *356*, 1012–1024. [CrossRef]
7. Liu, Y.; Sun, D.; You, H.; Chung, J.S. Corrosion resistance properties of organic-inorganic hybrid coatings on 2024 aluminum alloy. *Appl. Surf. Sci.* **2005**, *246*, 82–89. [CrossRef]
8. Borsellino, C.; Di Bella, G.; Ruisi, V.F. Adhesive joining of aluminium AA6082: The effects of resin and surface treatment. *Int. J. Adhes. Adhes.* **2009**, *29*, 36–44. [CrossRef]
9. Prolongo, S.G.; Ureña, A. Effect of surface pre-treatment on the adhesive strength of epoxy-aluminium joints. *Int. J. Adhes. Adhes.* **2009**, *29*, 23–31. [CrossRef]
10. Saleema, N.; Sarkar, D.K.; Paynter, R.W.; Gallant, D.; Eskandarian, M. A simple surface treatment and characterization of AA 6061 aluminum alloy surface for adhesive bonding applications. *Appl. Surf. Sci.* **2012**, *261*, 742–748. [CrossRef]
11. Boutar, Y.; Naïmi, S.; Mezlini, S.; Ali, M.B.S. Effect of surface treatment on the shear strength of aluminium adhesive single-lap joints for automotive applications. *Int. J. Adhes. Adhes.* **2016**, *67*, 38–43. [CrossRef]
12. Sarkar, D.K.; Farzaneh, M.; Paynter, R.W. Superhydrophobic properties of ultrathin rf-sputtered Teflon films coated etched aluminum surfaces. *Mater. Lett.* **2008**, *62*, 1226–1229. [CrossRef]
13. Burokas, V.; Martushene, A.; Bikul'Chyus, G.; Ruchinskene, A. Aluminum alloy etching in phosphoric acid solutions. *Russ. J. Appl. Chem.* **2009**, *82*, 1835–1839. [CrossRef]
14. Chambers, B. Etching of aluminum alloys by ferric ion. *Met. Finish.* **2000**, *98*, 26–29. [CrossRef]
15. Branch, L.C. Bright dipping aluminum. *Met. Finish.* **1998**, *96*, 24–29. [CrossRef]
16. Oh, H.J.; Lee, J.H.; Ahn, H.J.; Jeong, Y.; Park, N.J.; Kim, S.S.; Chi, C.S. Etching characteristics of high-purity aluminum in hydrochloric acid solutions. *Mater. Sci. Eng. A* **2007**, *448–451*, 348–351. [CrossRef]
17. Zhu, H. Etching Behavior of Aluminum Alloy Extrusions. *JOM* **2014**, *66*, 2222–2228. [CrossRef]
18. Wu, R.; Chao, G.; Jiang, H.; Hu, Y.; Pan, A. The superhydrophobic aluminum surface prepared by different methods. *Mater. Lett.* **2015**, *142*, 176–179. [CrossRef]
19. Branzoi, V.; Golgovici, F.; Branzoi, F. Aluminium corrosion in hydrochloric acid solutions and the effect of some organic inhibitors. *Mater. Chem. Phys.* **2003**, *78*, 122–131. [CrossRef]
20. Wahab, F.M.A.E.; Khedr, M.G.A.; Din, A.M.S.E. Effect of anions on the dissolution of Al in acid solutions. *J. Electroanal. Chem. Interfacial Electrochem.* **1978**, *86*, 383–393. [CrossRef]

21. Wu, R.; Liang, S.; Pan, A.; Yuan, Z.; Tang, Y.; Tan, X.; Guan, D.; Yu, Y. Fabrication of nano-structured super-hydrophobic film on aluminum by controllable immersing method. *Appl. Surf. Sci.* **2012**, *258*, 5933–5937. [CrossRef]

22. Kamaraj, A.B.; Shaw, V.; Sundaram, M.M. Novel Fabrication of Un-coated Super-hydrophobic Aluminum via Pulsed Electrochemical Surface Modification. *Procedia Manuf.* **2015**, *1*, 892–903. [CrossRef]

23. Adelkhani, H.; Nasoodi, S.; Jafari, A.H. A study of the morphology and optical properties of electropolished aluminum in the Vis-IR region. *Int. J. Electrochem. Sci.* **2009**, *4*, 238–246.

24. Feng, L.; Zhang, H.; Mao, P.; Wang, Y.; Ge, Y. Superhydrophobic alumina surface based on stearic acid modification. *Appl. Surf. Sci.* **2011**, *257*, 3959–3963. [CrossRef]

25. Liu, W.; Sun, L.; Luo, Y.; Wu, R.; Jiang, H.; Chen, Y.; Zeng, G.; Liu, Y. Facile transition from hydrophilicity to superhydrophilicity and superhydrophobicity on aluminum alloy surface by simple acid etching and polymer coating. *Appl. Surf. Sci.* **2013**, *280*, 193–200. [CrossRef]

26. Saleema, N.; Sarkar, D.K.; Gallant, D.; Paynter, R.W.; Chen, X.G. Chemical nature of superhydrophobic aluminum alloy surfaces produced via a one-step process using fluoroalkyl-silane in a base medium. *ACS Appl. Mater. Interfaces* **2011**, *3*, 4775–4781. [CrossRef] [PubMed]

27. Wrzecionko, E.; Minařík, A.; Smolka, P.; Minařík, M.; Humpolíček, P.; Rejmontová, P.; Mráček, A.; Minaříková, M.; Gřundělová, L. Variations of Polymer Porous Surface Structures via the Time-Sequenced Dosing of Mixed Solvents. *ACS Appl. Mater. Interfaces* **2017**, *9*, 6472–6481. [CrossRef] [PubMed]

28. Flemming, R.G.; Murphy, C.J.; Abrams, G.A.; Goodman, S.L.; Nealey, P.F. Effects of synthetic micro- and nano-structured surfaces on cell behavior. *Biomaterials* **1999**, *20*, 573–588. [CrossRef]

materials

Article

Development of Compact High-Voltage Power Supply for Stimulation to Promote Fruiting Body Formation in Mushroom Cultivation

Katsuyuki Takahashi [1,2,*], Kai Miyamoto [1], Koichi Takaki [1,2] and Kyusuke Takahashi [3]

1   Faculty of Science and Engineering, Iwate University, Morioka, Iwate 020-8551, Japan;
    g0317144@iwate-u.ac.jp (K.M.); takaki@iwate-u.ac.jp (K.T.)
2   Agri-Innovation Center, Iwate University, Morioka, Iwate 020-8550, Japan
3   Morioka Forest Association, Morioka, Iwate 028-4132 Japan; sotoyama@smile.ocn.ne.jp
*   Correspondence: ktaka@iwate-u.ac.jp; Tel.: +81-19-621-6460

Received: 21 November 2018; Accepted: 1 December 2018; Published: 5 December 2018

Abstract: The compact high-voltage power supply is developed for stimulation to promote fruiting body formation in cultivating *L. edodes* and *Lyophyllum deeastes Sing.* mushrooms. A Cockcroft-Walton (C-W) circuit is employed to generate DC high-voltage from AC 100 V plug power for the compact, easy handling and high safety use in the hilly and mountainous area. The C-W circuit is connected to high-voltage coaxial cable which works for high-voltage transmission and for charging up as energy storage capacitor. The output voltage is around 50 kV with several microseconds pulse width. The dimension and weight of the developed power supply are $0.4 \times 0.47 \times 1$ m^3 and 8.1 kg, respectively. The effect of the high-voltage stimulation on enhancement of fruiting body formation is evaluated in cultivating *L. edodes* and *Lyophyllum deeastes Sing.* mushrooms using the developed compact high-voltage power supply. The conventional Marx generator is also used for comparison in effect of high-voltage stimulation for fruiting body formation. *L. edodes* is cultivated with hosting to natural logs and the pulsed high voltage is applied to the cultivated natural logs. The substrate for *Lyophyllum deeastes Sing.* cultivation consists of sawdust. The results show that the fruiting body formation of mushrooms of *L. edodes* for four cultivation seasons and that of *Lyophyllum deeastes Sing.* for two seasons both increase approximately 1.3 times higher than control group in terms of the total weight. Although the input energy per a pulse is difference with the generators, the improvement of the fruit body yield mainly depends on the total input energy into the log. The effect for promotion on fruiting body formation by the developed compact high-voltage power supply is almost same that by the conventional Marx generator.

Keywords: pulse power; electrical stimulation; electric field; mushroom; *L. edodes*; *Lyophyllum deeastes Sing*

---

1. Introduction

The application of a pulsed high voltage to improve the yield in edible mushroom cultivation has also been attempted by some research groups. The fruiting capacity of shiitake mushroom (*L. edodes*; *L. edodes*) was remarkably promoted by applying a pulsed high voltage to log wood [1–3]. This effect was also recognized in *L. edodes* fruiting on a mature sawdust-based substrate [4,5]. The fruit body (sporocarp) yield in the electrically stimulated substrate was observed to be 1.7 times more than that in the spontaneous fruiting substrate control [6]. This effect was also recognized in the sporocarp formation of edible mushrooms: *Grifola frondosa, Pholiota nameko, Flammulina velutipes, Hypsizygus marmoreus, Pleurotus ostreatus, Pleurotus. eryngii* and *Agrocybe cylindraceas* [7–9]. Sporocarp yield, that is, fruit body formation in the electrically stimulated substrate, was observed to be 130–180% greater than that in the spontaneous fruiting substrate control [6]. The pulsed high-voltage stimulation technique was also applied to ectomycorrhizal fungi, which form associations with some types of wood, such as *Laccaria laccata* and *Tricholoma matsutake* [9,10].

Many types of electrical power supplies have been employed to provide electrical stimulation. A large-scale 1 MV high-voltage impulse generator was used to stimulate *L. edodes* log wood [1]. High-voltage AC was used to stimulate an *L. edodes* sawdust substrate [4]. Inductive energy storage (IES) pulsed power generators have favorable features for mushroom-cultivating applications, for example, they are compact, cost effective, light and have high voltage amplification compared with capacitive energy storage generators such as the impulse generator [9,10]. The yield of *L. edodes* fruiting bodies was improved with high-voltage stimulation generated by the IES pulsed power generators [2,3]. The effect of the pulsed voltage stimulation on some other types of mushroom such as *P. nameko* and *Lyophyllum decastes* (*L. decastes*) was also confirmed using an IES generator developed for the improvement of mushroom yield [6,7]. As a result of these studies, the total harvested weight from log wood and/or sawdust substrates for mushroom cultivation increased by applying a pulsed voltage as an electrical stimulation.

The hilly and mountainous area is suitable for the farmland of mushroom production because of its abundant forest resources and the significant overnight temperature changes. The method of pulsed voltage stimulation has been attracting attention as a promising technology that replaces the conventional stimulation methods such as the immersing water and the beating mushroom logs and improve the working efficiency in the hill and mountains. On the other hands, the electrical power supplies for pulsed voltage stimulation, such as Marx and IES pulsed power generators, has a heavy weight, a large size and a low safety because of its high power, large charging energy and high voltage, which is the major obstacle in a practical use. In this study, a Cockcroft-Walton (C-W) circuit is developed and employed to generate DC high-voltage from AC 100 V plug power as a compact and easy-handling high-voltage power supply for pulsed voltage stimulation. The promotion of mushroom production is affected by electric parameters such as applied voltage, pulse width and input energy. In the present experiment, the influence of the electric parameters on the mushroom production is evaluated using two types of power supply, C-W circuit and a conventional Marx generator [11]. The experiments are conducted on the mushroom production using two different fruiting types, Shiitake (*L. edodes*) mushroom and Hatakeshimeji (*Lyophyllum deeastes Sing.*) mushroom. The mushrooms are cultivated at a farmland in the hilly and mountainous area.

2. Experimental Setup

2.1. Pulsed Power Generators

Figure 1 shows circuit diagram and photograph of high voltage pulsed power supply based on Cockcroft-Walton circuit (Green techno, Kanagawa, Japan; GM100) [12,13]. The circuit is consisted of an AC/DC converter, a DC/AC converter, 12 stages of ceramic capacitors and diodes, a charging capacitor, a 100 MΩ charging resistor and a spark gap switch. The ceramic capacitors have a capacity of several hundred pF. The DC/AC converter consists of a high voltage transformer driven by a resonance circuit and its output voltage of DC/AC converter is 6.2 kV with frequency of 25 kHz. The charging capacitor consists of a 2.6 m coaxial cable with the capacitance of 130 pF (50 pF/m). The AC/DC and DC/AC converters, C-W circuit, the charging capacitor and the charging resistor are inside of the box as shown in Figure 1b, which is filled by a resin for insulation. Figure 2 shows the charging voltage to the charging capacitor. Although the charging time depends on number of the stages and the frequency, the capacitor is charged during approximately 230 ms after turning the spark gap switch on because the output current of the DC/AC converter is limited.

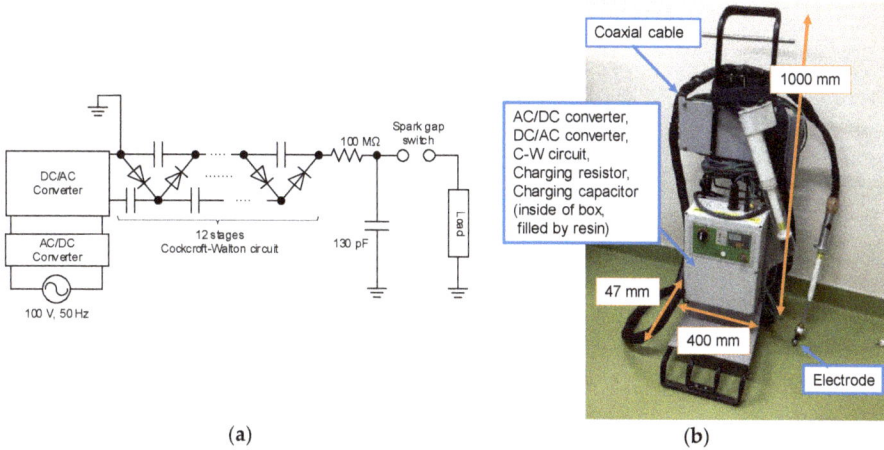

(a)

(b)

Figure 1. Circuit diagram (a) and photograph (b) of C-W circuit.

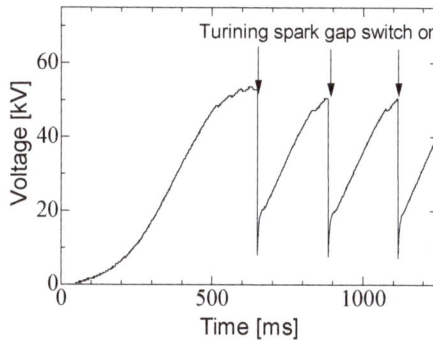

Figure 2. Waveforms of output voltage of C-W circuit during charging without load.

Figure 3 shows the circuit diagram and photograph of pulsed power generator based on Marx generator [8,11]. The Marx generator consists of 4 energy storage 0.22 μF capacitors (Maxwell, 31160), charging resistors (1 and 5 MΩ) connected to the capacitors and the spark gap switches. The capacitors are charged up using a high voltage DC power supply (Gamma high voltage research, RR3-5R/100) up to 12.5 kV. The charging time is required for approximately 10 s because of the output current limit. After charging up the capacitors, a spark gap switch is manually closed. When a spark gap switch is closed, the other switches are sequentially closed automatically and the connection of capacitors is changed from parallel to series. The voltage is stepped up and is applied to the load.

Although the sizes of the Marx generator (1.0 m × 0.45 m × 0.45 m) and C-W circuit (0.4 m × 0.47 m × 1.0 m) are almost same; however, the weights of them are 39.4 kg and 8.1 kg, respectively. Therefore, the handling of C-W circuit in the farmland in hilly and mountainous areas is much easier than that of Marx generator.

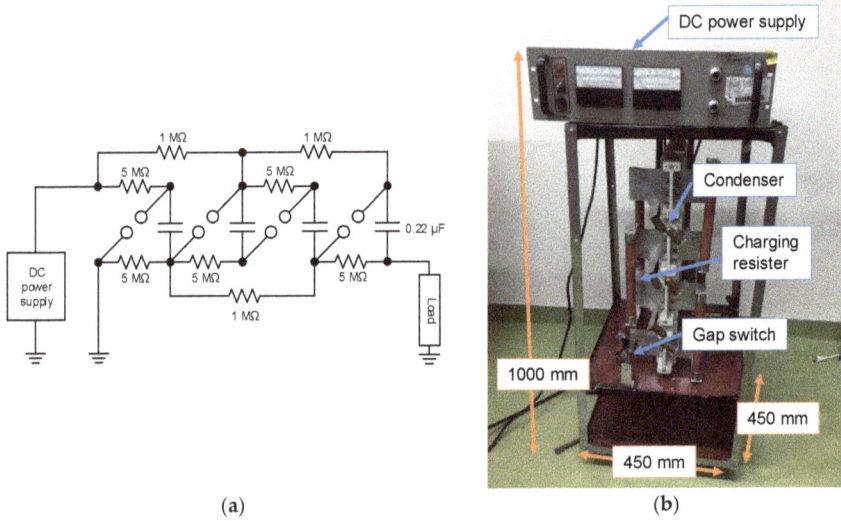

(a)　　　　　　　　　　　　　　　　(b)

Figure 3. Circuit diagram (**a**) and photograph (**b**) of Marx generator.

2.2. Electrical Stimulation to L. edodes

The cultivating mushroom, *L. edodes*, is inoculated on natural logs of *Quercus crispula Blume* two years before the experiment. The strain of the fruiting type is Mori#290 (Mori. Co. Ltd., Gunma, Japan). The dimensions of shiitake mushroom logs with a length of 0.9 m and a diameter of about 0.1 m. The logs are covered with a blackout curtain to maintain the moisture content in the logs hosting the mushroom hyphae. After two years incubation, the blackout curtain is unveiled and the logs are placed side by side under environment as shown in Figure 4.

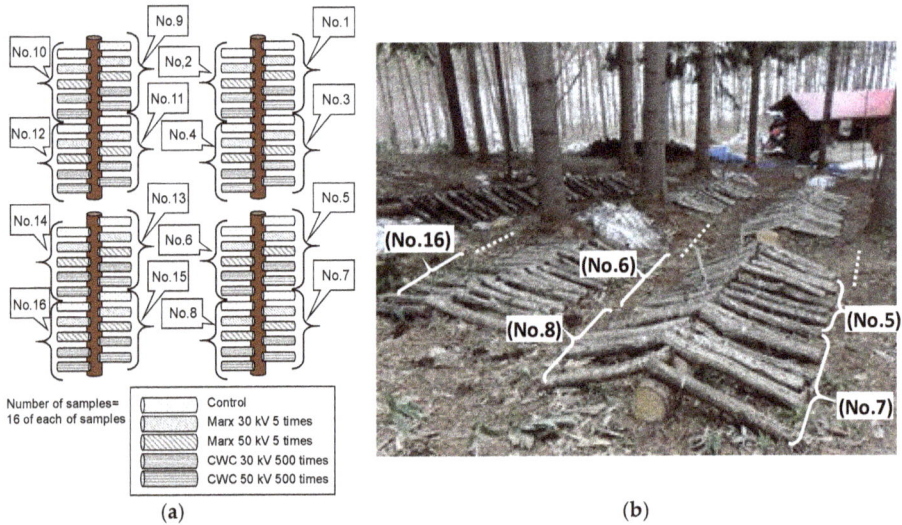

(a)　　　　　　　　　　　　　　　　(b)

Figure 4. Arrangement (**a**) and photograph (**b**) of *L. edodes* logs for cultivation.

Mushroom fruits body production varies among logs, which makes the evaluation difficult. Therefore, it is needed to reduce the influence of variation on the evaluation. In the experiments, the total 80 logs are divided into 4 pulsed voltage stimulated groups and a control group without pulsed voltage to make the average amount of mushroom production of each group almost same after 1st flash. The number of logs for each stimulated group and a control group is 16 logs and numbered from 1 to 16. After the 1st flash, the logs are alternately rearranged as shown in Figure 4a to reduce the influence of arrangement positions.

The pulsed voltage is applied to the logs 1 month before the date that mushroom fruit body is usually expressed. Since the impedance of the logs is affected by the moisture content of wood, the pulsed voltage is applied when that day and its previous day are not rained. The fruit body of mushrooms can be cropped from the logs in every two seasons, spring and autumn, over two years. Therefore, the experiments are conducted for 4 seasons, from 15 May to 20 June in 2017 (1st flush), from 22 September to 22 November in 2017 (2nd flush), from 4 April to 11 June in 2018 (3rd flush) and from 17 September to 9 November in 2018 (4th flush). Figure 5 shows the experimental setup for pulse application to the logs. To apply the pulse voltages to logs, the electrode plate was installed at both ends of logs placed on an insulator of acrylic. The pulsed voltages are applied to the logs at the first day of 2nd and 4th flush seasons using the C-W circuit and the Marx generator. The total input energy into the logs is controlled by the amplitude and the number of the applying. Four groups are stimulated by the pulsed voltages with the different amplitudes, 30 kV and 50 kV, for each generator. The number of pulses is fixed at 500 times in the case of C-W circuit and 5 times in the case of Marx generator. Because the mechanical stress to the mushroom hypha could be affected to the mushroom production, the logs in the control group are set the experimental setup without the applying voltage. The fruit bodies of mushroom are cropped when their pileus is 80% opened, which is suitable to be in the market.

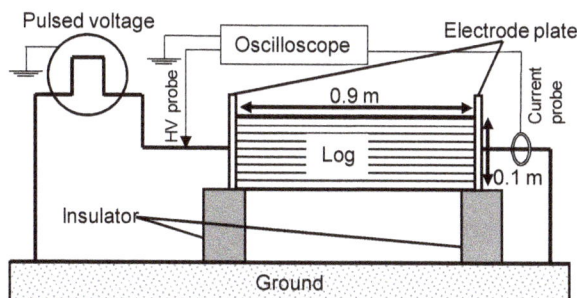

Figure 5. Experimental setup for pulsed voltage stimulation to the *L. edodes* logs.

2.3. Electrical Stimulation to Lyophyllum deeastes Sing

The substrate for *Lyophyllum deeastes Sing.* cultivation consists of sawdust from *Cryptomeria japonica* produced by Kamiyotsuba agricultural cooperative (Kami, Miyagi, Japan). The strain of the fruiting type is Miyagi LD-2 (Tsukidate bio service. Co. Ltd., Miyagi, Japan). The dimensions of the sawdust substrate are 0.12 m × 0.2 m × 0.1 m and it has a cuboid-block shape. The weight of the substrate was 2.5 kg ± 200 g. *Lyophyllum deeastes Sing.* fungus are inoculated on the block and the incubated for 50–60 days under the temperature of 22–23 deg-C with a relative humidity of 65–70%. The blocks are stimulated by the pulsed voltage after the incubation. The pulsed voltage was applied to a needle electrode with a 4 mm diameter driven into the block to a depth of 50 mm, as shown in Figure 6, using C-W circuit. The total input energy into the blocks are controlled by the amplitude and number of the applying voltage. Four groups are stimulated by the pulsed voltages with the different amplitudes, 30 kV and 50 kV and the different numbers of pulses, 100 times and 500 times. The number of blocks for each group is 16 and numbered from 1 to 16.

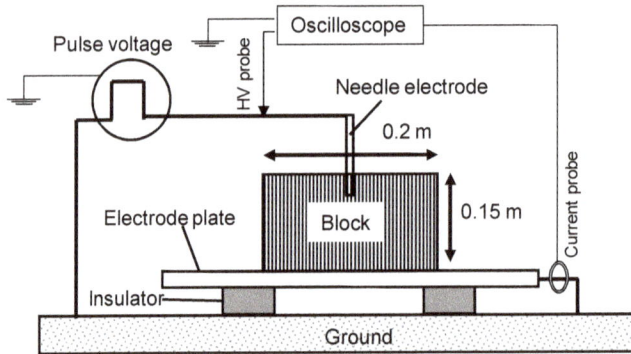

Figure 6. Experimental setup for pulsed voltage stimulation to the *Lyophyllum deeastes Sing.* sawdust block.

After the stimulation, the blocks are buried under the soil with the unburied upper surface as shown in Figure 7. The blocks are alternately arranged as shown in Figure 7a to reduce the influence of the arrangement positions. The fruit bodies of mushroom are cropped when their pileus is 80% opened, which is suitable to be in the market.

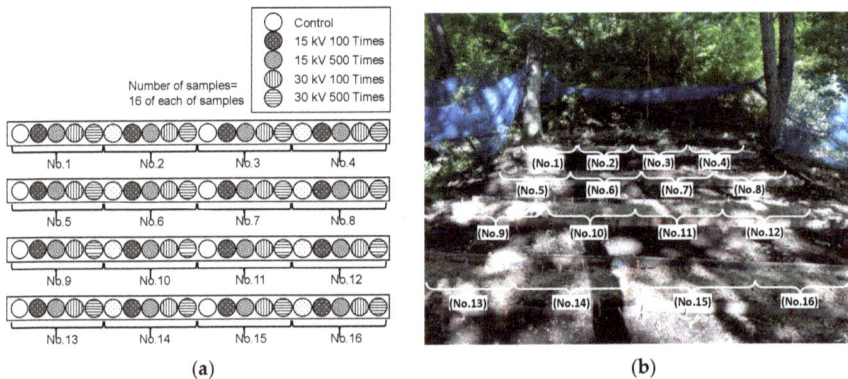

Figure 7. Arrangement (**a**) and photograph (**b**) of the *Lyophyllum deeastes Sing.* sawdust block for cultivation.

3. Results

3.1. Electrical Stimulation to Logs and Cropping Fruits Body of L. edodes

Figure 8a,b shows the typical waveforms of the applied voltage and output current to the shiitake mushroom logs using the C-W circuit and the Marx generator in 4th flush season. When the gap switch of the circuits is shortened, the voltage charged at the capacitors is applied to the log and then the voltage exponentially decays. The impedance of the log is calculated from the waveforms and is 2.67 kΩ with a standard deviation of 0.64 kΩ in 2nd flush season and 5.29 kΩ with a standard deviation of 1.97 kΩ in 4th flush season. The differences of the impedance could be caused by the moisture contents of the logs and a decay with a hypha filled in the log. The time constant in the cases of C-W circuit and Marx generator in the case of 2nd flush season are approximately 1.2 µs with a standard deviation of 0.28 µs and 190 µs with a standard deviation of 51 µs, respectively and those in the case of 4th flush season are approximately 150 µs with a standard deviation of 0.61 µs and 410 µs with a standard deviation of 0.41 ms, respectively. Although the impedances and the time constants are difference, the total input energy into the logs are almost same in the two seasons. In the case of C-W circuit, high voltage pulses with maximum voltage of 30 kV and 50 kV are applied for 500 times and the total input energy are 60 J and 148 J, respectively. In the case of Marx generator, the high voltage pulses with maximum voltage of 30 kV and 50 kV are applied to the cultivation log for 5 times and the total input energy are 127 J and 345 J, respectively. Assuming that the electric field in the log is uniform, the electric field inside log in the case of 30 kV and 50 kV is 34 and 56 kV/m, respectively.

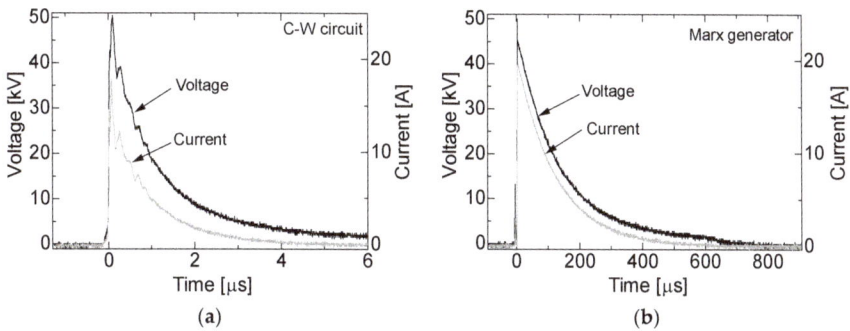

Figure 8. Typical waveforms of applied voltage and current to the *L. edodes* logs using (a) C-W circuit and (b) Marx generator.

Figure 9a–c shows the diurnal change of the accumulated weight of fruitbody of shiitake mushroom in three seasons, 2nd flush, 3rd flush and 4th flush. In the 2nd flush (Figure 9a), the accumulated of fruit body in the case of applying voltage is higher than that in the control group with the harvest duration. The yield of fruit body in the cases of the stimulate groups of 30 kV and 50 kV using the C-W circuit and 30 kV and 50 kV using the Marx generator are 1.15, 1.38, 1.49 and 1.66 times higher than the control group. In the 3rd flush, the pulsed voltages are not applied to logs (Figure 9b), the yield of fruit body does not increase in the cases of stimulated groups in comparison with 2nd flush. The yield of fruit body in the cases of the stimulated groups of 30 kV and 50 kV using the C-W circuit and 30 kV using the Marx generator are 1.04, 1.14 and 1.04 times higher than the control group in the 3rd flush. The yield in the case of the stimulated group of 50 kV using the Marx generator is much lower than other groups. Generally, the yield depends on the yield in previous flush, which could cause the decrease of the yield. In the 4th flush, the fruit bodies are cropped from only the stimulated groups of 30 kV using C-W circuit and 30 and 50 kV using the Marx generator.

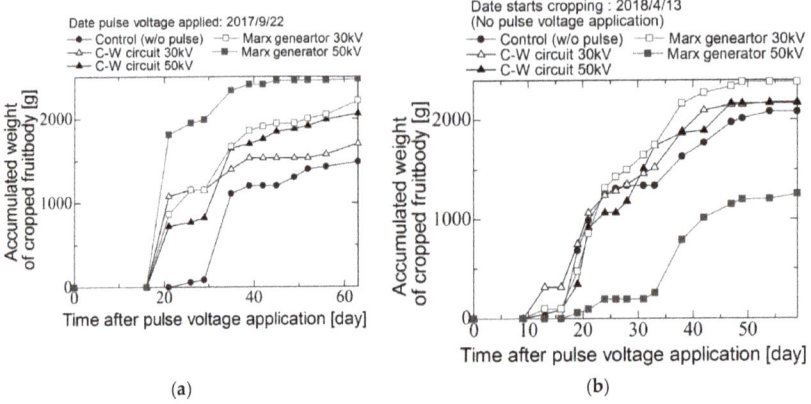

(a)

(b)

Figure 9. *Cont.*

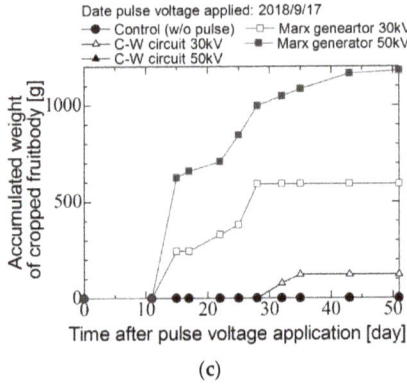

(c)

Figure 9. Diurnal change of the accumulated weight of fruitbody of *L. edodes* in (**a**) 2nd flush, (**b**) 3rd flush and (**c**) 4th flush.

Figure 10 shows the average weight of fruit body cropped per a log, cropped for 4 seasons. The error bars represent the standard error. Since the logs are divided into 5 groups stimulated groups after 1st flash, the amount of the weight of fruit body is almost same. The average weight of fruit body is improved by applying pulse voltages and increased with increasing total input energy into the log. The average weight of fruit body in the case of the Marx generator is approximately 1.3 times higher than that in the control group. The results show that the improvement of the fruit body yield mainly depends on the total input energy into the log.

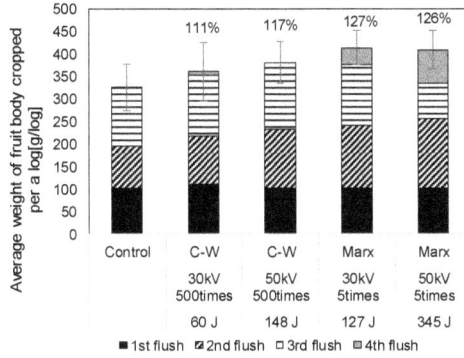

Figure 10. Average yield of fruit body fruitbody of *L. edodes* per a log for 4 flushes.

3.2. Electrical Stimulation to Blocks and Cropping Fruits Body of Lyophyllum deeastes

Figure 11 shows waveforms of applied voltage and output current to the *Lyophyllum deeastes* mushroom block. The resistivity of the bed is 45 Ωm and the impedance of the block is calculated from the waveforms and is 0.35 kΩ with a standard deviation 0.12 kΩ. Because the coaxial cable in the C-W circuit acts as a transmission line, the waveforms of applied voltage and output current are distorted and do not have an exponential shape by the forward and backward transmitted waves [14]. Figure 12 shows the electric field distribution analyzed by the finite element method (Ansoft Maxwell 2D). The analysis results show that the electric field inside of the block is concentrated at the tip of the needle and is ranged from 18 to 360 kV/m with the applied voltage of 30 kV. The input energy per a pulse in the cases of 15 and 30 kV applied voltages are 54 mJ and 27 mJ, respectively. The total input energy in the case of 15 kV is 5.4 J for 100 times pulses and 27 J for 500 times pulses and that of 30 kV is 27 J for 100 times pulses and 160 J for 500 times pulses.

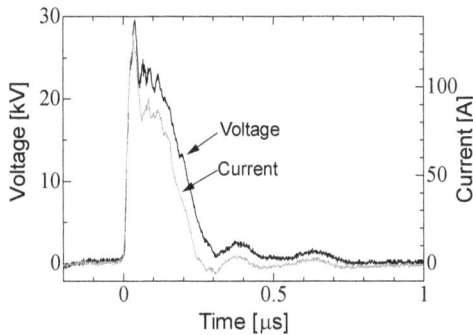

Figure 11. Waveforms of applied voltage and current to the *Lyophyllum deeastes Sing.* Sawdust block using. C-W circuit.

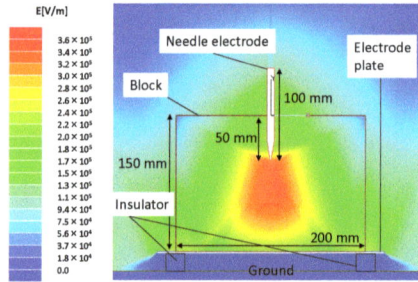

Figure 12. Electric field distribution inside of *Lyophyllum deeastes Sing.* Sawdust block for applied voltage of 30 kV.

Figure 13a–c shows diurnal change of accumulated weight of fruitbody of *Lyophyllum deeastes* mushroom from 27 August to 25 October in 2017 and from 13–21 June in 2018. The pulsed voltage is applied to the logs at the first day using the C-W circuit. Figure 14 shows the average weight of fruit body cropped per a block for two flush seasons. The error bars represent the standard error. The average weight of fruit body was improved by applying pulse voltages and increased with increasing total input energy into the block.

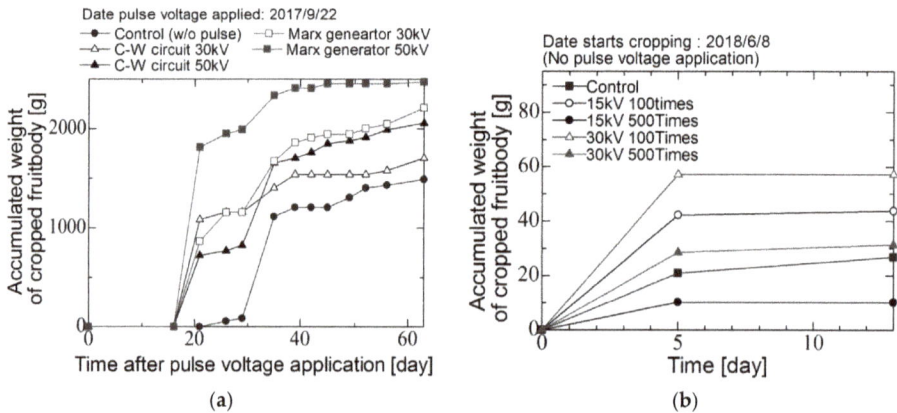

(a)

(b)

Figure 13. Diurnal change of the accumulated weight of fruitbody of *Lyophyllum deeastes Sing* in (**a**) 1st flush and (**b**) 2nd flush.

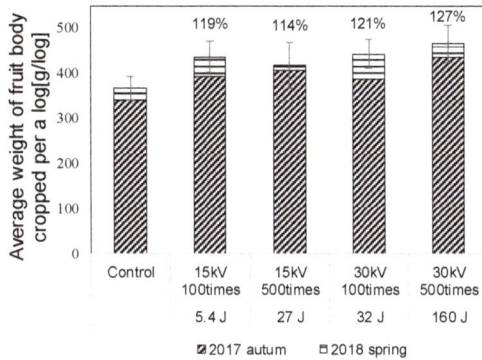

Figure 14. Average yield of fruit body fruitbody of *Lyophyllum deeastes Sing* per a log for 2 flush seasons.

4. Discussion

When the pulsed voltages are applied to the logs, the mushroom hyphae are subjected to an electric field. When the frequency component of the applied pulse voltage is less than several MHz, the membrane of the cell, rather than the inside of the cell, is mainly subjected to the electric field [15]. The hyphae are accelerated and displaced according to the electric field by the electrostatic force such as a Coulomb force [16], which could induce a physical stress on the hyphae. It has been suggested that some genes encoding enzymes such as laccase and protease [17–19] could be upregulated by the physical stress [1,5] in the same manner as other physical stresses such as scrapping of surface hyphae, which induces fruit body formation [16,20]. Since the physical stress relates to the fruit body formation, the flush is accelerated and the amount of cropped fruitbody is increased in the seasons that the logs are stimulated by the voltage pulses as shown in Figure 10.

The total *L. edodes* mushroom yield cropped from the logs is improved approximately 1.3 times from the control group by stimulating. The *Lyophyllum deeastes Sing*. mushroom cropped from the logs is also improved about 1.2 times. It has been reported that *Lyophyllum deeastes Sing*. yield is improved as same level using Marx-IES circuit. These results show the effect for promotion on fruiting body formation by the developed compact high-voltage power supply is almost same that by the conventional Marx generator.

In the economic aspect, the production improvement of 1.2 to 1.3 times using electric stimulation directly increases the farmer's income. The electric power consumption of the high voltage pulsed power supply based on C-W circuit for the operation of the electrical stimulation is measured using an electric power monitor (SANWA SUPPLY, TAP-TST7) and is less than 40 W, which shows that the energy cost is low enough to be negligible. The time cost for the operation and the work load of the electrical stimulation could be much lower than the traditional stimulation methods such as a beating and a shaking. Furthermore, the acceleration of the flush as shown in Figures 9 and 13 could reduce the total time cost for a cropping period, which could enhance the work efficiency. Therefore, the electrical stimulation has a high potential for the farmer's management improvement.

5. Conclusions

The C-W circuit is developed and employed to generate DC high-voltage from AC 100 V plug power as a compact and easy-handling high-voltage power supply for pulsed voltage stimulation. The influence of the electric parameters on the mushroom production is evaluated using two types of power supply, C-W circuit and a conventional Marx generator. The weight of the C-W circuit is approximately 5 times lower than the Marx generator. The handling of C-W circuit in the farmland in hilly and mountainous areas is much easier than that of Marx generator. The experiments are conducted on the mushroom production using two different fruiting types, Shiitake (*L. edodes*) mushroom and

Hatakeshimeji (*Lyophyllum deeastes Sing.*) mushroom. The fruiting body formation of mushrooms of *L. edodes* for four cultivation seasons and that of *Lyophyllum deeastes Sing.* for two seasons both increase approximately 1.3 times higher than control group in terms of the total weight. Although the input energy per a pulse is difference with the generators, the improvement of the fruit body yield mainly depends on the total input energy into the log. The effect for promotion on fruiting body formation by the developed compact high-voltage power supply is almost same that by the conventional Marx generator.

Author Contributions: Katsuyuki Takahashi (K.T.), K.M. and Koichi Takaki (K.T.) conceived and designed the experiments; K.M. and Kyusuke Takahashi (K.T.) performed the experiments; K.M. analyzed the data; Katsuyuki Takahashi (K.T.) and Koichi Takaki (K.T.) wrote the paper.

Funding: This work was supported by a Grant-in-Aid for Scientific Research (A) from the Japan Society for the Promotion of Science, Grant Number 15H02231.

Acknowledgments: The author would like to thank Yuichi Sakamoto at Iwate Biotechnology Research Center for his valuable comments and discussions. The author would also like to thank Yutaka Shida at the Iwate University technical staff.

Conflicts of Interest: The authors declare no conflict of interest.

References

1. Jitsufuchi, Y.; Jitsufuchi, Y.; Yamamoto, M. *Research for Improvement of Lentinula edodes Cultivation: Application of Electric Stimulation for Mushroom Cultivation*; Kyushu Electrical Co.: Fukuoka, Japan, 1987. (In Japanese)
2. Tsukamoto, S.; Maeda, T.; Ikeda, M.; Akiyama, H. Application of pulsed power to mushroom culturing. In Proceedings of the 14th IEEE International Pulsed Power Conference, Dallas, TX, USA, 15–18 June 2003; Volume 2, pp. 1116–1119.
3. Takaki, K.; Yamazaki, N.; Mukaigawa, S.; Fujiwara, T.; Kofujita, H.; Takahashi, K.; Narimatsu, M.; Nagane, K. Fruit body formation of basidiomycete by pulse voltage stimulations. *Front. Appl. Plasma Technol.* **2009**, *2*, 61–64.
4. Kudo, S.; Mitobe, S.; Yoshimura, Y. Electric stimulation multiplication of *Lentinulus edodes*. *J. Inst. Electrost. Jpn.* **1999**, *23*, 186–190.
5. Ohga, S.; Iida, S.; Koo, C.-D.; Cho, N.-S. Effect of electric impulse on fruit body production on *Lentinula edodes* in the sawdust-based substrate. *Mushroom Sci. Biotechnol.* **2001**, *9*, 7–12.
6. Ohga, S.; Cho, N.S.; Li, Y.; Royse, D.J. Utilization of pulsed power to stimulate fructification of edible mushroom. *Mushroom Sci.* **2004**, *16*, 343–352.
7. Takaki, K.; Kanesawa, K.; Yamazaki, N.; Mukaigawa, S.; Fujiwara, T.; Takahasi, K.; Yamasita, K.; Nagane, K. Effect of pulsed high-voltage stimulation on Pholiota nameko mushroom yield. *Acta Phys. Pol. Ser. A* **2009**, *115*, 953–956. [CrossRef]
8. Takaki, K.; Kanesawa, K.; Yamazaki, N.; Mukaigawa, S.; Fujiwara, T.; Takahasi, K.; Yamasita, K.; Nagane, K. Improvement of edible mushroom yield by electric stimulations. *J. Plasma Fusion Res. Ser.* **2009**, *8*, 556–559.
9. Ohga, S.; Iida, S. Effect of Electric Impulse on sporocarp formation of ectomycorrhizal fungus *Laccaria laccata* in Japanese red pine plantation. *J. For. Res.* **2001**, *6*, 37–41. [CrossRef]
10. Islam, F.; Ohga, S. The response of fruit body formation on *Tricholoma matsutake* in situ condition by applying electric pulse stimulator. *ISRN Agron.* **2012**, *2012*, 462724:1–462724:6. [CrossRef]
11. Mankowski, J.; Kristiansen, M. A review of short pulse generator technology. *IEEE Trans. Plasma Sci.* **2000**, *28*, 102–108. [CrossRef]
12. Takaki, K.; Takahashi, K.; Ueno, T.; Akiyama, M.; Sakugawa, T. Design and practice of pulsed power circuit. *J. Plasma Fusion Res.* **2011**, *87*, 202–215.
13. Kobougias, I.C.; Tatakis, E.C. Optimal design of a half-wave Cockcroft-walton voltage multiplier with minimum total capacitance. *IEEE Trans. Power Electron.* **2010**, *25*, 2460–2468. [CrossRef]
14. Rossi, J.O.; Tan, I.H.; Ueda, M. Plasma implantation using high-eneryg ions an short high voltage pulses. *Nucl. Instrum. Method Phys. Res. B* **2006**, *242*, 328–331. [CrossRef]
15. Buescher, E.S.; Schoenbach, K.H. Effects of submicrosecond, high intensity pulsed electric fields on living cells—Intracellular electromanip- ulation. *IEEE Trans. Dielectr. Electr. Insul.* **2003**, *10*, 788–794. [CrossRef]

16. Takaki, K.; Takahashi, K.; Sakamoto, Y.; Plant and Mushroom Development. Chap. 7 High-voltage methods for mushroom fruit-body developments. In *Physical Methods for Stimulation of Plant and Mushroom Development*; El-Esawai, M.A., Ed.; IntechOpen Ltd.: London, UK, 2018; pp. 95–113.

17. Miyazaki, Y.; Sakuragi, Y.; Yamazaki, T.; Shishido, K. Target Genes of the Developmental Regulator PRIB of the Mushroom Lentinula edodes Target Genes of the Developmental Regulator PRIB of the Mushroom *Lentinula edodes*. *Biosci. Biotechnol. Biochem.* **2004**, *68*, 1898–1905. [CrossRef] [PubMed]

18. Muraguchi, H.; Fujita, T.; Kishibe, Y.; Konno, K.; Ueda, N.; Nakahori, K.; Kamada, T. The *exp1* gene essential for pileus expansion and autolysis of the inky cap mushroom *Coprinopsis cinerea* (*Coprinus cinereus*) encodes an HMG protein. *Fungal Genet. Biol.* **2008**, *45*, 890–896. [CrossRef] [PubMed]

19. Nakade, K.; Watanabe, H.; Sakamoto, Y.; Sato, T. Gene silencing of the Lentinula edodes lcc1 gene by expression of a homologous inverted repeat sequence. *Microbiol. Res.* **2001**, *166*, 484–493. [CrossRef] [PubMed]

20. Takaki, K.; Yoshida, K.; Saito, T.; Kusaka, T.; Yamaguchi, R.; Takahashi, K.; Sakamoto, Y. Effect of electrical stimulation on fruit body formation in cultivating mushrooms. *Microoganisms* **2014**, *2*, 58–72. [CrossRef] [PubMed]

materials

MDPI

Article

Green Procedure to Manufacture Nanoparticle-Decorated Paper Substrates

Werner Schlemmer [1], Wolfgang Fischer [1], Armin Zankel [2], Tomislava Vukušić [3], Gregor Filipič [4], Andrea Jurov [4,5], Damjan Blažeka [6], Walter Goessler [7], Wolfgang Bauer [1], Stefan Spirk [1,*] and Nikša Krstulović [6,*]

[1] Institute of Paper-, Pulp- and Fibre Technology (IPZ), Graz University of Technology, Inffeldgasse 23, 8010 Graz, Austria; werner.schlemmer@gmx.at (W.S.); wolfgang.johann.fischer@gmail.com (W.F.); wolfgang.bauer@tugraz.at (W.B.)

[2] Institute of Electron Microscopy and Nanoanalysis (FELMI), Steyrergasse 17, 8010 Graz, Austria; armin.zankel@felmi-zfe.at

[3] Faculty of Food Technology and Biotechnology, University of Zagreb, Pierottijeva 6, 10000 Zagreb, Croatia; tvukusic@pbf.hr

[4] Jožef Stefan Institute, Jamova 39, Ljubljana 1000, Slovenia; Gregor.Filipic@ijs.si (G.F.); andrea.jurov@ijs.si (A.J.)

[5] Jožef Stefan International Postgraduate School, Jamova 39, Ljubljana 1000, Slovenia

[6] Institute of Physics, Bijenička 46, 10000 Zagreb, Croatia; dblazeka@ifs.hr

[7] Institute of Chemistry, University of Graz, Universitaetsplatz 1, 8010 Graz, Austria; walter.goessler@uni-graz.at

* Correspondence: stefan.spirk@tugraz.at (S.S.); niksak@ifs.hr (N.K.); Tel.: +43-316-873-30763 (S.S.); +385-1-4698-803 (N.K.)

Received: 30 October 2018; Accepted: 26 November 2018; Published: 29 November 2018

Abstract: For this study, a paper impregnated with silver nanoparticles (AgNPs) was prepared. To prepare the substrates, aqueous suspensions of pulp fines, a side product from the paper production, were mixed with AgNP suspensions. The nanoparticle (NP) synthesis was then carried out via laser ablation of pure Ag in water. After the sheet formation process, the leaching of the AgNPs was determined to be low while the sheets exhibited antimicrobial activity toward Escherichia coli (*E. coli*).

Keywords: silver nanoparticles; laser ablation in liquids; laser synthesis of colloidal nanoparticles solution; nanoparticle-impregnated paper; antimicrobial activity; fiber fines; sheet forming; vacuum filtration

1. Introduction

Cellulose, a biopolymer consisting of β-D 1→4 linked glucose, is the main component in many industrial products such as pulp and paper. In recent years, the development of micro- and nanostructured materials (e.g., nanocellulose and microfibrillated cellulose) was a boost to cellulose science and new applications for industrial purposes have been proposed. One way to create the new materials is to include inorganic matter having particular properties. In this context, the incorporation of functional inorganic nanoparticles (NPs) (e.g., antimicrobial, conductive, photoactive, catalytic, magnetic, Raman active) into different types of cellulosic substrates has been described for many cellulosic materials [1–9]. In principle, there are different ways of implementing nanoparticles into these materials. One approach is the so called in situ synthesis, where the nanoparticle precursor, typically a metal salt or metal citrate, is added to a cellulosic material. The metal ions coordinate to the cellulose macromolecules and a chemical reaction is induced to convert the salt into either metal, metal oxide, or metal sulfide nanoparticles, which are formed in close spatial proximity to the cellulose material [10]. In this way, silver and gold nanoparticle-decorated fibers have been prepared from AgNO₃ and HAuCl₄, in the presence of different types of reduction agents [11,12]. While NaBH₄ was frequently used in older studies to reduce the metal salts, in recent years the use of environmentally friendly reduction agents has become more popular. Here, either glucose or polysaccharides with reducing end groups (i.e., aldehydes) were used to generate silver and gold nanoparticles [13–17]. In particular, polysaccharides are interesting since they not only act as reduction agents but also serve as electrosteric stabilizers, preventing the aggregation of the nanoparticles and providing suspensions which can be stable for several months [18]. In addition, monodisperse silver nanoparticles (AgNPs) can be prepared using a pulsed sono-electrochemical technique from silver citrate where poly (vinyl pyrrolidone) can serve as a stabilizer [19].

A different route involves the application of the ready-made nanoparticles to a cellulosic material, and then to perform a processing step. If the nanoparticles are stabilized by a polymer shell, covalent or physical binding onto the cellulosic substrate can be additionally performed. For example, silver and gold nanoparticles have been synthesized and encapsulated by sulfated chitosan showing antithrombogenic behavior [3,20]. After immobilization on surfaces, simultaneous antimicrobial and anticoagulant surfaces were obtained.

The common element in these methods is that a metal salt precursor is required which must be converted to the desired nanoparticles, very often using elevated temperatures that give a particular type of nanoparticle shape (e.g., spheres, rods). In most cases, nanoparticles contain either capping agents or polymer shells to prevent agglomeration and to obtain stable colloidal suspensions. However, the largest drawback of many wet-chemical routes is their yield which is hardly ever reported in scientific papers. In order to overcome limitations and to make the production of nanoparticles scalable and environmentally friendly, laser ablation in liquids (LAL) has been proposed recently for nanoparticle preparation [21]. This technique is based on the process of pulsed laser ablation of a target (metals, metal oxides, nitrides, etc.) immersed in a liquid [22–24]. There are many advantages of adopting this technique as compared to standard ones, but its potential has still not been fully exploited. It is known as a "green synthesis" technique since additional chemicals are not required [25] while the formation of any by-products is prevented. In principle, pure nanoparticles can be obtained that do not contain any residues on the surface, except oxidation products (e.g., oxides). The LAL technique is not limited by the choice of materials because any metal target or target made of other materials (composites, isolators, conductive materials, semiconductors, organic materials, ceramics, catalytic, hybrid materials, and magnetic or paramagnetic materials) can be used for the synthesis of colloidal nanoparticle suspensions. Therefore, a wide variety of liquids can be used for tailoring the nanoparticle properties [26,27]. Laser pulses can additionally generate, de-agglomerate, fragmentate, re-shape, and reduce the size of the initially formed nanoparticles either by secondary laser interaction (post-irradiation) [28–31] or by double-pulse LAL [32–34].

In this paper, the authors explored the use of LAL for preparing AgNPs in water and integrating them into a paper-based substrate, consisting of fine cellulosic microparticles. Sheets from such particles resemble to a significant extent those made of microfibrillated cellulose, a recently proposed material for many applications such as barrier coatings and packaging materials. However, the price of the material is much lower since it is a large-scale underutilized material in cellulose processing industries. The outline of the paper is as follows: after a description of the synthesis/characterization of the NPs by laser ablation, the procedure to incorporate nanoparticles into sheets made from fines is described. Finally, the properties of the substrates in terms of antimicrobial activity as well as the leaching of the AgNPs from the material are evaluated.

2. Materials and Methods

2.1. Laser Synthesis of Nanoparticles

A colloidal solution of AgNPs was synthesized using pulsed laser ablation of a Ag plate (purity 99.99%, thickness 1 mm, Kurt J. Lesker, Jefferson Hills, PA, USA) immersed in a cuvette filled with 25 mL of deionized water. Laser ablation was conducted with a Nd:YAG laser (Quantell, Brilliant, Les Ulis, France) using the following specifications: pulse duration 5 ns, wavelength 1064 nm, output energy 290 mJ, and repetition rate 5 Hz. The target was fixed on the holder in order to drill craters during ablation. The laser beam was focused by a 10 cm lens onto the target surface. The laser pulse energy in front of the target was 120 mJ while the diameter of a focused pulse on the target surface was 1 mm yielding a laser fluence of 15 J/cm^2. The thickness of a water layer above the target was kept constant at 1.5 cm during the experiment in order to keep the ablation efficiency constant [35]. A detailed scheme of the experimental setup for laser ablation in water is shown in [36].

The morphology of AgNPs was observed using a transmission electron microscope (TEM, FEI Tecnai G2 20 Twin, Thermo Fischer Scientific, Waltham, MA, USA). Obtained TEM pictures were used to calculate size distribution of the AgNPs.

After the ablation, the crater created on the Ag target was studied by an optical microscope (Leitz, Leica Aristomet, reflective illumination mode, Wetzlar, Germany) in order to determine the crater's volume for the assessment of nanoparticle concentration in a procedure described in [37]. The procedure regarding the crater volume determination is described in detail in [38]. It was found that an ablation volume of 23.8·10^6 μm^3 yielded a 250 μg mass of ablated material. Under the assumption that most of the ablated material is transferred into nanoparticles, its mass is also the total mass of synthesized nanoparticles. Therefore, the production rate of nanoparticles is 10 μg per mL as ablation is performed in 25 mL of water. From the known size distribution and the crater volume, the concentration of AgNPs is determined to be in the order of 10^{10} mL^{-1}.

A colloidal solution of laser-synthesized AgNPs was analyzed using a spectrophotometer (Perkin Elmer, Lambda 25, Waltham, MA, USA) to assess the UV-Vis absorption spectrum. Zeta-potential (Zetasizer, Malvern Instruments, Worcestershire, Great Britain) of Ag colloidal solution was measured as −50 mV, indicating a colloidal solution of high stability (the solution is stable for months).

AgNP-impregnated fines sheets were analyzed by scanning electron microscope (SEM, Jeol JSM-7600F, Tokyo, Japan). The electron accelerating voltage was set to 10 kV. Before the imaging, samples were coated with a thin layer of amorphous carbon (PECS 682) to prevent charge accumulation on the surface. Observation mode was set to low-energy secondary electron detection. Energy-dispersive X-ray spectroscopy (EDS) was done on the same instrument by an INCA Oxford 350 EDS SDD detector with an accelerated potential of 15 kV.

2.2. Hand Sheet Formation

All tests were performed using primary fines separated from never-dried, bleached sulfite pulp (mixture of spruce and beech). In order to prepare hand sheets from pure fines, a vacuum filtration method described in [39] was used. A defined amount of fines (0.24 g, dry weight) was diluted with deionized water to reach a solid content of 0.24 wt %. This suspension was stirred at 450 rpm for at least 2 h. After stirring, 2 mL of AgNP colloidal solution (10^{10} particles mL^{-1}) was added to the suspension and stirred for 5 min. After the addition of AgNPs, the sheets were formed by vacuum filtration using a Britt Dynamic Drainage Jar (Frank PTI, Birkenau, Germany). The scheme used for AgNP-impregnated sheet forming is shown in Figure 1.

The Britt Dynamic Drainage Jar was equipped with a supporting plate, a 500 mesh screen (hole diameter 20 m), two filter papers, and a nitrocellulose membrane (DAWP29325 from Merck Chemicals and Life Science GesmbH, Darmstadt, Germany) with a pore size of 0.65 m. The major advantage of using the Britt Dynamic Drainage Jar as compared to a Büchner funnel was to improve the fines sheet formation. In particular, the sandwich-like setup prevented the loss of fine cellulosic material. After the filtration step, the membrane with the fines/MFC sheet on top was pre-dried in a Rapid-Köthen sheet dryer (Frank PTI, Birkenau, Germany) for about 20 s at 93 °C under vacuum. Then, the membrane was peeled off, and the neat fines/MFC sheets were dried for 10 min in the Rapid-Köthen sheet dryer. The sheets were then stored in a climate room at 23 °C and 50% RH for least 12 h prior to testing.

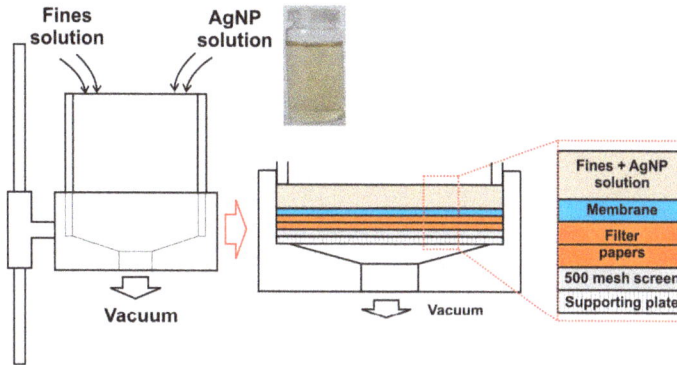

Figure 1. Scheme of the manufacturing process of preparing silver nanoparticle (AgNP)-impregnated sheets.

From the known concentration of AgNPs, volume of colloidal AgNPs added to the suspension, and sheet dimension (10 cm in diameter), the density of impregnated AgNPs could be calculated. The number density of impregnated AgNPs obtained was in the order of 10 nanoparticles per μm^2.

2.3. Antimicrobial Test

For the determination of antimicrobial activity, *E. coli* MG1655-K12 was used (donated by the Laboratory for Biology and Microbial Genetics, Faculty of Food Technology and Biotechnology, University of Zagreb, Zagreb, Croatia). The bacterial suspension was prepared by inoculating 20 µL of *E. coli* K12 in 10 mL of nutrient broth (Biolife, Milan, Italy). This bacterial suspension was incubated at 37 °C for 24 h to create bacteria in the stationary growth phase. The incubated bacterial suspension was centrifuged (Tehtnica, Centric 150, Domel, Železniki, Slovenia) at 4000 rpm for 10 min at room temperature. Harvested cells were washed three times and re-suspended in phosphate buffer saline (PBS) and sterile water solution.

Paper impregnated with AgNPs was inoculated using 10–100 µL of the selected microorganism and placed in an Eppendorf tube containing phosphate buffer. The aliquot sample was incubated 0 h, 1 h, 4 h, 6 h, and 24 h, respectively, after the addition of *E. coli* K12 at a temperature of 37 °C. After incubation, the number of microorganisms (CFU/mL) was determined by the standard dilution method on nutrient agar (Biolife, Milan, Italy). As a control, the selected microorganism (100 µL) was inoculated into 900 µL of phosphate buffer, incubated, and counted by the number of augmented cells. All experiments were analyzed three times (as the repetition of three experiments) and the final results were the mean values of three determinations. The results were reported as log colony forming units per milliliter (log CFU/mL).

2.4. Leaching Tests and Analysis of the Solutions

To investigate the leaching of AgNPs upon contact with water, the paper substrates were shaken in ultrapure water and samples of the liquid were taken after 5 min, 10 min, 30 min, 1 h, 2 h, 5 h, 8 h, and 24 h. The total Ag concentration in these extracts was determined after acidification with nitric acid (10% v/v) with an Agilent 7700cx ICPMS at m/z 107. National Institute of Standards and Technology (NIST) validated Standard Reference Materials (SRM) including 1640a "Trace elements in water" was used for quality control. The Ag concentration of each extract was determined from three parallel measurements. From these results, the extracted amount of Ag in µg was calculated taking the extraction volume into account. The amount of leached mass was directly calculated from the inductively coupled plasma mass spectrometry (ICP-MS) measurements, whereas the relative leaching was determined with respect to the initially added amount of colloidal AgNPs (1 or 2 mL which corresponded to the addition of 10 or 20 µg of AgNPs, respectively).

3. Results

AgNPs feature surface plasmons (SP) activated by the interaction of conductive electrons and an external electromagnetic field, that is, via irradiation with light. If these photons have the right frequency (i.e., the plasmonic frequency), the SPs are excited into a resonant state, oscillating with the highest possible amplitude. This plasmonic frequency depends on the size, shape, chemical composition, and environment of the NPs. From the frequency (wavelength) dependency on the intensity (i.e., absorbance) of the colloidal NPs, which can be assessed by UV-Vis spectroscopy, one can roughly estimate the NP size in solution.

The UV-Vis spectrum and the size distribution of the LAL-synthesized AgNPs are shown in Figure 2a,b. The UV-Vis spectrum of the AgNPs synthesized by laser ablation shows a distinct maximum at 400 nm, typical for a surface plasmon resonance of Ag nanoparticles with a dimension of a few tens of nanometers [10]. This rough estimation is further supported by TEM analysis (inset of Figure 2a), which, in addition, reveals that the AgNPs have a spherical shape. The AgNPs feature an average dimension of 28 nm and their size distribution is relatively broad (FWHM of 20 nm; fitting by a log-normal function (black line)). The LogN fit is often used to describe the size distribution of nanoparticles synthesized from the gaseous phase which is the case in the nanoparticle synthesis process using LAL. It applies whenever particle growth depends on diffusion and drift of atoms to a growth zone of nanoparticles [40]. The final distribution is determined by the available growth time of the nanoparticles [41]. The formation of AgNPs synthesized by LAL is described by dynamic formation mechanisms which include a diffusion slow-growth (diffusion coalescence) process [42–44].

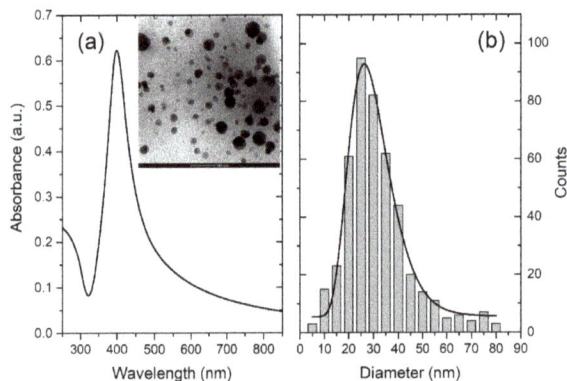

Figure 2. (**a**) UV-Vis photoabsorption spectrum of colloidal AgNPs (Inset: TEM image of AgNPs) and (**b**) Size distribution of AgNPs obtained by TEM imaging (black line: Lognormal fit).

After characterization, these nanoparticles were implemented into a sheet forming process. Suspensions containing the AgNPs and the paper fines were mixed and a stable colloidal suspension was obtained. This suspension was then applied to a Britt Jar process in order to avoid any loss of fine cellulosic material as described in [38]. Afterwards, the sheets were transferred to a drying system and hand sheets were obtained.

Figure 3 depicts SEM images of such AgNP-impregnated hand sheets at different magnification. In Figure 3a,b, distinct AgNPs located at the surface of the sheets can be observed. Identification of the AgNPs was further performed by EDS measurements (shown in the Supplementary Material). Figure 3c,d depicts the same images but recorded with different detectors. They reveal AgNP agglomerates (which are very rare) embedded beneath the sheet surface, covered with much smaller fines fibers. AgNPs in (a) and (b) are in direct contact, whereas those in (c) and (d) are in close contact with bacteria in antimicrobial testing experiments. In Figure 3a,b, one can identify only a few AgNPs per μm^2 on the sheet surface. It implies that AgNPs are impregnated in the whole volume of the sheet as their density is around 10 AgNPs per μm^2. The authors assume that AgNPs are impregnated into sheets by physical entrapment. It can be clearly seen that the AgNPs are hardly agglomerated, which relates to the processing conditions. There, the AgNP colloidal suspensions are added to the fines suspensions under rigorous stirring which ensures an even distribution of the AgNPs in the fines suspension and consequently in the formed sheets.

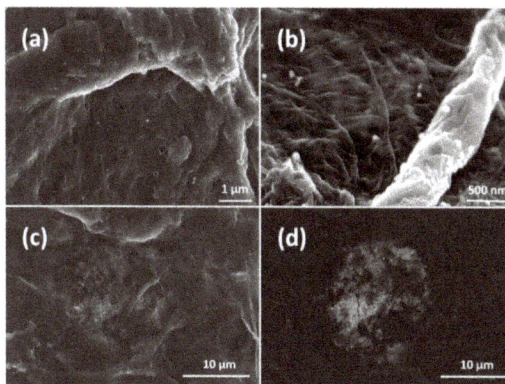

Figure 3. (**a**,**b**) are SEM images of AgNP-impregnated sheets at different magnifications. (**c**,**d**) are the same images recorded with different detectors to show impregnated agglomerates.

These papers have been subjected to antibacterial tests using *E. coli* K12. As indicated in Figure 4, the impregnated materials are highly active toward *E. coli* K12. Figure 4 shows the AgNPs' antimicrobial activity during 24 h, where it can be seen that the AgNPs' antibacterial activity is increasing in time reaching up to a 4 log reduction rate.

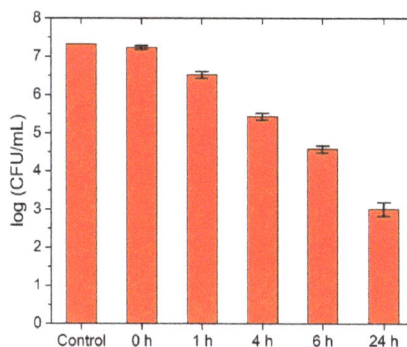

Figure 4. Antimicrobial test of AgNP-impregnated paper on *E. coli* . Please note the logarithmic scale on the y-axis.

In order to investigate the leaching behavior of the AgNPs, the substrates were prepared with two different loadings of silver via the addition of 1 and 2 mL of the colloidal NP solution to the fines dispersion. The filtrate after formation of the sheets was colorless indicating that the AgNPs had been quantitatively incorporated into the sheets. Considering the total amount of AgNPs in the dispersion, this correlates to an AgNP content of 10 and 20 µg, respectively, in the sheets, which corresponds to 42 and 84 µg Ag per gram of paper. The relative mass of leaching AgNPs as well as the total leached mass of AgNPs after exposure to water under rigorous shaking are depicted in Figure 5. The amount of leached AgNPs was measured with ICP-MS after 5 min, 10 min, 30 min, 1 h, 2 h, 5 h, 8 h, and 24 h of substrate immersion in water and vigorous shaking. Relative leaching is determined with respect to the total amount of AgNPs in dispersion. It can be seen that the amount of leached AgNPs increases smoothly up to 8 h spent in immersion and under shaking conditions as expected. After 8 h the leaching is significantly increased for the denser solution, whereas after 24 h the leaching drops due to the degradation of the paper samples upon long exposure to water environment and shaking, points which will be discussed later.

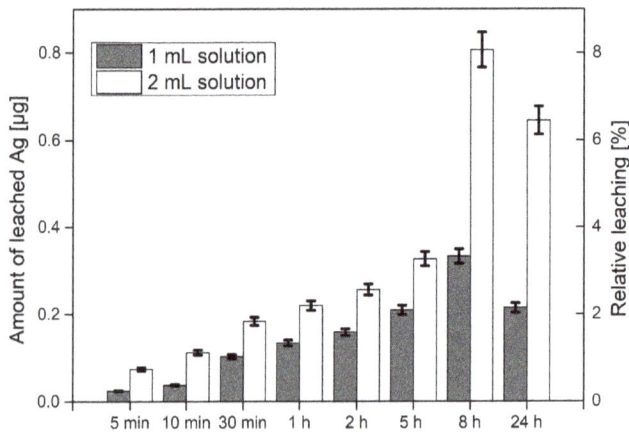

Figure 5. Amount of leached AgNPs from sheets and relative leaching with respect to the AgNPs used for the sheet formation. For the tests, 1 and 2 mL of the colloidal Ag dispersion (10 µg/mL Ag) were added to the fines suspension.

4. Discussion

The use of LAL for the generation of AgNPs which are then implemented into a microstructured paper substrate is a step toward a scalable approach to implement these nanoparticles into cellulosic materials. There are several advantages of using this approach: the amount of applied nanoparticles can be precisely controlled, which is unlikely with any of the other methods. This is important in order to avoid overdosing and to equip materials with the required amount of nanoparticles to prevent bacterial growth, which is beneficial from both an ecological and an economic point of view. Almost any type of bulk material can be used to generate nanoparticles by LAL, and the use of reduction and capping agents can be avoided. In the case of AgNPs, a thin oxide layer results in a negatively charged surface (= −50 mV) making them stable in a colloidal solution according to the Derjagin-Landau-Verwey-Overbeek (DLVO) theory [45]. On the other hand, paper fines represent an underutilized stream in the paper industry and technology, exhibiting a few positive and several negative properties in the course of the paper manufacturing process. One of their drawbacks in paper production, namely their tendency to strongly interact with colloids, is exploited in the case of nanoparticles. After mixing the AgNPs with paper fines suspensions, the colored NPs are homogeneously distributed in the fines' suspensions. Any precipitation of the AgNPs has not been

Materials **2018**, *11*, 2412

observed, while aggregates on the fines surface are very rare. A certain interaction of these AgNPs with the fines must be occurring, because no removal of the AgNPs was observed during the sheet formation process, since the filtrates did not exhibit any color. In contrast, the sheets were slightly yellowish.

The silver concentrations in the study's system were rather small with only 10 AgNPs per m^2. Nonetheless, the AgNP-decorated sheets exhibited antimicrobial activity which was the authors' main goal. There were two types of particles: those that were in direct contact with the bacteria during testing and those that were buried inside the paper network, thereby slowly providing silver ions over time. The action mechanism of silver nanoparticles on bacteria is still not completely clear and fully understood, but the concentration, shape, and size of AgNPs are known to have a significant effect on inactivation effectiveness [46,47]. After the penetration of AgNPs inside the bacteria, NPs interact with intracellular materials, DNA loses its replication ability, and cellular proteins are inactivated [48].

Recent investigations into antimicrobial activity have indicated that bacterial growth is not suppressed by affecting the maximum growth rate but rather by extending the lag phase, that is, the time bacteria spend without replicating and adapting to a new environment. A continuous release of Ag$^+$ ions exhibits higher antimicrobial effects than adding a specific amount of Ag$^+$ at the beginning of the experiment [49]. Further, electrically generated Ag$^+$ has better antimicrobial properties than other silver-based compounds [50–52].

Upon immersion in water under vigorous shaking, a small fraction of the AgNPs leached from the paper. The determined concentration increased slowly for a period of 8 h, whereas the leaching was directly correlated with the amount of particles in the paper. After 8 h of exposure to water, the paper network disassembled, leading to a burst of released silver into the solution. One might expect that these levels would rise for the 24 h sample, but the silver levels in the solution were smaller than after 8 h. Most likely, the AgNPs were readsorbing and reattaching into fines which were abraded from the paper substrates during the shaking process.

5. Conclusions

The authors proposed a simple, fast, and environmentally friendly method to fabricate AgNP-impregnated paper fines sheets with antimicrobial activity. The method was based on the standard route for paper sheet formation to which a step was added—the addition of colloidal nanoparticles. The method worked for any type of colloidal nanoparticle solution synthesized by LAL, whereas a highly stable nanoparticle solution (high zeta-potential) was required for better dispersion into the paper solution and thus better homogeneity in the final product.

Further improvements in the antimicrobial activity will be assessed by optimizing the nanoparticle concentration and using different types of LAL nanoparticles. Future development of the method will be realized in terms of the application of a broad range of nanoparticles with unique properties (Cu, Ti, metal oxides, two-component) for different applications (catalysis, photoactivity, sensors).

Supplementary Materials: The following are available online at http://www.mdpi.com/1996-1944/11/12/2412/s1.

Author Contributions: W.S. analyzed the data and partially wrote the article. W.F., D.B., W.B., and N.K. performed experiments with sheet formation and AgNP impregnation. A.Z., G.F., and A.J. performed SEM and EDX measurements and analysis. S.S. and N.K. wrote the main body of the paper and analyzed the data. D.B. and N.K. synthesized and analyzed the AgNPs. T.V. performed antimicrobial tests. W.G. performed leaching tests.

Funding: This work has been partially supported by project IP-11-2013-2753 funded by the Croatian Science Foundation and by Austrian Academy of Science under Joint Excellence in Science and Humanities fund.

Acknowledgments: N.K. acknowledge the COST Action TD 1208 for fruitful discussions. *E. coli* K12 was donated by the Laboratory for Biology and Microbial Genetics, Faculty of Food Technology and Biotechnology, University of Zagreb (Zagreb, Croatia).

Conflicts of Interest: The authors declare no conflict of interest. The founding sponsors had no role in the design of the study; in the collection, analyses, or interpretation of data; in the writing of the manuscript; and in the decision to publish the results.

References

1. Reishofer, D.; Rath, T.; Ehmann, H.M.; Gspan, C.; Dunst, S.; Amenitsch, H.; Plank, H.; Alonso, B.; Belamie, E.; Trimmel, G.; et al. Biobased Cellulosic–CuInS$_2$ Nanocomposites for Optoelectronic Applications. *ACS Sustain. Chem. Eng.* **2017**, *5*, 3115–3122. [CrossRef]

2. Breitwieser, D.; Kriechbaum, M.; Ehmann, H.M.A.; Monkowius, U.; Coseri, S.; Sacarescu, L.; Spirk, S. Photoreductive generation of amorphous bismuth nanoparticles using polysaccharides-Bismuth-cellulose nanocomposites. *Carbohydr. Polym.* **2015**, *116*, 261–266. [CrossRef] [PubMed]

3. Breitwieser, D.; Spirk, S.; Fasl, H.; Ehmann, H.M.A.; Chemelli, A.; Reichel, V.E.; Gspan, C.; Stana-Kleinschek, K.; Ribitsch, V. Design of simultaneous antimicrobial and anticoagulant surfaces based on nanoparticles and polysaccharides. *J. Mat. Chem. B* **2013**, *1*, 2022–2030. [CrossRef]

4. Sahoo, K.; Biswas, A.; Nayak, J. Effect of synthesis temperature on the UV sensing properties of ZnO-cellulose nanocomposite powder. *Sens. Actuators A* **2017**, *267*, 99–105. [CrossRef]

5. Spiridonov, V.V.; Panova, I.G.; Makarova, L.A.; Afanasov, M.I.; Zezin, S.B.; Sybachin, A.V.; Yaroslavov, A.A. The one-step synthesis of polymer-based magnetic γ-Fe$_2$O$_3$/carboxymethyl cellulose nanocomposites. *Carbohydr. Polym.* **2017**, *177*, 269–274. [CrossRef] [PubMed]

6. Van Rie, J.; Thielemans, W. Cellulose–gold nanoparticle hybrid materials. *Nanoscale* **2017**, *9*, 8525–8554. [CrossRef] [PubMed]

7. Croes, S.; Stobberingh, E.E.; Stevens, K.N.J.; Knetsch, M.L.W.; Koole, L.H. Antimicrobial and Anti-Thrombogenic Features Combined in Hydrophilic Surface Coatings for Skin-Penetrating Catheters. Synergy of Co-Embedded Silver Particles and Heparin. *ACS Appl. Mater. Interfaces* **2011**, *3*, 2543–2550. [CrossRef] [PubMed]

8. Shrivastava, S.; Bera, T.; Singh, S.K.; Singh, G.; Ramachandrarao, P.; Dash, D. Characterization of Antiplatelet Properties of Silver Nanoparticles. *ACS Nano* **2009**, *3*, 1357–1364. [CrossRef] [PubMed]

9. Moram, S.S.B.; Byram, C.; Shibu, S.N.; Chilukamarri, B.M.; Soma, R.V. Ag/Au Nanoparticle-Loaded Paper-Based Versatile Surface-Enhanced Raman Spectroscopy Substrates for Multiple Explosives Detection. *ACS Omega* **2018**, *3*, 8190–8201. [CrossRef]

10. Reishofer, D.; Ehmann, H.M.; Amenitsch, H.; Gspan, C.; Fischer, R.; Plank, H.; Trimmel, G.; Spirk, S. On the formation of Bi$_2$S$_3$-cellulose nanocomposite films from bismuth xanthates and trimethylsilyl-cellulose. *Carbohydr. Polym.* **2017**, *164*, 294–300. [CrossRef] [PubMed]

11. Schlücker, S. SERS Microscopy: Nanoparticle Probes and Biomedical Applications. *Chem. Phys. Chem.* **2009**, *10*, 1344–1354. [CrossRef] [PubMed]

12. Taajamaa, L.; Rojas, O.J.; Laine, J.; Yliniemi, K.; Kontturi, E. Protein-assisted 2D assembly of gold nanoparticles on a polysaccharide surface. *Chem. Commun.* **2013**, *49*, 1318–1320. [CrossRef] [PubMed]

13. Zhicong, M.; Zilin, G.; Ruoxia, C.; Xiaoqing, Y.; Zhiqiang, S.; Wei, G. Surface-Bioengineered Gold Nanoparticles for Biomedical Applications. *Curr. Med. Chem.* **2018**, *25*, 1–25.

14. Coseri, S.; Spatareanu, A.; Sacarescu, L.; Rimbu, C.; Suteu, D.; Spirk, S.; Harabagiu, V. Green synthesis of the silver nanoparticles mediated by pullulan and 6-carboxypullulan. *Carbohydr. Polym.* **2015**, *116*, 9–17. [CrossRef] [PubMed]

15. Donati, I.; Travan, A.; Pelillo, C.; Scarpa, T.; Coslovi, A.; Bonifacio, A.; Sergo, V.; Paoletti, S. Polyol Synthesis of Silver Nanoparticles: Mechanism of Reduction by Alditol Bearing Polysaccharides. *Biomacromolecules* **2009**, *10*, 210–213. [CrossRef] [PubMed]

16. Dahl, J.A.; Maddux, B.L.S.; Hutchison, J.E. Toward Greener Nanosynthesis. *Chem. Rev.* **2007**, *107*, 2228–2269. [CrossRef] [PubMed]

17. Huang, H.; Yang, X. Synthesis of polysaccharide-stabilized gold and silver nanoparticles: A green method. *Carbohydr. Res.* **2004**, *339*, 2627–2631. [CrossRef] [PubMed]

18. Breitwieser, D.; Moghaddam, M.M.; Spirk, S.; Baghbanzadeh, M.; Pivec, T.; Fasl, H.; Ribitsch, V.; Kappe, C.O. In situ preparation of silver nanocomposites on cellulosic fibers—Microwave vs. conventional heating. *Carbohydr. Polym.* **2013**, *94*, 677–686. [CrossRef] [PubMed]

19. Jiang, L.-P.; Wang, A.-N.; Zhao, Y.; Zhang, J.-R.; Zhu, J.-J. A novel route for the preparation of monodisperse silver nanoparticles via a pulsed sonoelectrochemical technique. *Inorg. Chem. Commun.* **2004**, *7*, 506–509. [CrossRef]

20. Ehmann, H.M.A.; Breitwieser, D.; Winter, S.; Gspan, C.; Koraimann, G.; Maver, U.; Sega, M.; Köstler, S.; Stana-Kleinschek, K.; Spirk, S.; Ribitsch, V. Gold nanoparticles in the engineering of antibacterial and anticoagulant surfaces. *Carbohydr. Polym.* **2015**, *117*, 34–42. [CrossRef] [PubMed]

21. Yan, G. *Laser Ablation in Liquids: Principles and Applications in the Preparation of Nanomaterials*; Pan Stanford Publishing: Singapore, 2012.

22. Barcikowski, S.; Mafuné, F. Trends and Current Topics in the Field of Laser Ablation and Nanoparticle Generation in Liquids. *J. Phys. Chem. C* **2011**, *115*, 4985–4985. [CrossRef]

23. Kang, H.W.; Lee, H.; Welch, A.J. Laser ablation in liquid confinement using a nanosecond laser pulse. *J. Appl. Phys.* **2008**, *103*, 083101. [CrossRef]

24. Zhang, D.; Gökce, B.; Barcikowski, S. Laser Synthesis and Processing of Colloids: Fundamentals and Applications. *Chem. Rev.* **2017**, *117*, 3990–4103. [CrossRef] [PubMed]

25. Besner, S.; Kabashin, A.V.; Winnik, F.M.; Meunier, M. Ultrafast laser based 'green' synthesis of non-toxic nanoparticles in aqueous solutions. *Appl. Phys. A* **2008**, *93*, 955–959. [CrossRef]

26. Dolgaev, S.I.; Simakin, A.V.; Voronov, V.V.; Shafeev, G.A.; Bozon-Verduraz, F. Nanoparticles produced by laser ablation of solids in liquid environment. *Appl. Surf. Sci.* **2002**, *186*, 546–551. [CrossRef]

27. Tarasenko, N.V.; Butsen, A.V. Laser synthesis and modification of composite nanoparticles in liquids. *Quantum Electron.* **2010**, *40*, 986–1003. [CrossRef]

28. Tsuji, T.; Watanabe, N.; Tsuji, M. Laser induced morphology change of silver colloids: Formation of nano-size wires. *Appl. Surf. Sci.* **2003**, *211*, 189–193. [CrossRef]

29. Tsuji, T.; Okazaki, Y.; Higuchi, T.; Tsuji, M. Laser-induced morphology changes of silver colloids prepared by laser ablation in water Enhancement of anisotropic shape conversions by chloride ions. *J. Photochem. Photobiol. A* **2006**, *183*, 297–303. [CrossRef]

30. Zeng, H.; Yang, S.; Cai, W. Reshaping Formation and Luminescence Evolution of ZnO Quantum Dots by Laser-Induced Fragmentation in Liquid. *J. Phys. Chem. C* **2011**, *115*, 5038–5043. [CrossRef]

31. Giorgetti, E.; Giammanco, F.; Marsili, P.; Giusti, A. Effect of Picosecond Postirradiation on Colloidal Suspensions of Differently Capped AuNPs. *J. Phys. Chem. C* **2011**, *115*, 5011–5020. [CrossRef]

32. Burakov, V.S.; Tarasenko, N.V.; Butsen, A.V.; Rozantsev, V.A.; Nedel'ko, M.I. Formation of nanoparticles during double-pulse laser ablation of metals in liquids. *Eur. Phys. J. Appl. Phys.* **2005**, *30*, 107–112. [CrossRef]

33. DeGiacomo, A.; DeBonis, A.; Dell'Aglio, M.; De Pascale, O.; Gaudiuso, R.; Orlando, S.; Santagata, A.; Senesi, G.S.; Taccogna, F.; Teghil, R. Laser Ablation of Graphite in Water in a Range of Pressure from 1 to 146 atm Using Single and Double Pulse Techniques for the Production of Carbon Nanostructures. *J. Phys. Chem. C* **2011**, *115*, 5123–5130. [CrossRef]

34. Dell'Aglio, M.; Gaudiuso, R.; El Rashedy, R.; De Pascale, O.; Palazzo, G.; De Giacomo, A. Collinear double pulse laser ablation in water for the production of silver nanoparticles. *Phys. Chem. Chem. Phys.* **2013**, *15*, 20868–20875. [CrossRef] [PubMed]

35. Krstulovic, N.; Shannon, S.; Stefanuik, R.; Fanara, C. Underwater-laser drilling of aluminum. *Int. J. Adv. Manuf. Technol.* **2013**, *69*, 1765–1773. [CrossRef]

36. Krstulovic, N.; Umek, P.; Salamon, K.; Capan, I. Synthesis of Al-doped ZnO nanoparticles by laser ablation of ZnO:Al_2O_3 target in water. *Mater. Res. Express* **2017**, *4*, 105003. [CrossRef]

37. Krstulovic, N.; Salamon, K.; Budimlija, O.; Kovac, J.; Dasovic, J.; Umek, P.; Capan, I. Parameters optimization for synthesis of Al-doped ZnO nanoparticles by laser ablation in water. *Appl. Surf. Sci.* **2018**, *15*, 916–925. [CrossRef]

38. Krstulovic, N.; Milosevic, S. Drilling enhancement by nanosecond–nanosecond collinear dual-pulse laser ablation of titanium in vacuum. *Appl. Surf. Sci.* **2010**, *256*, 4142–4148. [CrossRef]

39. Fischer, W.J.; Mayr, M.; Spirk, S.; Reishofer, D.; Jagiello, L.A.; Schmiedt, R.; Colson, J.; Zankel, A.; Bauer, W. Pulp Fines—Characterization, Sheet Formation, and Comparison to Microfibrillated Cellulose. *Polymers* **2017**, *9*, 366. [CrossRef]

40. Kiss, L.B.; Söderlund, J.; Niklasson, G.A.; Granqvist, C.G. New approach to the origin of lognormal size distributions of nanoparticles. *Nanotechnology* **1999**, *10*, 25–28. [CrossRef]

41. Söderlund, J.; Kiss, L.B.; Niklasson, G.A.; Granqvist, C.G. Lognormal size distributions in particle growth processes without coagulation. *Phys. Rev. Lett.* **1998**, *80*, 2386–2388. [CrossRef]

42. Mafuné, F.; Kohno, J.; Takeda, Y.; Kondow, T. Formation and Size Control of Silver Nanoparticles by Laser Ablation in Aqueous Solution. *J. Phys. Chem. B* **2000**, *104*, 9111–9117. [CrossRef]

43. Zeng, H.; Du, X.-W.; Singh, S.C.; Kulinich, S.A.; Yang, S.; He, J.; Cai, W. Nanomaterials via Laser Ablation/Irradiation in Liquid: A Review. *Adv. Funct. Mater.* **2012**, *22*, 1333–1353. [CrossRef]

44. Zhang, D.S.; Liu, J.; Liang, C.H. Perspective on how laser-ablated particles grow in liquids. *Sci. China-Phys. Mech. Astron.* **2017**, *60*, 074201. [CrossRef]

45. Lyklema, J.; van Leeuwen, H.P.; Minor, M. DLVO-theory, a dynamic re-interpretation. *Adv. Colloid Interface Sci.* **1999**, *83*, 33–69. [CrossRef]

46. Agnihotri, S.; Mukherji, S.; Mukherji, S. Size-controlled silver nanoparticles synthesized over the range 5–100 nm using the same protocol and their antibacterial efficacy. *RSC Adv.* **2014**, *4*, 3974–3983. [CrossRef]

47. Sondi, I.; Salopek-Sondi, B. Silver nanoparticles as antimicrobial agent: A case study on E. coli as a model for Gram-negative bacteria. *J. Colloid Interface Sci.* **2004**, *275*, 177–182. [CrossRef] [PubMed]

48. Morones, J.R.; Elechiguerra, J.L.; Camacho, A.; Holt, K.; Kouri, J.B.; Ramírez, J.T.; Yacaman, M.J. The bactericidal effect of silver nanoparticles. *Nanotechnology* **2005**, *16*, 2346–2353. [CrossRef] [PubMed]

49. Liau, S.; Read, D.; Pugh, W.; Furr, J.; Russell, A. Interaction of silver nitrate with readily identifiable groups: Relationship to the antibacterialaction of silver ions. *Lett. Appl. Microbiol.* **1997**, *25*, 279–283. [CrossRef] [PubMed]

50. Berger, T.J.; Spadaro, J.A.; Bierman, R.; Chapin, S.E.; Becker, R.O. Antifungal properties of electrically generated metallic ions. *Antimicrob. Agents Chemother.* **1976**, *10*, 856–860. [CrossRef] [PubMed]

51. Berger, T.J.; Spadaro, J.A.; Chapin, S.E.; Becker, R.O. Electrically generated silver ions: Quantitative effects on bacterial and mammalian cells. *Antimicrob. Agents Chemother.* **1976**, *9*, 357–358. [CrossRef] [PubMed]

52. Vukusic, T.; Shi, M.; Herceg, Z.; Rogers, S.; Estifaee, P.; Mededovic Thagard, S. Liquid-phase electrical discharge plasmas with a silver electrode for inactivation of a pure culture of Escherichia coli in water. *Innov. Food Sci. Emerg. Technol.* **2016**, *38*, 407–413. [CrossRef]

materials

MDPI

Article

Lectins at Interfaces—An Atomic Force Microscopy and Multi-Parameter-Surface Plasmon Resonance Study

Katrin Niegelhell [1], Thomas Ganner [2], Harald Plank [2], Evelyn Jantscher-Krenn [3] and Stefan Spirk [1,*]

[1] Institute for Paper-, Pulp- and Fiber Technology, Graz University of Technology, Inffeldgasse 23, 8010 Graz, Austria; k.niegelhell@gmx.at

[2] Institute for Electron Microscopy and Nanoanalysis, Graz University of Technology, Steyrergasse 17, 8010 Graz, Austria; Thomas.ganner@a1.net (T.G.); harald.plank@felmi-zfe.at (H.P.)

[3] Department of Obstetrics and Gynecology, Medical University of Graz, Auenbruggerplatz 14, 8036 Graz, Austria; evelyn.jantscher-krenn@medunigraz.at

* Correspondence: stefan.spirk@tugraz.at; Tel.: +43-316-873-30763

Received: 27 September 2018; Accepted: 17 November 2018; Published: 22 November 2018

Abstract: Lectins are a diverse class of carbohydrate binding proteins with pivotal roles in cell communication and signaling in many (patho)physiologic processes in the human body, making them promising targets in drug development, for instance, in cancer or infectious diseases. Other applications of lectins employ their ability to recognize specific glycan epitopes in biosensors and glycan microarrays. While a lot of research has focused on lectin interaction with specific carbohydrates, the interaction potential of lectins with different types of surfaces has not been addressed extensively. Here, we screen the interaction of two specific plant lectins, Concanavalin A and Ulex Europaeus Agglutinin-I with different nanoscopic thin films. As a control, the same experiments were performed with Bovine Serum Albumin, a widely used marker for non-specific protein adsorption. In order to test the preferred type of interaction during adsorption, hydrophobic, hydrophilic and charged polymer films were explored, such as polystyrene, cellulose, $N,-N,-N$-trimethylchitosan chloride and gold, and characterized in terms of wettability, surface free energy, zeta potential and morphology. Atomic force microscopy images of surfaces after protein adsorption correlated very well with the observed mass of adsorbed protein. Surface plasmon resonance spectroscopy studies revealed low adsorbed amounts and slow kinetics for all of the investigated proteins for hydrophilic surfaces, making those resistant to non-specific interactions. As a consequence, they may serve as favorable supports for biosensors, since the use of blocking agents is not necessary.

Keywords: lectin; bovine serum albumin; adsorption; cellulose thin film; polystyrene; gold; surface plasmon resonance spectroscopy

1. Introduction

Lectins are a diverse group of carbohydrate binding proteins featuring at least one non-catalytic domain that reversibly binds to specific mono- or oligosaccharides [1]. These sugar binding proteins are commonly classified in terms of their source (i.e., plants, fungi, animals), or carbohydrate specificity (e.g., glucose/mannose, galactose, sialic acid or fucose) [2,3]. Lectins are critical for cell communication and signaling in many physiologic and pathophysiologic processes. The versatile structure of lectins results in a large diversity of properties, reaching from anti-insect, anti-tumor, immunomodulatory, antimicrobial to HIV-I reverse transcriptase inhibitor activities [4]. Many human pathogens (viral, bacterial or protozoan) employ lectins to bind to glycans displayed on the host's cell surfaces and thus, initiate adhesion and infection. Escherichia coli, for instance, binds to mannosides and the Influenza virus attaches via sialic acid residues on the host's cell surfaces. Aberrant cell surface glycosylation is a hallmark of tumor cells, and lectin interactions with tumor-specific glycan epitopes can promote tumor growth and immune modulation [5,6]. Therefore, lectins are subject to extensive studies in the fields of infectious diseases and cancer research as potential drug targets or therapeutic agents, as well as diagnostic and prognostic tools [2,7].

Other applications of lectins are their use in structural glycan analysis. The carbohydrate binding affinity of lectins is exploited for the detection of glycans or glycan containing molecules. Lectin microarrays are applied to separate, isolate and identify mono-, oligo- or polysaccharides, glycoproteins and glycolipids. Additionally, lectins are employed in biosensors to analyze lectin-carbohydrate interactions, such as specificity, affinity and kinetics [3]. When it comes to biosensors, protein adsorption is a very critical factor, since non-specific interactions of the protein with the substrate influence sensitivity and selectivity. Therefore, blocking agents are employed to minimize those factors [8].

Many parameters are affecting the adsorption behavior of proteins, among others the nature of the substrates, such as hydrophilicity or hydrophobicity and surface morphology. Therefore, fundamental adsorption studies assist to predict the behavior of proteins in the environment of a certain substrate e.g., used in a biosensor. Despite the countless number of lectin applications, there are only a few studies concerning non-specific adsorption of lectins. For instance, Amim et al. investigated the effect of the use of amino-terminated substrates for cellulose ester films and the concomitant change of surface free energy on the lectin-carbohydrate interaction [9] and Zemla et al. determined the preferred adsorption of lectins on parts of phase separated polymer thin films [10].

In this study, we examine the adsorption behavior of two lectins, Ulex Europaeus Agglutinin-I (UEA-I), a fucose binding lectin that is extracted from common gorse, and Concanavalin A (Con A), a lectin with mannose and glucose specificity extracted from jack bean [11,12]. Their adsorption behavior onto the different surfaces was compared to that of Bovine Serum Albumin (BSA), which is a widely used marker for non-specific protein interaction. The interaction capacity of the proteins with substrates of various kinds, such as hydrophilic, hydrophobic and charged (positively and negatively), was tested in real time by means of multi-parameter surface plasmon resonance spectroscopy (MP-SPR) in order to determine not only the adsorbed amount, but also the adsorption kinetics. The herein presented results give insight into the type of interaction that governs the adsorption behavior of these specific proteins.

2. Materials and Methods

Materials. Trimethylsilyl cellulose (TMSC, Avicel, $M_w = 185,000$ g·mol^{-1}, $M_n = 30,400$ g·mol^{-1}, PDI = 6.1 determined by GPC in chloroform) with a DS_{Si} value of 2.8 was purchased from TITK (Rudolstadt, Germany). Chloroform (99.3%), Toluene (99.9%), disodium phosphate heptahydrate (Na_2HPO_4 $7H_2O$), sodium dihydrogen phosphate monohydrate ($NaH_2PO_4 \cdot H_2O$), hydrochloric acid (37%), sodium chloride (Ph.Eur.), sodium hydroxide (99%), polystyrene (PS, $M_w = 35,000$ g·mol^{-1}), Bovine Serum Albumin (lyophilized powder, $\geq$96%, 66.5 kDa), Ulex Europaeus Agglutinin (lyophilized powder $\geq$80%, 63 kDa) and Concanavalin A (Type IV, lyophilized powder, 110 kDa) were purchased from Sigma Aldrich and used as received. *N,N,N*-trimethyl chitosan chloride (TMC, $M_w = 90$ kDa, medical grade, $D_{Acetylation}$: 32%, DS_{Me3+Cl}: 66%) was purchased from Kitozyme, Belgium. Silicon wafers were cut into 1.5 × 1.5 cm^2. SPR gold sensor slides (CEN102AU) were purchased from Cenibra, Bramsche, Germany. Milli-Q water (resistivity = 18.2 Ω^{-1}·cm^{-1}) from a Millipore water purification system (Millipore, Burlington, MA, USA) was used for contact angle and zeta-potential measurements and SPR investigations.

Substrate Cleaning and Film Preparation. Prior to spin coating, SPR gold sensor slides/silicon wafers were immersed in a "piranha" solution containing H_2O_2 (30 wt.%)/H_2SO_4 (1:3 *v/v*) for 10 min.

Then substrates were extensively rinsed with Milli-Q water and blow dried with N_2 gas. TMSC was dissolved in chloroform by stirring over night at room temperature and filtered through 0.45 μm PVDF filters. 120 μL of TMSC (1 wt.%) solution were deposited onto the substrate and then rotated for 60 s at a spinning speed of 4000 rpm and an acceleration of 2500 rpm·s^{-1}. For converting TMSC into pure cellulose, the sensors/wafers were placed in a polystyrene petri-dish (5 cm in diameter) containing 3 mL of 10 wt.% hydrochloric acid (HCl). The dish was covered with its cap and the films were exposed to the vapors of HCl for 15 min. The regeneration of TMSC to cellulose was verified by ATR-IR (Figure A1) and water contact angle (Figure A2) measurements as reported elsewhere [13,14]. PS was dissolved in toluene by stirring over night at room temperature and filtered through 0.45 μm PVDF filters afterwards. 120 μL of PS (1 wt.%) solution were deposited onto the substrate and then rotated for 30 s at a spinning speed of 3000 rpm and an acceleration of 4500 rpm·s^{-1}. TMC films were prepared by adsorption of TMC (1 mg·ml^{-1} dissolved in water, ionic strength was adjusted to 150 mM NaCl and pH value was adjusted to pH 7) onto cellulose substrates at a flow rate of 50 μL·min^{-1} for 5 min. TMC adsorption was monitored by MP-SPR.

Infrared Spectroscopy. IR spectra were attained by an Alpha FT-IR spectrometer (Bruker, Billerica, MA, USA) using an attenuated total reflection (ATR) attachment. Spectra were obtained in a scan range between 4000 to 400 cm^{-1} with 48 scans and a resolution of 4 cm^{-1}. The data were analyzed with OPUS 4.0 software.

Profilometry. Film thicknesses were acquired with a DETAK 150 Stylus Profiler from Veeco (Plainview, USA). The scan length was set to 1000 μm over a duration of 3 s. Measurements were performed with a force of 3 mg, a resolution of 0.333 μm per sample and a measurement range of 6.5 μm. A diamond stylus with a radius of 12.5 μm was used. Samples were measured after scratching the film (deposited on a silicon wafer). The resulting profile was used to calculate the thickness of different films. All measurements were performed three times.

Contact Angle (CA) and Surface Free Energy (SFE) Determination. Static contact angle measurements were performed with a Drop Shape Analysis System DSA100 (Krüss GmbH, Hamburg, Germany) with a T1E CCD video camera (25 fps) and the DSA1 v 1.90 software. Measurements were done with Milli-Q water and di-iodomethane using a droplet size of 3 μL and a dispense rate of 400 μL·min^{-1}. All measurements were performed at least 3 times. CAs were calculated with the Young-Laplace equation and SFE was determined with the Owen-Wendt-Rabel-Kaelble (OWRK) method [15–17].

Atomic Force Microscopy—AFM. Surface characterization was done in ambient atmosphere at room temperature using two Multimode Quadrax MM and FastScanBio AFMs (both Bruker Nano, Billerica, MA, USA). While the former was operated with an NCH-VS1-W cantilever (NanoWorld AG, Neuchâtel, Switzerland, SUI) with force constants around 42 N·m^{-1}, the latter used FastScan A cantilever (Bruker Nano, Billerica, MA, USA) with force constants around 18 N·m^{-1}. Data analyses were done with the software packages Nanoscope (V7.30r1sr3, Veeco) and Gwyddion (V2.50). Image processing and in particular roughness analysis used line and/or plane fitting procedures together with cross-sectional analyses to remove curved and/or tilted background. No additional filters were used to prevent influence on data analyses. Root mean square (R_q) values were derived from multiply selected area statistics to exclude unusually large particles (min. 3 images per sample were fully analyzed). Typical variation from area to area in the same, and in different images, vary less than 0.3 nm, which allows one to specify an accuracy range of ±0.2 nm for all Rq values.

Zeta Potential Measurements. The zeta potential measurements were performed by using a commercial electrokinetic analyzer (SurPASS™3, Anton Paar GmbH, Graz, Austria). For each sample, two zeta potential/pH value functions have been measured in 0.001 M KCl solution. For statistical reasons, four streaming potentials were measured at each pH value. The mean value of these data were used to calculate the potential/pH function.

Multi Parameter Surface Plasmon Resonance Spectroscopy—MP-SPR. MP-SPR spectroscopy was accomplished with an SPR Navi 210 from Bionavis Ltd., Tampere, Finland, equipped with two

Materials **2018**, *11*, 2348

different lasers (670 and 785 nm, respectively) in both measurement channels, using gold coated glass slides as substrate (gold layer 50 nm, chromium adhesion layer 10 nm). All measurements were performed using a full angular scan (39–78°, scan speed: $8° \cdot s^{-1}$).

Gold sensor slides coated with the investigated thin films were mounted in the SPR, equilibrated with water and then with 10 mM PBS with an ionic strength of 100 mM NaCl at pH 5.5/7.4 (The pH value of the buffers was adjusted with 0.1 M hydrochloric acid or 0.1 M NaOH.). After equilibration, protein at a concentration of 0.1 $mg \cdot mL^{-1}$ (dissolved in the same buffer used for equilibration) is introduced into the flow cell. The protein is pumped through the cell with a flow rate of 50 $\mu L \cdot min^{-1}$ over a period of 5 min. After rinsing with buffer, the shift of SPR angle was determined and used to evaluate the amount of adsorbed protein. After protein adsorption all samples were rinsed with Milli-Q water and dried in a stream of N_2 gas. All experiments have been performed in three parallels.

Protein adsorption was quantified according to Equation (1), which considers the dependence of the angular response of the surface plasmon resonance in dependence of the refractive index increment (dn/dc) of the adsorbing layer [18].

$$\Gamma = \frac{\Delta\Theta \times k \times d_p}{dn/dc} \tag{1}$$

For thin layers (<100 nm), $k \times d_p$ can be considered constant and can be obtained by calibration of the instrument by determination of the decay wavelength l_d. For the SPR Navi 210 used in this study, $k \times d_p$ values are approximately 1.09×10^{-7} cm/° (at 670 nm) and 1.9×10^{-7} cm/° (at 785 nm) in aqueous systems. For proteins, dn/dc in water-based buffer systems was reported 0.187 $cm^3 \cdot g^{-1}$, which was used to calculate the amount of adsorbed masses [19]. For TMC, the dn/dc value of chitosan (0.192 $cm^3 \cdot g^{-1}$) [20] was used for the calculation of adsorbed mass.

3. Results and Discussion

In order to provide a variety of interaction possibilities for the proteins, hydrophobic polystyrene (PS), gold and hydrophilic substrates, such as negatively charged cellulose and positively charged *N*,-*N*,-*N*-trimethyl chitosan (TMC) were chosen as substrates for this adsorption study. Prior to adsorption experiments, the materials were characterized in terms of film thickness, surface free energy and morphology. As gold substrates, cleaned SPR sensor slides consisting of 50 nm gold deposited on a glass substrate with an adhesion layer of chromium in between (as reported from manufacturer), were used. Cellulosic substrates were prepared from spin coating trimethylsilyl cellulose (TMSC, Hsinchu City, Taiwan) and subsequent regeneration to cellulose by treatment with HCl vapors, which yielded thin films with a thickness of 30 ± 2 nm as determined by stylus profilometry measurements. Polystyrene films were spin coated as well, leading to film thicknesses of 58 ± 1 nm. The thickness of the positively charged TMC substrate could not be determined, since the substrate was prepared by adsorption of TMC onto cellulose resulting in thicknesses that were too low for detection with stylus profilometry. The different substrates were then subjected to atomic force microscopy (Figure 1). The high root-mean-square-RMS roughness (R_q = 4.3 nm) of the gold substrate is caused by the cleaning procedure with piranha, a very harsh treatment that removes all of the adventitious carbon that was adsorbed from the atmosphere [21]. The cellulosic and TMC substrate display similar RMS roughness (R_q = ca. 2 nm) originating from homogeneous TMC adsorption, thereby forming a thin layer on the cellulose film. The PS thin films show the lowest RMS roughness (R_q = 0.6 nm). All of the substrates are very homogenous and free of any visible contamination or pin-holes.

50/50/50/8 nm

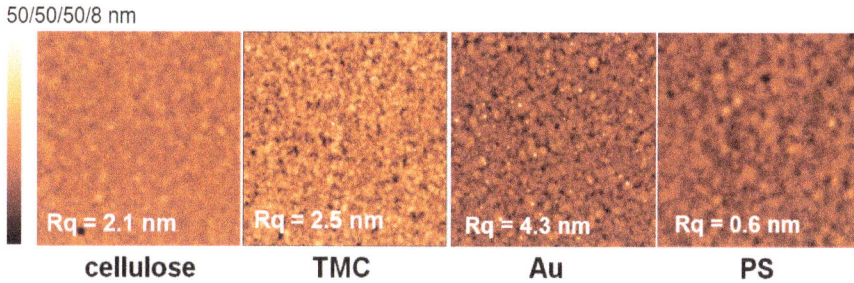

Figure 1. Atomic force microscopy height images (3×3 μm^2) of the different substrates and corresponding RMS roughness (R_q). All images are 3×3 μm^2 while Z scales are 50 nm for cellulose, *N,N,N*-trimethyl chitosan chloride (TMC) and Au and 8 nm for polystyrene (PS) to visualize surface features.

As mentioned above, the positively charged substrate was prepared by adsorption of TMC onto cellulose thin films. Modification of cellulose substrates with TMC as an approach to control nonspecific protein adsorption behavior (using BSA) was already reported earlier [22,23] and the appropriate adsorption conditions for preparation of the cationic TMC substrates were adopted from these studies. In this work, TMC adsorption was monitored by multi-parameter surface plasmon resonance spectroscopy (MP-SPR) and zeta potential measurements (Figure 2). First, we observed a steady equilibration (rinsing with buffer) signal with MP-SPR, associated with a negative zeta potential (ca. -27 mV) for the pure cellulose film. Upon injection of the TMC solution, the SPR-angle increased and the zeta potential changed to positive values indicating deposition of TMC on the surface. Loosely bound material was clearly removed upon rinsing. However, the zeta potential of the adsorbed TMC layer shifted when rinsed with buffer to higher values (from 35 mV to 38 mV), which might be due to a change in conformation of the adsorbed layer. After adsorption, the MP-SPR sensogram showed a change of SPR-angle of 0.05°, which corresponds to an adsorbed amount of 0.6 ± 0.05 mg·m^{-2}. For comparison, cationic starches with a similar charge density as TMC adsorbed to a much higher extent (1.2 mg·m^{-2}) as shown recently [24].

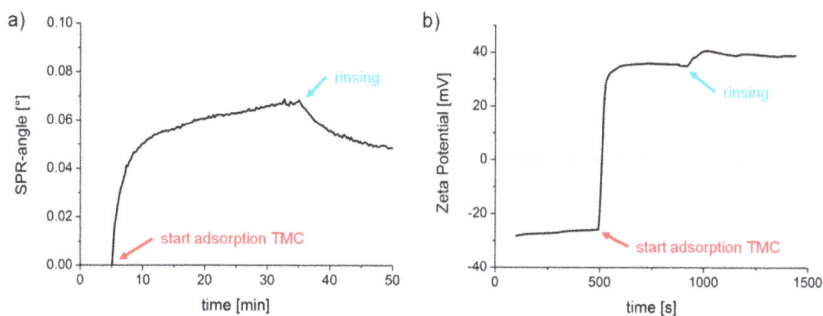

Figure 2. Adsorption of TMC on cellulose thin films. (**a**) Multi-parameter surface plasmon resonance spectroscopy (MP-SPR) sensogram measured at 785 nm, (**b**) zeta potential measurements.

Compared to the other substrates used in this study, TMC is the only one featuring a positive zeta potential. According to literature, the employed and cleaned gold surface displays a negative zeta potential above pH 5 [25], and PS also exhibits negative surface charge (-20 to -30 mV close to pH 7) [26]. The negative zeta potential for the cellulose thin films used in this study is supported by values reported in the literature on cellulosic fibers (-13 mV to -17 mV at pH 4.7–7.2) [22].

The surface free energies (SFE) of the substrates were calculated from static contact angle measurements (Figure A3) and are presented in Figure A4. Cellulose and TMC surfaces both display a hydrophilic character. Although TMC displays a higher zeta potential than cellulose, cellulose shows higher SFE and larger polar contributions than TMC, which could be attributed to the different conformation of the adsorbed polymer in the dry state during contact angle measurements compared to the wet state in the zeta potential determination. PS exhibits, as expected, a hydrophobic surface without any significant polar contribution to the SFE. The lowest SFE of the investigated substrates is presented by the gold substrate. It is important to note, that the gold substrates were immediately used after the cleaning procedure for the adsorption experiments and for the other characterization tests. Thereby, it is guaranteed that the determined SFE is representative of all the samples in this work.

After proper characterization of the substrates, the adsorption behavior of the different proteins was monitored by MP-SPR and the adsorbed amounts were calculated by the change in SPR-angle (Figures 3 and 4). All of the examined proteins did adsorb to the least extent on the cellulose surface and to the highest on PS. This can be attributed to the apolar nature of the PS substrate leading to hydrophobic effects between protein and substrate. In general, hydrophobic effects in proteins are very common and are influencing the folding of proteins in aqueous environments. In such environments, the hydrophobic moieties are buried inside the protein minimizing their free energy. At hydrophobic surfaces, rearrangements of the proteins can take place by exposing the hydrophobic moieties towards that surface. This may even lead to denaturation of the protein, if the degree of interaction is very high. It is widely accepted that this process is mainly governed by entropic contributions rather than enthalpy, unless specific interactions come into play [27]. However, it should be noted here, that also non-hydrophobic effects may contribute to entropy, such as changes in low energy vibrational states.

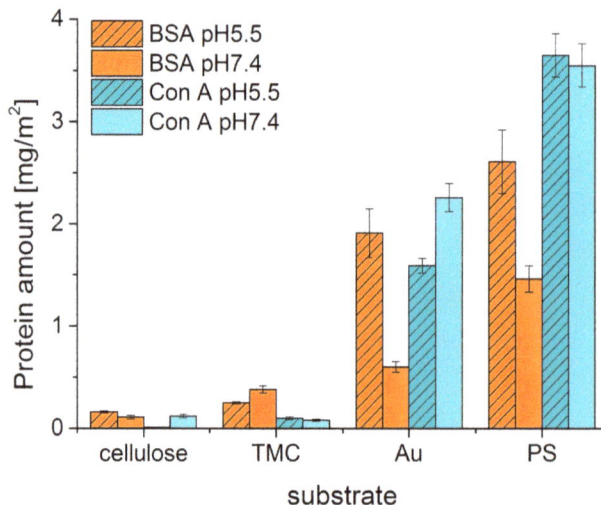

Figure 3. Comparison of adsorbed amount of Bovine Serum Albumin (BSA) and ConA calculated from the change in SPR-angle for different substrates at two pH values.

There is less electrostatic repulsion between surface and protein in the case of PS than for cellulose, still the hydrophobic effect overrules the electrostatic attractions as seen by comparison of PS and TMC. At the investigated pH values, all of the proteins are negatively charged, because the pH values are above the isoelectric points (IEP$_{BSA}$ = pH 4.7, IEP$_{Con A}$ = pH 4.5–5.5, IEP$_{UEA-I}$ = pH 4.8) [28,29]. Enhanced protein adsorption was observed when adsorbing proteins onto TMC modified cellulose compared to pure cellulose. Since TMC is positively charged, more electrostatic attraction takes place, whereas in the case of negatively charged cellulose, proteins are rather compelled at the investigated pH values [30]. The low protein adsorption on cellulose surfaces can be further rationalized by their high water content (ca 60 wt.%). Proteins are highly hydrated molecules as well and any removal of water will lead to a reduction in entropy. However, upon protein adsorption, water needs to be removed from the protein in order to irreversibly adsorb on the surface. Since this is, as we stated above, entropically unfavorable, cellulose surfaces (as well as nearly all highly swollen surfaces) are rather resistant towards non-specific protein deposition and fouling [31].

In general, the highest extent of protein adsorption is reached at the pH value near the isoelectric point, where the proteins exhibit a zero net charge. The balance of positive and negative charges leads to reduced solubility at pH 5.5 for all three proteins investigated, whereas at pH 7.4 the proteins are negatively charged, which increases solubility and causes smaller adsorbed amounts onto the surfaces. This effect is highly pronounced for BSA on all the examined surfaces, except for the TMC substrate. Since TMC is positively charged, it prefers the interaction with the negatively charged BSA at pH 7.4 rather than the more or less neutral BSA at pH 5.5. As for Con A, adsorption onto hydrophilic substrates was extremely low; for cellulose at pH 5.5 it was not even detectable. UEA-I was only investigated at pH 5.5 and showed the highest interaction capacity of all proteins and all pH values with all types of surfaces (Figure 4). Another factor affecting solubility, and subsequently protein deposition, is the aggregation of the proteins in solution, which may take place upon a change in pH value. For ConA, the dimers present at a pH of 5.5 are transformed into tetramers in solution at a pH value larger than 6 [32]. As a consequence, the solubility at the interface is reduced leading to larger deposited amounts in the case of non-specific interaction, which is indeed the observation for the Au and—to some extent—for the PS surfaces. For the latter, the additional complication is the rather large amount of deposited ConA, corresponding to a multilayer. It is known that upon the growth of such protein multilayers, at a certain point the surface reaches saturation and no more protein is adsorbed beyond this limit.

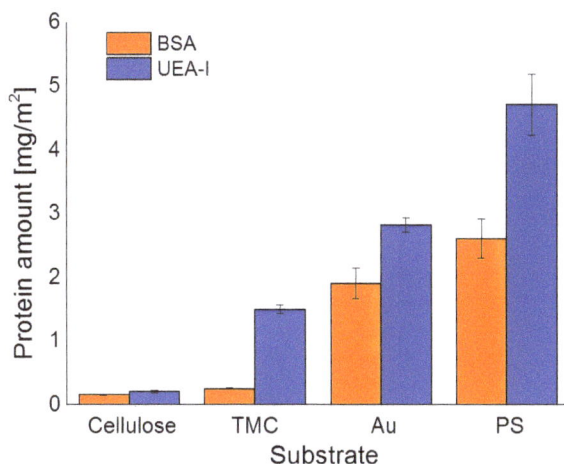

Figure 4. Comparison of adsorbed amount of BSA and UEA-I calculated from the change in SPR-angle for different substrates at pH 5.5.

The sensograms, as shown in Figure 5, give an insight into the adsorption behavior observed in real-time and thereby allow for making statements on the kinetics. BSA adsorbs extremely fast (steep slope and quickly reaching an equilibrium) at pH 7.4 onto PS and cellulose, whereas adsorption is rather slow at pH 5.5. The interactions with PS and gold are strong since no material is removed upon rinsing. Only minor adsorption of Con A is detectable on the cellulose and TMC surface. However, extremely fast adsorption onto PS and gold is monitored indicating a high affinity to the substrates. At pH 7.4

only small amounts detached during rinsing, whereas at pH 5.5 an overshoot effect occurs, which is in the case of proteins usually explained by the so-called rollover model describing a reorientation of end-on into side-on adsorbed proteins [33]. The same effect is observed for UEA-I adsorption onto PS. Adsorption of UEA-I onto TMC and cellulose is very slow, not even reaching an equilibrium in the observed timeframe.

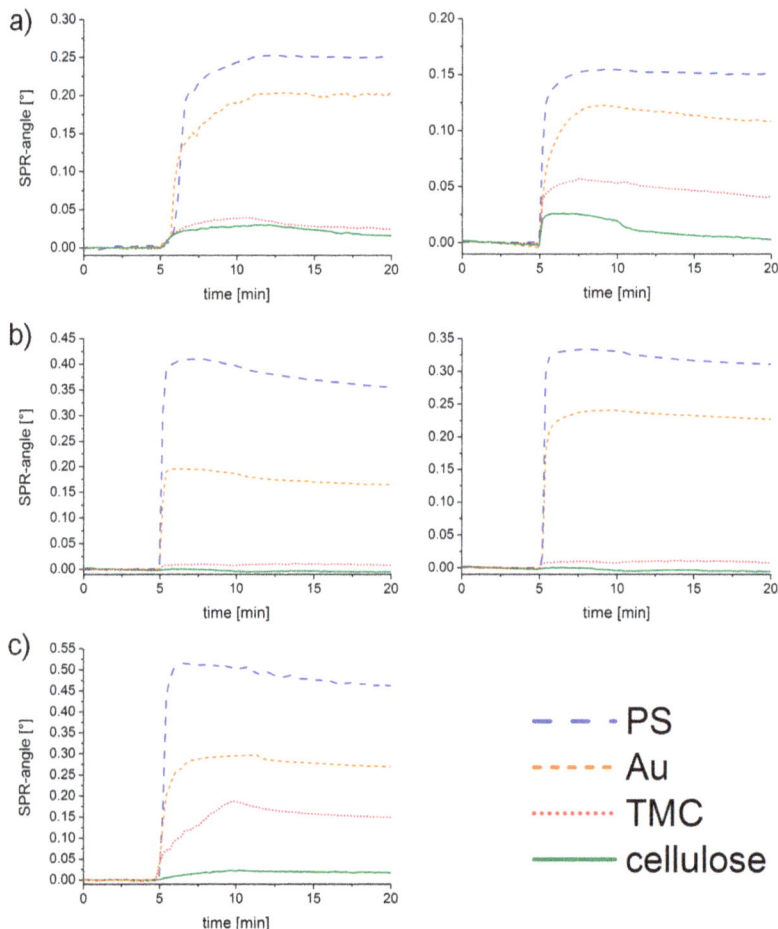

Figure 5. Sensograms measured by MP-SPR at 785 nm for different proteins; (**a**) BSA, (**b**) Con A and (**c**) Ulex Europaeus Agglutinin-I (UEA-I) at a pH value of 5.5 (**left**) and a pH value of 7.4 (**right**).

All of the surfaces were rinsed with water, dried and measured with atomic force microscopy (AFM) directly after protein adsorption. The images (Figures 6 and 7) depict the adsorbed amount obtained by MP-SPR. For some surfaces there is hardly any change in surface topography, because the adsorbed amount was too low to be detected by AFM. In general, the more protein adsorbed on the surfaces, the lower the roughness of the surfaces was. This is an indication for preferable adsorption into valleys/pores of the substrates. However, for the extremely flat surface of PS (R_q = 0.6 nm), it is vice versa meaning that the roughness increases upon protein deposition. There, the proteins form island like features that fuse into a patch like morphology with increasing adsorbed amount before full coverage is achieved [34]. This is represented best by comparing the AFM images of BSA adsorbed onto PS. At pH 5.5 we observe nearly full coverage with a roughness of 1.8 nm, whereas at pH 7.4 islands of BSA lead to a higher roughness (2.8 nm).

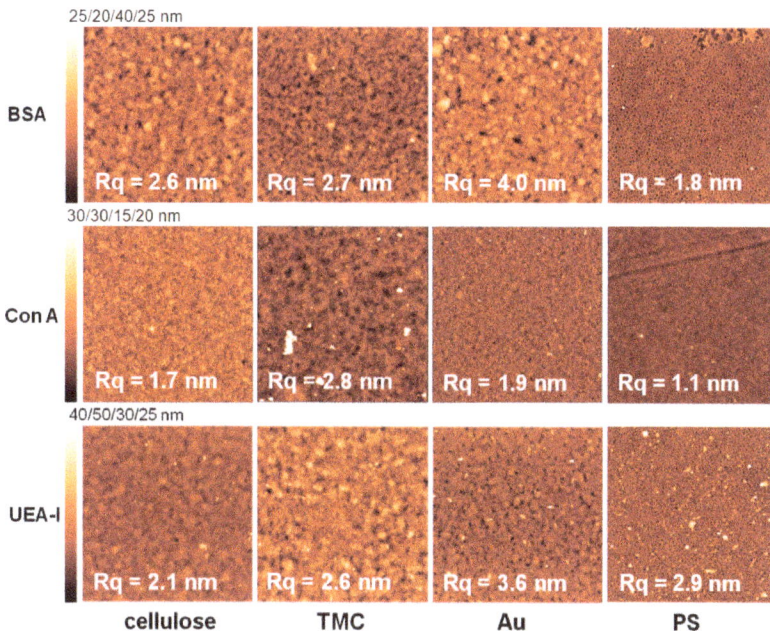

Figure 6. Atomic force microscopy height images of different substrates after protein adsorption at pH 5.5. All images are 3×3 μm^2 while Z scales were individually adapted as specified above the scale bars in each row.

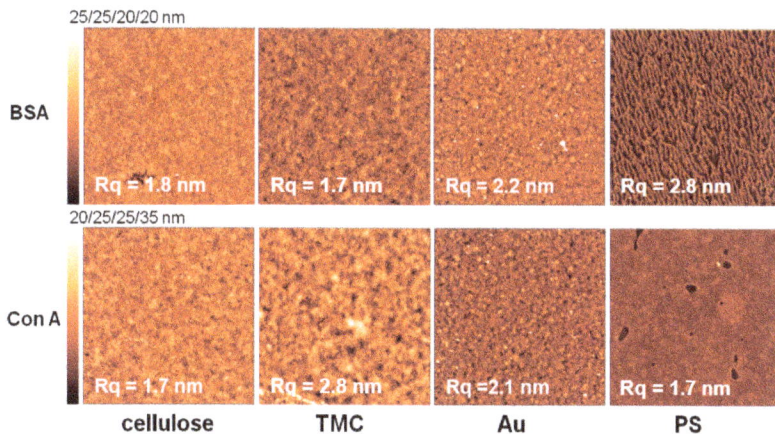

Figure 7. Atomic force microscopy height images of different substrates after protein adsorption at pH 7.4. All images are 3×3 μm^2 while Z scales were individually adapted as specified above the scale bars in each row.

4. Conclusions

The results of this adsorption study can be rationalized in the following way. The adsorption behavior of the examined lectins is comparable to BSA in terms of affinity to substrates of different types. The largest adsorbed amounts and fastest kinetics were observed on the PS surface indicating that hydrophobic effects govern the attraction of the investigated proteins to the substrate, which in turn are mostly driven by entropic contributions. The preferred adsorption onto gold is most likely enhanced by interactions of the thiol groups of the proteins (e.g., methionin for ConA), because of the good interaction capacity of sulfur and gold. The affinity to the hydrophilic substrates was exceptionally low, even when positive charges were introduced by adsorbing TMC. In addition, both types of polysaccharide layers are very prone to swelling (water contents up to 60 wt.%), which impedes protein adsoprtion via entropy since some of the water must be removed from the system in order to accomplish for protein deposition. Although Con A is a mannose/glucose binding lectin, which could interact with the glucose residues from cellulose, no adsorption was detected, probably due to the small number of available end groups of the cellulose.

In conclusion, the binding interactions of BSA, UEA-I and Con A are primarily based on hydrophobic effects, therefore hydrophilic substrates, such as cellulose and TMC, compared to for instance PS, offer huge advantages for the utilization in biosensor development. They are not only stemming from renewable resources, but when used as a support material they are resistant to non-specific protein adsorption thereby avoiding the introduction of blocking agents. As a consequence, they are highly suitable to be used for a variety of lectin-based arrays. Future research will focus on interactions of human milk oligosaccharide with specific lectins immobilized on polysaccharide surfaces, which is an important topic concerning the health of breastfed new-born children.

Author Contributions: Data curation, K.N.; Investigation, K.N., T.G. and H.P.; Methodology, K.N., T.G., H.P. and E.J.-K.; Supervision, S.S.; Validation, S.S.; Visualization, H.P.; Writing—original draft, K.N., H.P., E.J.-K. and S.S.

Funding: This research was partly funded by the People Programme (Marie Curie Actions—Career Integration Grants of the European Union's Seventh Framework Programme (FP7/2007-2013) under REA grant agreement No. 618158 and Marie Curie Actions—International Incoming Fellowship under grant agreement No. 627056).

Acknowledgments: Claudia Payerl is gratefully acknowledged for zeta potential measurements of the TMC coated cellulose films.

Conflicts of Interest: The authors declare no conflict of interest.

Appendix

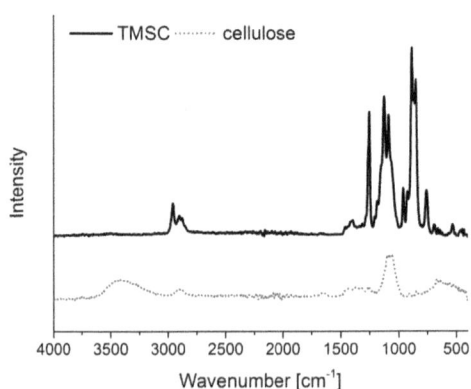

Figure A1. ATR-IR spectra of a TMSC thin film before and after regeneration. Conversion to cellulose upon HCl treatment is proven.

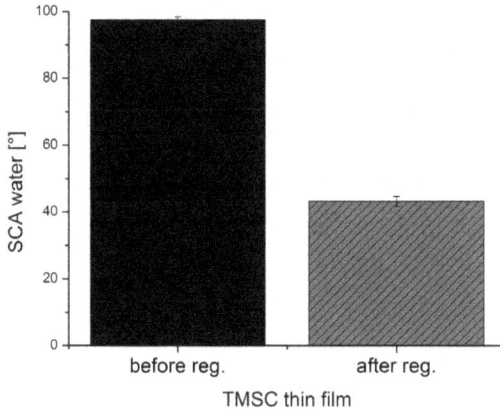

Figure A2. Static water contact angles of a TMSC thin film before and after regeneration to cellulose.

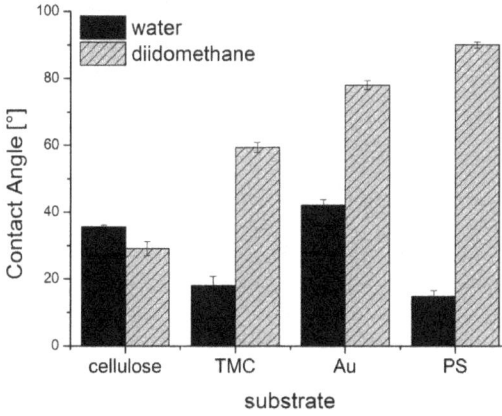

Figure A3. Contact angles measured with water and diiodomethane of different substrates used in the adsorption study. This data were used to calculate the surface free energies of the substrates by the OWRK method.

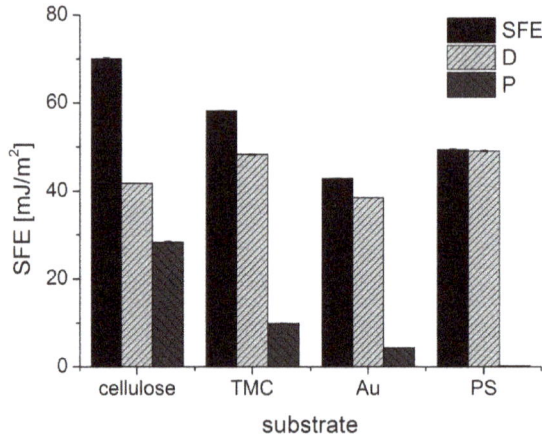

Figure A4. Surface free energies (SFE) and disperse (D) and polar (P) contributions to the SFE of different substrates used in the adsorption study.

References

1. Rini, J.M. Lectin Structure. *Annu. Rev. Biophys. Biomol. Struct.* **1995**, *24*, 551–577. [CrossRef] [PubMed]
2. Kumar, K.; Chandra, K.; Sumanthi, J.; Reddy, G.; Shekar, P.; Reddy, B. Biological role of lectins: A review. *J. Orofac. Sci.* **2012**, *4*, 20–25. [CrossRef]
3. Safina, G. Application of surface plasmon resonance for the detection of carbohydrates, glycoconjugates, and measurement of the carbohydrate-specific interactions: A comparison with conventional analytical techniques. A critical review. *Anal. Chim. Acta* **2012**, *712*, 9–29. [CrossRef] [PubMed]
4. Lam, S.K.; Ng, T.B. Lectins: Production and practical applications. *Appl. Microbiol. Biotechnol.* **2011**, *89*, 45–55. [CrossRef] [PubMed]
5. Taniguchi, N.; Kizuka, Y. Chapter Two—Glycans and Cancer: Role of N-Glycans in Cancer Biomarker, Progression and Metastasis, and Therapeutics. In *Advances in Cancer Research*; Drake, R.R., Ball, L.E., Eds.; Academic Publisher: Waltham, MA, USA, 2015; Volume 126, pp. 11–51.
6. Pinho, S.S.; Reis, C.A. Glycosylation in cancer: Mechanisms and clinical implications. *Nat. Rev. Cancer* **2015**, *15*, 540. [CrossRef] [PubMed]
7. Hamid, R.; Masood, A.; Wani, I.H.; Rafiq, S. Lectins: Proteins with Diverse Applications. *J. Appl. Pharmaceut. Sci.* **2013**, *3*, S93–S103.
8. Jeyachandran, Y.L.; Mielczarski, J.A.; Mielczarski, E.; Rai, B. Efficiency of blocking of non-specific interaction of different proteins by BSA adsorbed on hydrophobic and hydrophilic surfaces. *J. Coll. Interf. Sci.* **2010**, *341*, 136–142. [CrossRef] [PubMed]
9. Amim, J.; Petri, D.F.S. Effect of amino-terminated substrates onto surface properties of cellulose esters and their interaction with lectins. *Mater. Sci. Eng. C* **2012**, *32*, 348–355. [CrossRef]
10. Zemła, J.; Lekka, M.; Raczkowska, J.; Bernasik, A.; Rysz, J.; Budkowski, A. Selective Protein Adsorption on Polymer Patterns Formed by Self-Organization and Soft Lithography. *Biomacromolecules* **2009**, *10*, 2101–2109. [CrossRef] [PubMed]
11. Goldstein, I.J.; Hollerman, C.E.; Smith, E.E. Protein-Carbohydrate Interaction. II. Inhibition Studies on the Interaction of Concanavalin A with Polysaccharides. *Biochemistry* **1965**, *4*, 876–883. [CrossRef] [PubMed]
12. Matsumoto, I.; Osawa, T. Purification and characterization of a Cytisus-type anti-H(O) phytohemagglutinin from Ulex europeus seeds. *Arch. Biochem. Biophys.* **1970**, *140*, 484–491. [CrossRef]
13. Kontturi, E.; Thüne, P.C.; Niemantsverdriet, J.W. Cellulose Model Surfaces Simplified Preparation by Spin Coating and Characterization by X-ray Photoelectron Spectroscopy, Infrared Spectroscopy, and Atomic Force Microscopy. *Langmuir* **2003**, *19*, 5735–5741. [CrossRef]
14. Ehmann, H.M.A.; Werzer, O.; Pachmajer, S.; Mohan, T.; Amenitsch, H.; Resel, R.; Kornherr, A.; Stana-Kleinschek, K.; Kontturi, E.; Spirk, S. Surface-Sensitive Approach to Interpreting Supramolecular

Rearrangements in Cellulose by Synchrotron Grazing Incidence Small-Angle X-ray Scattering. *ACS Macro Lett.* **2015**, *4*, 713–716. [CrossRef]

15. Owens, D.K.; Wendt, R.C. Estimation of the surface free energy of polymers. *J. Appl. Polym. Sci.* **1969**, *13*, 1741–1747. [CrossRef]

16. Rabel, W. Einige Aspekte der Benetzungstheorie und ihre Anwendung auf die Untersuchung und Veränderung der Oberflächeneigenschaften von Polymeren. *Farbe Lack* **1971**, *77*, 997–1005.

17. Kaelble, D.H. Dispersion-Polar Surface Tension Properties of Organic Solids. *J. Adhes.* **1970**, *2*, 66–81. [CrossRef]

18. De Feijter, J.A.; Benjamins, J.; Veer, F.A. Ellipsometry as a tool to study the adsorption behavior of synthetic and biopolymers at the air–water interface. *Biopolymers* **1978**, *17*, 1759–1772. [CrossRef]

19. Robeson, J.L.; Tilton, R.D. Spontaneous Reconfiguration of Adsorbed Lysozyme Layers Observed by Total Internal Reflection Fluorescence with a pH-Sensitive Fluorophore. *Langmuir* **1996**, *12*, 6104–6113. [CrossRef]

20. Nguyen, S.; Hisiger, S.; Jolicoeur, M.; Winnik, F.M.; Buschmann, M.D. Fractionation and characterization of chitosan by analytical SEC and 1H NMR after semi-preparative SEC. *Carbohydr. Polym.* **2009**, *75*, 636–645. [CrossRef]

21. Niegelhell, K.; Leimgruber, S.; Grießer, T.; Brandl, C.; Chernev, B.; Schennach, R.; Trimmel, G.; Spirk, S. Adsorption Studies of Organophosphonic Acids on Differently Activated Gold Surfaces. *Langmuir* **2016**, *32*, 1550–1559. [CrossRef] [PubMed]

22. Ristić, T.; Mohan, T.; Kargl, R.; Hribernik, S.; Doliška, A.; Stana-Kleinschek, K.; Fras, L. A study on the interaction of cationized chitosan with cellulose surfaces. *Cellulose* **2014**, *21*, 2315–2325. [CrossRef]

23. Mohan, T.; Ristic, T.; Kargl, R.; Doliska, A.; Köstler, S.; Ribitsch, V.; Marn, J.; Spirk, S.; Stana-Kleinschek, K. Cationically rendered biopolymer surfaces for high protein affinity support matrices. *Chem. Comm.* **2013**, *49*, 11530–11532. [CrossRef] [PubMed]

24. Niegelhell, K.; Chemelli, A.; Hobisch, J.; Griesser, T.; Reiter, H.; Hirn, U.; Spirk, S. Interaction of industrially relevant cationic starches with cellulose. *Carbohydr. Polym.* **2018**, *179*, 290–296. [CrossRef] [PubMed]

25. Giesbers, M.; Kleijn, J.M.; Cohen Stuart, M.A. The Electrical Double Layer on Gold Probed by Electrokinetic and Surface Force Measurements. *J. Coll. Interf. Sci.* **2002**, *248*, 88–95. [CrossRef] [PubMed]

26. Kirby, B.J.; Hasselbrink, E.F. Zeta potential of microfluidic substrates: 2. Data for polymers. *Electrophoresis* **2004**, *25*, 203–213. [CrossRef] [PubMed]

27. Lombardo, S.; Eyley, S.; Schütz, C.; van Gorp, H.; Rosenfeldt, S.; Van den Mooter, G.; Thielemans, W. Thermodynamic Study of the Interaction of Bovine Serum Albumin and Amino Acids with Cellulose Nanocrystals. *Langmuir* **2017**, *33*, 5473–5481. [CrossRef] [PubMed]

28. Böhme, U.; Scheler, U. Effective charge of bovine serum albumin determined by electrophoresis NMR. *Chem. Phys. Lett.* **2007**, *435*, 342–345. [CrossRef]

29. Entlicher, G.; Koštíř, J.V.; Kocourek, J. Studies on phytohemagglutinins. VIII. Isoelectric point and multiplicity of purified concanavalin A. *Biochim. Biophys. Acta* **1971**, *236*, 795–797. [CrossRef]

30. Privalov, P.L.; Gill, S.J. The hydrophobic effect: A reappraisal. *Pure Appl. Chem.* **1989**, *61*, 1097–1104. [CrossRef]

31. Norde, W.; Lyklema, J. Why proteins prefer interfaces. *J. Biomater. Sci. Polymer Edn.* **1991**, *2*, 183–202. [CrossRef]

32. Senear, D.F.; Teller, D.C. Thermodynamics of concanavalin A dimer-tetramer self-association: Sedimentation equilibrium studies. *Biochemistry* **1981**, *20*, 3076–3083. [CrossRef] [PubMed]

33. Rabe, M.; Verdes, D.; Seeger, S. Understanding protein adsorption phenomena at solid surfaces. *Adv. Colloid Interface Sci.* **2011**, *162*, 87–106. [CrossRef] [PubMed]

34. Wangkam, T.; Yodmongkol, S.; Disrattakit, J.; Sutapun, B.; Amarit, R.; Somboonkaew, A.; Srikhirin, T. Adsorption of bovine serum albumin (BSA) on polystyrene (PS) and its acid copolymer. *Curr. Appl. Phys.* **2012**, *12*, 44–52. [CrossRef]

materials

MDPI

Article

Physical and Morphological Changes of Poly(tetrafluoroethylene) after Using Non-Thermal Plasma-Treatments

Jorge López-García [1,*], Florence Cupessala [2], Petr Humpolíček [1] and Marian Lehocký [1]

[1] Centre of Polymer Systems, University Institute, Tomas Bata University in Zlín, Trida Tomase Bati 5678, 76001 Zlín, Czech Republic; humpolicek@utb.cz (P.H.); lehocky@utb.cz (M.L.)

[2] École supérieure de Chimie Organique et Minérale ESCOM, Allée du Réseau Jean Marie Buckmaster, 60200 Compiègne, France; florence.cupessala05@gmail.com

* Correspondence: lopez@utb.cz; Tel.: +420-776-862-375

Received: 26 September 2018; Accepted: 16 October 2018; Published: 17 October 2018

Abstract: A commercial formulation of poly(tetrafluoroethylene) (PTFE) sheets were surface modified by using non-thermal air at 40 kHz frequency (DC) and 13.56 MHz radiofrequency (RF) at different durations and powers. In order to assess possible changes of PTFE surface properties, zeta potential (ζ), isoelectric points (IEPs) determinations, contact angle measurements as well as Scanning Electron Microscopy (SEM) and Atomic Force Microscopy (AFM) imaging were carried out throughout the experimentation. The overall outcome indicated that ζ-potential and surface energy progressively changed after each treatment, the IEP shifting to lower pH values and the implicit differences, which are produced after each distinct treatment, giving new surface topographies and chemistry. The present approach might serve as a feasible and promising method to alter the surface properties of poly(tetrafluoroethylene).

Keywords: Poly(tetrafluoroethylene); Teflon; plasma treatment; zeta potential; surface energy; contact angle measurement

---

1. Introduction

One straightforward strategy to modify certain surface properties without altering polymer bulk properties is by using non-thermal plasma technologies, such as corona, dielectric barrier, radiofrequency and microwave discharges. Furthermore, plasma treatment is a very versatile technique, since various carrier gases may be employed, giving unique features to the treated material [1–5]. Poly(tetrafluoroethylene) (PTFE), known commercially as Teflon® is a type of fluorinated polymer formed by a succession of molecules of two fluorine atoms (F) and one of carbon (C), its chemical structure is similar to polyethylene; instead of having carbon and hydrogen atoms, the latter are replaced by fluorine atoms as shown in Figure 1.

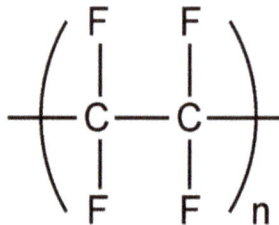

Figure 1. Chemical structure of poly(tetrafluoroethylene) (PTFE).

The strong cohesive force between fluorine and carbon makes poly(tetrafluoroethylene) an inert and nonstick material. Except for other PTFE-like molecules, there is no other molecule that adheres to it [6–8]. One of PTFE's properties is thermal stability, since it is one of the most thermostable plastic materials; for instance it does not show any decomposition above 250 °C, keeping most of its properties. PTFE's thermal conductivity coefficient does not change with temperature, and is relatively high, and thus it is a good insulator. The mixture of PTFE with other materials like glass fibers or carbon increases its thermal conductivity and its resistance to chemical agents. Regarding resistance to atmospheric agents and light, it has been shown that PTFE specimens do not drastically change their properties after being exposed to extreme environment conditions. This inert material reacts only with fluorinated hydrocarbons and fluorinated oils at high temperatures (above 300 °C) causing some swelling and dissolution, which may be reversible. The molecular configuration of PTFE gives the surface a high anti-adhesion. Hence, PTFE surfaces are unwettable and are deemed as a super hydrophobic material [9–13]. PTFE withstands elevated temperatures, and unlike typical thermoplastics, its viscosity above the melting point is so high that PTFE may not be processed by traditional methods, such as extrusion or injection molding. For this reason, PTFE components are manufactured by means of special compression molding and sintering techniques to create blocks, sheets and rods. Recently though, some modified PTFE materials may already be processed as thermoplastics by using traditional techniques as well as electrospinning, which is one of the simplest ways to produce the polymer fibers with nano-sized porous media that has a high surface area per unit volume [14–17].

The aim of this contribution was to make a comprehensive assessment of the effect of air plasma treatment on PTFE commercial sheets by using 40 kHz (DC) and 13.56 MHz (RF) plasma discharges at different plasma duration and power inputs. This outcome is supported by surface probe techniques, such as Scanning Electron Microscopy (SEM), Atomic Force Microscopy (AFM), zeta potential analysis and contact angle measurements. The extensive and indispensable use of PTFE in our daily life underlies the motivation of choosing this material as the target of this work.

2. Materials and Methods

2.1. Materials

Poly(tetrafluoroethylene) (PTFE) was obtained from Dow Chemical Company, (Midland, MI, USA). Distilled-deionized water was retreated in a Simplicity UV® unit, Milipore S.A.S, Molsheim, France, equipped with dual wavelength 185/254 nm UV lamp. The ultra-pure water was used for all experiments and solutions with a water resistivity of 18.2 MΩ at 25 °C and total organic carbon content lower than 5 ppb. Potassium chloride KCl 99%, sodium hydroxide NaOH 98%, anhydrous ethylene glycol $C_2H_6O_2$ 99.8% and diiodomethane CH_2I_2 99% were purchased from Sigma-Aldrich, (Saint Louis, Missouri) USA. Hydrochloric acid HCl 35% was obtained from Penta, Prague, Czech Republic. The reactants were used as received without any further purification.

Both sides of 5 × 5 × 0.1 cm PTFE foils were exposed to non-thermal air plasma by using the following plasma reactors: Pico (Diener electronic, Ebhausen, Germany) with a cylindrical chamber of 150-mm inner diameter Ø and 320 mm length operated at a frequency of 40 kHz, hereinafter (DC). Pico (Diener electronic, Ebhausen, Germany) Ø 150 mm and 320 mm length operated at a frequency of 13.56 MHz, referred to as (RF). The power inputs were 10, 20 and 50 W. The treatment durations were 0, 1, 2, 5, 10 and 20 min respectively, and the pressure in every experiment was 40 Pa. Once the treatment was completed, the specimens were withdrawn from the plasma reactor and immediately used for the next experiments. The foils were stored in a vacuum desiccator (MERCI S.R.O, Brno, Czech Republic) with stopcock, porcelain plate and cobalt chloride ($CoCl_2$) (MERCI S.R.O, Brno, Czech Republic) indicating silica gel (MERCI S.R.O, Brno, Czech Republic).

2.2. Surface Wettability Assessment

Wettability of the samples was evaluated by contact angle measurement before and immediately after each modification. The sessile drop method was employed for this purpose on a Surface Energy Evaluation (SEE) system equipped with a CCD camera (Advex Instruments, Brno, Czech Republic). Deionized water, ethylene glycol and diiodomethane were used as testing liquids at 22 °C and 60% relative humidity. The droplets volume was set to 5 µL for all experiments. Every representative contact angle value was an average of 10 independent measurements. The substrate surface free energy was evaluated by using the acid-basic model.

2.3. Electrokinetic Analysis

The ζ-potential of the sample surfaces was determined by using a SurPASS electrokinetic analyzer (Anton Paar GmbH, Graz, Austria) with a clamping rectangular measuring cell as the one shown in Figure 2. Streaming current and streaming potential measurement methods for flat solid surfaces were used (Anton Paar GmbH, Graz, Austria). The measurements were performed with 0.001M KCl as an electrolyte solution. The pH range was within 2–6 and adjusted by adding either NaOH 0.05M or HCl 0.05M.

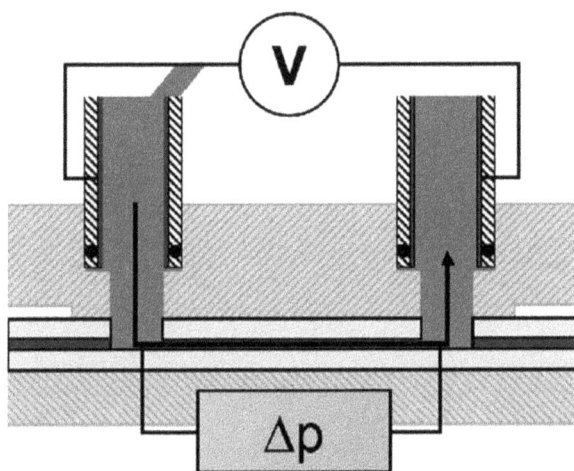

Figure 2. Schematic representation of a clamping cell for the determination of ζ-potential.

2.4. Topographical Evaluation

The surface morphology was evaluated by using an Atomic Force Microscope (AFM) Solver PRO (NT-MDT, Moscow, Russia). The surfaces were analyzed with standard Si cantilever with a constant force of $10 \, \text{N} \cdot \text{m}^{-1}$ and at resonance frequency of 170 kHz. In order to obtain a reliable result, the average surface roughness was obtained from different spots of the samples. The Scanning Electron Microscope (SEM) was carried out by using a Nova NanoSEM 450 (FEI, Brno, Czech Republic) with Schottky field emission electron source operated at acceleration voltage ranging from 200 V to 30 kV and low-vacuum SED (LVD) detector. A coating with a thin layer of gold was performed by a sputter coater SC 7640 (Quorum Technologies, Lewes, UK).

3. Results

3.1. Surface Energy Evaluation

In order to estimate the extent of plasma treatment at different treatment times and powers, the PTFE specimens were assessed by contact angle measurements. Figures 3 and 4 depict the surface energy variation with respect to the plasma duration of either DC or RF plasma treatments.

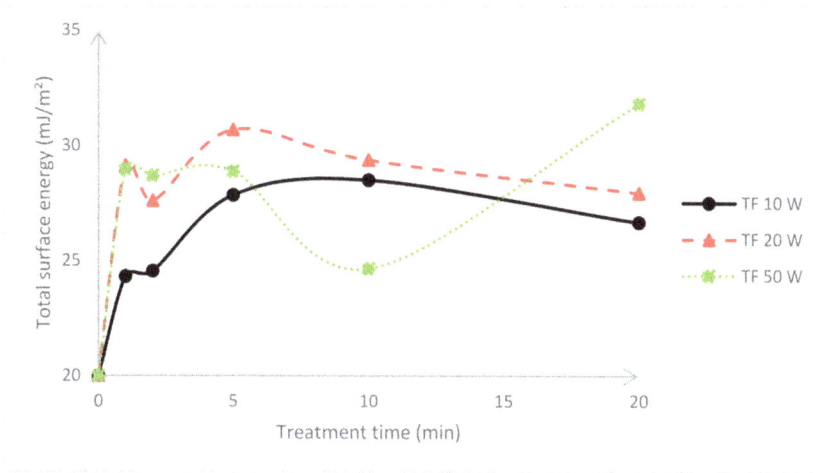

Figure 3. Surface energy of untreated and plasma-treated PTFE after using 40 kHz frequency (DC).

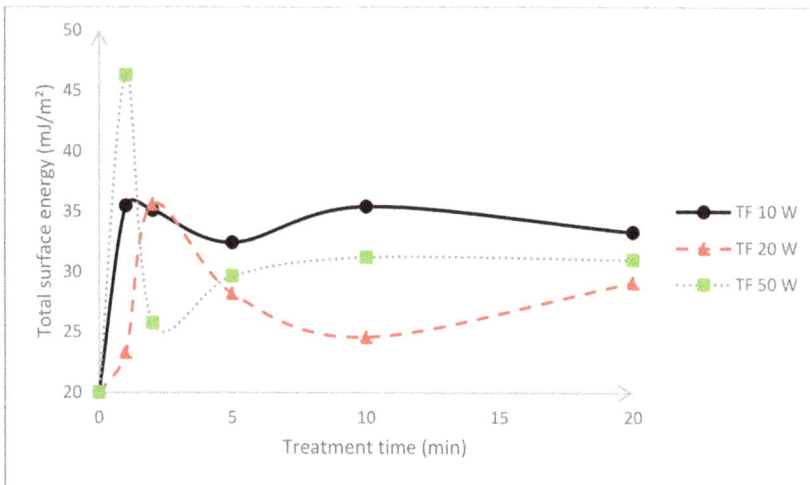

Figure 4. Comparison of the PTFE surface energy after radiofrequency (RF) plasma treatment with different plasma durations and power inputs.

With regard to the surface energy and surface wettability assessments, the contact angles of water, ethylene glycol and diiodomethane are listed in Table 1.

Table 1. Contact angles of PTFE samples after using DC and RF plasma treatments. The Lifshitz-Van der Waals/acid-base (LW/AB) theory was employed to obtain the total surface energy.

Sample		Contact Angle (°) After DC			Contact Angle (°) After RF		
Plasma Duration (min)	Power Input (W)	θ_w	θ_e	θ_d	θ_w	θ_e	θ_d
0	0	108.9	90.7	75.2	108.9	90.7	75.2
1	10	85.0	69.7	68.2	83.1	70.4	57.4
1	20	81.9	68.3	62.6	88.1	74.2	70.3
1	50	77.6	62.6	60.6	75.2	69.0	51.8
2	10	82.0	70.4	68.9	83.1	76.1	63.5
2	20	78.7	69.7	67.2	81.7	74.6	62.6
2	50	81.4	66.8	62.1	82.5	67.3	65.1
5	10	74.2	66.8	66.9	86.9	78.6	65.7
5	20	81.9	69.1	61.8	83.0	62.7	66.4
5	50	80.0	63.1	59.8	78.7	68.1	63.7
10	10	81.5	61.4	66.8	87.6	78.4	61.9
10	20	81.4	71.8	65.8	81.0	67.3	67.7
10	50	88.6	70.1	70.0	75.7	67.2	63.3
20	10	80.9	63.4	70.5	83.6	73.6	62.2
20	20	77.6	69.3	67.1	78.8	67.1	63.3
20	50	78.5	73.5	67.2	81.8	59.8	59.9

3.2. Surface Charge Appraisal

The surface charge with respect to the plasma duration was appraised by electrokinetic analysis, where the ζ-potential is an indicator for charge formation at the solid-liquid interface and the surface charge is generated by the interaction of the solid surface with the electrolyte solution. Likewise, the isoelectric points (IEPs) are defined as the pH at which a substance has a net charge of zero, or at which it is at its minimum ionization. Figure 5 shows the trend of ζ-potential versus the pH and the isoelectric points of untreated PTFE as well as the treated sample after 20 min 50W DC and RF. For this experiment, only the longest durations and the highest power input were compared to untreated PTFE.

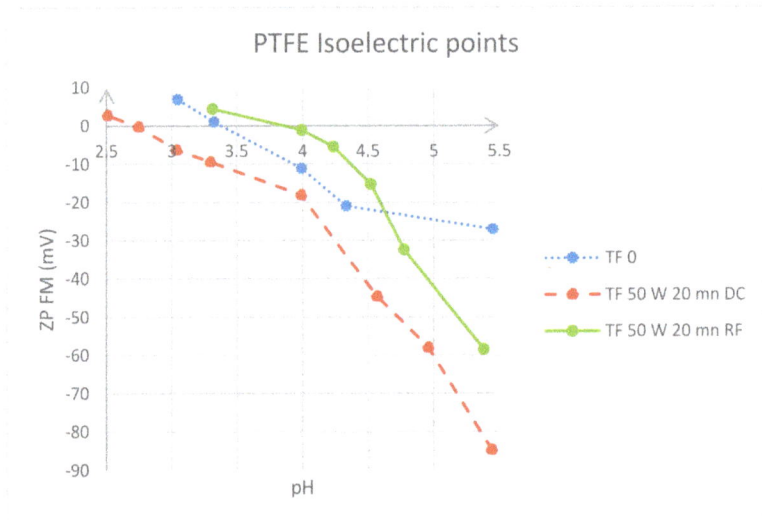

Figure 5. ζ-potential as a function of pH in aqueous solution of 0.001 M potassium chloride, and Isoelectric points (IEPs) of untreated and treated PTFE films.

3.3. Topographical Assessment

Concerning the topographical patterning, the scanning electron microscopy (SEM) images along with the AFM ones of the examined specimens are presented in Figure 6. In addition, the connection between exposure time, mass change and by extension surface roughness is shown in Figure 7.

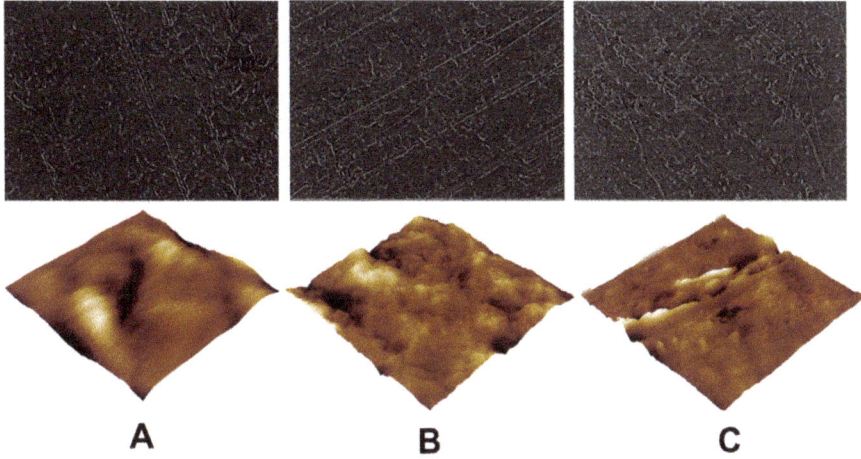

A   **B**   **C**

Figure 6. 2D SEM and 3D AFM images of: (**A**) Untreated, (**B**) RF plasma treated, and (**C**) DC plasma treated PTFE samples.

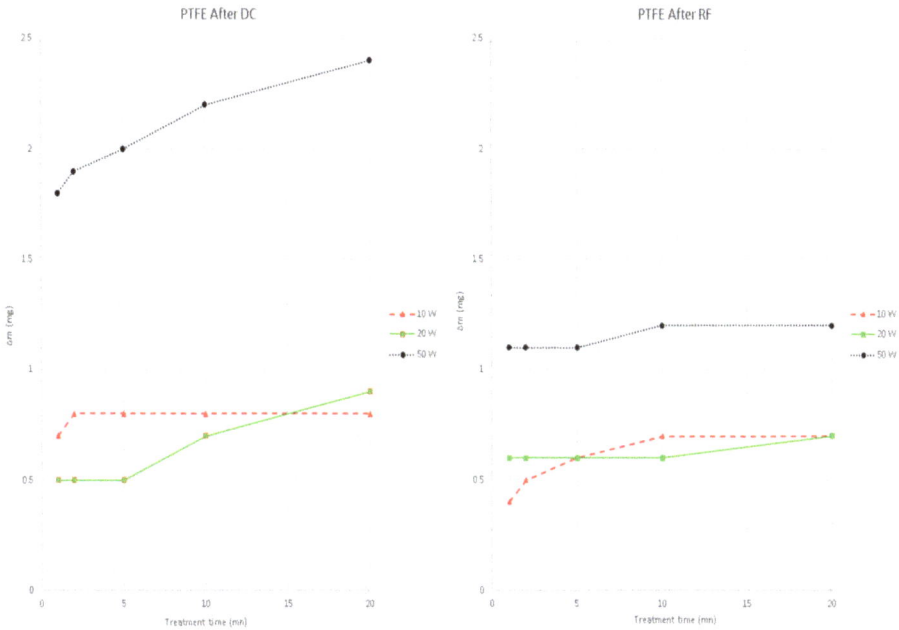

Figure 7. Effect of plasma treatment on the mass of the treated films.

4. Discussion

Solid surfaces may be classified into two basic groups, hydrophilic (wettable with water and high surface energy) and hydrophobic (not wettable with water and low surface energy). The contact angles of the employed liquids on the studied surfaces diminished. For instance, the contact angle of deionized water decreased within the range 74–88°, which indeed indicates a surface wettability change. The typical water contact angle value of poly(tetrafluoroethylene) (PTFE) is ≥105 and its surface energy is typical 20 mJ/m^2. The superhydrophobic character of PTFE plummeted after using air plasma treatment, demonstrating the capability of plasma treatment for surface modification [18–20]. Oxygen-containing (as air) plasmas increase the surface energy and introduce polar (O-containing) moieties. This phenomenon is a consequence of breaking bonds and free radicals formation that once the samples are withdrawn from the plasma reactor trigger the reaction between atmospheric oxygen and free radicals. The hydrophobic character of PTFE was altered making the specimens more hydrophilic after each treatment [21,22].

PTFE has a relatively low surface energy and its total surface energy increased regardless of the treatment. In the case of DC, the highest values correspond to the foils exposed to 20 min treatment with a power input of 50 W; nevertheless, there is no drastic surface energy change if shorter plasma durations are applied, and 1 min of treatment is enough for a cost-efficient surface modification [23–25]. This may be confirmed with the radiofrequency experiments, where it is evident that 1 min is sufficient as the further treatment or power input merely had an effect on surface energy. This phenomenon may be associated with chemical saturation after 1 min of plasma treatment. This is in agreement with previous studies, where short plasma duration is a rather cost-effective treatment. It should be noted that longer treatments might provoke thermal degradation that potentially damage the previously obtained surface attributes [26–28]. The treatment efficiency is intrinsically connected with experimental parameters of the plasma reactors; for example, pursuant to the plasma reactors' supplier, the kHz machines are more robust, provide more uniform treatments, and are more efficient than any other commercial machines. The efficiency factor of 40-kHz frequency (DC) is 80%, whereas 13.56-MHz (RF) has the lowest efficiency factor of commercial plasma reactors, close to 50% [29,30].

As seen from the electrokinetic analysis, which was performed to evaluate the surface charge and to determine the isoelectric points of the studied sample, all the plots have negative values, and these negative charges increased after 20 min RF plasma treatment; conversely, the curve of 20 min RF shifted towards higher pH values. PTFE treated under DC holds the most negative ζ-potential values, which correspond to the highest surface charge. The untreated PTFE showed a value around pH 3.4, which coincides with the IEP found in literature [31]. The studied samples had negative charges indicating the presence of chemical groups which may be deprotonated. Poly(tetrafluoroethylene) is an inert and stable polymer, and its backbone may not be deprotonated; therefore, it may be assumed that new chemical groups were incorporated and strong monomer fragmentation occurred during the plasma treatment. The new surface-functionalities may act as electron donors (Lewis-base), which may explain why the curves drop to negative numbers [32].

With respect to the SEM images, both the untreated sample and the treated ones have wavy areas, and all studied sheets possessed scratches. These anomalies may come from the processing line or an inadequate storage. Nevertheless, the untreated film is relatively smoother and its morphology is more uniform than the treated ones. These alterations may be observed in more detail with the Atomic Force Microscopy (AFM) microphotographs, where the surface topography changes following the PTFE films' exposure to air plasma treatment are more visible. The treated samples depict relatively rougher morphologies, with the sample treated under DC power being the roughest, with etched features and irregularly shaped textures compared with the untreated film. This may be substantiated in Figure 7, where the extent of plasma treatment was assessed with respect to the loss of weight. It may be noticed that the increase of Δm is proportional to the plasma duration and the power input. Hence, higher power input leads to greater weight loss. The foils treated for a longer period have rougher surfaces and underwent higher loss of mass. In fact, the generated pattern on the plasma treated samples stems from the competition between ablation and functionalization. It has been seen throughout this study that 1min under 40 kHz frequency is the most efficient in terms of surface modification, and this information is in agreement with previous studies, where, as it was mentioned above, the kHz machines are more robust, providing more uniform treatment, and are more efficient than other machines [33].

5. Conclusions

We conclude that the non-thermal plasma sources used in this work are suitable for the surface modification of poly(tetrafluoroethylene). The superhydrophobic character of PTFE is transformed after air plasma treatment. Different treatment times and powers were employed, and as far as this contribution is concerned, plasma treatment at 40 kHz is the most efficacious system and it is in agreement with previous studies. Short duration plasmas are cost-efficient methods to enhance surface properties without causing any negative impact on the treated samples. Surface charge and surface energy have been succinctly characterized by surface probe techniques. All the results demonstrate how the surface charge is gradually changed, and provide the moment that chemical saturation and thermal degradation begin. Surface energy increases with increasing treatment time. Physical and chemical changes are clearly manifested by the rise of surface charge after RF plasma treatment; likewise, the isoelectric point of treated PTFE after DC plasma treatment is lower than the untreated one. The microphotographs illustrate the surface morphology and the etching effects of the treatment, which are corroborated by the Δmass of the treated specimens. This contribution underlines the use of plasma treatment as a reliable tool for surface modification and functionalization.

Author Contributions: F.C. carried out all the experiments. J.L., P.H. and M.L. designed the experiments; make the data collection and analysis. All authors approved the final manuscript.

Funding: The authors would like to express their sincere gratitude to the Czech Science Foundation (project 17-10813S) for financial support.

Acknowledgments: The authors are grateful to Tomas Bata University in Zlín, Czech Republic (UTB) and the École supérieure de chimie organique et minérale, Compiègne, France (ESCOM) for supporting us.

Conflicts of Interest: The authors declare that no conflict of interest.

References

1. Desmet, T.; Morent, R.; De Geyter, N.; Leys, C.; Schacht, E.; Dubruel, P. Nonthermal plasma technology as a versatile strategy for polymeric biomaterials surface modification: A review. *Biomacromolecules* **2009**, *10*, 2351–2378. [CrossRef] [PubMed]
2. Morent, R.; De Geyter, N.; Desmet, T.; Dubruel, P.; Leys, C. Plasma Surface Modification of Biodegradable Polymers: A Review. *Plasma Process Polym.* **2011**, *8*, 171–190. [CrossRef]
3. Puliyalil, H.; Recek, N.; Filipic, G.; Cekada, M.; Jerman, I.; Mozetic, M.; Thomas, S.; Cvelbar, U. Mechanisms of hydrophobization of polymeric composites etched in CF$_4$ plasma. *Surf. Interface Anal.* **2017**, *49*, 334–339. [CrossRef]
4. Gotoh, K.; Shohbuke, E.; Ryu, G. Application of atmospheric pressure plasma polymerization for soil guard finishing of textiles. *Text. Res. J.* **2018**, *88*, 1278–1289. [CrossRef]
5. Vesel, A.; Mozetic, M. New developments in surface functionalization of polymers using controlled plasma treatments. *J. Phys. D Appl. Phys.* **2017**, *50*, 293001. [CrossRef]
6. Khedkar, J.; Negulescu, I.; Meletis, E.I. Sliding wear behavior of PTFE composites. *Wear* **2002**, *252*, 361–369. [CrossRef]
7. Pricea, D.M.; Jarratt, M. Thermal conductivity of PTFE and PTFE composites. *Thermochim. Acta* **2002**, *392*, 231–236. [CrossRef]
8. Buerkle, M.; Asai, Y. Thermal conductance of Teflon and Polyethylene: Insight from an atomistic, single-molecule level. *Sci. Rep.* **2017**, *7*, 41898. [CrossRef] [PubMed]
9. Suh, J.; Bae, D. Mechanical properties of poly(tetrafluoroethylene) composites reinforced with graphene nanoplatelets by solid-state processing. *Compos. Part* **2016**, *95*, 317–323. [CrossRef]

10. Morent, R.; De Geyter, N.; Verschuren, J.; De Clerck, K.; Kiekens, P.; Leys, C. Non-thermal plasma treatment of textiles. *Surf. Coat. Technol.* **2008**, *202*, 3427–3449. [CrossRef]

11. Svorcik, V.; Reznickova, A.; Kolska, Z.; Slepicka, P.; Hnatowicz, V. Variable surface properties of PTFE foils. *e-Polymers* **2010**, *133*, 1–6.

12. Preocanin, T.; Selmani, A.; Lindqvist-Reis, P.; Heberling, F.; Kallaya, N.; Lutzenkirchen, J. Surface charge at Teflon/aqueous solution of potassium chloride interfaces. *Colloids Surf. A* **2012**, *412*, 120–128. [CrossRef]

13. Zanini, S.; Barni, R.; Della Pergola, R.; Riccardi, C. Modification of the PTFE wettability by oxygen plasma treatments: Influence of the operating parameters and investigation of the ageing behavior. *J. Phys. D Appl. Phys.* **2014**, *47*, 325202. [CrossRef]

14. Kang, D.H.; Kang, H.W. Surface energy characteristics of zeolite embedded PVDF nanofiber films with electrospinning process. *Appl. Surf. Sci.* **2016**, *387*, 82–88. [CrossRef]

15. Jia, J.; Kang, G.; Zou, T.; Li, M.; Zhou, M.; Cao, Y. Sintering process investigation during polytetrafluoroethylene hollow fibre membrane fabrication by extrusion method. *High Perform. Polym.* **2017**, *29*, 1069–1082. [CrossRef]

16. Jurczuk, K.; Galeski, A.; Morawiec, J. Effect of poly(tetrafluoroethylene) nanofibers on foaming behavior of linear and branched polypropylenes. *Eur. Polym. J.* **2017**, *88*, 171–182. [CrossRef]

17. Kang, D.H.; Kang, H.W. Advanced electrospinning using circle electrodes for freestanding PVDF nanofiber film fabrication. *Appl. Surf. Sci.* **2018**, *455*, 251–257. [CrossRef]

18. Lopez-Garcia, J. Wettability analysis and water absorption studies of plasma active polymeric materials. In *Non-Thermal Plasma Technology for Polymeric Materials: Applications in Composites, Nanostructured Materials and Bio-Medical Fields*, 1st ed.; Thomas, S., Mozetic, M., Cvelbar, U., Spatenka, P., Praveen, K.M., Eds.; Elsevier: Amsterdam, The Netherlands, 11 October 2018; Chapter 9. ISBN 978-018131527.

19. Bursikova, V.; Stahel, P.; Navratil, Z.; Bursik, J.; Janca, J. *Surface Energy Evaluation of Plasma Treated Materials by Contact Angle Measurement*, 1st ed.; Masaryk University: Brno, Czech Republic, 1 January 2004; pp. 1–70. ISBN 80-210-3563-3.

20. Garcia, J.L.; Asadinezhad, A.; Pachernik, J.; Lehocky, M.; Junkar, I.; Humpolicek, P.; Saha, P.; Valasek, P. Cell proliferation of HaCaT keratinocytes on collagen films modified by argon plasma treatment. *Molecules* **2010**, *15*, 2845–2856. [CrossRef] [PubMed]

21. Garcia, J.L.; Pachernik, J.; Lehocky, M.; Junkar, I.; Humpolicek, P.; Saha, P. Enhanced keratinocyte cell attachment to atelocollagen thin films through air and nitrogen plasma treatment. *Prog. Colloid Polym. Sci.* **2011**, *138*, 89–94. [CrossRef]

22. Recek, N.; Resnik, M.; Motaln, H.; Lah-Turnsek, T.; Augustine, R.; Kalarikkal, N.; Thomas, S.; Mozetic, M. Cell Adhesion on Polycaprolactone Modified by Plasma Treatment. *Int. J. Polym. Sci.* **2016**, *1*, 1–9. [CrossRef]

23. Vesel, A. Modification of polystyrene with a highly reactive cold oxygen plasma. *Surf. Coat. Technol.* **2010**, *205*, 490–497. [CrossRef]

24. Vesel, A.; Mozetic, M. Surface modification and ageing of PMMA polymer by oxygen plasma treatment. *Vacuum* **2012**, *86*, 634–637. [CrossRef]

25. Ozaltin, K.; Lehocky, M.; Humpolicek, P.; Pelkova, J.; Saha, P. A new route of fucoidan immobilization on low density polyethylene and its blood compatibility and anticoagulation activity. *Int. J. Mol. Sci.* **2016**, *17*, 908. [CrossRef] [PubMed]

26. Asadinezhad, A.; Lehocky, M.; Saha, P.; Mozetic, M. Recent Progress in Surface Modification of Polyvinyl Chloride. *Materials* **2012**, *5*, 2937–2959. [CrossRef]

27. Vesel, A.; Zaplotnik, R.; Modic, M.; Mozetic, M. Hemocompatibility properties of a polymer surface treated in plasma containing sulfur. *Surf. Interface Anal.* **2016**, *48*, 601–605. [CrossRef]

28. Stoleru, E.; Zaharescu, T.; Hitruc, E.G.; Vesel, A.; Ioanid, E.G.; Coroaba, A.; Safrany, A.; Pricope, G.; Lungu, M.; Schick, C.; et al. Lactoferrin-immobilized surfaces onto functionalized PLA assisted by the gamma-rays and nitrogen plasma to create materials with multifunctional properties. *ACS Appl. Mater. Interfaces* **2016**, *46*, 31902–31915. [CrossRef] [PubMed]

29. Lopez-Garcia, J.; Primc, G.; Junkar, I.; Lehocky, M.; Mozetic, M. On the hydrophilicity and water resistance effect of styrene-acrylonitrile copolymer treated by CF_4 and O_2 plasmas. *Plasma Process. Polym.* **2015**, *12*, 1075–1084. [CrossRef]

30. Lopez-Garcia, J.; Lehocky, M.; Humpolicek, P.; Novak, I. On the correlation of surface charge and energy in non-thermal plasma-treated polyethylene. *Surf. Interface Anal.* **2014**, *46*, 625–629. [CrossRef]

31. Kolska, Z.; Makajova, Z.; Kolarova, K.; Kasalkova, N.; Trostova, S.; Reznickova, A.; Siegel, J.; Svorcik, V. *Electrokinetic Potential and Other Surface Properties of Polymer Foils and Their Modifications*, 1st ed.; Yilmaz, F., Ed.; IntechOpen: London, UK, 23 January 2013; Chapter 8. [CrossRef]

32. Lopez-Garcia, J.; Bilek, F.; Lehocky, M.; Junkar, I.; Mozetic, M.; Sowe, M. Enhanced printability of polyethylene through air plasma treatment. *Vacuum* **2013**, *95*, 43–49. [CrossRef]

33. Mozetic, M.; Ostrikov, K.; Ruzic, D.N.; Curreli, D.; Cvelbar, U.; Vesel, A.; Primc, G.; Leisch, M.; Jousten, K.; Malyshev, O.B.; et al. Recent advances in vacuum sciences and applications. *J. Phys. D Appl. Phys.* **2014**, *47*, 153001. [CrossRef]

materials

MDPI

Article

Reinforcement of Stearic Acid Treated Egg Shell Particles in Epoxy Thermosets: Structural, Thermal, and Mechanical Characterization

Ahmer Hussain Shah [1,2], Yuqi Zhang [1], Xiaodong Xu [1,*], Abdul Qadeer Dayo [1,3], Xiao Li [1], Shuo Wang [1] and Wenbin Liu [1]

[1] Key Laboratory of Superlight Material and Surface Technology, Ministry of Education, College of Materials Science and Chemical Engineering, Harbin Engineering University, Harbin 150001, China; ahmer.shah@buitms.edu.pk (A.H.S.); zhangyuqi7@hrbeu.edu.cn (Y.Z.); abdul.qadeer@buitms.edu.pk (A.Q.D.); shawn@hrbeu.edu.cn (X.L.); wangshuo123@hrbeu.edu.cn (S.W.); liuwenbin@hrbeu.edu.cn (W.L.)

[2] Department of Textile Engineering, Balochistan University of Information Technology, Engineering and Management Sciences, Quetta 87300, Pakistan

[3] Department of Chemical Engineering, Balochistan University of Information Technology, Engineering and Management Sciences, Quetta 87300, Pakistan

* Correspondence: xuxiaodong@hrbeu.edu.cn; Tel.: +86-451-82568191

Received: 12 August 2018; Accepted: 20 September 2018; Published: 1 October 2018

Abstract: This work reports the modification of egg shell (ES) particles by using stearic acid (SA) and their reinforcement in the epoxy matrix. The ES treatment via SA was optimized, the optimum conditions for concentration, temperature, and time were found to be 2.5%, 85 °C, and 50 min, respectively. The untreated ES (UES) and treated ES (TES) particles were characterized by Fourier transform infrared spectroscopy (FTIR), differential scanning calorimetry (DSC), X-ray diffraction (XRD), scanning electron microscope (SEM), particle size distribution, and contact angle. FTIR confirmed the chemical modification of SA on ES surface and DSC reflects an endothermic peak at 240 °C. XRD reveal a decrease in crystal size and crystallinity, while contact angle increases to 169° from 42°. The SEM observations clearly reflect a distinct decrease and separation of small domains of ES particles thus improving an increased surface area. Afterwards, the UES and TES particles were reinforced in epoxy at 15 and 20 weight (wt.) % loading. The tensile tests confirmed a 22% increase in elongation as compared to pure epoxy due to the hydrogen bonding between TES particles and matrix. The lowest brittleness was recorded for TES/epoxy composites on 20 wt % loading. The TGA confirmed the improved thermal stabilities at 20 wt % loading of TES particles in matrix, the improvements in $T_{5\%}$, $T_{10\%}$, and $T_{20\%}$ values were recorded as 33, 26, and 21 °C higher than the corresponding values for neat matrix. The TES/epoxy composites on 20 wt % showed 41% increase in storage modulus as compared to the pristine epoxy, and cross-link density reaches to 2.71×10^{-3} from 1.29×10^{-3} mol/cm^3 for neat matrix. The decline in *tan δ* height and improvement in T_g were also observed. The best adhesion effectiveness was recorded for TES/epoxy composites. This simple and economical modification technique can enhance the application of ES particles in various polymeric coating and composites applications.

Keywords: egg shell; stearic acid; modification; particle characterization; epoxy composites; dynamic mechanical analysis; adhesion effectiveness

1. Introduction

In the modern world, scientists and process industries are focusing on the use of biomaterials, either as raw materials or as a phase reinforcement to improve the properties of material.

Epoxies—having various structures depending on the hardener used—are cheap materials and widely used in number of industries. Aliphatic and aromatic amine are commonly used hardeners for epoxies curing, the aliphatic amine type hardeners are cheap and can cure epoxy at room temperature, but their properties are lower as compared to the aromatic amine type hardeners. However, both hardeners showed the similar behavior for brittleness, the high brittleness is considered as major associated drawback [1]. The polymer structure has poor resistance to crack propagation and low impact strength along with low thermal stability. One way to overcome the toughness of the thermosets like epoxies is to reinforce a material either in fiber or particle forms both at macro and nano levels [2–5]. The reinforcement phase physically interacts and/or chemically reacts with epoxy or hardener depending on the treatment [6].

The egg shell (ES) is a bio-waste generated in enormous quantity, as eggs are part of daily meal around the world, the chemical composition include 95% calcite embedded with an organic matrix [7]. China is in the first place for the production and consumption of eggs having 40% share in world consumption [8]. Table 1 shows the composition of ES particles [9]. The ES particles have been used in different applications after various treatments [10–15]. The carbonized ES particles (at 500 °C for 3 h) reinforced metal matrix composites show increased tensile strength, hardness and fatigue strength on 12.5 wt % ES loading but toughness was reduced [16]. The micro size ES particles reinforced epoxy composites showed a 72% improvement in toughness at 5 wt % ES loading reinforced composites. However, on higher amount (>5 wt %) the toughness decreases and reaches to the pure material [17]. Reduction in toughness was the result of agglomeration of particles at higher loading amount. The nano ES particles epoxy composites on 1 to 5 wt % ES were studied by Mohan and Kanny, the best results for the tensile and impact strength were observed on 2 wt % ES reinforcement, while on higher loading the decline was observed in the properties values due to serious agglomeration [18]. In the nano composites, the aliphatic room temperature hardener was used and lower amount of ES was loaded, but the results were much better than the micro size particle composite [16,17]. However, the associated cost for the size decrease and heat treatment of particles are not in favor for commercial application.

Table 1. Chemical composition of ES particles [9].

Composition	Amount (%)
Calcium carbonate	98.2
Magnesium	≈0.9
Phosphorous	≈0.9
Carbon, zinc, manganese, iron, and copper	<0.1

As the particles size decreases, the particles agglomeration at higher loading increases. The agglomeration can be avoided or reduced by effective surface treatment with suitable modifying agents, which improve the dispersion of fillers in matrix [19]. The calcium carbonate (CC) is the major component of ES particles, and an easy and effective surface modifications technique for calcium carbonate by reacting with stearic acid (SA) is reported [20,21]. SA is a cheap organic acid having hydrophilic head and oleophobic tail [22]. Several reports are reported for increased performance of composites by reinforcing CC treated with SA. Lam et al. [23] reported that SA treatment of CC improved the dispersion and adhesion of polypropylene/CC composites, which improved the mechanical and thermal stability of composites. In another study by Osman et al. [24], it was found that the amount necessary to gain high level of dispersion depends on the amount of SA coating. Monolayer coating of CC particles achieves more beneficial effects as compared to bilayer formation, if the amount of SA increases on surface. From literature analysis it is believed that SA treatment of ES particles may be adopted as a good technique to improve reinforcing affects in composites. At this stage, it is therefore necessary to optimize the process of ES treatment such that to prevent the formation of bilayer on ES surface. We believe that this method can effectively modify the ES particles' surface, and ultimate results of modification can enhance the reinforcement effects [25].

In this work, we report a simple and economical modification method for the ES particles by using stearic acid (SA), the treatment parameters were optimized. The efficacy of treatment was studied by FTIR, XRD, DSC, SEM, and water contact angle. The untreated and SA treated ES particles were reinforced in epoxy at higher loading i.e., 15 and 20 wt %, and their impacts on the tensile, thermal, and thermomechanical properties were evaluated along with structural changes by FTIR analysis.

2. Materials and Methods

2.1. Materials

ES were purchased from the local food market at Harbin, China. Epoxy (diglycidylether of bisphenol-A (DGEBA), E-51) having 184–195 g/eq EEW (epoxy equivalent weight), triethylenetetramine (TETA), stearic acid (99%, molecular weight 284.48), and dimethylbenzene (xylene) were procured from Sigma–Aldrich, Shanghai, China. Sodium hydroxide (NaOH) was purchased from Sigma–Aldrich with a 99.99% purity grade. Ethanol and deionized water were used throughout the study.

2.2. Treatment

The inner membrane of ES was manually separated; afterwards, shells were boiled in water for 30 min and overnight oven dried at 80 °C. The ES particles size was reduced by grinding in a home grinder and particles passed 500 mesh were used in the study. Later, the impurities from particles were removed by 30 min soaking in ethanol at 15:1 liquor ratio, ultrasonicated and stirred. After that, ES particles (100 g) were dipped in a SA solution i.e., water 500 mL + 250 mL of 0.014 mol NaOH and 0–3% varied concentration of SA, mixture was stirred at 300 rpm at various temperatures (25 to 105 °C at 20 °C ramp) and time periods (20 to 60 min at 10 min interval). The particles were separated and overnight vacuum dried at 80 °C.

2.3. Process Optimization

The process optimization of modification of ES particles was carried out by studying four parameters—i.e., sediment volume, oil-wet coefficient (O_w), active ratio % (A_d), and viscosity ratio (V_R). The procedures are defined below:

1. 1.0 g of ES sample was placed in a cylinder containing 30 mL of absolute ethanol with a stopper at the top. The cylinder was shaken well up and down for 3 min and kept at room temperature for 15 min. Settling volume V_{set} (mL/g) was calculated as per

$$V_{set} = \frac{V_{sed}}{m} \quad (1)$$

where V_{sed} is the sediment volume in mL, m is the mass of sample in grams, and V_{set} is volume per gram of sediment.

2. 0.2 g of the ES particles were taken and poured into a beaker containing 50 mL of deionized water. A measured quantity of methanol (V) was added drop wise to make the surface particles setting down. Oil-wet coefficient (O_w) was calculated by using the equation

$$O_w = \frac{V}{V + 50} \times 100 \quad (2)$$

3. Fixed amount of ES powder was placed in a separatory funnel to which 200 mL of water was added and shaken manually for 1 min. The mixture was kept still for 20–30 min so that the stratification of ES particles becomes obvious. After clear stratification of layers, the settled ES particles were separated from the layered. The layered particles were carefully taken and kept at

105 °C in vacuum oven to constant weight (≤0.1 mg). Active ratio (A_d) was calculated by the equation [26].

$$A_d = \frac{M_f}{M_t} \times 100 \tag{3}$$

where A_d is the active ratio %, M_f is mass of floating part, and M_t is total mass of the sample.

4. 10% ES particles solution with distilled water was flowed through a viscometer at atmospheric pressure and temperature, similar was repeated for pristine distilled water. The ratio of 10% ES particles solution viscosity to pure water viscosity was taken as viscosity ratio (V_R).

After the process optimization, following characterization methods were used to distinguish between untreated egg shell (UES) and treated egg shell (TES) particles.

2.4. Composite Fabrication

50 g of E51 resin was taken in a glass beaker and heated at 80 °C; the 15 or 20 wt % of UES or TES particles were added in the beaker and vigorously stirred at 300 rpm for 30 min by using a mechanical mixer. Later, for the better dispersion of particles in resin, the mixture was ultrasonicated for 30 min, and cooled down to room temperature. The stoichiometric amount of TETA hardener was added in the mixture and stirred for further 10 min. The mixture was poured into the desired testing sample dimension moulds. The samples were cured at room temperature for 24 h and post cured at 90 °C for 2 h.

2.5. Measurements and Analyses

Fourier transform infrared (FTIR) spectrum analysis was applied to study the chemical structure of UES and TES particles and epoxy composites by casting a thin film. A small amount of sample powder was mixed with KBr and film was prepared and examined on PerkinElmer Spectrum 100 spectrometer (Waltham, MA, USA). The spectra were recorded at 4 cm^{-1} resolution in the range of 4000–400 cm^{-1}.

X-ray diffraction (XRD) measurements were conducted on an X'Pert High Score PW3209 diffractometer for the investigation of crystal structure and other impurities. The tests were performed at room temperature from 20° to 80° of 2θ. The measurements were further used to analyze UES and TES crystal sizes by using Scherrer equation (Equation (4)). The samples were prepared by uniformly spreading the powder on a quartz sample holder.

$$L = \frac{K\lambda}{\beta \cos\theta} \tag{4}$$

where L is the crystallite size, λ is the wavelength of X-ray in nanometers (nm) which is taken as monochromatized CuKα radiation = 0.154184 nm, K is crystallite shape constant (0.9), β is diffraction peak profile at half maximum height in radians, and θ is the dominant peak of 2θ of material in radians.

Differential scanning calorimetry (DSC) measurements were performed on differential scanning calorimeter model TA Q200, TA Instruments, New Castle, DE, USA, under 50 mL/min steady flow rate of nitrogen. The 5.0 mg of sample was carefully weighed in a DSC sample pan at room temperature. The experiments were conducted from 50 to 300 °C at 20 °C/min heating rate.

Morphology and surface characteristics were investigated via scanning electron microscope (SEM) (CamScan SU 8010, Oxford Instruments, Oxford, UK) at 20 kV. The samples for SEM testing were collected carefully and sealed in glass bottles. The SEM copper plate was covered by conductive resin tape and particles/composites were distributed on the tape and gold coated.

Contact angle measurement of UES and TES particles was evaluated by sessile drop method. A thin pallet of both the powders was made by compression and dropping water droplet on it. The test was measured at 25 °C. The particle sizes were calculated on Malvern ZETASIZER nano series, UK, by dispersing them in a water media at 25 °C temperature.

Thermogravimetric analysis (TGA) was performed on TA Instruments Q50 at a heating rate of 20 °C/min from 40 to 800 °C under a constant (50 mL/min) nitrogen flow rate.

Tensile tests of the composites were performed on Universal Instron 4467, Instron, Norwood, MA, USA, machine at a cross head speed of 2 mm/min and the load cell capacity of 30.0 kN as per standard ASTM D-683. Dumbbell shape specimens were cast in a mold containing small amount of silicon oil for easy removal of samples after cure. The working gauge length of specimen was 12 mm. A total of five specimens were tested for each category and the average values are reported.

Thermomechanical properties of the composites were estimated under bending mode testing on a dynamic mechanical analyzer (DMA) model Q800 from TA Instruments, USA. $30 \times 10 \times 2$ mm^3 polished specimens were loaded in a single cantilever mode, and examined under a nitrogen atmosphere from 50 to 150 °C at 3 °C/min heating rate. The sinusoidal strain was applied with 1 Hz frequency by air.

3. Results

3.1. Optimization of SA Treatment

The effects of concentration of SA, treatment temperature, and time on the ES particles were optimized by evaluating the sediment volume, oil-wet coefficient (O_w), active ratio % (A_d), and viscosity ratio (V_R). The results are summarized in Table 2.

Table 2. Process optimization of SA treatment of ES particles.

SA Concentration (%)	Treatment Temperature (°C)	Treatment Time (min)	V_{set} (mL/g)	O_w (%)	A_d (%)	V_R
0			0.590	0.223	2.3	32.1
1			0.699	0.242	65.4	1.71
1.5	70	40	0.892	0.310	85.7	1.68
2			0.990	0.473	94.1	1.64
2.5			1.186	0.456	96.3	1.58
3			1.201	0.468	96.2	1.59
	25		0.317	0.295	92.3	1.72
	45		0.332	0.285	94.2	1.63
2.5	65	40	0.425	0.305	96.1	1.60
	85		0.428	0.468	96.3	1.58
	105		0.425	0.451	96.2	1.59
		20	0.291	0.447	91.9	1.67
		30	0.376	0.452	92.1	1.63
2.5	85	40	0.428	0.468	95.9	1.58
		50	1.178	0.472	96.3	1.58
		60	1.189	0.450	96.2	1.57

V_{set} is an important parameter to measure the dispersion of particles in a liquid. If the particles have good dispersion, they have large volume and will not aggregate and rate of settling will be slow. On the contrary, if the powder has poor dispersion, the sedimentation rate will be faster with lower volume. A direct relation was seen between V_{set} and optimized parameters—i.e., concentration, temperature, and time. It is however sure that an equilibrium volume is reached after a certain time [27], therefore the effect was seen for all process conditions at 15 min time interval. The porosity and surface roughness change along with hydrophobicity of ES particles must have a direct influence on changing the O_w value. In order to find out the effects of modification parameters, the increase of O_w value reflects more attachment of SA on ES particles. While after reaching a certain quantity of concentration, the high dipole–dipole interaction of molecules decreases the surface attachment because of higher molecular motion which negatively influences the surface attachment. Surface treatment of ES particles also changes the active ratio % in liquid. A_d was calculated to see the behavior of particles behavior after modification. Due to the difference in surface tension, well treated, and modified ES floats while poorly interacted particles sink in water. It is reported that before the formation of bilayer on surface, the particles have increasing trend of this ratio which becomes constant after reaching a certain value [26]. The surface becomes hydrophobic as a result of increase in active ratio % due to uniform arrangement of alkyl chains on ES surface. Similar hydrophobic behavior of the ES particles were observed by He and coworkers during the preparation of super hydrophobic material from ES for the oil/water separation [28]. Treatment of particles also affects the viscosity of liquid due to change of surface energy and change of particles size. Well treated ES particles disperse well and hence reduce the viscosity of the solution. The effect of different parameters also shows a difference in V_R values but the difference is small.

The optimized conditions from the obtained results were found to be a 2.5% concentration, 85 °C treatment temperature, and 50 min treatment time. At these processing conditions, the TES particles have highest V_{set}, O_w, A_d, and lowest V_R values.

3.1.1. Structural Characteristics by FTIR

The chemical structure before and after treatment of ES particles was examined by FTIR analysis as depicted in Figure 1. It can be seen that UES particles represent characteristic peaks at 1435, 875, and 714 cm^{-1}, representing the asymmetric stretch, out-of-plane bend and in-plane bending vibrations of CO_3^{2-} ion, respectively [29,30]. The absorption bands situated in between 3000 to 2500 cm^{-1} were identified as organic matter and a broad absorption band at 3287 cm^{-1} was due to the stretching vibration of structural H_2O [31]. The peak positioned at 2505 cm^{-1} can be assigned to acidic hydrogen group –OH stretching [32]. The carbonyl (C=O) stretching vibrations of SA can be observed in the range of 1800 to 1700 cm^{-1}. After treatment with SA, it is evident that asymmetric stretch peak of CO_3^{2-} almost vanishes and out-of-plane bend CO_3^{2-} decreases in intensity to a greater extent, indicating CO_3^{2-} on the surface of UES particles react with SA. A clear peak is also observed at 1746 cm^{-1} for C=O stretching in TES which is absent in UES particles, representing the presence of SA ion RCOO$^-$. On the other hand, the stretch vibration of structural water of UES was broadened and shifted from 3287 cm^{-1} to 3418 cm^{-1} in TES; which shows a considerable change of structural water in TES particles. The peak of acidic hydrogen group (–OH) stretching also increases in TES which can be attributed to chemical reaction of SA with ES particles. Based on these observations, the proposed reaction is shown in Scheme 1.

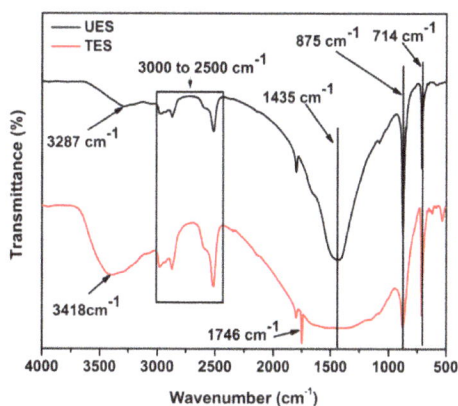

Figure 1. FTIR spectra of UES and TES particles.

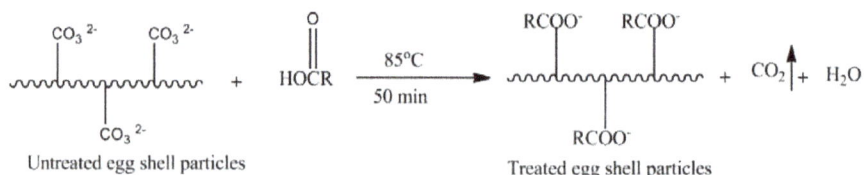

Scheme 1. Chemical reaction of SA treatment of ES particles.

3.1.2. XRD

The XRD of UES and TES particles are illustrated as Figure 2. The UES particles exhibit characteristic peaks at 2θ regions of 23, 29, and 35–48, which is consistent with the crystalline nature of calcite of ES particles in the literature [33]. These peaks represent the high ordered crystalline phase of calcite in ES particles. Comparing the XRD of UES and TES particles, we can easily find that the characteristics diffraction peaks are at same positions, which confirmed that after modification new phases were not generated. This represents that only TES surface was modified without affecting the internal crystalline structure. However, it can be observed easily that the relative peaks intensities decreased, which confirms the decrease of crystallinity after the SA treatment. The decrease of crystallinity can be attributed to the covering of ES surface with SA modifier [34]. These XRD results are in good agreement with FTIR results.

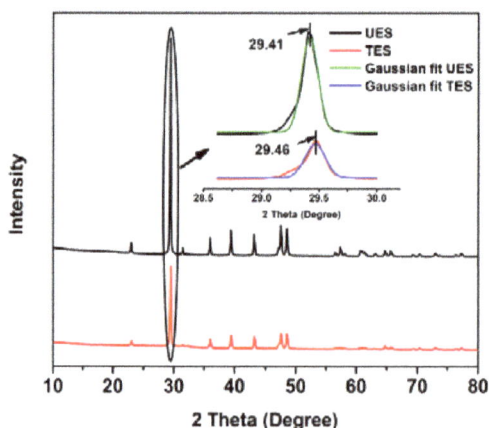

Figure 2. XRD of UES and TES particles.

The effects of SA treatment on the crystallite size were evaluated by using Scherrer equation (Equation (4)) and shown in Table 3. The dominant peak plot was fitted according to Gaussian function.

Table 3. XRD crystallite size parameters of UES and TES particles.

Sample	Dominant Peak 2θ	cos θ (rad)	Root Mean Square	β	L (nm)
UES	29.41	0.016888	0.979	0.17131	47.9
TES	29.46	0.016886	0.964	0.21113	38.9

We can easily observe the decrease in crystallite size after SA treatment, which suggests the increased surface area for TES particles. The reason could be attributed to the removal of surface impurities during the treatment, as SA removes the proteins and organic matter without affecting the calcite phase. This is important observation for the TES particles application as a calcium carbonate mineral in various drugs, medicated implants or polymers with improved absorption, dissolution, flow ability, solubility, and bio-availability.

3.1.3. DSC

The DSC was employed for the detection of phase transformation of ES particles on heating; plots are produced as Figure 3. For UES particles, few peaks around 60 to 150 °C were detected; these can be related to free water and bio-impurities degradation. For TES particles, an obvious endothermic peak was detected at 240 °C. The endothermic reactions occurred due to volatilization of the molecules, whereas exothermic reaction occurred due to the formation of charring (solid residue) [35,36]. From this analysis, it can be apparent that the surface of TES was modified by chemical bonding of SA to the UES surface to a greater extent.

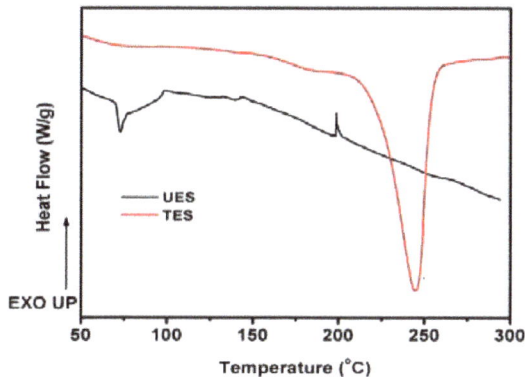

Figure 3. DSC curves of UES and TES particles.

3.1.4. Morphology

The SEM images of unwashed egg shell, UES, and TES particles at three different magnifications were studied to evaluate the morphological changes on the treatment, SEM are shown as Figure 4.

The gummy net structure having an approximately 1 μm diameter confirms the appearance of proteins, collagen, and sulfated polysaccharides materials in unwashed ES sample, Figure 4a–c, and it is not possible to use unwashed ES as filler in polymer industry. After ethanol washing the UES particles, Figure 4d–f, sufficient removal of proteins and impurities was observed, and gummy material was removed to a good extent and definite irregular shaped particles were observed. On higher magnification, Figure 4e,f, the effect of ethanol washing was clearly visible; however, the particles were agglomerated in bunches even after washing. After the SA treatment, Figure 4g–i, the ES particles agglomerates were dispersed and size was reduced to a greater extent. On high magnification Figure 4h,i, the morphology clearly reflects the breaking of particles with improved edges which means that SA removed the gummy texture and impurities to a greater extent and changed the surface roughness. This improved roughness has a prime importance in TES use in polymers which improve the adhesion of particles with matrix, and ultimately improve the properties of composites.

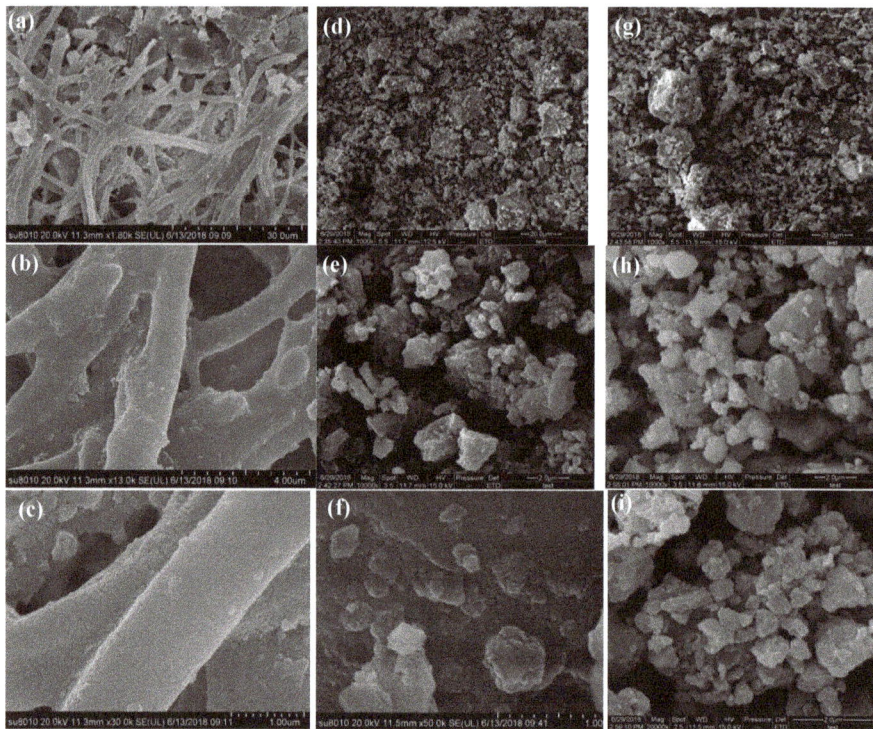

Figure 4. Morphological images of unwashed (**a–c**), untreated (**d–f**), and treated (**g–i**) egg shell particles.

3.1.5. Contact Angle

The hydrophobic characteristics of the ES surface are measured by the contact angle in Figure 5. Water dropping onto UES (Figure 5a) resulted in no firm shape, and water was distributed into the pellet more quickly. The contact angle of water on UES particles was very small (42°) representing a hydrophilic nature of UES particles. The contact angle of TES particles was 169° (Figure 5b), which indicates a very strong hydrophobic surface of the TES particles due to SA modification [37].

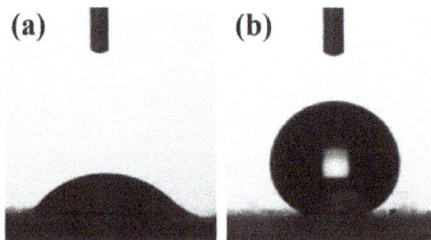

Figure 5. Contact angle of UES (**a**) and TES (**b**).

3.1.6. Particle Size and Distribution

The UES and TES particle sizes after dispersing them in water were calculated at 25 °C temperature. The relative intensities were fitted by using the Guassian fit are shown in Figure 6 and corresponding data is represented in Table 4. It is evident from the plot that UES particles have broader distribution range (500 to 5000 nm) and larger mean particle size (1521 nm). In comparison with UES particles, the TES particles have significantly narrower distribution after SA treatment. The distribution range and mean particle size was recorded at 200–2000 nm and 683 nm, respectively. It is evident from the produced particle size results that SA treatment removed the bio organic impurities to a greater extent and reduces the particle size which is in good agreement with SEM analysis. Moreover, the polydispersity index (PDI) calculated from the plot data also showed decline in the values, which confirms the improved size distribution due to SA reaction on surface causing uniform particle size.

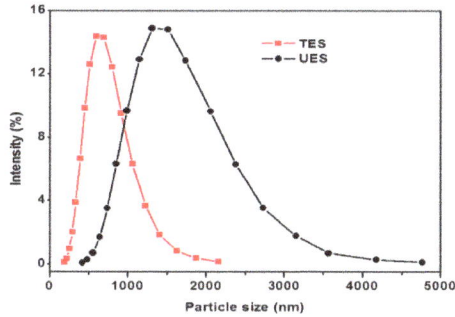

Figure 6. Particles size distribution of UES and TES.

Table 4. Particle size and polydispersity index (PDI) of aqueous dispersion of UES and TES.

Sample	Mean Particle Size (nm)	PDI
UES	1521	0.43
TES	683	0.39

3.2. Effects of UES and TES on Epoxy Filled Composites

The epoxy/UES and epoxy/TES composites were prepared on 15 and 20 wt % loading of particles. The effects of particles loading on the epoxy matrix were studied by evaluating the chemical structure, thermomechanical, mechanical, and thermal properties.

3.2.1. FTIR

The chemical structure of UES and TES filled epoxy composites were examined by FTIR and plotted as Figure 7.

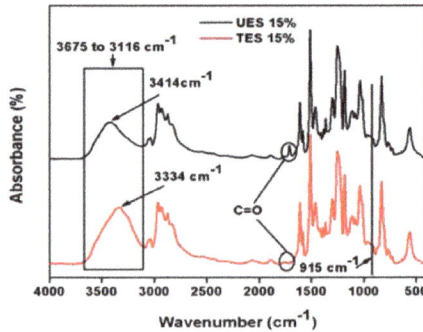

Figure 7. The FTIR spectra of 15% UES and 15% TES filled epoxy composites.

It is important to highlight that there are two main parameters to describe the changes in the chemical structure for UES and TES particles reinforced composites; i.e., shifting in wavelength and intensity variation. The wavelength shift represents either hydrogenation or dehydrogenation if shifted to lower or higher wavelength, respectively [38]. The intensity variation represents the bonds formation or elimination [39]. The spectra of both composites represent characteristics peaks of epoxy. The oxirane group peak located at 915 cm^{-1} corresponds to C–O stretching. The absence of this peak represents that composites were completely cured at given curing conditions. The –OH stretching vibrations are located in the region at 3675 to 3116 cm^{-1}. As the reaction progresses, this region between 3675 to 3116 cm^{-1} develops in the structure. The carbonyls stretch (C=O) due to ES particles can be seen at 1710 cm^{-1} and (–OH) peak at 3414 cm^{-1} in UES composites. In TES composites (–OH), the stretching peak is shifted to lower wave number 3334 cm^{-1} while no peak is detected for carbonyl (C=O) stretch. The absence of carbonyl peak reflects the dissociation of formed surface carbonyls with epoxy and is also reported [40]. The area of hydroxyl (–OH) region intensity is also broadened for TES composites. This observation clearly reflects that TES particles are attached with increased hydrogen bonding with epoxy due to greater functionality of TES particles. The proposed chemical structure is shown in Scheme 2.

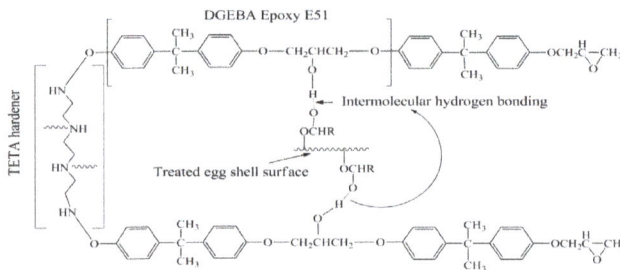

Scheme 2. Chemical structure of epoxy with TES particles.

3.2.2. Particles–Matrix Interaction

The interaction of UES and TES particles with epoxy matrix was examined by studying the variation in storage modulus (E'), loss modulus (E''), and damping factor (*tan δ*) with respect to temperature on a dynamic mechanical analysis (DMA). The plots for particles filled epoxy composites are shown in Figure 8 and the values are tabulated in Table 5.

The study of E', Figure 8A, shows that the E' values steady fall along with temperature increase. Two distinct regions, glassy and rubbery, were observed in all composites. In the glassy region, the high values of E' are due to highly immobile, close and tightly packed matrix components. The UES and TES particles filled composite has increased values of E' values in comparison to neat epoxy due to stiffer nature of particles. In the rubbery region, the 15 wt % UES particles filled composites showed a little higher modulus value due to the interaction of protein contained by the UES particles with epoxy, and they formed a chemically cross-link with epoxy [41]. The TES particles filled epoxy composites showed higher stiffness (E' at 50 °C) values, the stiffness for 15 and 20 wt % composite were 23 and 25% higher than the neat matrix value, respectively. In addition to this, a distinct increase in the rubbery region was also observed. This can be dedicated to the increased porosity, sufficient removal of bio-organic impurities, and reduced TES particles size after SA treatment, along with the chemical cross-linking of alkyl chains with epoxy. This behavior also confirms the improved cross-link points of TES particles in epoxy. The cross-link density was calculated by following equation [42] and values are already summarized in Table 5.

$$\rho = E'/3RT \tag{5}$$

where E' is the storage modulus in rubbery plateau (i.e., $T_g + 30$ °C), R is the universal gas constant (8.314472 J/K mol), and T (K) is the temperature in a rubbery plateau.

Table 5. DMA data and effectiveness coefficient C of neat epoxy; UES and TES filled epoxy composite.

Sample	Storage Modulus (MPa)	Loss Modulus Peak (°C)	tan δ Peak Height	tan δ Peak (°C)	Cross-Link Density (10^{-3} mol/cm³)	C 100	120	140
Pure	1822	99	0.94	111	1.29	-	-	-
15% UES	2051	98	0.78	109	2.14	0.97	0.63	0.55
15% TES	2254	101	0.58	113	2.63	0.73	0.49	0.42
20% UES	2325	97	0.60	110	1.51	0.95	0.62	0.54
20% TES	2578	101	0.44	114	2.71	0.69	0.31	0.25

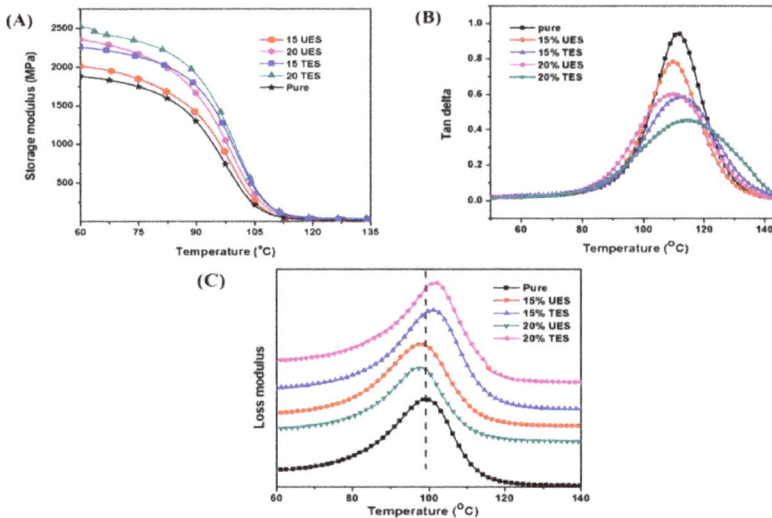

Figure 8. Storage modulus (**A**), *tan δ* (**B**) and loss modulus (**C**) of pure epoxy, UES and TES filled epoxy composites.

The TES particles filled composite exhibit higher number of cross-link density confirming the improved stiffness of the composite. We can observe that, at 15 wt % UES loading the cross-link density of composite was improved than the neat epoxy, but on 20 wt % UES loading the cross-link density value decreases to a greater extent, this shows agglomeration of particles in the composites. On the other hand, the 20 wt % TES filled composites showed an increase cross-link density value which is in consistent with E' results.

The *tan δ* plot of pure epoxy, UES and TES particles filled epoxy composites with respect to temperature are shown in Figure 8B. The neat epoxy showed increased *tan δ* height value as compared to UES and TES particles filled composites; this represents the weak interfaces and frictional damping behavior. The deformation energy is dissolute mostly at the interface of composite materials. If the particle amount, geometry and matrix are identical, the *tan δ* value can be used to study the interfacial properties of composites. The higher *tan δ* height means the composite has a tendency to dissipate more energy—i.e., poor interfacial adhesion results in higher damping values. Moreover, the broadening of *tan δ* is related to molecular relaxations at the interfacial region. The introduction of ES particles reduces the height and increase the broadening of *tan δ* values, the composite showed increase of molecular interaction and relaxations at the interfacial regions. The smallest and broadening damping curves were observed for TES particles filled composites as compared to the UES particles filled composites [43]. This confirmed that TES particles filled composites have higher load bearing capacity than the UES particles filled composites. Moreover, a positive shift in loss modulus values to higher temperature of TES filled composites composite also represents more energy retention in contrast to UES filled composites and pure epoxy as shown in Figure 8C, which is in consistent with the *tan δ* results.

The effectiveness coefficient (C) was also calculated to cross check the adhesion effectiveness of particles and matrix in the composites. It is the ratio of composite storage modulus (E') in glassy and rubbery region in relation to the neat resin, and calculated by using the equation [44]

$$C = \frac{E'_g / E'_r \; composite}{E'_g / E'_r \; resin} \tag{6}$$

Here E' represents storage modulus, while subscripts g and r represent glassy and rubbery regions, respectively.

The C for UES and TES particles filled composites was calculated at three different temperatures—i.e., 100, 120, and 140 °C—and values are shown in Table 5. The glassy region of composites and neat matrix was taken at 40 °C. The lower value of C represents more effectiveness for particle–matrix adhesion [45]. From the calculated values for C, we can conclude that the TES particles filled composite show lower values at all temperatures, which also cross verified the improved particle–matrix adhesion for TES particles than the UES particles.

3.2.3. Tensile Properties

The tensile stress–strain behavior of epoxy filled particle composites is shown in Figure 9 and the corresponding values are tabulated in Table 6.

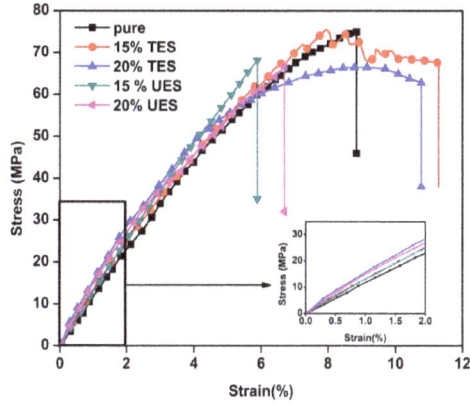

Figure 9. Tensile stress–strain curve of UES and TES filled epoxy composites on various compositions.

Table 6. Tensile and toughness values of composites.

Sample	Tensile Strength (MPa)	Elongation (%)	Toughness [1]	Brittleness (%Pa $\times$ 10^{-5})
Pure epoxy	74.91 $\pm$ 2.21	8.84 $\pm$ 0.15	405 $\pm$ 4.10	5.5
15% UES	68.31 $\pm$ 1.95	5.87 $\pm$ 0.09	211 $\pm$ 2.69	9.29
20% UES	66.44 $\pm$ 1.84	6.67 $\pm$ 0.11	207 $\pm$ 2.91	11.4
15% TES	67.97 $\pm$ 2.11	11.31 $\pm$ 0.18	575 $\pm$ 5.10	3.79
20% TES	63.12 $\pm$ 2.07	10.76 $\pm$ 0.12	491 $\pm$ 3.87	3.84

[1] area under stress strain curve value calculated by integration.

The UES particles filled epoxy composites showed semi ductile behavior, while TES particles filled epoxy composites showed increased ductility. It has been observed that on the macro size particles reinforcement ultimate strength of the composites was lower than the neat matrix [46]. On the other hand, the effective surface modification can change the behavior depending upon the bonding and size alteration. The tensile strength values at 15 and 20 wt % were recorded as 68.31 $\pm$ 1.95 and 66.44 $\pm$ 1.84 MPa, respectively, while the elongation was recorded as 5.87 $\pm$ 0.09 and 6.67 $\pm$ 0.11% for 15 and 20 wt % UES particles loading, respectively, the values for neat epoxy were 74.91 $\pm$ 2.21 MPa and 8.84 $\pm$ 0.15%. The highest elongation value (11.31 $\pm$ 0.18%) was recorded for 15 wt % TES filled epoxy composite, on 20 wt % TES loading the value reduced and recorded as 10.76 $\pm$ 0.12%. This enhancement in the elongation values can be explained by the reduced amount of protein and increased carbonyl groups due to SA treatment, improved dispersion of TES particle, reduced crystal size and increased hydrogen bonding as earlier discussed in FTIR. This employs the reduced brittleness (*B*) of the epoxy composites, the *B* was calculated by using the equation [47].

$$B = \frac{1}{\varepsilon \, E'} \tag{7}$$

Here ε is elongation at break and E' is storage modulus by DMA. The calculated *B* values, summarized in Table 6, shows that the loading of TES particles filled epoxy composites have very lower brittleness as compared to UES particles filled epoxy composites and pure epoxy. These results are consistent with the toughness value, which confirms that TES filled epoxy composites have increased load bearing capacity with improved bending.

3.2.4. TGA

Thermal stability of UES and TES particles filled epoxy composites, and neat epoxy were evaluated in an inert atmosphere and illustrated as Figure 10. The calculated data for the degradation temperatures, including $T_{5\%}$, $T_{10\%}$, and $T_{20\%}$ weight loss temperatures, residual mass, and information on the maximum degradation intensity from DTG curves for UES and TES particles filled epoxy composites are summarized in Table 7.

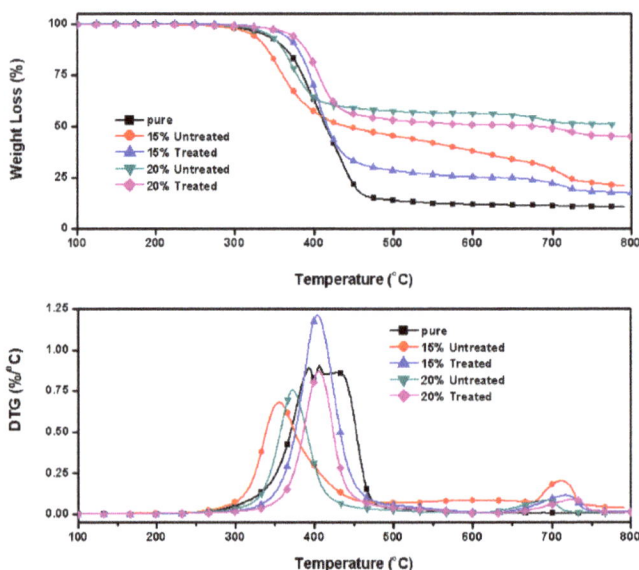

Figure 10. TGA and DTG curves of UES and TES filled epoxy composites on various compositions.

The initial degradation temperature (values of $T_{5\%}$, $T_{10\%}$, and $T_{20\%}$) for 15 wt % UES particles containing epoxy composites were much lower than the neat matrix; however, the 20 wt % UES particles epoxy composites showed a slightly higher values, which may be due to the improved cross links formed between proteins and epoxy as discussed in DMA section. These results are much different than the reported literature due to the alkyl structure attached on the TES particles [18]. The UES particles only increases the char yield values for the composites which were much higher than the neat matrix. However, the TES particles also increase the thermal stability of composites. The TES particles filled epoxy composites also showed the improved initial degradation temperature as the loading of TES particles increased. The DTG peak values for the TES/epoxy composites were moved to higher temperature.

Table 7. TGA and DTG data of pure, UES and TES composites investigated under nitrogen atmosphere.

Sample	$T_{5\%}$ (°C)	$T_{10\%}$ (°C)	$T_{20\%}$ (°C)	Residual Mass (%)	DTG (°C; %/°C)
Pure epoxy	337	359	379	10.4	394; 0.89
15% UES	321	337	353	20.5	356; 0.68
15% TES	361	375	390	17.1	403; 1.21
20% UES	342	357	372	50.8	372; 0.75
20% TES	370	385	400	44.7	405; 0.87

4. Conclusions

In the present work, the ES particles surface was modified with SA. The process was optimized on the basis of sediment volume, oil-wet coefficient, active ratio %, and viscosity ratio. The optimized process conditions were 2.5% concentration, 85 °C temperature, and 50 min treatment time. Afterwards, the UES and TES particles were characterized by conventional methods. The FTIR confirmed the attachment of SA on the particles surface, no phase transformation of calcite phase and reduced particle size were observed by XRD. The DSC showed the chemical bonding of SA to ES particles which increases the hydrophobicity of TES particles by increasing the contact angle to 169°. The morphological studies confirmed the removal of bio-impurities and formation of small particles. Later, the UES and TES particles were reinforced in the epoxy matrix at 15 and 20 wt %, and prepared composites were tested for FTIR, DMA, TGA, and tensile properties. The TES particles filled composites showed increased thermal stabilities, improved stiffness and higher T_g as compared to the neat matrix. The toughness value from the tensile stress–strain plot showed a remarkable increase of 42 and 22% for 15 and 20 wt % TES composites and elongation increase to 27 and 16%, respectively. Moreover, the loading of TES also reduced the brittleness. The effectiveness of particle adhesion was also verified by calculating effectiveness coefficient (C), the TES filled composites showed lowest C value in comparison to the UES filled composites, which proved that the TES particles have increased particle–matrix adhesion. The new simple and economical modification technique for the modification of ES particles can be useful in the dispersion of ES particles in functional polymers, coating applications, and various other high-performance polymers for improved performance characteristics of structure at comparatively high loading levels.

Author Contributions: Conceptualization, Methodology, and Writing—Original Draft Preparation, A.H.S.; Investigation and Data Curation, Y.Z.; Supervision, Project Administration, and Funding Acquisition, X.X.; Formal Analysis and Investigation, X.L. and S.W.; Writing—Review & Editing, A.Q.D. and W.L.

Funding: The research was funded by the National Natural Science Foundation of China (Project No. 21404027), the Open Research Fund of State Key Laboratory of Polymer Physics and Chemistry, Changchun Institute of Applied Chemistry, Chinese Academy of Sciences, and the Natural Science Foundation of Heilongjiang Province of China (No. QC2016060).

Acknowledgments: The authors highly acknowledge the efforts of Syed Salman Shah (Director PDMA, Karachi, Sindh, Pakistan) and Syed Sanaullah Shah (General Manager, SUPARCO, Karachi, Sindh, Pakistan) for the proof reading of manuscript.

Conflicts of Interest: The authors declare no conflicts of interest.

References

1. Xu, Y.-L.; Dayo, A.Q.; Wang, J.; Wang, A.-R.; Lv, D.; Zegaoui, A.; Derradji, M.; Liu, W.-B. Mechanical and thermal properties of a room temperature curing epoxy resin and related hemp fibers reinforced composites using a novel in-situ generated curing agent. *Mater. Chem. Phys.* **2018**, *203*, 293–301. [CrossRef]
2. Dayo, A.Q.; Zegaoui, A.; Nizamani, A.A.; Kiran, S.; Wang, J.; Derradji, M.; Cai, W.-A.; Liu, W.-B. The influence of different chemical treatments on the hemp fiber/polybenzoxazine based green composites: Mechanical, thermal and water absorption properties. *Mater. Chem. Phys.* **2018**, *217*, 270–277. [CrossRef]
3. Brostow, W.; Lobland, H.E.H. *Materials: Introduction and Applications*; Wiley: Hoboken, NJ, USA, 2016.
4. Dayo, A.Q.; Ma, R.-K.; Kiran, S.; Zegaoui, A.; Cai, W.-A.; Shah, A.H.; Wang, J.; Derradji, M.; Liu, W.-B. Reinforcement of economical and environment friendly Acacia catechu particles for the reduction of brittleness and curing temperature of polybenzoxazine thermosets. *Compos. Part A Appl. Sci. Manuf.* **2018**, *105*, 258–264. [CrossRef]
5. Zegaoui, A.; Derradji, M.; Medjahed, A.; Dayo, A.Q.; Wang, J.; Cai, W.-A.; Liu, W.-B. Tailoring the desired properties of dicyanate ester of bisphenol-A/bisphenol-A based benzoxazine resin by silane-modified acacia catechu particles. *React. Funct. Polym.* **2018**, *131*, 333–341. [CrossRef]
6. Yu, Z.-Q.; Wu, Y.; Wei, B.; Baier, H. Boride ceramics covalent functionalization and its effect on the thermal conductivity of epoxy composites. *Mater. Chem. Phys.* **2015**, *164*, 214–222. [CrossRef]

7. Rodriguez-Navarro, A.; Kalin, O.; Nys, Y.; Garcia-Ruiz, J.M. Influence of the microstructure on the shell strength of eggs laid by hens of different ages. *Br. Poult. Sci.* **2010**, *43*, 395–403. [CrossRef] [PubMed]

8. Production of Eggs, Hen, in Shell in China. Available online: http://www.fao.org/faostat/en/#data/QL/visualize (accessed on 30 June 2018).

9. MacNeil, J. Separation and utilization of waste eggshell. In Proceedings of the International Egg Commission Annual Production and Marketing Conference, Toronto, ON, Canada, 30 September 1997.

10. Montilla, A.; Castillo, M.D.; Sanz, M.L.; Olano, A. Egg shell as catalyst of lactose isomerisation to lactulose. *Food Chem.* **2005**, *90*, 883–890. [CrossRef]

11. Tangboriboon, N.; Khongnakhon, T.; Kittikul, S.; Kunanuruksapong, R.; Sirivat, A. An innovative CaSiO₃ dielectric material from eggshells by sol–gel process. *J. Sol-Gel Sci. Technol.* **2010**, *58*, 33–41. [CrossRef]

12. Pundir, C.S.; Bhambi, M.; Chauhan, N.S. Chemical activation of egg shell membrane for covalent immobilization of enzymes and its evaluation as inert support in urinary oxalate determination. *Talanta* **2009**, *77*, 1688–1693. [CrossRef] [PubMed]

13. Takehira, K.; Shishido, T.; Shoro, D.; Murakami, K.; Honda, M.; Kawabata, T.; Takaki, K. Preparation of egg-shell type Ni-loaded catalyst by adopting "Memory Effect" of Mg–Al hydrotalcite and its application for CH₄ reforming. *Catal. Commun.* **2004**, *5*, 209–213. [CrossRef]

14. Saeb, M.R.; Rastin, H.; Nonahal, M.; Paran, S.M.R.; Khonakdar, H.A.; Puglia, D. Cure kinetics of epoxy/chicken eggshell biowaste composites: Isothermal calorimetric and chemorheological analyses. *Prog. Org. Coat.* **2018**, *114*, 208–215. [CrossRef]

15. Rivera, E.M.; Araiza, M.; Brostow, W.; Castaño, V.M.; Diaz-Estrada, J.R.; Hernández, R.; Rodríguez, J.R. Synthesis of hydroxyapatite from eggshells. *Mater. Lett.* **1999**, *41*, 128–134. [CrossRef]

16. Dwivedi, S.P.; Sharma, S.; Mishra, R.K. A comparative study of waste eggshells, CaCO₃, and SiC-reinforced AA2014 green metal matrix composites. *J. Compos. Mater.* **2017**, *51*, 2407–2421. [CrossRef]

17. Ji, G.; Zhu, H.; Qi, C.; Zeng, M. Mechanism of interactions of eggshell microparticles with epoxy resins. *Polym. Eng. Sci.* **2009**, *49*, 1383–1388. [CrossRef]

18. Mohan, T.P.; Kanny, K. Thermal, mechanical and physical properties of nanoegg shell particle-filled epoxy nanocomposites. *J. Compos. Mater.* **2018**. [CrossRef]

19. Ulus, H.; Üstün, T.; Eskizeybek, V.; Şahin, Ö.S.; Avcı, A.; Ekrem, M. Boron nitride-MWCNT/epoxy hybrid nanocomposites: Preparation and mechanical properties. *Appl. Surf. Sci.* **2014**, *318*, 37–42. [CrossRef]

20. Cao, Z.; Daly, M.; Clémence, L.; Geever, L.M.; Major, I.; Higginbotham, C.L.; Devine, D.M. Chemical surface modification of calcium carbonate particles with stearic acid using different treating methods. *Appl. Surf. Sci.* **2016**, *378*, 320–329. [CrossRef]

21. Zhang, J.; Liu, Z.; Liu, J.; Liu, E.; Liu, Z. Effects of seed layers on controlling of the morphology of ZnO nanostructures and superhydrophobicity of ZnO nanostructure/stearic acid composite films. *Mater. Chem. Phys.* **2016**, *183*, 306–314. [CrossRef]

22. Jun, G.L.G. Concentration of stearic acid in monolayers adsorbed from solution. *Nature* **1959**, *184*, 1139.

23. Lam, T.D.; Hoang, T.V.; Quang, D.T.; Kim, J.S. Effect of nanosized andsurface-modified precipitated calcium carbonate on properties ofCaCO₃/polypropylene nanocomposites. *Mater. Sci. Eng. A* **2009**, *501*, 87–93. [CrossRef]

24. Osman, M.A.; Atallah, A.; Suter, U.W. Influence of excessive filler coating on thetensile properties of LDPE–calcium carbonate composites. *Polymer* **2004**, *45*, 1177–1183. [CrossRef]

25. El-Nahhal, I.M.; Elmanama, A.A.; Amara, N.; Qodih, F.S.; Selmane, M.; Chehimi, M.M. The efficacy of surfactants in stabilizing coating of nano-structured CuO particles onto the surface of cotton fibers and their antimicrobial activity. *Mater. Chem. Phys.* **2018**, *215*, 221–228. [CrossRef]

26. Mihajlović, S.; Daković, A.; Sekulić, Z.; Jovanovic, V.; Vučinić, D. Influence of the modification method on the surface adsorption of stearic acid by natural calcite. *J. Serb. Chem. Soc.* **2009**, *67*, 1–19.

27. Pukanszky, B.; Fekete, E. Aggregation tendency of particulate fillers: Determination and consequences. *Period. Polytech. Chem. Eng.* **1998**, *42*, 167–186.

28. He, J.; He, J.; Yuan, M.; Xue, M.; Ma, X.; Hou, L.; Zhang, T.; Liu, X.; Qu, M. Facile fabrication of eco-friendly durable superhydrophobic material from eggshell with oil/water separation property. *Adv. Eng. Mater.* **2018**. [CrossRef]

29. LeGeros, R.Z.; LeGeros, J.P.; Trautz, O.R.; Klein, E. Spectral properties of carbonate in carbonate-containing apatites. In *Developments in Applied Spectroscopy*; Grove, E.L., Perkins, A.J., Eds.; Springer: Boston, MA, USA, 1970; pp. 3–12.

30. Engin, B.; Demirtas, H.; Eken, M. Temperature effects on egg shells investigated by XRD, IR and ESR techniques. *Radiat. Phys. Chem.* **2006**, *75*, 268–277. [CrossRef]

31. Michel, V.; Ildefonse, P.; Morin, G. Assessment of archaeological bone and dentine preservation from Lazaret cave (Middle Pleistocene) in France. *Palaeogeogr. Palaeoclimatol. Palaeoecol.* **1996**, *126*, 109–119. [CrossRef]

32. Bhaumik, R.; Mondal, N.K.; Das, B.; Roy, P.; Pal, K.C.; Das, C.; Banerjee, A.; Datta, J.K. Eggshell powder as an adsorbent for removal of fluoride from aqueous solution: Equilibrium, kinetic and thermodynamic studies. *E-J. Chem.* **2012**, *9*, 1457–1480. [CrossRef]

33. Hassan, T.A.; Rangari, V.K.; Jeelani, S. Value-Added biopolymer nanocomposites from waste eggshell based CaCO$_3$ nanoparticles as fillers. *Sustain. Chem. Eng.* **2014**, *2*, 706–717. [CrossRef]

34. Jiang, L.X.; Jiang, L.Y.; Xu, L.J.; Han, C.T.; Xiong, C.D. Effect of a new surface-grafting method for nano-hydroxyapatite on the dispersion and the mechanical enhancement for poly(lactide-co-glycolide). *Express Polym. Lett.* **2014**, *8*, 133–141. [CrossRef]

35. Kabir, M.M.; Wang, H.; Lau, K.T.; Cardona, F. Effects of chemical treatments on hemp fibre structure. *Appl. Surf. Sci.* **2013**, *276*, 13–23. [CrossRef]

36. Ball, R.; McIntosh, A.; Brindley, J. Feedback processes in cellulose thermal decomposition: Implications for fire-retarding strategies and treatments. *Combust. Theory Model.* **2004**, *8*, 281–291. [CrossRef]

37. Deepika, S.K.; Hait, Y.C. Optimization of milling parameters on the synthesis of stearic acid coated CaCO$_3$ nanoparticles. *J. Coat. Technol. Res.* **2014**, *11*, 273–282. [CrossRef]

38. Zhang, W.; Dehghani-Sanij, A.A.; Blackburn, R.S. IR study on hydrogen bonding in epoxy resin–silica nanocomposites. *Prog. Nat. Sci.* **2008**, *18*, 801–805. [CrossRef]

39. Anwar, A.; Elfiky, D.; Ramadan, A.M.; Hassan, G.M. Effect of γ-irradiation on the optical and electrical properties of fiber reinforced composites. *Radiat. Phys. Chem.* **2017**, *134*, 14–18. [CrossRef]

40. Chai, Z.-Y.; Xie, Z.; Zhang, P.; Ouyang, X.; Li, R.; Gao, S.; Wei, H.; Liu, L.-H.; Shuai, Z.-J. High impact resistance epoxy resins by incorporation of quadruply hydrogen bonded supramolecular polymers. *Chin. J. Polym. Sci.* **2016**, *34*, 850–857. [CrossRef]

41. Riew, C.K.; Gillham, J.K. *Rubber-Modified Thermoset Resins*; American Chemical Society: Washington, DC, USA, 1984; Volume 208, p. 388.

42. Koleske, J.V. *Paint and Coating Testing Manual: Of the Gardner-Sward Handbook Paint Testing Manual*; ASTM: West Conshohocken, PA, USA, 1995.

43. Militký, J.; Jabbar, A. Comparative evaluation of fiber treatments on the creep behavior of jute/green epoxy composites. *Compos. Part B Eng.* **2015**, *80*, 361–368. [CrossRef]

44. Pothan, L.A.; Oommen, Z.; Thomas, S. Dynamic mechanical analysis of banana fiber reinforced polyester composites. *Compos. Sci. Technol.* **2003**, 283–293. [CrossRef]

45. Idicula, M.; Malhotra, S.K.; Joseph, K.; Thomas, S. Dynamic mechanical analysis of randomly oriented intimately mixed short banana/sisal hybrid fibre reinforced polyester composites. *Compos. Sci. Technol.* **2005**, *65*, 1077–1087. [CrossRef]

46. Singh, V.K. Mechanical behavior of walnut (*Juglans* L.) shell particles reinforced bio-composite. *Sci. Eng. Compos. Mater.* **2014**, *22*, 383–390. [CrossRef] [PubMed]

47. Brostow, W.; Hagg Lobland, H.E.; Khoja, S. Brittleness and toughness of polymers and other materials. *Mater. Lett.* **2015**, *159*, 478–480. [CrossRef]

materials

MDPI

Article

Tribological Performance of Microhole-Textured Carbide Tool Filled with CaF$_2$

Wenlong Song [1,*], Shoujun Wang [1], Yang Lu [2,*] and Zixiang Xia [1]

[1] Department of Mechanical Engineering, Jining University, Qufu 273155, China; shoujun0531@163.com (S.W.); xiazixiang168@163.com (Z.X.)
[2] Department of Mechanical Engineering, Shandong University, Jinan 250061, China
* Correspondence: wlsong1981@163.com (W.S.); luyang@mail.sdu.edu.cn (Y.L.)

Received: 7 August 2018; Accepted: 31 August 2018; Published: 7 September 2018

Abstract: To enhance the friction and wear performance of cemented carbide, textured microholes were machined on micro Electron Discharge Machining (EDM) on the tool rake face, and Calcium Fluoride (CaF$_2$) powders were burnished into the microholes. The friction and wear characteristics of the microhole-textured tool filled with CaF$_2$ were investigated using sliding friction tests and dry cutting tests. Results exhibited that the working temperature could affect the tribological performance of the microhole-textured tool filled with CaF$_2$ due to the temperature-sensitive nature of CaF$_2$. There is no obvious lubrication effect for the textured tool filled with CaF$_2$ at room temperature, while it was shown to be more effective in improving tribological property at a cutting speed of higher than 100 m/min with a corresponding to cutting temperature of 450 °C. The possible mechanisms for the microhole-textured tool filled with CaF$_2$ were discussed and established.

Keywords: microhole-textured tool; CaF$_2$; micro-EDM; tribological properties

1. Introduction

Due to the properties of high surface hardness, good thermal stability, outstanding chemical inertness and excellent wear resistance, cemented carbide has been widely applied in engineering applications [1,2], such as machining tools, engine components, mechanical seal parts and bearing modules. However, without a cutting fluid during the cutting process, the carbide tool will be subjected to more severe friction and wear, leading to the increase of the cutting temperature, abrasive wear and adhesions, and hence the reduction of service life. Accordingly, considerable efforts have been made to improve the cutting performance of carbide tools, i.e., optimal carbide geometries and cutting parameters [1], cryogenic minimum quantity lubrication (MQL) [2], subzero treatment [3,4], thermal treatment [5], and surface coatings such as TiN, TiCN, TiAlN, TiAlSiN and CrSiCN, etc. [6–13]. With superior hardness and chemical stability, the coated cutting tools have significantly promoted the application of carbide inserts.

In recent years, surface textures on sliding surfaces have been utilized to enhance the friction and wear performance, and have been applied in many fields such as bearing rings, engine cylinder blocks and cutting tools [14–17]. The surface texturing is beneficial to entrap the wear debris, supply lubricant, and enhance the load capacity with fluid lubrication [18–22], which may effectively reduce the friction and wear of the sliding surface. The literature has increasingly been investigating the role of the textured surface in cutting tools. Lei et al. [23] made an array of microholes on the tool surface to carry out the lubrication, and cutting forces were found to be reduced by 10–30% in the turning of hardened steel. Xiong et al. [24] machined surface texturing filled with molybdenum disulfide (MoS$_2$) on nickel based composite materials. The sliding tests against alumina balls were carried out with both a textured surface and a non-textured surface. Results showed that the average coefficient of friction and wear rate were decreased for the textured composite filled with MoS$_2$ compared to that of the un-textured ones. Kawasegi et al. [25] showed that the texture by a femtosecond laser in the tool-chip contacting zone improved the tribological characteristics owing to the reduction of friction and wear. Deng et al. [26] made micro-textures with various arrays on the carbide tool surface, and MoS$_2$ were burnished into the texturing. Results exhibited that the cutting forces, cutting heat and the friction coefficient for the tools with the micro-texture were significantly reduced compared with the smooth ones, and textures with elliptical arrays were superior to those with perpendicular or parallel arrays. They also reported that the textured carbide tools deposited with WS$_2$ and TiAlN coatings effectively improved the dry cutting capability [27–29].

However, the previous research on surface textures primarily focused on the combination of sulfide additives like MoS$_2$ and WS$_2$. The sulfides start to oxidize as the operating temperature reaches 450–550 °C, and gradually lose the function of lubrication in higher temperatures [30–34]. Compared with the sulfides, Calcium fluoride (CaF$_2$) is a widely utilized solid lubricant at high temperatures. The average friction coefficient of CaF$_2$ decreases gradually with the increasing of temperature from about 450 °C, and it still exhibits an excellent lubricating effect at a temperature of 1000 °C [34]. Then CaF$_2$ solid lubricant can be used as an addition in the fabrication of ceramic material to improve the friction characteristics. For example, the Al$_2$O$_3$/TiC based ceramic tool material with the combinations of CaF$_2$ powder possessed excellent lubricating performance especially in high-speed turning, owing to elevated temperature [35,36]. Therefore, CaF$_2$ was an effective and economical solid lubricant to improve the tribological properties in cutting tools. However, there are few studies on the wear resistance of the textured tool combined with CaF$_2$ lubricants [37], and this area therefore still needs a systematic and comprehensive study.

The aim of this paper is to present the friction behavior and wear mechanisms of the microhole-textured carbide tool with a combination of micro-EDM and CaF$_2$ lubricants at different temperatures. Sliding tests at lower speeds and cutting tests at higher speeds were implemented using the textured carbide tool and the conventional (untextured) one, while the coefficient of friction, cutting forces, temperature, workpiece surface quality and tool wear was analyzed and compared. Based on the test results, the tribological properties were studied and the corresponding possible reasons for performance improvement were proposed. This study may provide a method of combining surface texturing and CaF$_2$ to expand the application of cemented carbide.

2. Experimental Procedures

2.1. Fabrication of the Microhole-Textured Tool Filled with CaF$_2$

In this work, carbide insert (WC/TiC/Co) with the size of 16 mm × 16 mm × 4.5 mm was utilized as the test sample. The physical mechanical performance and composition are listed in Table 1. Microholes were then fabricated on the carbide surface using a micro-EDM system (DZW-10, Lunan Machine Tool Co., Ltd., Tengzhou, China). The processing was accomplished with a capacitance of 4.45 nF and voltage of 125 V. Figure 1 shows the scanning electron microscope (SEM) images and corresponding dispersive X-ray (EDX) component analysis on the microholes. To store more lubricants and catch more debris, the average diameter of the microhole was 150 ± 10 μm, the depth was 200 ± 5 μm, and the distance between the micro-holes of the samples for sliding friction test and for cutting tests was 350 μm and 300 μm, respectively. As shown in Figure 1f, the affected layer of the microhole was just 2.5 ± 0.5 μm, which could ignore the influence of EDM on the substrate mechanical properties. The EDX composition analysis was performed before and after fabrication of the micro-EDM indicated in Figure 1d,e. The results obtained are given in Table 2. It indicated that the oxygen and Cu elements were also detected alongside the elements of carbide substrate, and it was clear that these two elements were created and attached to the surface of the microholes during the EDM process.

Table 1. Properties of the cemented carbide material.

Composition (wt. %)	Density (g/cm³)	Hardness (GPa)	Flexural Strength (MPa)	Young's Modulus (GPa)	Thermal Expansion Coefficient (10^{-6}/K)	Poisson's Ratio
WC + 15%TiC + 6%Co	11.5	15.5	1130.0	510	6.51	0.25

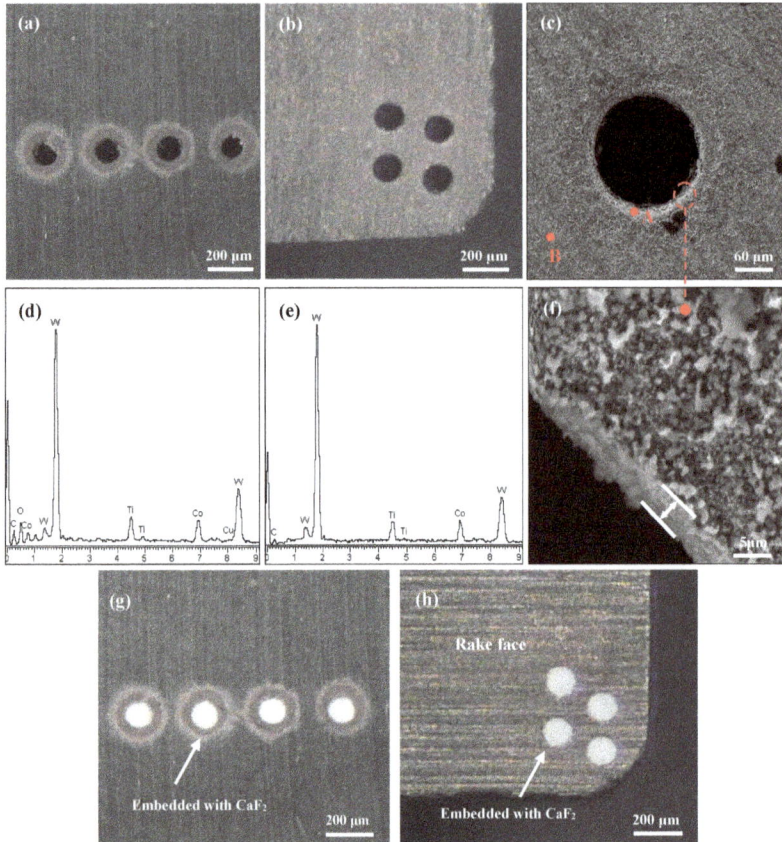

Figure 1. Micrographs of the microholes on the carbide surface: (**a**,**b**) a sample embedded without CaF_2 for sliding friction test and for cutting test, (**c**) enlarged micrograph corresponding to the micro-hole in (**b**,**d**,**e**) corresponding EDX composition analysis of point A and B in (**c**,**f**) enlarged micrograph corresponding to the micro-hole in (**c**,**g**,**h**) sample embedded with CaF_2 for the sliding wear test (SC1) and for the cutting test (SCT1).

Table 2. Element compositions analysis of the cemented carbide before and after EDM.

Element Content	Before EDM (wt. %)	After EDM (wt. %)
C	11.41	20.97
O	0	10.6
Ti	5.57	5.31
Co	8.18	5.84
Cu	0	0.37
W	74.84	56.91
Total	100%	100%

CaF$_2$ powders with an average diameter of 40 nm were manually embedded into the microholes to form microhole-textured tool with combination of CaF$_2$. The micrograph of textured carbide filled with CaF$_2$ for sliding wear test (SC1) and for cutting tests (SCT1) are shown in Figure 1g,h.

2.2. Friction Tests

Sliding tests of the microhole-textured SC1 sample and conventional smooth one (SC2) were executed using a ball-on-plate tribometer (UMT-2, CETR, Campbell, CA, USA). The schematic diagram of the frictional tester and tribometer are shown in Figures 2a and 2b, respectively. The above sample was a WC carbide ball with a hardness of HRA90 and a diameter of 9.5 mm. The sample below was a WC/TiC/Co carbide sample. The sample below was in a fixed position, and the above ball did linear reciprocal sliding against the counterpart. The tests were implemented with the following parameters: Sliding velocity = 2–10 mm/s, normal force = 10–70 N, stroke sliding = 8 mm and sliding time = 15 min. The worn regions of the specimen were tested using a scanning electron microscope (SEM) and an energy dispersive X-ray (EDX).

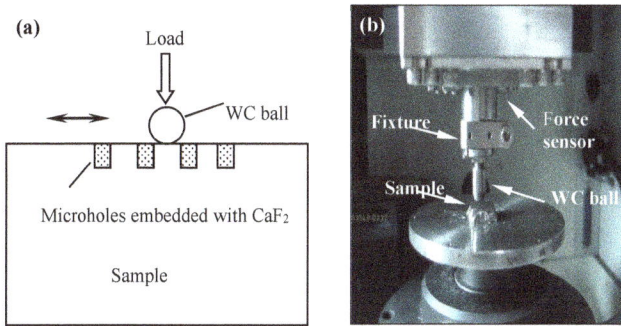

Figure 2. (**a**) Schematic diagram of frictional tester; (**b**) The ball-on-plate tribometer.

2.3. Cutting Tests

Cutting experiments were implemented with a CA6140 turning machine (Syms, Shenyang, China) including a conventional fixture with the following parameters: Clearance angle $\alpha_o = 8°$, rake angle $\gamma_o = 8°$, side cutting edge angle $k_r = 45°$, inclination angle $\lambda_s = 2°$. AISI 1045 quenched steel with a surface hardness of HRC 36–42 was selected as machined material. Cutting tools were utilized with the SCT1 tool and the conventional untextured one (SCT2), and cutting coolant was not applied during the machining experiment. The processing conditions were shown as follows: Cut depth $a_p = 0.2$ mm, feed rate $f = 0.1$ mm/r, cutting speed $v = 60$–180 m/min, and cutting time 5 min. Each condition was repeated three times.

Figure 3 presents the setup for the cutting experiment. Cutting forces were evaluated using a KISTLER piezoelectric 9275A quartz dynamometer (Dijia, Chongqing, China). Cutting temperature was measured using a TH5104R infrared thermography (TH5104R, NEC, Tokyo, Japan). The machining quality of workpiece was obtained with a surface profilometer (TR200, SDCH Co., Ltd., Beijing, China) and the sampling length for each test was about 10 mm. The average measurements of the thrice-conducted tests were presented and compared. The micrographs of the worn carbides were observed through SEM (INCA Penta FETXS, Oxford, UK), and the compositions on the corresponding area were analyzed via EDX (D8 ADVANCE, Bruker, Karlsruhe, Germany).

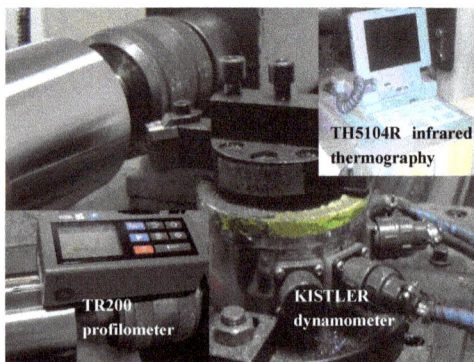

Figure 3. Experimental setup in dry cutting of hardened steel.

3. Results and Discussion

3.1. Friction Test and Surface Wear

Figures 4 and 5 exhibit the friction coefficient of the two kinds of samples in reciprocating sliding wear tests at different sliding speeds and loads. It was evident that there was no marked difference in the friction coefficient between the SC1 and SC2 samples during the test duration. The friction coefficient of the SC2 sample stabilized at about 0.24–0.27, and the SC1 sample possessed a friction coefficient of 0.23–0.26. The average friction coefficient decreased with the increase of the speed, and it increased with the load.

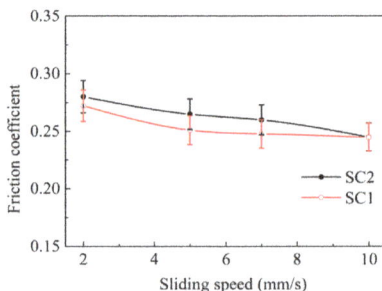

Figure 4. Friction coefficient of the sliding couple at different speeds (load = 50 N).

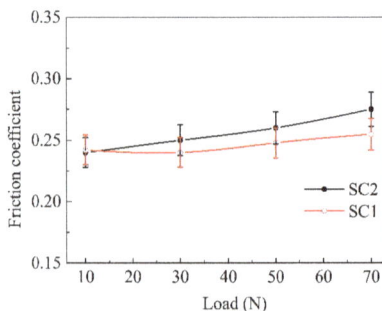

Figure 5. Average friction coefficient of the sliding couple at different loads (speed = 7 mm/s).

Figure 6 shows the SEM micrographs and EDX analysis on the worn track of SC2 sample after 15 min friction test. There was clear abrasive wear on the wear surface, which was characterized as mechanical plowing and scratched appearance. The EDX composition analysis of point A Figure 6c indicated that there were W, Ti and Co elements on the wear face.

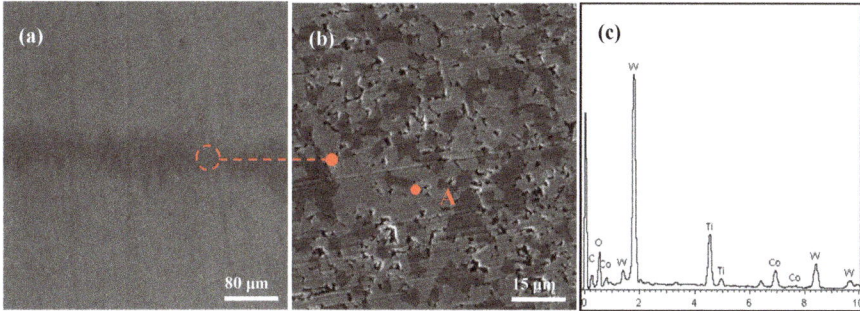

Figure 6. SEM micrographs of the worn surface of the conventional SC2 sample after 15 min sliding friction at the speed of 7 mm/s and load of 50 N: (**a**) SEM micrograph of the wear scar; (**b**) enlarge micrograph corresponding to (**a**); (**c**) EDX composition analysis of point A in (**b**).

Figure 7 exhibits the surface topographies and composition analysis on the wearing area of SC1 sample. No clear abrasive wear can be observed on the friction track between two micro-holes, and large amounts of adhering materials were clearly observed on the sliding track. The EDX analysis in Figure 7c confirmed that the adhesives were CaF_2 powders, which indicated that CaF_2 powders were smeared and transferred to the sliding surface from the micro-holes by friction extrusion. Once a thin layer of CaF_2 was created on the surface, the sliding pairs were separated by the CaF_2 film, which was beneficial for reducing wear. As a result, the SC1 substrate surface exhibited smaller wear than that of the SC2. However, CaF_2 powder kept a brittle state at normal temperature [34], and it acted as the abrasive particle in the process of friction, which led to a high coefficient of friction.

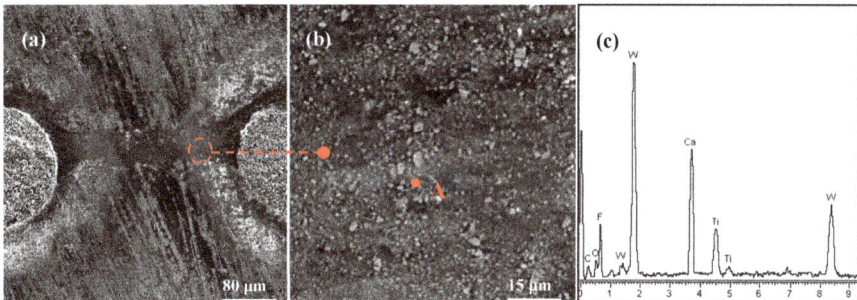

Figure 7. SEM micrographs and EDX composition analysis of the worn surface between two microholes of the SC1 sample after 15 min sliding friction at the speed of 7 mm/s and load of 50 N: (**a**) SEM micrograph of the wear scar; (**b**) enlarge micrograph corresponding to (**a**); (**c**) corresponding EDX composition analysis of point A in (**b**).

3.2. Cutting Performance

3.2.1. Cutting Forces

Figure 8 shows the values of three components of cutting force under different speeds with the SCT1 and SCT2 in machining experiments. From the figure, the turning forces were mainly inversely proportional to the variations of speed. Cutting speed was found to affect the changing rule of force for the SCT1. At a cutting speed of lower than 100 m/min, the three cutting force components for the SCT1 were decreased by about 10–15% compared to the SCT2; while at a cutting speed of higher than 100 m/min, the three cutting force components for the SCT1 were decreased by about 15–25%.

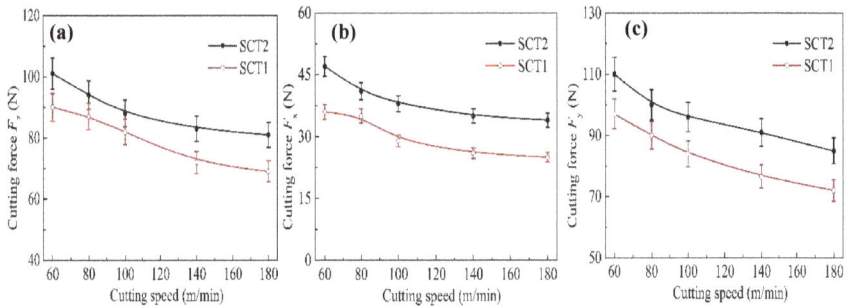

Figure 8. Cutting forces at different speeds with SCT1 and SCT2 in dry cutting of hardened steel (**a**) main force F_z, (**b**) axial thrust force F_x, and (**c**) radial thrust force F_y (a_p = 0.2 mm, f = 0.1 mm/r, cutting time 5 min).

3.2.2. Cutting Temperature

The highest temperature of the chip near the cutting edge was determined by the infrared thermal imaging system in the dry cutting of hardened steels. Figure 9 presents the machining heat energy distribution of chip with SCT1 at turning velocity of 100 m/min, and the highest temperature was about 450 °C under such conditions.

The variation temperatures of chip with cutting speed are plotted and shown in Figure 10. It was clear that the temperature with the two tools rose with the turning speed increasing, and it exceeded 450 °C as cutting speed exceeded 100 m/min. The temperature of chip with SCT1 was reduced apparently in comparison with that of the SCT2.

The experimental results also showed that cutting speed affected temperature variation of the SCT1. The temperature of chip with the SCT1 was decreased by 5–10% with speeds lower than 100 m/min; while the cutting temperature was reduced by 10–20% with speeds higher than 100 m/min.

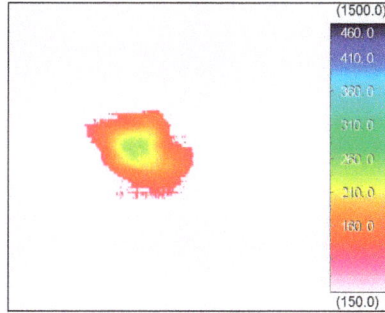

Figure 9. Cutting temperature distribution of chip with SCT1 at speed of 100 m/min in dry cutting hardened steels (a_p = 0.2 mm, f = 0.1 mm/r).

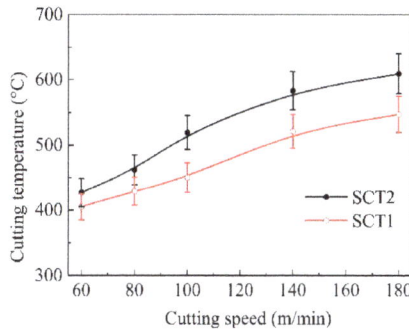

Figure 10. Cutting temperature with SCT1 and SCT2 in dry cutting of hardened steel at different cutting speeds (a_p = 0.2 mm, f = 0.1 mm/r, cutting time 5 min).

3.2.3. Average Friction Coefficient on the Rake Face

The average coefficient of friction μ between the rake face and the chip can be expressed as the formula below [38]:

$$\mu \;=\; \tan(\beta) \;=\; \tan(\gamma_0 + \arctan(F_y/F_z)) \tag{1}$$

where β is angle of friction, F_z is primary cutting force, F_y is radial thrust force and γ_0 is front rake angle.

Figure 11 presents the variation of the friction coefficient on the cutting tool rake face with machining speed. As indicated in the figure, it could be considered that the SCT1 owned improved surface lubricity on the rake face. Under same machining conditions, the average value of friction coefficient for SCT1 was obviously smaller than that of SCT2 at a cutting speed of higher than 100 m/min; yet there was a relatively small decrease in friction coefficient at a speed of lower than 100 m/min.

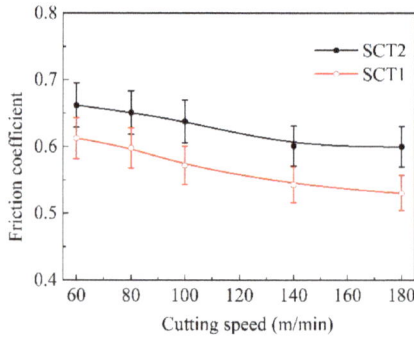

Figure 11. Friction coefficient at the tool-chip interface of SCT1 and SCT2 at different cutting speeds (a_p = 0.2 mm, f = 0.1 mm/r, cutting time 5 min).

3.2.4. Surface Roughness of Machined Workpiece

Figure 12 indicates the average roughness of the machining surface along with the change of cutting speed. The surface roughness result was an average value of three measurements at different positions. The surface roughness of two kinds of tools exhibited a declining trend with speed increases, and the surface roughness value of SCT1 reduced slightly compared to the value of SCT2.

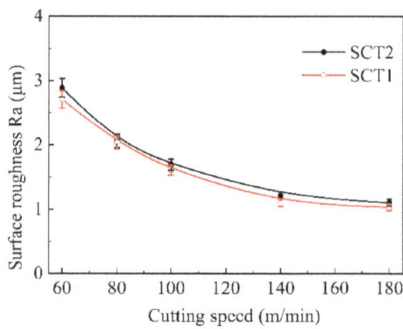

Figure 12. Surface roughness of machined workpiece with SCT1 and SCT2 at different cutting speeds (a_p = 0.2 mm, f = 0.1 mm/r).

3.2.5. Wear Properties

Figure 13 indicates the change of flank wear rate of two tested tools with machining speed. It was evident that the wear of flank was increased with the enhancement of the cutting speed, and the value of flank wear for the SCT1 was lower than the smooth SCT2. This suggests that the microhole-textured tool filled with CaF$_2$ was conducive to enhancing the wear resistance of the flank face, especially at cutting speeds of higher than 100 m/min.

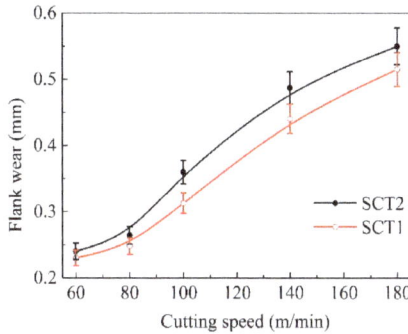

Figure 13. Flank wear of SCT1 and SCT2 tool in dry cutting of hardened steel (a_p = 0.2 mm, f = 0.1 mm/r, cutting time 5 min).

To better evaluate the friction performance and wear mechanism of the tested tools, the wear micrograph and surface component on the worn area for the SCT1 and SCT2 were investigated with SEM and XRD, as indicated in Figures 14 and 15. From Figure 14, significant abrasive wear and evident mechanical ploughs can be found at the flank face (Figure 14a) and rake face (Figure 14b,c) of the SCT2, and clear adhesion material attached to the surface near the tool edge can be observed. The corresponding EDX composition analysis (Figure 14d,e) showed that Fe element was also detected in addition to the elements of WC/Ti/Co carbide tool substrate. This was clearly due to severe friction and chip adhesion that took place on the tool face. The continuous chip friction, attachment and detachment of chip to the tool surface may exacerbate the wear of the flank face, rake face and cutting edge.

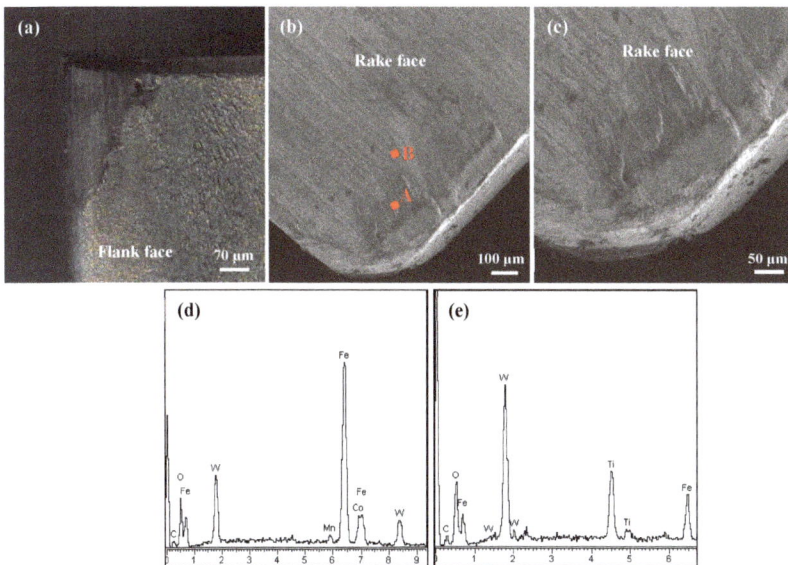

Figure 14. SEM micrographs and EDX component analysis of tool surface of SCT1 after 5 min dry cutting at speed of 100 m/min: (**a**) Micrograph of the worn flank face; (**b**) SEM micrograph of the worn rake face, (**c**) enlarge micrograph corresponding to (**b**,**d**,**e**) corresponding EDX component analysis of point A and B in (**b**).

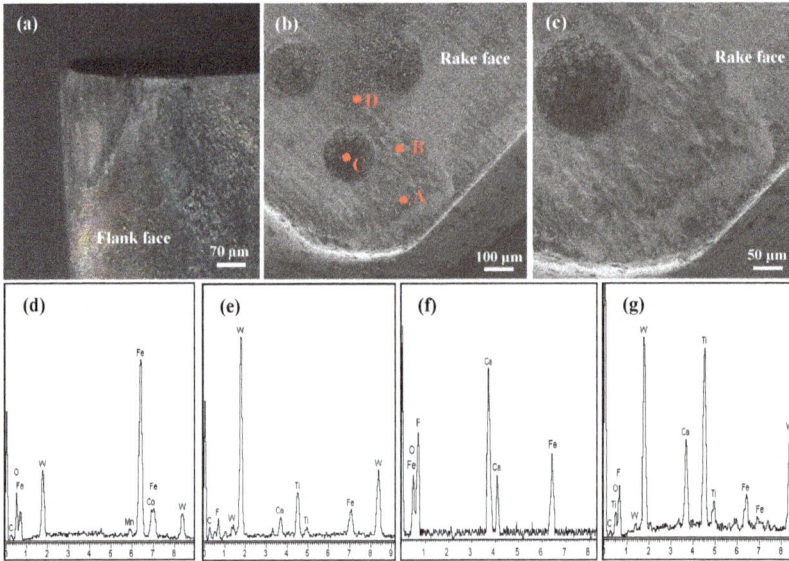

Figure 15. SEM micrographs and EDX component analysis of SCT1surfac after 5 min dry cutting at speed of 100 m/min: (**a**) Micrograph of worn flank face; (**b**) SEM micrograph of worn rake face, (**c**) enlarge micrograph corresponding to (**b,d–g**) corresponding EDX component analysis of point A, B, C and D in (**b**).

As shown in Figure 15, there were also many plows and types of adhering material, but the flank wear, rake wear and edge wear of the SCT1 were mild in contrast to the SCT2. The corresponding surface composition measurements on the wear face are shown in Figure 15d–g. It can be determined that there were iron materials of workpiece in the microholes (Figure 15f) and on the tool face (Figure 15d,e,g), and the CaF$_2$ lubricants were dragged out from the microholes and applied on the tool face in cutting process (Figure 15e,g). The results identified that a continuous and/or discontinuous CaF$_2$ lubricating layer had been produced on the friction track of the SCT1, which was conducive to reducing the wear of cutting edge and tool surface. In addition, the microholes were beneficial to entrap the wear debris and in doing so slow down abrasion and adhesion of the workpiece on the tool face, and supply more CaF$_2$ lubricants.

As shown in the figures above, it could be considered that the main wear mechanisms of rake face for the samples were abrasive and adhesive wear, and abrasive wear was the main wear mechanism of the flank face.

4. Discussion

The test results showed that the microhole-textured tool filled with CaF$_2$ was ideal for the enhancement of tribological performance. The mechanisms responsible for the improvement of tribological properties of the SCT1 are discussed next.

4.1. Cutting Forces

During practical machining, the average frictional force F_f on tool rake face can be expressed as [39]:

$$F_f = a_w l_f \bar{\tau}_c = a_w l_f (k\tau_c + (1-k)\tau_f) \tag{2}$$

where F_f is frictional force on tool surface, $\bar{\tau}_c$ is the average shear strength of tool surface, l_f is the superficial tool-chip contact length, a_w is the width of cutting, k is the ratio of effective contact length to superficial contact length, τ_c is the shear strength of the machined workpiece, and τ_f is the shear strength of lubricating film on the tool surface.

Then, the three force components of similar oblique cutting shown in Figure 16 can be determined as follows [40,41]:

$$F_z = F_r \cos(\beta - \gamma_o) = \frac{F_f}{\sin\beta}\cos(\beta - \gamma_o) = a_w l_f (k\tau_c + (1-k)\tau_f)\left(\sin\gamma_o + \frac{\cos\gamma_o}{\tan\beta}\right) \quad (3)$$

$$F_x = F_r \sin(\beta - \gamma_o)\cos(\psi_r + \psi_\lambda) = a_w l_f (k\tau_c + (1-k)\tau_f)\left(\cos\gamma_o - \frac{\sin\gamma_o}{\tan\beta}\right)\cos(\psi_r + \psi_\lambda) \quad (4)$$

$$F_y = F_r \sin(\beta - \gamma_o)\sin(\psi_r + \psi_\lambda) = a_w l_f (k\tau_c + (1-k)\tau_f)\left(\cos\gamma_o - \frac{\sin\gamma_o}{\tan\beta}\right)\sin(\psi_r + \psi_\lambda) \quad (5)$$

where F_x is the axial force, F_r is the cutting force component of the shear plane, ψ_r is the approach angle, ψ is the shear angle, and ψ_λ is the angle of chip flow.

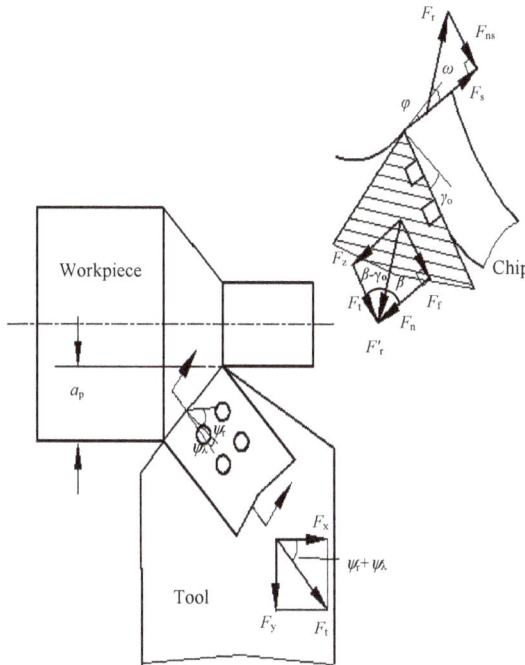

Figure 16. Simplified model of oblique cutting.

Equations (2)–(5) demonstrate that the three cutting force components have a variation of direct proportion with the average shear strength $\bar{\tau}_c$ and tool–chip contact length l_f. The thermal expansion coefficient of CaF_2 (18.38×10^{-6}/K) is significantly larger than that of the tool substrate (6.21×10^{-6}/K). The CaF_2 powders could be dragged out from the microholes because of high cutting temperature and chip friction, and attach to the rake face unevenly. Then a continuous and/or discontinuous CaF_2 lubricating layer may form on the tool surface, which is consistent with the results obtained in Figure 15. The sliding condition between chip and tool is converted from dry friction to boundary friction. That is, the tool substrate endures the load, the fiction occurs on the lubricating film, and self-lubricity is achieved. The lubricating model of the SCT1 in machining process is shown in Figure 17. Supposing that ratio k is 0.8, the ratio of contact length covered with CaF_2 layer is about 0.2. Owing to the much smaller shear stress of CaF_2, the average shear strength $\bar{\tau}_c$ of the SCT1 will be reduced by about 20 percent. Then the three cutting force components will be decreased by about 20% based on Equations (3)–(5). At the same time, forming a CaF_2 lubricating layer between tool-chip can lead to a reduction of chip distortion and angle of friction [40], which is beneficial to a further decrease of cutting forces. Thus, the formation of CaF_2 lubricating layer on the tool surface can efficiently decrease the cutting forces.

Furthermore, the microholes on the tool surface can reduce the tool-chip contact length l_f as indicated in Figure 17, and the effective contact length l_a can be calculated as:

$$l_a = l_f - nd \tag{6}$$

where d is the microhole diameter and n is the microhole quantity in the effective contact area.

As the initial length is 0.8 mm, and there are two microholes of 0.15 mm in diameter at the contact area, the effective length l_a will change to 0.5 mm (Figure 16), and the three cutting force components will be decreased by about 37.5% without loss in mechanical properties according to Equations (3)–(5), due to the decrease of the tool-chip contact area.

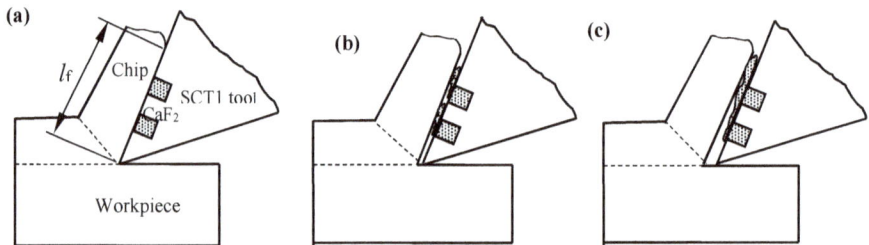

Figure 17. Lubricating model of SCT1 tool in machining process: (a) cutting beginning; (b) CaF_2 solid lubricant piled out; (c) film forming.

4.2. Cutting Temperature

The heat produced in the metal machining process mainly consists of three components [40]: The elastic-plastic deformation of chip on the shear plane, the friction between tool rake face and chip, and the friction between tool flank face and processed surface of workpiece. As a rule, the friction of tool flank has little influence and can be neglected, and the cutting heat can be simplified in calculation. A schematic diagram of the cutting heat distribution is presented in Figure 18, and the average temperature of cutting tool ($\bar{\theta}_{tt}$) and chip ($\bar{\theta}_t$) can be expressed as [40,41]:

$$\bar{\theta}_t = \bar{\theta}_s + \bar{\theta}_f = \theta_0 + \frac{R_1 \tau_s \cos\gamma_o}{c_1 \rho_1 (\sin(2\phi - \gamma_o) + \sin\gamma_o)} + 0.7524 R_2 \bar{\tau}_c \sqrt{\frac{k_2 v a_w l_f}{c_2 \rho_2 \zeta}} \tag{7}$$

$$\bar{\theta}_{tt} = \bar{\theta}_s + \bar{\theta}_{ft} = \theta_0 + \frac{R_1 \tau_s \cos\gamma_o}{c_1 \rho_1 (\sin(2\phi - \gamma_o) + \sin\gamma_o)} + 0.7524(1 - R_2) \bar{\tau}_c \sqrt{\frac{k_2 v a_w l_f}{c_2 \rho_2 \zeta}} \tag{8}$$

where $\bar{\theta}_s$ is shear plane temperature of the chip, $\bar{\theta}_{ft}$ and $\bar{\theta}_f$ is the temperature increase of the cutting tool and chip, respectively, caused by friction of the tool and chip, R_1 is the proportion between the heat of chip and the whole heat caused by chip deformation, c_1 is heat capacity of the chip as the temperature is $(\theta_0 + \bar{\theta}_s)/2$, ρ_1 is the workpiece density, ϕ is the shearing angle, τ_s is the workpiece shear strength, ζ is the chip deformation coefficient, θ_0 is ambient temperature, R_2 is the proportion between the heat of chip and the whole heat produced by severe tool-chip friction, k_2 is thermal diffusivity coefficient of chip as temperature is $(2\bar{\theta}_s + \bar{\theta}_f)/2$, c_2 and ρ_2 is the heat capacity and density of the chip respectively as the temperature is $(2\bar{\theta}_s + \bar{\theta}_f)/2$.

According to Equations (7) and (8), the average cutting temperature of the cutting tool and chip are both positively correlated with the shear strength $\bar{\tau}_c$ and tool-chip contact length l_f. Owing to the reduced shear strength and contact area, the cutting temperature of the SCT1 goes down in comparison with that of SCT2; meanwhile, from Lee and Shaffer shear angle formula [38], the decreased friction angle β can bring about the increase of shear angle ϕ, which is also propitious for the decrease of chip temperature on the shear plane. This is well consistent with the cutting temperature results shown in Figure 10.

Furthermore, the CaF$_2$ solid lubricant had a smaller thermal conductivity (9.17 W/(m·K)) compared to the carbide insert (33.47 W/(m·K)). Once a continuous and/or discontinuous CaF$_2$ film is created on the carbide surface, the thin film could act as a thermal barrier to prevent the heat transfer to the carbide substrate, which is propitious to further lower tool temperature and tool wear.

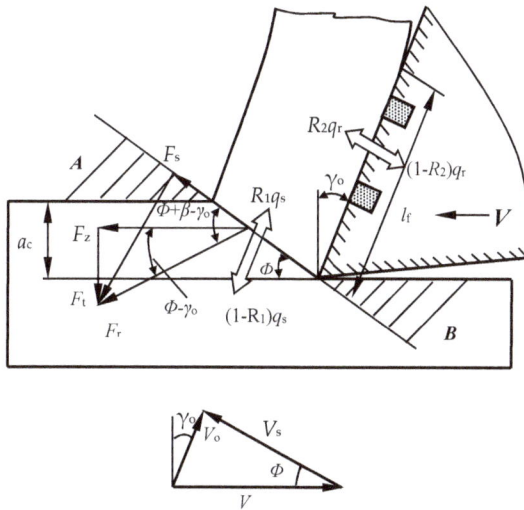

Figure 18. Schematic diagram of the heat distribution during cutting process.

4.3. Average Friction Coefficient at The Sliding Interface

The average friction coefficient between two elastic contact surfaces in sliding can be represented as [39]:

$$\mu = \tan\beta = \frac{F_f}{P} = \frac{\bar{\tau}_c A_r}{\sigma_b A_r} = \frac{\bar{\tau}_c}{\sigma_b} = \frac{k\tau_c + (1-k)\tau_f}{\sigma_b} \qquad (9)$$

where P is the normal load, A_r is the actual contact area, and σ_b is the compressive yield limit of tool substrate materials.

It can be seen that a CaF$_2$ lubricating layer attached to the tool surface conduces to lower friction coefficient for the SCT1 by Equation (9); meanwhile, the microholes on the rake face by reasonable design can supply more lubricant Figure 15e,g and entrap more wear debris of chip Figure 15c, which are conducive to the reduction of friction coefficient. This was in accordance with the variation of the friction coefficient obtained in Figure 11.

Service temperature had an obvious influence on the tribological performance of the microhole-textured carbide filled with CaF$_2$. This is because the CaF$_2$ solid lubricant is a wonderful lubricating material suitable for high temperatures, it can effectively carry out lubrication in the range of 450–700 °C, and still maintains good lubricating performance even at a temperature of 1000 °C [34]. However, if the working temperature drops below 400–450 °C, CaF$_2$ begins to transit from ductile to brittle mode, and the average coefficient of friction increases gradually for lower temperatures. Therefore, the textured carbide embedded with CaF$_2$ powders can more efficiently implement lubrication at a higher cutting speed with corresponding to a higher temperature (Figures 8–15), and result in improved cutting performance. But at a lower cutting speed, the textured carbide exhibits a relatively poor lubricating performance, and even loses the lubricating effect at room temperature, which has been confirmed by the sliding tests and cutting tests as shown in Figures 4–13.

Future investigations will be carried out on the lifetime of textured tools under different test conditions (speed, load, temperature, etc.), and will seek to determine what is the cutting performance without a lubricant supply after a long period of service.

5. Conclusions

The study presented the tribological properties of a microhole-textured carbide tool filled with CaF$_2$. The friction performance and antiwear mechanism of the textured carbide tool during sliding friction tests and dry machining tests were investigated and studied. The main conclusions are as follows:

1. There was no significant change in the friction coefficient of the conventional microhole-textured carbide filled with CaF$_2$ (SC1) and an conventional one (SC2) in sliding tests with WC ball.
2. Compared with the untextured carbide tool (SCT2), the microhole-textured carbide tool filled with CaF$_2$ (SCT1) was effective in promoting machining performance. The tool rake face revealed adhesion and abrasive wear, and flank face indicated severe abrasive wear.
3. Service temperature was found to affect the tribological performance of the textured carbide, which was probably due to the sensitivity of CaF$_2$ solid lubricant to the cutting temperature. At machining speeds higher than 100 m/min, corresponding to temperature of 450 °C, the textured carbide improved the tribological performance compared to the untextured carbide; while at machining speeds lower than 100 m/min, the tribological properties of the textured carbide were only slightly improved in comparison with the smooth one, and it lost the lubrication effect at room temperature.
4. The reasons of performance improvement for the textured tool were as follows: Owing to high cutting heat and friction, CaF$_2$ powders may be drawn out of the microhole textures, adhere to the tool surface and create an uneven CaF$_2$ layer on the rake face, which is propitious to reducing cutting forces, cutting temperature, friction coefficient and tool wear. On the other hand, the microhole textures at the tool face could lower the tool-chip contact length and entrap workpiece debris, which is beneficial to increasing machining performance.

Author Contributions: W.S. conceived and designed the experiments; S.W. and Z.X. performed the experiments; Y.L. analyzed the data; W.S. and Y.L. wrote the paper.

Funding: This research was funded by the Key Research and Development Program of Shandong Province (Grant No. 2017GGX203007), Scientific Research Planning Project of Shandong Province (Grant No. J16LB02), Postdoctoral Innovative Projects of Shandong Province (Grant No. 201603028), and China Postdoctoral Science Foundation (Grant No. 2016M592181).

Acknowledgments: This work was supported by Scientific Research Foundation of Jining University.

Conflicts of Interest: The authors declare no conflict of interest.

References

1. Ai, X. *High Speed Machining Technology*; National Defense Industry Press: Beijing, China, 2004.
2. Liu, X.L. *Guide for Selection of CNC Cutting Tools*; China Machine Press: Beijing, China, 2014.

3. Özbek, N.A.; Çiçek, A.; Gülesin, M.; Özbek, O. Investigation of the effects of cryogenic treatment applied at different holding times to cemented carbide inserts on tool wear. *Int. J. Mach. Tools Manuf.* **2014**, *86*, 34–43. [CrossRef]

4. Akıncıoğlu, S.; Gökkaya, H.; Uygur, İ. The effects of cryogenic-treated carbide tools on tool wear and surface roughness of turning of Hastelloy C22 based on taguchi method. *Int. J. Adv. Manuf. Technol.* **2016**, *82*, 303–314. [CrossRef]

5. Serdyuk, Y.D.; Semizhon, O.A.; Prokopiv, N.M.; Petasyuk, G.A.; Kharchenko, O.V.; Omel'Chuk, T.V. The influence of thermal compression treatment parameters on quality characteristics and wear mechanisms of T5K10 carbide inserts in rough turning. *J. Superhard Mater.* **2011**, *33*, 120–128. [CrossRef]

6. Haubner, R.; Lessiak, M.; Pitonak, R.; Köpf, A.; Weissenbacher, R. Evolution of conventional hard coatings for its use on cutting tools. *Int. J. Refract. Met. Hard Mater.* **2016**, *62*, 210–218. [CrossRef]

7. Xiao, B.J.; Chen, Y.; Dai, W.; Kwork, K.Y.; Zhang, T.F.; Wang, Q.M.; Wang, C.; Kim, K.H. Microstructure, mechanical properties and cutting performance of AlTiN coatings prepared via arc ion plating using the arc splitting technique. *Surf. Coat. Technol.* **2017**, *311*, 98–103. [CrossRef]

8. Zheng, G.; Li, L.; Li, Z.; Gao, J.; Niu, Z. Wear mechanisms of coated tools in high-speed hard turning of high strength steel. *Int. J. Adv. Manuf. Technol.* **2018**, *94*, 4553–4563. [CrossRef]

9. Lorenzo-Martin, C.; Ajayi, O.; Erdemir, A.; Wei, R. Tribological performance of quaternary CrSiCN coatings under dry and lubricated conditions. *Wear* **2017**, *376*, 1682–1690. [CrossRef]

10. Chowdhury, M.S.I.; Chowdhury, S.; Yamamoto, K.; Beake, B.D.; Bose, B.; Elfizy, A.; Cavelli, D.; Dosbaeva, G.; Aramesh, M.; Fox-Rabinovich, G.S.; et al. Wear behaviour of coated carbide tools during machining of Ti6Al4V aerospace alloy associated with strong built up edge formation. *Surf. Coat. Technol.* **2017**, *313*, 319–327. [CrossRef]

11. Tazehkandi, A.H. Cutting forces and surface roughness in wet machining of Inconel alloy 738 with coated carbide tool. *Proc. Inst. Mech. E Part B J. Eng. Manuf.* **2016**, *230*, 215–226. [CrossRef]

12. Armarego, E.J.A.; Verezub, S.; Samaranayake, P. The effect of coatings on the cutting process, friction, forces and predictive cutting models in machining operations. *Proc. Inst. Mech. E Part B J. Eng. Manuf.* **2017**, *216*, 347–356. [CrossRef]

13. Yan, P.; Chen, K.; Wang, Y.; Zhou, H.; Peng, Z.; Jiao, L.; Wang, X.B. Design and performance of property gradient ternary nitride coating based on process control. *Materials* **2018**, *11*, 758. [CrossRef] [PubMed]

14. Bhardwaj, V.; Pandey, R.K.; Agarwal, V.K. Experimental investigations for tribo-dynamic behaviors of conventional and textured races ball bearings using fresh and MoS$_2$ blended greases. *Tribol. Int.* **2016**, *113*, 149–168. [CrossRef]

15. Hua, X.; Sun, J.; Zhang, P.; Ge, H.; Fu, Y.; Ji, J.; Yin, B. Research on discriminating partition laser surface micro-texturing technology of engine cylinder. *Tribol. Int.* **2016**, *98*, 190–196. [CrossRef]

16. Xing, Y.; Deng, J.; Wu, Z.; Liu, L.; Huang, P.; Jiao, A. Analysis of tool-chip interface characteristics of self-lubricating tools with nanotextures and WS$_2$/Zr coatings in dry cutting. *Int. J. Adv. Manuf. Technol.* **2018**, *97*, 1637–1647. [CrossRef]

17. Song, W.; Deng, J.; Zhang, H.; Yan, P.; Zhao, J.; Ai, X. Performance of a cemented carbide self-lubricating tool embedded with MoS$_2$ solid lubricants in dry machining. *J. Manuf. Process.* **2011**, *13*, 8–15. [CrossRef]

18. Wang, M.; Zhang, C.; Wang, X. The wear behavior of textured steel sliding against polymers. *Materials* **2017**, *10*, 330. [CrossRef] [PubMed]

19. Lin, N.; Qiang, L.; Zou, J.; Guo, J.; Li, D.; Yuan, S.; Ma, Y.; Wang, Z.; Wang, Z.; Tang, B. Surface texturing-plasma nitriding duplex treatment for improving tribological performance of AISI 316 stainless steel. *Materials* **2016**, *9*, 875. [CrossRef] [PubMed]

20. Basnyat, P.; Luster, B.; Muratore, C.; Voevodin, A.A.; Haasch, R.; Zakeri, R.; Kohli, P.; Aouadi, S.M. Surface texturing for adaptive solid lubrication. *Surf. Coat. Technol.* **2008**, *203*, 73–79. [CrossRef]

21. Bonse, J.; Kirner, S.V.; Griepentrog, M.; Spaltmann, D.; Krüger, J. Femtosecond laser texturing of surfaces for tribological applications. *Materials* **2018**, *11*, 801. [CrossRef] [PubMed]

22. Segu, D.Z.; Si, G.C.; Choi, J.H.; Kim, S.S. The effect of multi-scale laser textured surface on lubrication regime. *Appl. Surf. Sci.* **2013**, *270*, 58–63. [CrossRef]

23. Lei, S.; Devarajan, S.; Chang, Z. A study of micropool lubricated cutting tool in machining of mild steel. *J. Mater. Process Technol.* **2009**, *209*, 1612–1620. [CrossRef]

24. Li, J.; Xiong, D.; Dai, J.; Huang, Z.; Tyagi, R. Effect of surface laser texture on friction properties of nickel-based composite. *Tribol. Int.* **2010**, *43*, 1193–1199. [CrossRef]
25. Kawasegi, N.; Sugimori, H.; Morimoto, H.; Morita, N.; Hori, I. Development of cutting tools with microscale and nanoscale textures to improve frictional behavior. *Precis. Eng.* **2009**, *33*, 248–254. [CrossRef]
26. Deng, J.; Wu, Z.; Lian, Y.; Qi, T.; Cheng, J. Performance of carbide tools with textured rake face filled with solid lubricants in dry cutting processes. *Int. J. Refract. Met. Hard Mater.* **2012**, *30*, 164–172. [CrossRef]
27. Meng, R.; Deng, J.; Liu, Y.; Duan, R.; Zhang, G. Improving tribological performance of cemented carbides by combining laser surface texturing and W-S-C solid lubricant coating. *Int. J. Refract. Met. Hard Mater.* **2018**, *72*, 163–171. [CrossRef]
28. Lian, Y.; Mu, C.; Wang, L.; Yao, B.; Deng, J.; Lei, S. Numerical simulation and experimental investigation on friction and wear behaviour of micro-textured cemented carbide in dry sliding against TC4 titanium alloy balls. *Int. J. Refract. Met. Hard Mater.* **2018**, *73*, 121–131. [CrossRef]
29. Zhang, K.; Deng, J.; Meng, R.; Lei, S.; Yu, X. Influence of laser substrate pretreatment on anti- adhesive wear properties of WC/Co-based TiAlN coatings against AISI 316 stainless steel. *Int. J. Refract. Met. Hard Mater.* **2016**, *57*, 101–114. [CrossRef]
30. Song, W.; Wang, Z.; Deng, J.; Zhou, K.; Wang, S.; Guo, Z. Cutting temperature analysis and experiment of Ti-MoS_2/Zr-coated cemented carbide tool. *Int. J. Adv. Manuf. Technol.* **2017**, *93*, 799–809. [CrossRef]
31. Renevier, N.M.; Oosterling, H.; König, U.; Dautzenberg, H.; Kim, B.J.; Geppert, L. Performance and limitations of MoS_2/Ti composite coated inserts. *Surf. Coat. Technol.* **2003**, *172*, 13–23. [CrossRef]
32. Kao, W.H. Tribological properties and high speed drilling application of MoS_2-Cr coatings. *Wear* **2005**, *258*, 812–825. [CrossRef]
33. Renevier, N.M.; Hamphire, J.; Fox, V.C. Advantages of using self-lubricating, hard, wear-resistant MoS_2-based coatings. *Surf. Coat. Technol.* **2001**, *142*, 67–77. [CrossRef]
34. Shi, M.S. *Solid Lubricating Materials*; China Chemical Industry Press: Beijing, China, 2000.
35. Deng, J.; Cao, T.; Yang, X. Self-lubrication of sintered ceramic tools with CaF_2 additions in dry cutting. *Int. J. Mach. Tools Manuf.* **2006**, *46*, 957–963. [CrossRef]
36. Deng, J.; Cao, T.; Ding, Z. Tribological behaviors of hot-pressed Al_2O_3/TiC ceramic composites with the additions of CaF_2 solid lubricants. *J. Eur. Ceram. Soc.* **2006**, *26*, 1317–1323. [CrossRef]
37. Song, W.; Deng, J.; Wu, Z.; Zhang, H.; Yan, P.; Zhao, J.; Xing, A. Cutting performance of cemented-carbides-based self-lubricated tool embedded with different solid lubricants. *Int. J. Adv. Manuf. Technol.* **2011**, *52*, 477–485. [CrossRef]
38. Chen, R.Y. *Metal-Cutting Principles*; China Machine Press: Beijing, China, 2005.
39. Wen, S.Z. *Principles of Tribology*; Tinghua University Press: Beijing, China, 2012.
40. Usui, E. *Mechanical Metal Processing*; China Machine Press: Beijing, China, 1982.
41. Nakayama, K. *Theory of Metal Cutting*; China Machine Press: Beijing, China, 1985.

materials

MDPI

Article

The Effect of Plasma Pretreatment and Cross-Linking Degree on the Physical and Antimicrobial Properties of Nisin-Coated PVA Films

Zuzana Kolarova Raskova [1,*, Pavel Stahel [2], Jana Sedlarikova [1,3], Lenka Musilova [1], Monika Stupavska [2] and Marian Lehocky [1,3]**

[1] Centre of polymer systems, Tomas Bata University, Trida Tomase Bati 5678, 76001 Zlin, Czech Republic; sedlarikova@utb.cz (J.S.); musilova@utb.cz (L.M.); lehocky@utb.cz (M.L.)
[2] Department of Physical Electronics, Faculty of Science, Masaryk University, Kotlarska 267/2, 63711 Brno, Czech Republic; pstahel@physics.muni.cz (P.S.); 119414@mail.muni.cz (M.S.)
[3] Department of Fat, Surfactants and Cosmetics Technology, Faculty of Technology, Tomas Bata University, Vavrečkova 275, 76001 Zlin, Czech Republic
* Correspondence: zraskova@cps.utb.cz; Tel.: +420-724-333-988

Received: 13 July 2018; Accepted: 13 August 2018; Published: 16 August 2018

Abstract: Stable antimicrobial nisin layers were prepared on the carrying medium-polyvinyl alcohol (PVA) films, crosslinked by glutaric acid. Surface plasma dielectric coplanar surface barrier discharge (DCSBD) modification of polyvinyl alcohol was used to improve the hydrophilic properties and to provide better adhesion of biologically active peptide-nisin to the polymer. The surface modification of films was studied in correlation to their cross-linking degree. Nisin was attached directly from the salt solution of the commercial product. In order to achieve a stable layer, the initial nisin concentration and the following release were investigated using chromatographic methods. The uniformity and stability of the layers was evaluated by means of zeta potential measurements, and for the surface changes of hydrophilic character, the water contact angle measurements were provided. The nisin long-term stability on the PVA films was confirmed by tricine polyacrylamide gel electrophoresis (SDS-PAGE) and by antimicrobial assay. It was found that PVA can serve as a suitable carrying medium for nisin with tunable properties by plasma treatment and crosslinking degree.

Keywords: antimicrobial film; nisin; physical properties; plasma treatment polyvinyl alcohol; surface characterization

---

1. Introduction

Antimicrobial surfaces and packaging have been receiving increasing attention over the last few decades. Except antimicrobial properties, high levels of biodegradability and biocompatibility are important attributes of polymer materials that are suitable for medical and pharmaceutical applications and the cosmetic and food industries. The polymeric drug delivery systems with controlled release and increased solubility of the drug are investigated. Drugs or antimicrobial agents can be incorporated directly into polymers or they are attached via the side chains often with peptides [1–5].

Among environmentally favorable polymers, a synthetic biodegradable polyvinyl alcohol (PVA) possessing excellent mechanical properties is one of the most important representatives. PVA became attractive also due to its inherent hydrophilicity, large swelling capacity, low cost, and simplicity of use. Its biodegradability in various microbial environments has been reported [6–10].

PVA provides many potential functional groups for the adhesion of antimicrobial and preserving agents, among which the natural based ones are the most promising.

In particular, the proteins have been extensively investigated as a coating material. However, the immobilization of proteins is often associated with their tendency to lose their biological activity [6,11–14]. The use of peptides offers a solution to this problem. The peptides can be obtained as a part of protein after their digestion or as lantibiotics/bacteriocins [7–10]. Nisin is an amphiphilic peptide (bacteriocin) with low molecular weight (approximately 3.4 kDa) that is an effective inhibitor of Gram-positive bacterial strains [9,10,13–16], which are produced by lactic acid bacterial strain *Lactococcus lactis* during the fermentation process. Nisin has been studied as a promising preserving agent since the 1950s [17,18]. Moreover, nisin (and another bacteriocins) provide sufficient antimicrobial activity against bacterial strains that are resistant to conventional drugs. The antitumor activity of nisin is reported in several studies [19–21].

Antimicrobial properties of the films are based on peptide immobilization or incorporation and release. For the bioactivity of film/foils, peptide stability is very important. Many types of physical and chemical treatments have been employed to obtain uniform and stable surface treatment [18–22]. Atmospheric plasma surface activation can offer suitable functional groups for the binding of bioactive substances [21,23,24]. In many studies, the suitability of Dielectric Coplanar Surface Barrier Discharge (DCSBD) treatment was evaluated [25]. This discharge type was investigated with regard to industrial use, and the importance of dielectric barrier discharges for material processing has been continually increasing. This follows from the current trends in industrial applications, in which an effort to replace low-pressure plasma processing by the atmospheric-pressure systems can be observed. This discharge has been successfully used for polymer surface activation [24,26]. It operates in a non-thermal, uniform, glow plasma regime. Its advantage is manifold symmetry; therefore, it can be used also for large area surface modification [25,27]. Some studies have shown the significant effect of DCSBD on the selective functionalization of polymers; however, the DCSBD treatment on biodegradable PVA based polymers with regard to peptides adhesion on their surfaces, as well as on the peptide layer stability, was not reported before. The main advantage of choosing air as the process gas is that it is naturally present in the industrial production lines [26]. Contrary to other dielectric barrier discharges, the DCSBD is capable of generating a macroscopically uniform plasma layer with high power density and energetic efficiency when operated in air [27–29]. This enables fast and homogeneous in-line plasma treatment of materials at high-speeds.

The methods for setting up antimicrobial polymeric surfaces are still in developmental stages, and understanding the structure-function relationship of the material is important to the design and control of delivery systems. Although the surface modification for peptide and protein adhesion as well as the mechanism of adhesion were studied in many papers [30–44], the nisin stability, and thus its long-term antimicrobial effect, was not sufficiently reported. Therefore, the objective of this study is to evaluate the plasma treatment effect on the physico-chemical properties of PVA films as potential transport system for nisin. Also, nisin stability, its release from the surface, and the antibacterial properties of modified PVA films were investigated in correlation with crosslinking degree.

2. Materials and Methods

2.1. Materials

Polyvinyl alcohol (PVA, Mowiol 8-88), glutaric acid (GA), Polyethylene glycol (PEG) 2050, phosphate-buffered saline (PBS, pH 4.5), methanol, lactose, trifluoroacetic acid (TFAA), trichloracetic acid (TCA), formic acid, potassium chloride (KCl), ammonium bicarbonate (NH_4HCO_3), acrylamide, N,N'-methylene bisacrylamide, Coomasie brilliant blue R250, glutaraldehyde, molecular weight (MW) markers, nisin from lactococcus lactis standard (2.5%, balance sodium chloride and denatured milk solids) and bovine serum albumin (BSA), and Tween 80 were purchased from Sigma-Aldrich (St. Louis, MO, USA).

Tryptone soya agar (TSA), Nutrient Broth, MRS broth, and Mueller-Hinton agar (MHA) were obtained from HiMedia (Mumbai, India), bacterial strain *Staphylococcus aureus* CCM 4516 was purchased from Czech Collection of Microorganisms (Masaryk University in Brno, Czech Republic).

2.2. PVA Films Preparation

Prior to experiments, PVA films were prepared using solvent cast technique (10 wt. % of PVA, at 85 °C) from aqueous/HCl (0.02 M) solution. Glutaric acid (GA) was used as the crosslinking agent at different concentrations to obtain following crosslinking degrees: 0%, 5%, 10%, 20%, and 40% [3,45].

In most experiments, the PVA was prepared on microscopy glass (Fischer Sci, Pardubice, Czech Republic diameter: 26 × 26 × 1.1 mm). However, for the purpose of long-time electrokinetic measurement (titration curves), it was necessary to prepare PVA films on polyethylene (PE) foil substrate that prevented the PVA foil twisting under high pressure of electrolyte. The PE matrix was activated by 10 s plasma treatment [6,9,11] before the PVA dip-coating to improve the adhesion and homogeneity.

2.3. PVA Activation and Nisin Application

It is well known [2,9,13,16] that adhesion of peptides/nisin on many polymers is rather unsatisfactory, and so additional polymer activation step was carried out.

The PVA film surface was activated by two types of Dielectric Coplanar Surface Barrier Discharge (DCSBD), RPS600 and RPS 40 systems (Roplass s.r.o., Brno, Czech Republic). DCSBD consists of two comb-like electrodes printed on Al_2O_3 dielectric plate [15,25,26] in ambient air under the following condition sets: (1) power generator 14 kHz, surface power density 2 W/cm^2 (RPS 40) and (2) power generator 25 kHz, power density 7 W/cm^2 (RPS600). The treatment time was 10 s. The distance between the sample surface and plasma layer was 0.3 mm. The process was carried out in the dynamic regime to obtain homogeneous surface treatment.

On the surface of PVA films, 400 μL of commercial nisin solution was applied (nisin was dissolved in phosphate saline solution, the pH value was set at 4.5, and the nisin concentration was 525 μg/mL). It means that theoretical amount of nisin on the surface was 0.31 μg/mm^2. The samples were incubated for 20 min at room temperature, and then the surface was dried and shortly rinsed off by demineralized sterilized water and dried again.

As the finishing step, all samples were plasma-treated for 10 s to improve stability and sterility [19–21,23].

Just for comparison, a set of samples without the PVA plasma activation was prepared. They proved weak adhesion, homogeneity, and stability, which made them generally not applicable.

2.4. Release and Stability Analysis of Nisin

Nisin layers behavior in water (water resistance) was tested via nisin release from the prepared coatings that was carried out at preselected time intervals, while the PVA films were immersed into physiological solution for 1 month. The nisin concentrations in the eluate were analysed by reverse-phase HPLC using a C-18 column (5 μm, 4.6 × 250 mm, Waters) with pre-column Reprosil 100 C18 (5 μm, 50 × 4 mm, Watrex eluted with water/0.05% (v/v) TFA (eluent A) and acetonitrile/0.05% (v/v) TFA (eluent B) (gradient: 0–5 min, 20% eluent A; 5–20 min from 20% eluent A, to 80% eluent A) with UV detection at 220 nm. The injection volume was 25 μL, and the flow rate was of 0.6 ml/min. The amount of nisin was calculated by means of a calibration curve, when calibration solutions were prepared from 0.2 M filtered (the syringe filter with pore size 0.45 μm) solution.

The nisin release profile was tested also during long-term stability testing (90 days of storage in conditioned chamber at 25 °C and relative humidity (RH) of 54% of films). Simultaneously, the stability of released nisin was studied electrochemically using Tricine SDS-PAGE at 16% and 4% separating, and stacking acrylamide gel containing 3% bis-acrylamide, proteins, and peptides were visualized by staining solution: methanol/acetic acid/ water/Coomassie Blue G250 and by silver staining. After electrophoresis, the bands from the electrophoregrams were cut, and, after spot destaining, the peptide was extracted by 5% formic acid/acetonitrile (1:2) solution, and the nisin concentrations were analyzed by means of UV-VIS spectrophotometry with ELISA reader at a wavelength of 595 nm.

Besides, total nitrogen content eluted from the surface after films immersion (3 h into physiological solution) was measured by TOC/TN analyzer (TNM-L, Shimadzu, autosampler ASI-L, Shimadzu Corporation, Japan). Calibration curve was created using potassium nitrate solution in ultrapure water.

The results presented further are the average values from at least three measurements. Standard deviation value was always up to 10% of average value.

2.5. Antimicrobial Assays

To study the long-term stability of nisin coatings, the as-prepared samples were stored in conditioned chamber at 25 °C and relative humidity of 54%, while the measurements were performed also by antimicrobial testing, initially with freshly made coatings and with the coatings after their storage.

Ageing of these layers was monitored as a decrease of antibacterial activity of both released nisin and nisin surface by means of two types of testing.

Nisin in water activity was determined by agar diffusion method [42,43] using *Staphylococcus aureus* as the nisin sensitive microorganism. Nisin layers were immersed into demineralized sterile water, and eluate was used for testing. An aliquot of bacterial culture (10^8 cfu/mL) was applied with a sterile swab on the surface of the Mueller Hinton agar plate. The eluate (3 h into physiological solution) from the nisin coated samples was transferred into the bored wells of 8 mm in diameter (120 µl/well) and incubated at 35 °C for 24 h. Control samples were prepared as series of nisin solutions in demineralized and sterilized water, starting with 25 µg/mL concentration, that was used as supposed maximal nisin concentration on the surface. The antimicrobial activity was expressed as a diameter of microbial growth inhibition zone (*IZ*) around the sample well after the incubation, calculated as average of four independent measurements. Ageing effect, i.e., reduction of antibacterial activity (R_A) of released nisin with time, was calculated as ratio of *IZ* diameter after given time from their preparation and *IZ* diameter of freshly prepared sample, following the Equation (1).

$$R_A = \left(1 - \frac{IZ_t}{IZ_0}\right) \times 100 \ [\%] \tag{1}$$

in which IZ_t is the diameter of inhibition zone in mm related to nisin released from stored sample and IZ_0 is the diameter of inhibition zone in mm of nisin released from freshly prepared samples.

The antibacterial susceptibility testing of the nisin layer (surface) was covered by standards EN ISO 22196 *Measurement of antibacterial activity on plastics and other non-porous surfaces*. The specimens of untreated and treated blank material (without nisin) and untreated and treated material with nisin coating were used to measure viable cells after incubation immediately after inoculation and after incubation for 24 h. Prepared flat samples diameter was (25 ± 2) mm, and target concentration was 6×10^5 cfu/mL. On the surface, 0.4 mL of test inoculum was pipetted; then, it was covered with the sterile foil and gently pressed down. The Soybean casein digest broth with lecithin and polyoxyethylene sorbitan monooleate (SCDLP) broth was used for washing of specimens.

For each test specimen, the number of viable bacteria recovered was determined in accordance with Equation (2):

$$N = \frac{(100 \times C \times D \times V)}{A} \tag{2}$$

in which *N* is the number of viable bacteria recovered per cm^2 per test specimen; *C* is the average plate count for the duplicate plates; *D* is the dilution factor for the plates counted; *V* is the volume, in mL, of SCDLP added to the specimen; and *A* is the surface area, in mm^2, of the cover film

Antibacterial activity (bacteria reduction) was calculated as according to Equation (3)

$$R = U_t - A_t \tag{3}$$

in which *R* is the antibacterial activity; U_t is the average of the common logarithm of the number of viable bacteria, in cfu/cm^2, recovered from the blank test specimens after 24 h; and A_t is the average of the common logarithm of the number of viable bacteria, in cfu/cm^2, recovered from the treated test specimens after 24 h.

2.6. Surface Characterization

2.6.1. Wettability and Surface Energy Analysis

Contact angle measurements of PVA films (nisin coated, uncoated, plasma treated and untreated) were performed using Surface Energy Evaluation system (See System E, Advex instruments s.r.o., Brno, Czech Republic). Demineralized water and diiodomethane were used as reference liquids. The analysis of a drop (volume of each drop was 10 µL) on a film surface was done after 20 s in order to let the drop stabilize its shape. Each sample was measured five times, mean contact angle was calculated, and subsequently the solid, surface-free energy was determined. The total surface-free energies γ^{total} were calculated according to Owens approach that assumes that total surface-free energy is a sum of both dispersion and polar components, i.e., Lifschitz-van der Waals and electron acceptor/electron donor [19,21,22]. The results presented further are the average values from at least five measurements. Standard deviation value was always up to 10% of average value.

2.6.2. Elektrokinetic Characterization

Zeta potential of all samples was determined by SurPASS Instrument (Anton Paar-SurPass, Graz, Austria). The SurPASS instrument is an electrokinetic analyzer utilized for the investigation of the zeta-potential of macroscopic solids based on the streaming potential and/or the streaming current method. For the sample analysis, the adjustable gap cell was used in contact with the 1 mmol/dm^3 KCl electrolyte. All measurements were performed at laboratory temperature.

The samples with higher PVA crosslinking degree of 10%, 20%, and 40% were analyzed in pH range of 4.5 to 9. The not-crosslinked and low crosslinked (PVA crosslinking degree 5%) samples were not stable, so they could be analyzed only at one pH value. For each measurement, a pair of identical samples (prepared on PE foils) was fixed on two opposite sample holders (with a cross section of 20 × 10 mm^2 and gap between them of 100 µm). The instrument employs Smoluchowski's equation [44] to determine the electrophoretic mobility of the particle and subsequently convert it in the Zeta Potential (see Equation (4)).

$$\zeta = \frac{u\sigma}{\varepsilon\varepsilon_0}\frac{\Delta V}{\Delta p} \tag{4}$$

in which μ is viscosity; σ conductivity; ε, ε_0 permittivity; and linear fits to the pressure-voltage curve give zeta potential.

2.6.3. XPS Analysis

The XPS measurements were done using the ESCALAB 250Xi (ThermoFisher Scientific, Waltham, MA, USA) system equipped with 500 mm Rowland circle monochromator with microfocused Al Kα X-Ray source. An X-ray beam with 200 W power (650 microns spot size) was used. The survey spectra were acquired with pass energy of 50 eV and resolution of 1 eV. High-resolution scans were acquired with pass energy of 20 eV and resolution of 0.1 eV. In order to compensate the charges on the surface, an electron flood gun was used. Spectra were referenced to the hydrocarbon type C 1s component set at a binding energy of 284.8 eV. The spectra calibration, processing, and fitting routines were done using Avantage software (ThermoFisher Scientific, Waltham, MA, USA).

3. Results

3.1. Stability Analysis of Nisin

As it is well known, nisin is very unstable in pure form, without any supplements as a stabilizer. This fact can cause difficulties during the monitoring of its release and adhesion. For this reason, the additional techniques were used to verify the nisin stability and nisin presence due to total nitrogen content measurement.

Electrophoretic assay and subsequent peptide determination (according to Bradford method) [40,41,45] were performed to evaluate nisin stability. The results from Bradford assay were compared to the total nitrogen content (TN) measurement, and it was found that amount of nitrogen corresponds to the nisin decomposition products—amino acids and shorter peptides—that are present in the immersion solution.

Based on the release behavior, it was also estimated that nisin was attached to the surface.

The mean molar weight of the peptide mixture (from electrophoretic measurements) confirmed that the stability of nisin attached to the PVA surface is dependent on the cross-linking degree and on the presence or absence of the plasma treatment step. These results were also obtained from TN measurements. Nisin molecule ($C_{143}H_{230}N_{42}O_{37}S_7$) contains about 17% of nitrogen. Detected nitrogen (correlated to blank) was much higher than it would be if it corresponded to stable nisin measured by electrophoresis and HPLC (Table 1). This fact signifies that nisin undergoes decomposition, and nitrogen comes from present amino acids. Thus, amount of nisin that could be attached to the PVA surface was estimated from complementary methods. As it was confirmed later, by using antimicrobial activity testing and zeta potential measurement, the most stable (ageing resistant) nisin layer was observed at 20% PVA crosslinking. When 10% crosslinking was used, the nisin attachment (initial amount of nisin at the surface) was maximal, but the stability was lower than in the case of 20% crosslinking. On the surface of films with 40% crosslinking, the relative small amount of nisin was attached, and also its instability was observed. However, the antimicrobial activity testing (see Chapter 3.3.) of this film is still relatively high compared to estimated nisin concentration, which can be attributed to glutaric acid (crosslinker) activity [39,45].

Table 1. Attached nisin content (Bradford assay from electrophoresis) and nitrogen content (TN) on the PVA surface (PVA films were stored at 25 °C for 3 months, relative humidity 54%). Nisin content on untreated PVA films was not measurable, and SD was higher than calculated value.

PVA/Nisin-Buffer Crosslinking Degree (%)	Nisin Content ($\mu g/mm^2$)	TN ($\mu g/mm^2$)
0	1.8 ± 0.5	0.7
5	3.6 ± 1.5	4.3
10	31.4 ± 1.5	13.2
20	28.2 ± 1.5	9.7
40	19.3 ± 1.3	6.8

3.2. Nisin Release

The nisin release profile was monitored, after its adhesion on the surface, in comparison to the study [45], in which nisin release from PVA bulk was investigated.

The effect of PVA crosslinking degree on amount of nisin released and probably also adsorbed on PVA substrate can be clearly recognized in the release study presented in Figure 1. It is evident that crosslinking degree of PVA influences extent of nisin adsorption, as well as its release profile. Untreated PVA films reveal only minimum nisin adsorption. Then, apparently, nisin is released immediately after immersion in salt solution (it means after 1–24 h). It was found that nisin was strongly attached at the films of 10% and 20% crosslinking degree. Nisin was released slowly in small amounts. On the other hand, from the film with 0% or 5% crosslinking the whole attached amount of nisin was released immediately after immersion, as it was in the case of untreated films. Besides, PVA with low crosslinking dissolved very fast. Pure PVA (without crosslinking) and also untreated PVA films do not reach the corresponding nisin attachment. The highest nisin surface adsorption was found for the samples with 20% of crosslinking degree. But for 40% crosslinking degree, the nisin adhesion is unsatisfying. This could be explained by the change in physical properties of PVA. It was proved also in some previous study that increasing the crosslinking degree leads to a significant decrease of the solubility. Moreover, PVA with 20% of crosslinking degree exhibited more advantageous mechanical properties for further applications [45]. Additionally, plasma treatment was not very effective in this case. It means that more intensive initial introducing of crosslinking connections leads to apparent change of material properties and formations of inhomogeneity on the surface that are caused by higher inhomogeneity in the bulk, higher degree of crystallinity, inhomogeneous formation of crystalline phase of glutaric acid.

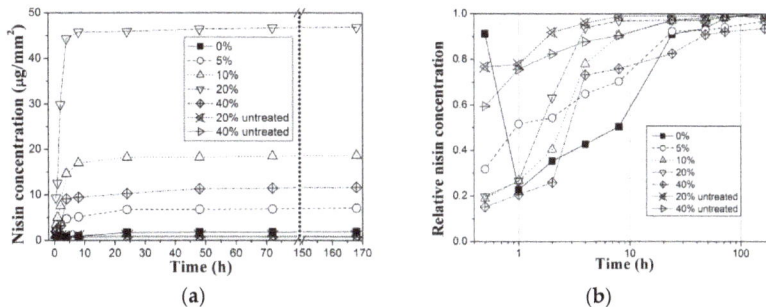

Figure 1. Nisin release from the PVA films after immersion in physiological solution, analysed by RP-HPLC: (**a**) cumulative concentration in $\mu g/mm^2$ and (**b**) related to total released amount.

3.3. Antimicrobial Assays

Anti-staphylococcal effectivity of nisin-coated PVA film was evaluated by antimicrobial susceptibility testing, with results shown in Figure 2a,b. Nisin in the PVA surface layer can be active against Gram positive (G+) bacterial strain *Staphylococcus aureus* even after 90 days of storage. However, nisin activity (and its ageing) is strongly dependent on the crosslinking degree (amount of GA) and on presence or absence of the plasma treatment step. The amount of GA influences not only bulk but also surface properties, and therefore the plasma treatment effectiveness. Only 35% reduction of antimicrobial activity after 90 days was observed in the case of nisin-coated PVA samples with 20% crosslinking degree, while almost 100% reduction of antimicrobial activity was observed in the case of 0% and 5% of crosslinking. It should be mentioned that synergic effect of the crosslinking agent, GA, cannot be neglected. Mainly in the samples with 40% crosslinking degree, modest antibacterial activity of GA in the systems has been found. However, this factor does not represent significant contribution to antibacterial activity of the PVA/GA/nisin samples [17,45]. Therefore, nisin is be supposed to be the principal antibacterial agent against G+ bacterial strains in the studied systems.

To compare it to HPLC analysis (Figure 1b), testing of released nisin was also performed by MHA diffusion testing, which is dependent on migration rate of antimicrobial agent and on its stability (Figure 2a). Antibacterial activity of released nisin from films that was stored at specific conditions for 90 days was determined according to inhibition zones of nisin in immersion solution.

Antimicrobial properties of PVA films depend on the amount of nisin that is attached to the PVA film and also on the nisin release from the PVA surface so the main aim should be to ensure maximal initial activity (surface concentration) of nisin and its stability in the layer. As was shown in antimicrobial testing, the crosslinking degree has an important influence on nisin adhesion.

It was not possible to calculate exact amount of nisin attached on the surface, but this amount can be estimated from the amount that is released from the surface.

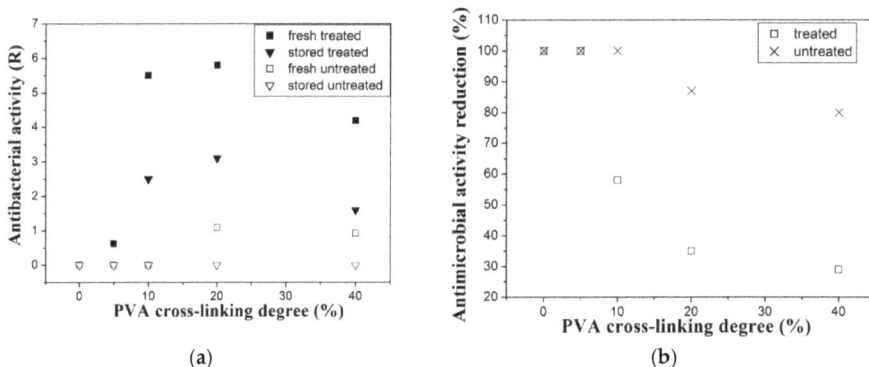

(a)   (b)

Figure 2. Antibacterial activity change of PVA films with attached nisin after 90 days of storage (**a**) after immersion in water solution for 3 h and (**b**) by ISO 22196 against *Staphylococcus aureus* (CCM 4516).

Nisin layers were tested as promising natural coating with anti-staphylococcal activity [6,9,11,17,18], and it was proved that even in combination with PVA surface, it can improve the antimicrobial activity of biocompatible films. Nevertheless, the key role lies in the plasma treatment, as was confirmed for other polymer materials in literature [16,19,22,29]. This corresponds also to our finding, when plasma treatment of our samples led to obtaining an effective layer with suitable nisin adhesion.

3.4. Wettability and Surface Energy Analysis

As it is shown in Figure 3, the PVA plasma treatment using the 2 W/cm^2 or 7 W/cm^2 power density resulted in essentially the same wettability. Nevertheless, the 7 W/cm^2 was finally chosen for the rest of experiments, since it was found that a decrease in water contact angle (WCA) of plasma-treated PVA films was sufficient for PVA wettability sustainment (WCA decrease is higher than 50% of the initial value).

Initial concentration of nisin (that was measured during nisin release) corresponded to nisin adhesion on PVA films, and it is in agreement with WCA measurements (Figure 4). The hydrophilic character (wettability) of PVA films significantly increased after the plasma treatment (see Figures 3 and 4). The γ^{AB} component of surface-free energy increases with plasma treatment, which indicates the presence of polar groups (attached via covalent bonds). Furthermore, the total surface energy increases, which enables nisin adhesion. With nisin immobilization on the surface, the γ^{LW} increases, thus the van der Waals interactions predominate. The different trend (WCA increase) with nisin adhesion demonstrates increase of contact angle, suggesting the hydrophobic nisin groups' attendance. While non-crosslinked PVA films and films with low crosslinking degree up to 5% show only slightly decreased wettability in comparison with untreated PVA films, samples with higher crosslinking degree (10 and 20%) show significant increase in WCA. This increase can be also caused by the surface roughness. With further increasing of crosslinking degree, the steep decrease of WCA was observed. This phenomenon can be explained by lower presence of free hydrophilic groups (OH, COO–) on the PVA surface, as was also confirmed by the previous study [45]. Maximal WCA of 84° was obtained for PVA with 20% of crosslinking. This can be attributed to protein surface adhesion when nisin was bonded to PVA substrate. It also reveals the optimal PVA crosslinking degree for the protein efficient surface immobilization. These results are in agreement with those in Sections 3.1 and 3.2.

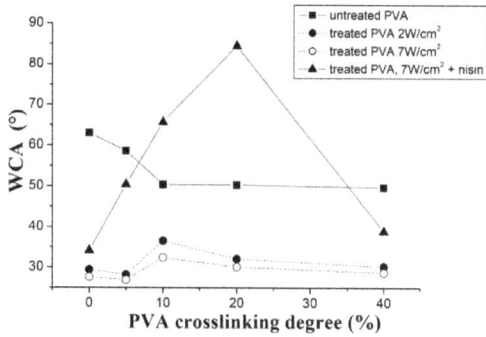

Figure 3. The water contact angle dependence as a function of PVA crosslinking degree; plasma treatment time was 10 s. Nisin was attached from PBS buffer.

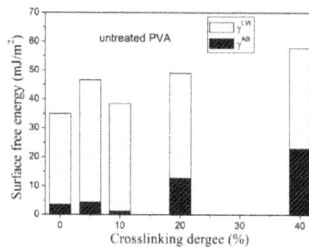

(a)

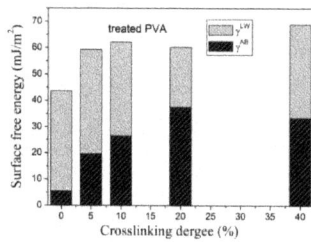

(b)

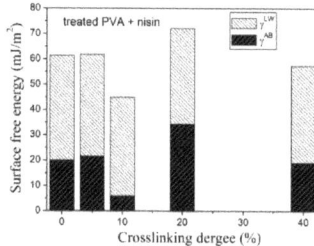

(c)

Figure 4. Surface-free energy evaluation for (**a**) untreated, (**b**) plasma-treated PVA films, and (**c**) films with attached nisin.

3.5. Electrokinetic Characterization

The stability study of coatings can be carried out, too, in correlation with the electrokinetic measurements. The surface charge on material is related to affinity-associated interactions. It can control degree of biomolecular affinity on a polymer surface [43,46–50]. Zeta potential is used for characterization of natural and synthetic polymers and in investigation of their hydrophilicity, which is important for their interactions with biomolecules, dyes, surfactants, etc. [30,31,42,44,50]. It is also important for research of modified surfaces (such as membranes, filters) or polymers modified by a plasma treatment or by plasma chemical coatings [32,50]. Water soluble polymers interact according to their functional group. Zeta potential of biodegradable polymer surface is not often evaluated in literature. In this paper, zeta potential, combined with pH of liquid phase, was used to characterize electrostatic interactions between the polymer surface and the immobilized bioactive peptide.

The zeta-potential measurement provides complementary information to WCA and surface-free energy evaluation. Results for zeta potential of treated and untreated films, and for the films with attached nisin, are shown in Figure 5. As it was expected, the zeta potential value increased after the plasma treatment (shift to more negative values). This behavior suggests that the surface is not stable; thus, it is susceptible to nisin attachment. Zeta potential of PVA films without nisin presents negative values in all cases. Lower (more negative) values were recorded in the case of the crosslinked PVA, mainly with higher content of glutaric acid (higher crosslinking degree). As zeta potential values are included inside the domain ± 30 mV, it means that the stability of measured surface is reduced; it means that the zeta potential measurements indicate the possibility of nisin adhesion. Surface charge becomes positive or less negative after nisin adhesion on PVA. Even the zeta potential values are dependent also on the surface roughness; it is clear that plasma-treated PVA show less negative zeta potential. Thus, it is confirmed that plasma treatment enhances the potential of nisin adsorption. In samples with nisin content, the shift to higher values was observed, reaching 5 mV at pH 4.5 in case of 20% of crosslinking degree. Moreover, as can be seen in Figure 5, the titration curve for PVA with 20% crosslinking shows the shift to the value close to zero at pH of 8 (pH of nisin is 8.3–8.5). All these findings signify the higher nisin adsorption and stability. The titration curves show the film high instability when the nisin is bonded, especially in the case of film with 10% of crosslinking.

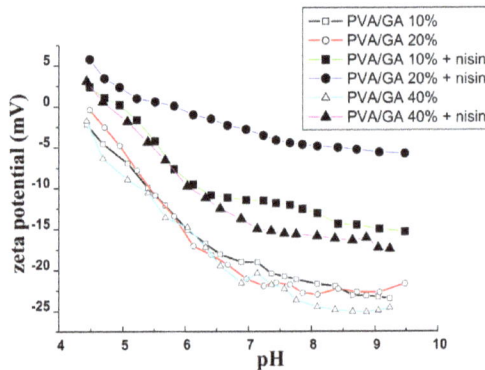

Figure 5. Zeta potential of treated PVA films with different crosslinking degree with and without adsorbed nisin as a function of pH values. Titration was performed by using 0.05 M NaOH. Non-crosslinked films and films with crosslinking up to 5% were not measured due to their low resistance to electrolyte.

3.6. XPS Analysis

Nisin attachment chemistry was estimated by XPS measurements. Surface composition of untreated and plasma treated samples is presented in Table 2. PVA films' ability to adsorb nisin after plasma treatment was investigated by analyzing changes in O and C peaks on the treated and untreated PVA films (PVA without crosslinking agent was used). The strong increase of O peak was observed on the surface of plasma treated samples (see Table 2).

Table 2. XPS elemental analysis of PVA samples.

PVA + 0% GA	C (at %)	O (at %)
untreated	68	32
treated 2 W/cm^2, 10 s	38	45
treated 7 W/cm^2, 10 s	30	51

The peak fitting routines were done in order to evaluate the changes in the bond structure and, therefore, explain the observed changes after plasma treatment. The high resolution C 1s peak was fitted with 6 principal components: C–C/C–H (binding energy at 284.7 eV), C–N (285.6 eV), C–O (286,3 eV), N–C–O (287.6 eV), C=O (288.1 eV), and O–C=O (289.2), as it is shown in Figure 6. The results of the fits are presented in Table 3.

As it was mentioned above, a significant decrease in WCA was observed after the plasma treatment. Hydrophilic character of PVA surface was achieved, because characteristic polar functional groups (mainly COO– groups) were introduced onto the PVA surface by the plasma treatment [10,13] (see Figure 6 and Table 3).

Table 3. Areas of C1s components.

Bonds (%)	Untreated	Treated, 7 W/cm^2	Treated with Nisin
C–C/C–H	41	31	21
C–N	34	36	19
C–O	13	15	47
C=O	12	2	2
N–C=O	0	5	11
O–C=O	0	11	0

The C1s peak fitting of plasma-treated samples shows O–C=O arising that can be attributed to the polar functional group formation. After nisin attachment, also C=O, C–O vs. C–C bonds presence was evaluated. Results (Table 3) show decrease of C–O and C–H bonds and appearance of N–C=O and O–C=O after nisin adhesion, while after final plasma treatment, the C–O and N–C=O bonds presence is higher. This can be caused by nisin attachment. On the other hand, the O–C=O are not present.

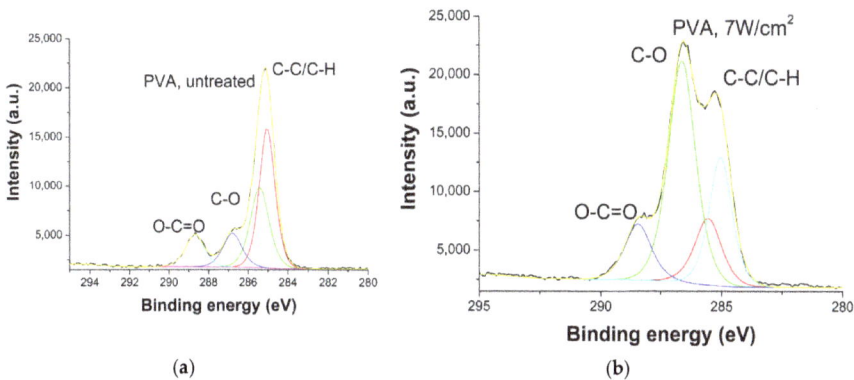

Figure 6. C1s high resolution peak fitting of (**a**) untreated PVA (0% crosslinking) and (**b**) plasma treated PVA (0% crosslinking).

4. Conclusions

Combination of coplanar Surface Barrier Discharge activation and cross-linking using environmentally friendly glutaric acid was applied to PVA to study its potential use as a carrier system for nisin adhesion. It was found that both mentioned factors have a crucial effect on the extent of nisin surface attachment and its release kinetics.

It was proved that crosslinking degree of polyvinyl alcohol plays an important role mainly in the range of 10–20%, which was evaluated as optimal for preparation of relatively water-stable peptide coatings that can sustain also the antibacterial properties with potential applicability in the medical field. According to the zeta potential measurement, the adhesion proceeded at acidic pH value, and surface inhomogeneity was observed, mainly in the case of 10% of crosslinking. This suggests that besides the bulk, also the surface properties of the polymer materials must be carefully controlled to achieve acceptable results.

Based on the results, DCSBD plasma treatment of PVA using 20% crosslinking with glutaric acid and nisin adhesion represents a novel and promising method for the design of biomaterials characterized by tunable stability and diffusion of bioactive compounds.

Author Contributions: Z.K.R. designed the study, participated in the sample preparation, HPLC analyses, electrophoresis, and antimicrobial testing; P.S. conducted the plasma treatment and samples preparation; J.S. carried out the antimicrobial and stability testing and WCA measurements; M.S. performed the XPS analysis; L.M. ensured zeta-potential measurement; and M.L. provided the skilled consultations and interpretation of results.

Funding: This work was supported by the grant 17-10813S of the Czech Science Foundation (Grant Agency of the Czech Republic).

Acknowledgments: We thank our colleagues from Faculty of Mathematics and Physics, Charles University in Prague for assistance with methodology and with the research. We would also like to show the gratitude to Professor Vit Kudrle from Masaryk university and Professor Vladimir Sedlarik from Centre of polymer systems for their support and comments on an earlier version of the manuscript.

Conflicts of Interest: The authors declare no conflict of interest. The founding sponsors had no role in the design of the study; in the collection, analyses, or interpretation of data; in the writing of the manuscript, and in the decision to publish the results.

References

1. Cho, D.; Hoepker, N.; Frey, M.W. Fabrication and characterization of conducting polyvinyl alcohol nanofibers. *Mater. Lett.* **2012**, *68*, 293–295. [CrossRef]
2. Bosco, R.; Edreira, E.U.; Wolke, J.G.; Leeuwenbugrh, C.G.; Van Den Beucken, J.; Jansen, J.A. Instructive coatings for biological guidance of bone implants. *Surf. Coat. Technol.* **2013**, *233*, 91–98. [CrossRef]
3. Hrabalikova, M.; Merchan, M.; Ganbold, S.; Sedlarik, V.; Valasek, P.; Saha, P. Flexible polyvinyl alcohol/2-hydroxypropanoic acid films: effect of residual acetyl moieties on mechanical, thermal and antibacterial properties. *J. Polym. Eng.* **2015**, *35*, 319–327. [CrossRef]
4. Yin, H.; Mix, R.; Friedrich, J.F. Combination of plasma-chemical and wet-chemical processes-a simple way to optimize interfaces in metal-polymer composites for maximal adhesion. *J. Adhes. Sci. Technol.* **2011**, *25*, 799.
5. Ducheyne, P.; Healy, K.; Dietmar, E.; Hutmacher, E.; Grainger, D.W.; Kirkpatrick, C.J. Comprehensive Biomaterials. Available online: https://www.elsevier.com/books/comprehensive-biomaterials/ducheyne/978-0-08-055302-3 (accessed on 13 August 2018).
6. Ryder, M.; Schilke, K.F.; Auxier, J.A.; McGuire, J.; Neff, J. Nisin adsorption to poly-ethylene oxide layers and its resistance to elution in the presence of fibrinogen. *J. Colloid Interface Sci.* **2010**, *350*, 194–199. [CrossRef] [PubMed]
7. Duan, J.; Park, S.I.; Daeschel, M.A.; Zhao, Y. Antimicrobialchitosan Lysozyme (CL) films and coatings for enhancingmicrobial safety of Mozzarella cheese. *Food Microbiol. Saf.* **2007**, *72*, 355–361.
8. Saraf, A.; Johnson, K.; Lind, M.L. Poly(vinyl) alcohol coating of the support layer of reverse osmosis membranes to enhance performance in forward osmosis. *Desalination* **2014**, *333*, 1–9. [CrossRef]
9. Xiang, C.; Taylor, A.G.; Hinestroza, J.P.; Frey, M.W. Controlled release of nonionic compounds from poly (lactic acid)/cellulose nanocrystal nanocomposite fibers. *J. Appl. Polym. Sci.* **2013**, *127*, 79–86. [CrossRef]

10. Karam, L.; Jama, C.; Dhulster, P.; Chibib, N. Study of surface interactions between peptides, materials and bacteria for setting up antimicrobial surfaces and active food packaging. *J. Mater. Environ. Sci.* **2013**, *4*, 798–821.

11. Resa, C.P.; Jagus, R.J.; Gerschenson, L.N. Effect of natamycin, nisin and glycerol on the physicochemical properties, roughness and hydrophobicity of tapioca starch edible films. *Mater. Sci. Eng.* **2014**, *40*, 281–287. [CrossRef] [PubMed]

12. Imran, M.; Klouj, A.; Revol-Junelles, A.M.; Desobry, S. Controlled release of nisin from HPMC, sodium caseinate, poly-lactic acid and chitosan for active packaging applications. *J. Food Eng.* **2014**, *143*, 178–185. [CrossRef]

13. Zasada, K.; Lukasiewitz-Atanasov, M.; Klysik, K.; Lewandowska, J.; Gzyl Malcher, B.; Malinowska, A. One-component ultrathin films based on poly (vinyl alcohol) as stabilizing coating for phenytoin-loaded liposomes. *Colloids Surf. B* **2015**, *135*, 133–142. [CrossRef] [PubMed]

14. Park, G.Y.; Park, S.J.; Choi, M.Y.; Koo, I.G.; Byun, J.H.; Hong, J.W.; Sim, J.Y.; Collins, G.J.; Lee, J.K. Atmospheric-pressure plasma sources for biomedical applications. *Plasma Sources Sci. Technol.* **2012**, *21*, 043001. [CrossRef]

15. Kim, K.; Lee, S.M.; Mishra, A.; Yeom, G. Atmospheric pressure plasmas for surface modification of flexible and printed electronic device. *Thin Solid Films* **2015**, *598*, 315–334. [CrossRef]

16. Donegan, M.; Dowling, D.P. Activation of PET using an RF atmospheric plasma system. *Surf. Coat. Technol.* **2013**, *234*, 53–59. [CrossRef]

17. Gubskaya, A.V.; Khan, L.J.; Valenzuela, L.M.; Lysniak, L.K. Ivestigating the release of a hydrophobic peptide from matrices of biodegradable polymers: An ittegrated method approach. *Polymer* **2013**, *54*, 3806–3820. [CrossRef] [PubMed]

18. Nisol, B.; Reniers, F. Challenges in the characteriyation of plasma polymers using XPS. *J. Electron. Spectrosc. Relat. Phenom.* **2015**, *200*, 311–331. [CrossRef]

19. Zuo, B.; Hu, Y.; Lu, X.; Zhang, S.; Fan, H.; Wang, X. Surface Properties of Poly (vinyl alcohol) Films Dominated by Spontaneous Adsorption of Ethanol and Governed by Hydrogen Bonding. *J. Phys. Chem. C* **2013**, *117*, 3396–3406. [CrossRef]

20. Schilke, K.F.; McGuire, J. Detection of nisin and fibrinogen adsorption on poly (ethylene oxide) coated polyurethane surfaces by time-of-flight secondary ion mass spectrometry (TOF-SI MS). *J. Colloid Interface Sci.* **2011**, *358*, 14–24. [CrossRef] [PubMed]

21. Kafi, K.; Magniez, K.; Fox, L.B. Surface properties relationship of atmospheric plasma treated jute composites. *Compos. Sci. Technol.* **2011**, *71*, 1692–1698. [CrossRef]

22. Siow, K.S.; Brichter, L.; Kumar, S.; Griesser, H.J. Plasma Methods for the Generation of Chemically Reactive Surfaces for Biomolecule Immobilization and Cell Colonization—A Review. *Plasma Process. Polym.* **2006**, *3*, 392–418. [CrossRef]

23. Bilek, F.; Krizova, T.; Lehocky, M. Preparation of active antibacterial LDPE surface through multistep physicochemical approach: I. Allylamine grafting, attachment of antibacterial agent and antibacterial activity assessment. *Colloid Surf. B Biointerfaces* **2011**, *88*, 440–447. [CrossRef] [PubMed]

24. Friedrich, F.J. *The Plasma Chemistry of Polymer Surfaces: Advanced Techniques for Surface Design*; Wiley-VCH: Weinheim, Germany, 2012.

25. Roth, J.R. *Industrial Plasma Engineering, Vol. II–Applications to Non-Thermal Plasma Processing (ISBN 7503 05444)*; Institute of Physics Publishing: Bristol, PA, USA, 2001.

26. Černák, M.; Hudec, I.; Kováčik, D.; Zahoranová, A. Diffuse Coplanar Surface Barrier Discharge and Its Applications for In-Line Processing of Low-Added-Value Materials. 2009. Available online: https://www.cambridge.org/core/journals/the-european-physical-journal-applied-physics/article/diffuse-coplanar-surface-barrier-discharge-and-its-applications-for-inline-processing-of-lowaddedvalue-materials/C072CA66231A6B78F2918F260101F39E (accessed on 13 August 2018).

27. ROPLASS (Robust Plasma Systems). Available online: http://http://www.roplass.cz/roplass-robust-plasma-systems (accessed on 4 May 2018).

28. Kogelschatz, U. Dielectric-barrier discharges: Their history, discharge physics and industrial applications. *Plasma Chem. Plasma Process.* **2003**, *23*, 1–46. [CrossRef]

29. Čech, J.; Hanusová, J.; Sťahel, P.; Černák, M. Diffuse coplanar surface barrier discharge in artificial air: statistical behaviour of microdischarges. *Open Chem.* **2015**, *13*, 528–540. [CrossRef]

30. Fuchs, S. Gelatin Nanoparticles as a Modern Platform for Drug Delivery-Formulation Development and Immunotherapeutic Strategies. Ph.D. Thesis, Ludwig-Maximilians-Universität München, Munich, Germany, 2010.

31. Imasaka, K.; Khaled, U.; Wei, S.; Suehiro, J. pH dependence of water-solubility of single-walled carbon nanotubes treated by microplasma in aqueous solution. *Electroanalysis* **2004**, 16.

32. Fang, D.L.; Chen, Y.; Xu, B.; Ren, K.; He, Z.Y.; He, L.L.; Lei, Y.; Fan, C.M.; Song, X.R. Development of Lipid-Shell and Polymer Core Nanoparticles with Water-Soluble Salidroside for Anti-Cancer Therapy. *Int. J. Mol. Sci.* **2014**, *15*, 3373–3388. [CrossRef] [PubMed]

33. Rachmawati, H.; Haryadi, B. The Influence of polymer structure on the physical characteristic of intraoral film containing BSA-loaded nanoemulsion. *J. Nanomed. Nanotechnol.* **2014**, *5*, 1. [CrossRef]

34. Cruz, E.F.; Zheng, Y.; Torres, E.; Li, W.; Song, W.; Burugapalli, K. Zeta potential of modified multi-walled carbon nanotubes in presence of poly (vinyl alcohol) hydrogel. *Int. J. Electrochem. Sci.* **2012**, *7*, 3577–3590.

35. Prombutara, P.; Kulwatthanasal, Y.; Supaka, N.; Samarala, I.; Chareonpornwattana, S. Production of nisin-loaded solid lipid nanoparticles for sustained antimicrobial activity. *Food Control* **2012**, *24*, 184–190. [CrossRef]

36. Malayoglu, U.; Tekin, K.C.; Shrestha, S. Influence of post-treatment on the corrosion resistance of PEO coated AM50B and AM60B Mg alloys. *Surf. Coat. Technol.* **2010**, *205*, 1793–1798. [CrossRef]

37. Weeks, M.D.; Subramanian, R.; Vaidya, A.; Mumm, D.R. Defining optimal morphology of the bond coat–thermal barrier coating interface of air-plasma sprayed thermal barrier coating systems. *Surf. Coat. Technol.* **2015**, *273*, 50–59. [CrossRef]

38. Chang, J.Y.; Godovsky, D.Y.; Han, M.J.; Hassan, C.M.; Kim, J.; Lee, B.; Lee, Y.; Peppas, N.A.; Quirk, R.P.; Yoo, T. *Biopolymers PVA Hydrogels, Anionic Polymerisation Nanocomposites*; Springer: Heidelberg/Berlin, Germany, 2000.

39. Alkan, C.; Gunther, E.; Hiebler, S.; Himpel, M. Complexing blends of polyacrylic acid-polyethylene glycol and poly(ethylene-co-acrylic acid)-polyethylene glycol as shape stabilized phase change materials. *Energy Convers. Manag.* **2012**, *64*, 364–370. [CrossRef]

40. Adamczyk, Z.; Nattich, M.; Wasilewska, M.; Zaucha, M. Colloid particle and protein deposition–Electokinetic studies. *Adv. Colloid Interface Sci.* **2011**, *168*, 3–28. [CrossRef] [PubMed]

41. Wiśniewski, J.R.; Gaugaz, F.Z. Fast and Sensitive Total Protein and Peptide Assays for Proteomic Analysis. *Anal. Chem.* **2015**, *87*, 4110–4116. [CrossRef] [PubMed]

42. Dorgan, K.M.; Wooderchak, W.L.; Wynn, D.P.; Karschner, E.L.; Alfaro, J.F.; Cui, Y.; Zhou, Z.S.; Hevel, J.M. An enzyme-coupled continuous spectrophotometric assay for S-adenosylmethionine-dependent methyltransferases. *Anal. Biochem.* **2006**, *350*, 249–255. [CrossRef] [PubMed]

43. Miyake, N.; Miura, T.; Sato, T.; Yoshinari, M. Effect of zeta potentials on bovine serum albumin adsorption on crown composite resin surfaces in vitro. *J. Biomed. Sci. Eng.* **2013**, *6*, 273–276. [CrossRef]

44. Sze, A.; Erickson, D.; Ren, L.; Li, D. Zeta-potential measurement using the Smoluchowski equation and the slope of the current–time relationship in electroosmotic flow. *J. Colloid Interface Sci.* **2003**, *261*, 402–410. [CrossRef]

45. Habalikova, M.; Holcapkova, P.; Suly, P.; Sedlarik, V. Immobilization of bacteriocin nisin into a poly(vinyl alcohol) cross-linked with non-toxic dicarboxylic acid. *J. Appl. Polym. Sci.* **2016**, *133*, 43674. [CrossRef]

46. Song, Y.W.; Shan, D.Y.; Han, E.H. High corrosion resistance of electroless composite plat- ing coatings on AZ91D magnesium alloys. *Electrochim. Acta* **2008**, *53*, 2135–2143. [CrossRef]

47. Belgacem, M.N.; Gandini, A. The surface modification of cellulose fibres for use as reinforcing elements in composite materials. *Compos. Interfaces* **2005**, *12*, 41–75. [CrossRef]

48. Cho, D.; Lee, S.; Frey, M.W. Characterizing zeta potential of functional nanofibers in a microfluidic device. *J. Colloid Interface Sci.* **2012**, *372*, 252–260. [CrossRef] [PubMed]

49. Salgın, S.; Salgın, U.; Bahadır, S. Zeta Potentials and Isoelectric points of biomolecules: The effects of ion types and ionic strengths. *Int. J. Electrochem. Sci.* **2012**, *7*, 12404–12414.

50. Balcão, V.M.; Costa, C.I.; Matos, C.M.; Moutinho, C.G.; Amorim, M.; Pintado, M.E.; Gomes, A.P.; Vila, M.M.; Teixeira, J.A. Nanoencapsulation of bovine lactoferrin for food and biopharmaceutical applications. *Food Hydrocoll.* **2013**, *32*, 425–431. [CrossRef]

materials

MDPI

Article

Increase in Strength and Fretting Resistance of Alloy 718 Using the Surface Modification Process

Auezhan Amanov [1,*], Rakhmatjon Umarov [1] and Tileubay Amanov [2]

[1]   Department of Mechanical Engineering, Sun Moon University, Asan 31460, Korea; umarov92@inbox.ru
[2]   Institute of Mechanics and Seismic Stability of Structures, Academy of Sciences,
      Tashkent 100125, Uzbekistan; amanov43@mail.ru
*   Correspondence: avaz2662@sunmoon.ac.kr; Tel.: +82-41-530-2892

Received: 9 July 2018; Accepted: 3 August 2018; Published: 6 August 2018

Abstract: This work comparatively investigated the strength (hardness, yield strength, dynamic elastic modulus, and surface residual stress), fretting failure, and corrosion resistance of the as-received and treated Ni-based superalloy Alloy 718. The goal of the current research is to improve the hardness, fretting wear, and corrosion resistances of Alloy 718 through the ultrasonic nanocrystal surface modification (UNSM) process with the aim of extending the lifespan of aircraft and nuclear components made of Alloy 718. The experimental results revealed that the surface hardness increased by about 32%, the fretting wear resistance increased by about 14%, and the corrosion resistance increased by about 18% after UNSM process. In addition, the UNSM process induced a tremendous high compressive surface residual stress of about -1324 MPa that led to an increase in yield strength and dynamic Young's modulus by about 14 and 9%, respectively. Grain size refinement up to ~50 nm after the UNSM process is found to be responsible for the increase in surface hardness as well. The depth of the effective layer generated by the UNSM process was about 20 μm. It was concluded that the UNSM process played a vital role in increasing the strength and enhancing the corrosion and fretting resistances of Alloy 718.

Keywords: Alloy 718; surface hardness; surface residual stress; grain size; fretting failure; corrosion

1. Introduction

Alloy 718 is mainly used in aerospace and nuclear applications in the range of operating temperature of 650–675 °C because of relatively excellent mechanical properties, fusion weldability, good, excellent oxidation resistance, corrosion resistance, and creep at elevated temperatures [1]. However, the relatively poor fretting wear resistance of Alloy 718 may restrict its use without any surface engineering [2]. Alloy 718 usually exhibits poor wear resistance, resulting in shortening the service life of components owing to the relatively low hardness in spite of its excellent properties such as corrosion, erosion, oxidation, etc. [3]. One of the main issues in the aerospace and nuclear industries is fretting damage, which can be derived from oscillating motion, and it may lead to high nominal stress between two fretting bodies, resulting in high wear or even fatigue limit [4]. Alloy 718 can be hardened by the presence of precipitation of γ' (Ni$_3$(Al,Ti)) and γ'' (Ni$_3$Nb) phases within a (face centered cubic) FCC structure. The latter phase can be transferred into Ni$_3$Nb—δ one at such a high temperature, where a softening can take place [5]. Also, δ phase (Ni$_3$Nb) is a vital phase, which precipitated in the range of 750~1020 °C [6,7]. In general, in Alloy 718, the precipitation of δ phase does not provide any strengthening mechanism, but it controls the microstructure [8]. It is well established that Alloy 718 can be directly used for various structural applications with no heat treatment thanks to the standard precipitation treatment leading to an increase in strength [9,10]. Lin et al. have reported earlier that a high amount of δ phase significantly decreased the strength and plasticity of Alloy 718 [7]. On the other hand, the balanced amount of δ phase may serve to alter the microstructure in terms of grain size refinement and dislocation impediment that led to an increase in strength [11]. However, the enhancement in strength of Alloy 718 by the presence of δ phase was not good enough to be used in real applications. Moreover, it is well known that the strength of Alloy 718 can be also substantially improved by various heat treatment processes. For instance, Chlebus et al. have investigated the heat treatment response to the microstructure and mechanical properties of Alloy 718 produced by additive manufacturing process [12]. In total, four series of heat treatments were performed at various annealing temperatures. It was found that the highest hardness of Alloy increased from 312 to 461 HV. Another study also claimed that the mechanical properties of Alloy 718 can be increased by annealing and hot isostatic pressing (HIP) + annealing processes, resulting in an increase in hardness by about 10 and 31%, respectively [13]. Raghavan et al. have also investigated the effect of heat treatment on mechanical properties of Alloy 718 [14]. The results revealed that increase in homogenization temperature coarsened the grain size that led to softening, but the strength increased remarkably after ageing treatment due to the precipitation hardening of Alloy 718.

In this regard, a number of investigations have been performed to increase the strength of Alloy 718 [15–17]. For instance, Zheng et al. have studied the influence of carbon (C) content on control in strength of Alloy 718 [18]. It was found that the amount of precipitation was reduced, and the grain size of carbide was refined with decreasing C content, leading to an increase in strength. Liu et al. have tried to control the mechanical properties of Alloy 718 using the electromagnetic stirring [EMS] process [19]. Interestingly, a tensile strength of Alloy 718 subjected to EMS process was found to be higher in comparison with that of the unprocessed Alloy 718 because of the increased surface hardness and induced compressive residual stress. In addition, Farber et al. have found the relationship between the mechanical properties and microstructure of four different combinations of Alloy 718: solution and ageing heat treatment, heat treatment along with hot isostatic pressing, heat treatment along with shot peening, and heat treatment together with hot isostatic pressing and shot peening. The results showed that even though the heat treated specimen exhibited better results in terms of surface hardness, the specimen subjected to heat treatment and hot isostatic pressing revealed similar results in terms of tensile strength and ductility, in which the surface hardness was found to be much lower compared to that of the heat treated one [8]. Surface modification methods such as shot peening and laser shock peening have also been widely used to increase the mechanical properties of Alloy 718. For example, Chamanfar et al. have used a shot peening process to increase the mechanical properties of Alloy 718 [20]. Gill et al. have analyzed the effect of laser shock peening on the hardness and residual stress [21]. Both surface modification methods increased the mechanical properties due to the severe plastic deformation of the surface as a result of inducing high strain and the transformation of tensile residual stress into compressive one.

In this current paper, Alloy 718 was treated by a relatively new ultrasonic nanocrystalline surface modification (UNSM) process to achieve high strength and excellent resistances to fretting and corrosion simultaneously. The main goal of the current research is to evaluate the impact of UNSM process in terms of surface roughness, surface hardness, surface residual stress, grain size, etc., on the increase in strength, fretting wear, and corrosion resistance that lead to an increase in lifespan of various components and parts made of Alloy 718 in both aerospace and nuclear industries. The effective depth of UNSM process was confirmed by cross-sectional observation. The impact of the UNSM process on the surface hardness and tensile behavior was obtained using hardness and tension tests, while the corrosion resistance was studied using the potentiodynamic polarization test method. Finally, the level of fretting failure was investigated by fretting wear tester under dry conditions.

2. Materials and Methods

2.1. Material

In this study, a number of wrought specimens made of solution-annealed Alloy 718 were used. The specimens were melted at VDM Metals (USA/FP), precipitation heat treated at 720 °C, and then air cooled. The grain size of specimens determined in accordance with a standard [22]. ASTM-E-112-13. The mechanical properties and chemical composition of Alloy 718 provided by the supplier are listed in Tables 1 and 2, respectively. Before and after UNSM process, all the specimens were washed in acetone ((CH₃)₂CO) for 5 min. Then, the optimized UNSM parameters listed in Table 3 were selected to be applied to Alloy 718. Detailed information about the UNSM process and its parameters are available elsewhere [23,24].

Table 1. Mechanical properties of Alloy 718.

Tensile Strength, MPa	Yield Strength, MPa	Elastic Modulus, GPa	Shear Modulus, GPa	Poisson's Ratio	Elongation, %
1150	950	211	77.2	0.294	25

Table 2. Chemical composition of Alloy 718 in wt %.

Fe	Cr	C	Ti	Mn	Si	Ni	S	P	Mo	Nb	Al
17.62	18.84	0.024	0.95	0.02	0.06	53.64	0.002	0.003	3.08	5.23	0.53

Table 3. UNSM process parameters.

Frequency, kHz	Amplitude, μm	Speed, mm/min	Load, N	Feed-Rate, μm	Ball Diameter, mm	Ball Material
20	50	2000	50	70	2.38	WC

2.2. Characterization

Before measuring the surface roughness and surface hardness, the specimens were ultrasonically cleaned in distilled water (H₂O) for 10 min to keep the surface clean. The average surface roughness (R_a) of the specimens was obtained using a surface profilometer (SJ-210, Mitutoyo, Kawasaki, Japan) with a scanning length 4 mm at a scan rate of 10 mm/min. The surface hardness was measured at a load of 300 gf for 10 s by micro-Vickers hardness tester (MVK E3, Mitutoyo, Japan). The indent points were randomly selected at a distance of 300 μm away from each other. The indent size was about 40–45 μm. The surface roughness and surface hardness measurements were repeated three times to give a good statistical representation of the results. To observe the difference in microstructure, the specimens were embedded on resin, and the conductive resin used to embed the metal sample was produced by press. The top surface of the specimens was mechanically polished using a silicon carbide (SiC) sandpaper with a grit in the range of 400 to 2000. The specimens were electrolytically etched in standard nital etchant solution (15% nitric acid and 85% ethanol) at 6 V for 10 s using an electropolisher-etcher (ElectroMetTM4, Buehler, Lake Bluff, IL, USA) to reveal the change in microstructure after UNSM process.

The mechanical properties were obtained using a tensile tester (Zwick/Roell Z010, Borchen, Germany) with increasing load from 0 to 10 kN at a loading rate of 5 mm/min. The dimensions of the tensile specimen are shown in Figure 1. The dynamic elastic modulus was measured with respect to temperature in the range of 100 to 400 °C by a resonant frequency. The temperature was controlled using an induction heating system with a medium frequency of 100 Hz. Tensile and dynamic Young's modulus tests were also repeated at least three times. Detailed information on measurement of dynamic Young's modulus is available elsewhere [25].

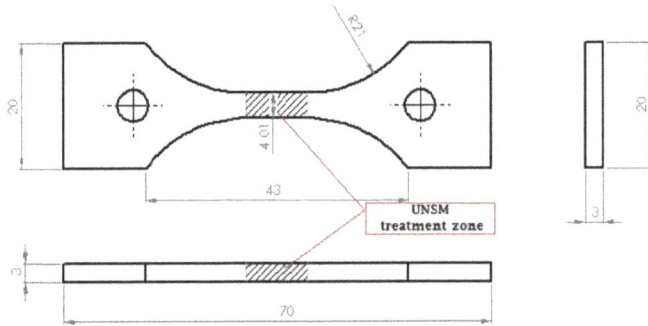

Figure 1. A schematic view of a tensile test specimen.

The fretting wear resistance was assessed by an Optimol SRV IV (Munich, Germany) tester at room temperature under dry conditions against SAE 52100 bearing steel ball with a diameter of 10 mm. The fretting wear test conditions are listed in Table 4.

Table 4. Fretting wear test conditions.

Frequency, Hz	Displacement Amplitude, μm	Normal Load, N	Test Duration, min	Hertzian Contact Stress, GPa
30	50	50	60	1.06

The corrosion resistance was evaluated by a potentiodynamic polarization tester (Solartron SI 1287, Gainesville, GA, USA) with three electrodes as a saturated calomel electrode (SCE), and platinum mesh, and specimen as reference. At a scanrate of 1 mV/s with a scan range of −0.5 V to +1.5 V in 3.5% NaCl solution corrosion resistance of as-received and UNSM-treated specimens was measured at room temperature. The exposed area of the specimen was 105.6 mm^2.

X-ray diffraction (XRD) pattern using a $\sin^2\psi$ method with a scanning speed 1/min and surface residual stress were obtained by an X-ray diffractometer (Bruker D8 Advance, Karlsruhe, Germany). The measurement area was 300 μm in diameter, and the penetration depth was 500 μm. The fretting wear rate was calculated using a fretting wear scar profile obtained by a laser scanning microscopy (VK-X100 Series, Keyence, Osaka, Japan). The effective depth of UNSM process was observed by field-emission (FE)-SEM images. It needs to be mentioned here that all the experimental tests and measurements were repeated three times to give a good statistical representation of the results.

3. Results and Discussion

3.1. Surface Roughness and Hardness

The surface roughness profile and surface hardness of the as-received and UNSM-treated specimens are shown in Figure 2. In fretting conditions, it is essential to have a rough surface; therefore, the average surface roughness (R_a) of the UNSM-treated specimen was intentionally modified to be in the range from 0.25 to 0.32 μm by creating high peaks and valleys, as illustrated in Figure 2a. It has previously been reported that a rougher surface can better avoid the initiation of fretting failure in comparison with the smoother one [26]. Meanwhile, it is important to mention that the UNSM process is also used as a surface smoothing process that achieves a very smooth surface with a surface roughness R_a of several nanometers [27]. In addition to surface roughness, the performance of the UNSM process on the surface hardness, which is an indication of resistance to deformation, was also evaluated as illustrated in Figure 2b. Obviously, the UNSM-treated specimen exhibited a higher surface hardness of 456 HV in comparison with the as-received one of 314 HV. The Hall-Petch expression explains the enhanced surface hardness of the UNSM-treated specimen, in which the surface hardness can be controlled by refining the coarse grain size [28]. It is well established that the UNSM process generates a nanocrystal layer up to a certain depth from the top surface, along with refined nano-grain size down to ~50 nm [29]. Moreover, an increase in surface hardness may be attributed to the work hardening by UNSM process as a result of the generated surface and subsurface severe plastic deformation (SPD). Alloy 718 can be also hardened by the presence of precipitation of γ' ($Ni_3(Al,Ti)$) and γ'' (Ni_3Nb) phases. The refined nano-grain size as a function of depth from the surface of Alloy 718 was reported previously, in which the finest grain of the top surface was about 30 nm [30].

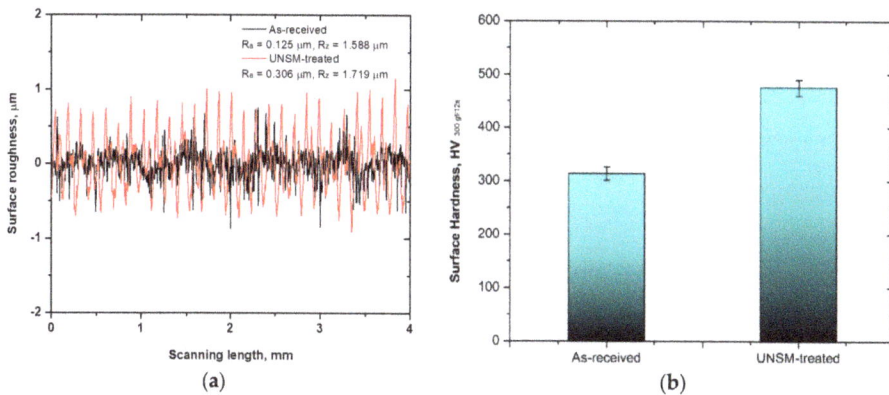

Figure 2. Surface roughness profile (**a**) and hardness (**b**) of the as-received and UNSM-treated specimens.

3.2. Tensile and Dynamic Elastic Modulus

The mechanical properties of the as-received and UNSM-treated specimens were evaluated by a tensile test, whereas the typical stress-strain curves are plotted in Figure 3a. It is clear that there is no significant difference in ultimate tensile strength (UTS) after the UNSM process, but the yield strength (YS) increased by about 14%. Unfortunately, the UNSM process lost the ductility of the as-received specimen by about 20%. The relationship between strength vs. ductility of engineering materials is still an important issue in Materials Science. Lu has suggested that a gradient nanostructured material can improve both the strength and ductility compared to the nanostructured one [31]. Hence, it is suggested to generate a gradient nanocrystal structure using the UNSM process, which may lead to a further increase in both the strength and ductility [32]. Metallic materials are ductile, and they have a relatively lower strength but a relatively higher toughness or energy to failure because of greater ductility/post yield deformation. Therefore, a gradient nanostructured Alloy 718 is needed to control the strength vs. ductility. In addition, dynamic elastic modulus, which is the ratio of stress to strain under vibratory conditions, as a function of temperature of the specimens, is presented in Figure 3b. The results showed that the dynamic elastic moduli of the as-received and UNSM-treated specimens were 187.24 and 199.79 GPa at a temperature of 100 °C, respectively. By increasing the temperature, the dynamic elastic modulus of the UNSM-treated specimen was higher throughout the range of temperatures compared to the as-received one, in which the amount of increase was reduced at higher temperature, as depicted in Figure 3b.

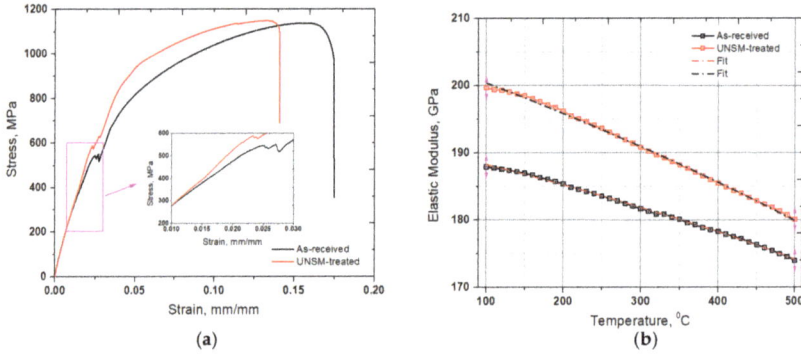

Figure 3. Stress-strain curve (**a**) and elastic modulus (**b**) change of the as-received and UNSM-treated specimens.

3.3. Surface Residual Stress and XRD Pattern

Figure 4a presents the surface residual stress measurement results obtained from the as-received and UNSM-treated specimens. The surface residual stress of the as-received specimen was found to be 0.178 GPa, which changed after the UNSM process to a value of about −1.324 GPa. One of the main influential factors in the ductility and fatigue of Alloy 718 is the compressive surface residual stress owing to detain crack initiation and propagation [33]. An XRD was utilized to analyze the grain size, micro-strain, and phase transformation alteration of the as-received and UNSM-treated specimens. In Figure 4b, it can be seen clearly that the intensity of both diffraction peaks of <200> and <220> reduced significantly, whereas the full width at half maximum (FWHM) widened after UNSM process. The FWHM of the primary <200> peak was found to be 0.49 and 0.61 for the as-received and UNSM-treated specimens, respectively. The grain size refinement played an important role in reducing the intensity and increasing the FWHM value, and also in increasing the dislocation density [34].

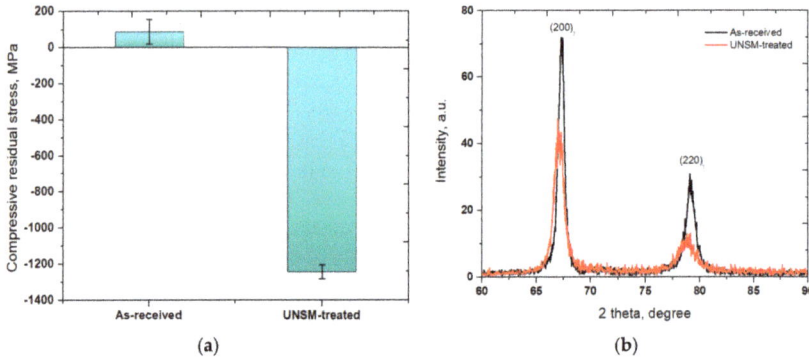

Figure 4. The surface residual stress measurement results (**a**) and XRD patterns (**b**) of the as-received and UNSM-treated specimens.

3.4. Cross-Sectional Observation by SEM

The cross-sectional SEM images of the as-received and UNSM-treated specimens are presented in Figure 5. It is obvious that the plastically deformed top surface with a thickness about 10 μm of the specimen after the UNSM process can be observed in Figure 5b. It can be also seen that the deformation occurred along the UNSM treatment direction, as indicated by yellow arrow. In general, once a material is subjected to SPD process, an increase in grain boundaries and refinement in grain size take place. It has been reported earlier that the UNSM process was able to refine grain size up to ~50 nm [29]. Several mechanisms of strengthening of Alloy 718 are available in the literature, in which grain size refinement is one of the possible mechanisms with which to increase strength [35]. It is announced earlier that the refined grain size using SPD methods less than 10 nm was found to be detrimental [36]. Also, slip or dislocation may occur along the grain boundary during SPD, in which the increased grain boundaries may block dislocation motion due to the orientations of neighboring grains. Hence, a nano-grained material has much higher hardness compared to coarse grained one because of the higher number of grain boundaries in nano-grained materials.

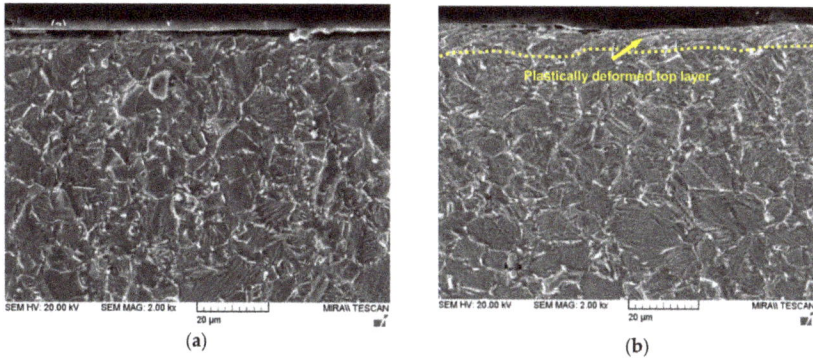

Figure 5. Cross-sectional SEM images of the as-received (**a**) and UNSM-treated (**b**) specimens.

3.5. Friction Coefficient and Fretting Wear Resistance

The friction coefficient as a function of fretting time of the specimens is shown in Figure 6a, in which the friction coefficient increased drastically at the start point of the test and then reduced gradually for the first 10–14 min, and it got stabilized throughout the testing time. In other words, the average friction coefficient of the as-received specimen reduced from 0.88 to 0.72, corresponding to 18% after the UNSM process. A high friction coefficient of the specimens at the start point of the test is attributed to the initial asperity contact that increases contact pressure significantly, since only a small number of asperities came into contact at the interface. Moreover, Figure 6b shows the evaluation of the fretting wear resistance based on the fretting wear scar profiles. It was revealed that the UNSM process reduced the fretting wear resistance significantly. The main reason for the reduction in the friction coefficient of the UNSM-treated specimen is the initial surface roughness, which reduced the true contact area between the surfaces of the specimen and the ball. It is worth mentioning here that high peaks and valley of rough surface may serve as additional space at the contact interface for wear debris formed during fretting oscillation. The improvement in wear resistance of the specimen subjected to UNSM process is due to the hardening and alteration of coarse grained structure into nano-grained one. Recently, the role of the UNSM process on the control of friction and fretting wear resistance of Inconel 690 alloy at 25 and 80 °C was reported in our previous publication [19]. It was reported that the friction behavior and fretting wear resistance of the UNSM-treated specimens were improved at both temperatures compared to the as-received ones. Another application of UNSM process on AlCrN coating with the aim of increasing the fretting resistance was reported earlier [37]. Interestingly, the UNSM process was capable of reducing the friction coefficient, which led to an enhanced fretting wear resistance.

The fretting wear scars formed on the surface of the specimens were characterized by SEM as presented in Figure 7a,b. It was found based on the SEM images that the diameter of fretting wear scar of the as-received and UNSM-treated specimens was about 1.202 and 1.048 mm in diameter, respectively. Besides the diameter of fretting wear scar, it is of interest to investigate the chemistry and wear mechanisms of fretting wear as well. Figure 7c,d presents the chemical mapping of the fretting wear scar. Basically, as presented in Figure 7e,f, the fretting wear scar was covered with Fe, which was transferred from the counterface. The EDS spectroscopy revealed that Ni composition in the fretting wear scar on the UNSM-treated specimen was found to be smaller than the as-received one as shown in Figure 7g,h due to the high amount of Fe transferred from the counter ball. Hence, the application of UNSM process was beneficial to controlling the friction behavior and the fretting wear resistance.

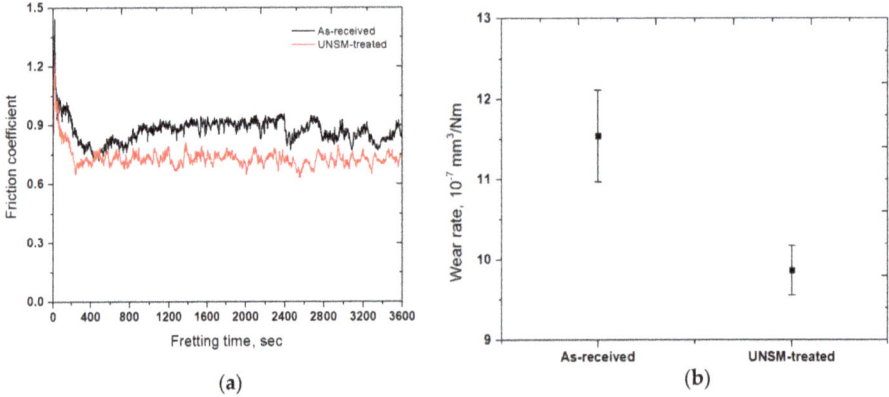

Figure 6. Friction coefficient (**a**) and fretting wear resistance (**b**) of the as-received and UNSM-treated specimens.

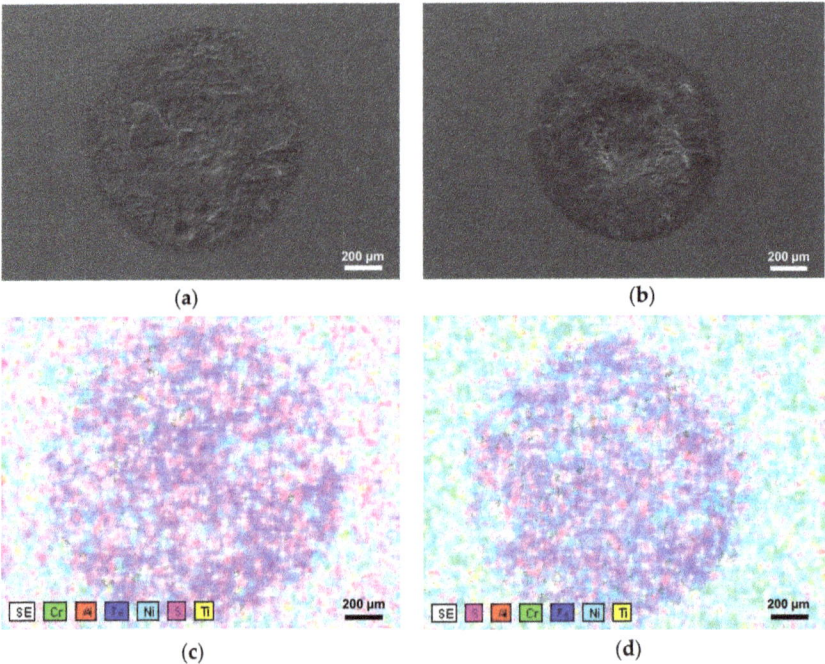

Figure 7. *Cont.*

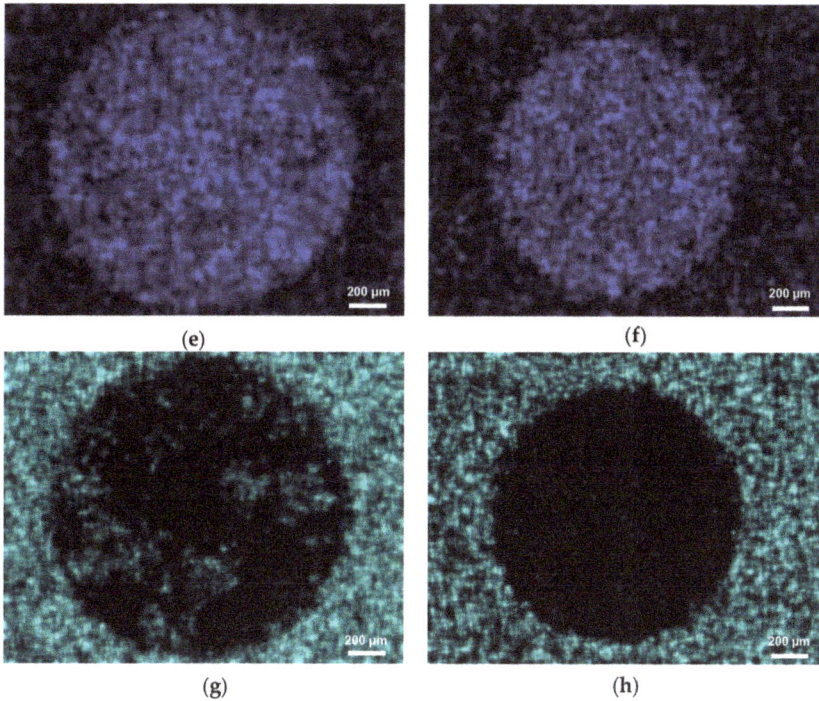

Figure 7. SEM images of the fretting wear scar formed on the surface of the as-received (**a**) and UNSM-treated (**b**) specimens. The composition mapping image of the fretting wear scar on the surface of the as-received (**c**) and UNSM-treated (**d**) specimens. Distribution of Fe (**e,f**), and Ni (**g,h**) over the fretting wear scar formed on the surface of the as-received (**a**) and UNSM-treated (**b**) specimens, respectively.

3.6. Corrosion Resistance

Electrochemical tests were performed to explore the effect of the UNSM process on the corrosion resistance of Alloy 718. The comparison of Tafel curves for the as-received and UNSM-treated specimens in 3.5% NaCl solution is presented in Figure 8. The corrosion potential and breakdown of Alloy 718 are −0.45 and 0.90 V_{SCE}, respectively. Clearly, it can be seen that the curves are divided into three zones as indicated by arrows. After UNSM process, the corrosion potential increased (E_{corr}), and the corrosion current density (i_{corr}) decreased in comparison with the as-received specimen, as shown in Figure 8. It is evident that the corrosion potential (E_{corr}) after the UNSM process shifted to a noble direction, while the corrosion current density (i_{corr}) shifted to the left side, corresponding to a higher resistance to corrosion compared than the as-received one. More negative corrosion potential and high corrosion current are indicative of lower resistance to corrosion. In other words, the smaller values of the corrosion current density indicate higher corrosion resistance, and the more positive the values of the corrosion potential are, the higher is the corrosion resistance [38]. Identified transpassive zone of the as-received specimen, as shown in Figure 8, could be probably due to anodic reaction.

The values of electrochemical test results are listed in Table 5. It was found that the corrosion potential and current density of the UNSM-treated specimen improved due to the inception of a dissolution on the anode, and galvanic coupling of cathodic and anodic area was sped up on the UNSM-treated specimen. The magnitude of corrosion potential is not the only property evaluating the corrosion resistance of Alloy 718. In general, the specimens with negative potential tend to have a lower resistance to corrosion. However, accelerated passivation inhibited the cathodic reaction, in which the motion of potential to negative corrosion takes place. No significant difference in χ^2 (chi-square) deviation was found, as shown in Table 5, in which the low values of the associated χ^2 indicate that the circuit was able to fit the experimental data accurately. In other words, the small values of chi-square indicates a better fit. The obtained results imply that the corrosion resistance of the UNSM-treated specimen increased compared to the as-received one. Lee et al. have studied the influence of shot peening and UNSM processes on the corrosion resistance of AISI304 stainless steel [39]. It was revealed that the UNSM-treated specimen with a smoother surface showed better corrosion resistance in comparison with the as-received and shot peened specimens. However, Hou et al. systematically investigated the influence of UNSM process on the corrosion behavior of AZ31B Mg alloy [40]. No improvement in corrosion behavior after UNSM process was found due to the increase in dislocation density induced by UNSM process. Reduction in surface roughness and change in microstructure after UNSM process can play an important role in increasing and decreasing the corrosion resistance. Hence, it is worth mentioning here that a rough surface roughness was prepared intentionally by UNSM process to improve the fretting wear resistance so that a rough surface could increase the corrosion resistance. Furthermore, Miyamoto overviewed the corrosion behavior of nano-grained materials using SDP methods and reported that there are some contradictory results [41]. Most of the reviewed papers reported that the refined grain size may increase the corrosion resistance following Ralston's rule [42].

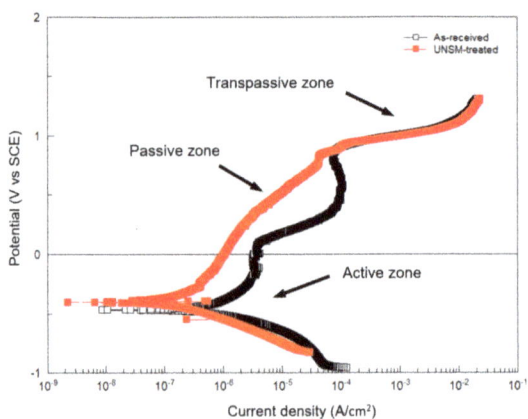

Figure 8. Potentiodynamic polarization curves of the as-received and UNSM-treated specimens.

Table 5. Electrochemical test results derived from polarization curves in 3.5% NaCl solution.

Specimens	E_{corr}, V	I_{corr}, A/cm^2	χ^2
As-received	−0.48244	2.2734×10^{-6}	8.439
UNSM-treated	−0.40361	1.1227×10^{-6}	9.864

4. Conclusions

The influence of UNSM process on the strength; hardness; tensile strength; dynamic elastic modulus; and surface residual stress, fretting, and corrosion resistances was investigated. It was found that the average surface roughness and the surface hardness were increased by about 58% and 27%, respectively. XRD diffraction pattern intensity and FWHM were reduced and broadened after the UNSM process, resulting in grain size refinement and generating a high surface compressive residual stress. Both the friction coefficient and the fretting wear were improved after the UNSM process. The corrosion resistance was also increased slightly after UNSM process. It was learned that an optimization of the UNSM process parameters is in need to further increase the resistance to corrosion. Accordingly, it can be summarized that the UNSM process played a vital role in increasing the strength, corrosion, and fretting resistances of Alloy 718 that are beneficial to extending the service lifespan of aircraft and nuclear components.

Author Contributions: A.A. and T.A. conceived and designed the experiments; A.A. and R.U. performed the experiments; A.A. analyzed the data; A.A. wrote the paper.

Funding: This study was supported by the Start-Up Research Project through the Ministry of Science, ICT, and Future Planning of Korea (NRF-2017R1C1B5017434). This study was also partially supported by the Korea Technology and Information Promotion Agency (TIPA) for Small and Medium Enterprises. Project (No. S2544322).

Conflicts of Interest: The authors declare no conflict of interest.

References

1. Zhu, L.; Xu, Z.; Gu, Y. Effect of laser power on the microstructure and mechanical properties of heat treated Inconel 718 superalloy by laser solid forming. *J. Alloys Compd.* **2018**, *746*, 159–167. [CrossRef]
2. Ott, E.A.; Groh, J.R.; Banik, A.; Dempster, I.; Gabb, T.P.; Helmink, R.; Liu, X.; Mitchell, A.; Sjoberg, G.P.; Wusatowska-Sarnek, A. *Superalloy 718 and Derivatives*; John Wiley & Sons, Inc.: New York, NY, USA, 2012.
3. Kumar, S.A.; Raman, S.G.; Narayanan, T.S.; Gnanamoorthy, R. Fretting wear behavior of surface mechanical attrition treated alloy 718. *Surf. Coat. Technol.* **2012**, *206*, 4425–4432. [CrossRef]
4. Lee, B.W.; Suh, J.J.H.; Lee, C.; Kim, T.G. Investigations on fretting fatigue in aircraft engine compressor blade. *Eng. Fail. Anal.* **2011**, *18*, 1900–1908. [CrossRef]
5. Jouiad, M.; Marin, E.; Devarapalli, R.S.; Cornier, J.; Ravaux, F.; Le Gall, C.; Franchet, J.M. Microstructure and mechanical properties evolutions of alloy 718 during isothermal and thermal cycling over-aging. *Mater. Des.* **2016**, *102*, 284–296. [CrossRef]
6. DuPont, J.N.; Lippold, J.C.; Kiser, S.D. *Welding Metallurgy and Weldability of Nickel-Base Alloys*; John Wiley & Sons, Inc.: New York, NY, USA, 2009.
7. Lin, Y.C.; Deng, J.; Jiang, Y.Q.; Wen, D.X.; Liu, G. Effects of initial δ phase on hot tensile deformation behaviors and fracture characteristics of a typical Ni-based superalloy. *Mater. Sci. Eng. A* **2014**, *598*, 251–262. [CrossRef]
8. Farber, B.; Small, K.A.; Allen, C.; Causton, R.J.; Nichols, A.; Simbolick, J.; Taheri, M.L. Correlation of mechanical properties to microstructure in Inconel 718 fabricated by direct metal laser sintering. *Mater. Sci. Eng. A* **2018**, *712*, 539–547. [CrossRef]
9. Damodaram, R.; Raman, S.G.; Prasad Rao, K. Microstructure and mechanical properties of friction welded alloy 718. *Mater. Sci. Eng. A* **2013**, *560*, 781–786. [CrossRef]
10. Mannan, S.; Pate, S.; deBarbadillo, J. Long term thermal stability of Inconel alloys 718, 706, 909, and Waspaloy at 593 °C and 704 °C. *Superalloys* **2000**, *10*, 449–458.
11. Dehmas, M.; Lacaze, J.; Niang, A.; Viguier, B. TEM study of high-temperature precipitation of delta phase in Inconel 718 alloy. *Adv. Mater. Sci. Eng.* **2011**, *2011*, 940634. [CrossRef]
12. Chlebus, E.; Gruber, K.; Kuznicka, B.; Kurzynowski, T. Effect of heat treatment on the microstructure and mechanical properties of Inconel 718 processed by selective laser melting. *Mater. Sci. Eng. A* **2015**, *639*, 647–655. [CrossRef]
13. Popovich, V.A.; Borisov, E.V.; Popovich, A.A.; Sufiiarov, V.S.; Masaylo, D.V.; Alzina, L. Impact of heat treatment on mechanical behavior of Inconel 718 processed with tailored microstructure by selective laser melting. *Mater. Des.* **2017**, *131*, 12–22. [CrossRef]
14. Raghavan, S.; Zhang, B.; Wang, P.; Sun, C.H.; Sharon Nai, M.L.; Li, T.; Wei, J. Effect of different heat treatments on the microstructure and mechanical properties in selective laser melted INCONEL 718 alloy. *Mater. Manuf. Process.* **2017**, *32*, 1588–1595. [CrossRef]
15. Wang, Z.; Guan, K.; Gao, M.; Li, X.; Chen, X.; Zeng, X. The microstructure and mechanical properties of deposited-IN718 by selective laser melting. *J. Alloys Compd.* **2012**, *513*, 518–523. [CrossRef]
16. Trosch, T.; Strosner, J.; Volkl, R.; Glatzel, U. Microstructure and mechanical properties of selective laser melted Inconel 718 compared to forging and casting. *Mater. Lett.* **2016**, *164*, 428–431. [CrossRef]

17. Tucho, W.M.; Cuvillier, P.; Sjolyst-Kverneland, A.; Hansen, V. Microstructure and hardness studies of Inconel 718 manufactured by selective laser melting before and after solution heat treatment. *Mater. Sci. Eng. A* **2017**, *689*, 220–232. [CrossRef]

18. Zheng, W.J.; Wei, X.P.; Song, Z.G.; Yong, Q.I.; Feng, H.; Xie, Q.C. Effects of carbon content on mechanical properties of Inconel 718 alloy. *J. Iron Steel Res.* **2015**, *22*, 78–83. [CrossRef]

19. Liu, F.; Cheng, H.; Yu, X.; Yang, G.; Huang, C.; Lin, X.; Chen, J. Control of microstructure and mechanical properties of laser solid formed Inconel 718 superalloy by electromagnetic stirring. *Opt. Laser Technol.* **2018**, *99*, 342–350. [CrossRef]

20. Chamanfar, A.; Monajati, H.; Rosenbaum, A.; Jahazi, M.; Bonakdar, A.; Morin, E. Microstructure and mechanical properties of surface and subsurface layers in broached and shot-peened Inconel-718 gas turbine disc fir-trees. *Mater. Charact.* **2017**, *132*, 53–68. [CrossRef]

21. Gill, A.S.; Telang, A.; Vasudevan, V.K. Characteristics of surface layers formed on Inconel 718 by laser shock peening with and without a protective coating. *J. Mater. Proc. Technol.* **2015**, *225*, 463–472. [CrossRef]

22. *Standard Test Methods for Determining Average Grain Size*; E112-13; ASTM International: West Conshohocken, PA, USA, 2014.

23. Amanov, A.; Umarov, R. The effects of ultrasonic nanocrystal surface modification temperature on the mechanical properties and fretting wear resistance of Inconel 690 alloy. *Appl. Surf. Sci.* **2018**, *441*, 515–529. [CrossRef]

24. Amanov, A.; Pyun, Y.S. Local heat treatment with and without ultrasonic nanocrystal surface modification of Ti-6AL-4V alloy: Mechanical and tribological properties. *Surf. Coat. Technol.* **2018**, *326*, 343–354. [CrossRef]

25. Mbrosia Co., Ltd. Available online: http://www.mbrosia.co.kr/ (accessed on 6 August 2018.

26. Anand Kumar, S.; Sundar, S.; Raman, S.G.S.; Kaul, R.; Ranganathan, K.; Bindra, K.S. Effects of laser peening on fretting wear behavior of alloy 718 fretted against two different counterbody materials. *Proc. Inst. Mech. Eng. Part J. Eng. Tribol.* **2017**, *231*, 1276–1288. [CrossRef]

27. Amanov, A.; Cho, I.S.; Pyun, Y.S. Microstructural evolution and surface properties of nanostructured Cu-based alloy by ultrasonic nanocrystalline surface modification technique. *Appl. Surf. Sci.* **2016**, *388*, 185–195. [CrossRef]

28. Cordero, Z.; Schuh, C.A.; Knight, B.E. Six decades of the Hall-Petch effect—A survey of grain size strengthening studies on pure metals. *Int. Mater. Rev.* **2016**, *61*, 495–512. [CrossRef]

29. Gill, A.; Telang, A.; Mannava, S.R.; Qian, D.; Pyun, Y.S.; Vasudevan, V.K. Comparison of mechanisms of advanced mechanical surface treatments in nickel-based superalloy. *Mater. Sci. Eng. A* **2016**, *576*, 346–353. [CrossRef]

30. Amanov, A.; Pyun, Y.S.; Kim, J.H.; Suh, C.M.; Cho, I.S.; Kim, H.D.; Wang, Q.; Khan, K. Ultrasonic fatigue performance of high temperature structural material Inconel 718 alloys at high temperature after UNSM treatment. *Fatigue Fract. Eng. Mater. Struct.* **2015**, *38*, 1266–1273. [CrossRef]

31. Lu, K. Making strong nanomaterials ductile with gradients. *Science* **2014**, *345*, 1455–1456. [CrossRef] [PubMed]

32. Chae, J.M.; Lee, K.O.; Amanov, A. Gradient nanostructured tantalum by thermal-mechanical ultrasonic impact energy. *Materials* **2018**, *11*, 452. [CrossRef] [PubMed]

33. Kattoura, M.; Telang, A.; Mannava, S.R.; Qian, D.; Vasudevan, V.K. Effect of ultrasonic nanocrystalline surface modification on residual stress, microstructure and fatigue behavior of ATI 718Plus alloy. *Mater. Sci. Eng. A* **2018**, *711*, 364–377. [CrossRef]

34. Oikawa, K.; Su, Y.H.; Tomota, Y.; Kawasaki, T.; Shinohara, T.; Kai, T.; Hiroi, K.; Zhang, S.Y.; Parker, J.D.; Sato, H.; et al. A comparative study of the crystallite size and the dislocation density of bent steel plates using Bragg-edge transmission imaging, TOF neutron diffraction and EBSD. *Phys. Procedia* **2017**, *88*, 34–41. [CrossRef]

35. Callister, W.D., Jr.; Rethwisch, D.G. *Materials Science and Engineering: An Introduction*, 9th ed.; Wiley & Sons, Inc.: Hoboken, NJ, USA, 2014.

36. Suryanarayana, C. Mechanical behavior of emerging materials. *Mater. Today* **2012**, *15*, 486–498. [CrossRef]

37. Cho, I.S.; Amanov, A.; Kim, J.D. The effects of AlCrN coating, surface modification and their combination on the tribological properties of high speed steel under dry conditions. *Tribol. Int.* **2015**, *81*, 61–72. [CrossRef]

38. Frankel, G.S. Fundamentals of Corrosion Kinetics. In *Active Protective Coatings New Generation Coatings for Metals*; Springer: Dordrecht, The Netherlands, 2016; pp. 17–32.

39. Lee, H.S.; Kim, D.S.; Jung, J.S.; Pyoun, Y.S.; Shin, K.S. Influence of peening on the corrosion properties of AISI304 stainless steel. *Corros. Sci.* **2009**, *51*, 2826–2830. [CrossRef]

40. Hou, X.; Qin, H.; Gao, H.; Mankoci, S.; Zhang, R.; Zhou, X.; Ren, Z.; Doll, G.L.; Martini, A.; Sahai, N.; et al. A systematic study of mechanical properties, corrosion behavior and biocompatibility of AZ31B Mg alloy after ultrasonic nanocrystal surface modification. *Mater. Sci. Eng. C* **2017**, *78*, 1061–1071. [CrossRef] [PubMed]

41. Moyomoto, H. Corrosion of ultrafine grained materials by severe plastic deformation, an overview. *Mater. Trans.* **2016**, *57*, 559–572. [CrossRef]

42. Ralston, K.D.; Birbilis, N. Effect of grain size on corrosion. *Corrosion* **2010**, *66*, 075005. [CrossRef]

materials

MDPI

Article

Surface Functionalization of Polyethylene Granules by Treatment with Low-Pressure Air Plasma

Hana Šourková [1,2], Gregor Primc [1,3] and Petr Špatenka [1,*]

[1] Faculty of Mechanical Engineering, Department of Materials Engineering, Center of Advanced Aerospace Technology, Czech Technical University in Prague, Karlovo náměstí 13, 121 35 Praha 2, Czech Republic; hana.sourkova@tul.cz (H.Š.); gregor.primc@ijs.si (G.P.)

[2] Faculty of Mechatronics, Informatics and Interdisciplinary Studies, Technical University of Liberec, Studentská 1402/2, 461 17 Liberec 1, Czech Republic

[3] Jozef Stefan Institute, Jamova cesta 39, 1000 Ljubljana, Slovenia

* Correspondence: petr.spatenka@fs.cvut.cz; Tel.: +420-607-516-901

Received: 20 April 2018; Accepted: 21 May 2018; Published: 25 May 2018

Abstract: Polyethylene granules of diameter 2 mm were treated with a low-pressure weakly ionized air plasma created in a metallic chamber by a pulsed microwave discharge of pulse duration 180 μs and duty cycle 70%. Optical emission spectroscopy showed rich bands of neutral nitrogen molecules and weak O-atom transitions, but the emission from N atoms was below the detection limit. The density of O atoms in the plasma above the samples was measured with a cobalt catalytic probe and exhibited a broad peak at the pressure of 80 Pa, where it was about 2.3×10^{21} m^{-3}. The samples were characterized by X-ray photoelectron spectroscopy. Survey spectra showed oxygen on the surface, while the nitrogen concentration remained below the detection limit for all conditions. The high-resolution C1s peaks revealed formation of various functional groups rather independently from treatment parameters. The results were explained by extensive dissociation of oxygen molecules in the gaseous plasma and negligible flux of N atoms on the polymer surface.

Keywords: polyethylene granules; low-pressure MW air plasma; optical emission spectroscopy; XPS; laser cobalt catalytic probe

1. Introduction

Surface properties of polymers are often inadequate, so they have to be modified prior to further treatment. A widely used technique for tailoring the surface properties of polymers is a brief exposure to non-equilibrium gaseous plasma. A recent survey indicates a variety of treatment conditions, and different authors reported results that are not always in agreement [1]. A possible reason for such discrepancies is the application of polymers in various forms and various grades. Recently, it was explained in detail that fibrous polymers might interact differently with gaseous plasma compared to interaction with flat materials, such as foils, due to the enormous surface-to-mass ratio [2]. Although numerous studies have been published in past decades, the exact mechanisms are still not well understood, so the interaction of plasma particles with polymer surfaces remains a hot topic [3]. A commonly used polymer is polyethylene, which comes in different forms and different grades. Recently, Orendáč et al. studied the modification of high-density polyethylene (HDPE) foils by argon plasma to create allyl and polyenyl radicals or dangling bonds [4]. Vartiainen et al. [5] applied gaseous plasma for the activation of low-density polyethylene, prior to coating it with a thin film of cellulose nanofibrils to make an effective barrier against oxygen and mineral oil residues. Zhao et al. [6] treated polypropylene fibrous membranes with weakly ionized plasma to activate the surface for better adhesion of a coating prepared by grafting with vinylimidazole acidic ionic liquids. Hu et al. treated a polyethylene surface with CO_2 plasma for carbene insertion [7]. Ozaltin et al. [8] used plasma pretreatment of low-density polyethylene (LDPE) before grafting of polymer brush of *N*-allylmethylamine, which acted as an antibacterial coating. Muzammil et al. [9] used a pulsed plasma technique for the deposition of hydrophobic coatings on low-density polyethylene. Van Vrekhem et al. [10] used an atmospheric pressure plasma jet for the treatment of ultrahigh-density polyethylene in order to improve the biological response of shoulder implants, and found improved adhesion to bone cement and an enhanced osteoblast proliferation. Lindner et al. [11] used a high-impedance corona discharge in air for the treatment of various polymers and laminates, and the best results in terms of laminate bond strength were obtained for low-density polyethylene. Popelka et al. [12] used corona discharge to enhance the surface hydrophilicity of composites prepared from linear low-density polyethylene and graphene nanoplatelets in order to develop materials suitable for electromagnetic interference shielding applications. The same group also applied the same discharge to obtain a significant increase in peel resistance in the linear low-density polyethylene/aluminum laminate [13]. This literature survey highlights important achievements reported in papers published only in 2018. Much work on polyethylene in different forms has been done in previous years using low-pressure microwave plasma: from the functionalization of low-density polypropylene with nitrogen plasma [14] and studies on the wetting properties of LDPE and polyethylene terephthalate (PET) [15] to the enhanced printability of polyethylene (PE) treated by air plasma [16]. Besides PE granules and foils, work has been also done on powders [17–19]. The complete survey is beyond the scope of this paper, but a general conclusion is that sometimes even a brief treatment of polyethylene causes improved surface wettability, which is a consequence of polar functional groups that appear on the surface upon plasma treatment. Usually, the surface is rich in oxygen-containing functional groups such as hydroxyl, carbonyl, and carboxyl/ester, even though the plasma was not generated in pure oxygen, but rather air, carbon dioxide, or even argon. The richness of oxygen-containing groups is usually explained by an excellent affinity of oxygen towards plasma-modified polymer surface. The authors used polymers in different forms, including foils and fibrous materials, but little work has been performed on the treatment of granules, which are the raw materials for synthesizing a variety of plastics. This paper reports on the surface modification of polyethylene granules treated with air plasma sustained in a chamber suitable for semi-industrial applications.

2. Materials and Methods

Samples of commercial polyethylene granules were treated in a plasma reactor (SurfaceTreat LA400, SurfaceTreat, a.s., Turnov, Czech Republic) suitable for the treatment of polymers on a semi-industrial scale. The PE granules were purchased from Sigma-Aldrich. The system's schematic is shown in Figure 1. The discharge chamber of dimensions approx. 40 cm × 40 cm × 40 cm was made from aluminum. The native oxide layer forms on the aluminum surface, leading to rather inert properties against different plasmas of molecular gases, since the coefficient for heterogeneous surface recombination is low for various atoms [20]. The chamber had the volume of about 60 L and was pumped with a Roots pump backed by a two-stage rotary pump through a high efficiency particulate air (HEPA) filter, a butterfly valve, and bellows. The discharge chamber was equipped with a system for the gas inlet: an adjustable flow controller that allowed flows of up to 2 standard liters per minute. Two Microwave (MW) source was placed on the top flange of the plasma system. The source was forced-air cooled. The rest of the experimental system was kept at room temperature. Gaseous plasma of high luminosity was concentrated to the region a few centimeters thick next to the upper flange where the microwave source was mounted. The radiation was less intensive and rather homogeneous in the rest of the discharge chamber. Pressure inside the chamber was measured with a Pirani gauge calibrated for dry air. The pressure was adjusted by changing the gas flow through the flow controller, which was also calibrated for dry air, and/or by changing the tilt of the butterfly valve.

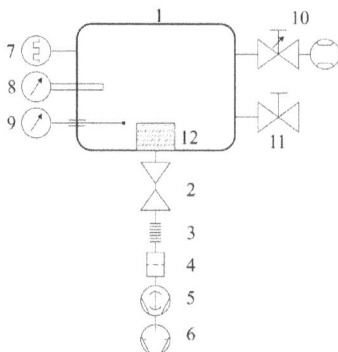

Figure 1. The schematic diagram of the vacuum of the small plasma reactor. 1—discharge chamber, 2—narrow tube with butterfly valve, 3—two-meter long bellows with diameter of 40 mm, 4—trap for fine powder, 5—Roots pump, 6—two-stage oil rotary pump, 7—Pirani vacuum gauge, 8—optical fiber with lens and spectrometer, 9—catalytic probe, 10—gas feeding flow measurement system, 11—air inlet valve, and 12—sample holder.

Plasma was characterized by optical emission spectroscopy (OES) and with a catalytic probe. An optical fibre was fixed to a window on a flange and connected to an optical spectrometer. We used an Avantes AvaSpec 3648 spectrometer (Avantes, Apeldoorn, The Netherlands). The device is based on the AvaBench 75 symmetrical Czerny Turner design, with a 3648-pixel charge-coupled device (CCD) detector and focal length of 75 mm. The range of measurable wavelengths is from 200 nm to 1100 nm, and the wavelength resolution is 0.5 nm. The spectrometer has a USB 2.0 interface, enabling high sampling rates of up to 270 spectra per second. It has a signal-to-noise ratio of 350:1. Integration time is adjustable from 10 μs to 10 min. At integration times of below 3.7 ms, the spectrometer itself performs internal averaging of spectra before transmitting them through the USB interface.

A catalytic probe for measuring the O-atom density was mounted at a flange and stretched 10 cm inside the discharge chamber in order to reveal the atom density away from the walls, where heterogeneous surface recombination might have influenced the probe signal. We used a professional laser-heated fiber-optics catalytic probe supplied by Plasmadis (Plasmadis Ltd., Ljubljana, Slovenia). The probe employed a catalytic tip made from pure cobalt with an oxide layer on the surface. The probe was sensitive to O atoms, but not to any N atoms that might be created in the gaseous plasma. For completeness of this paper, we shortly describe the main features as follows. The probe is made from a small catalytic tip, which is mounted onto an optical fibre. The other side of the fibre is connected to an electronic unit, which consists of an infrared (IR) radiation detector (for measuring tip temperature) and a laser (for heating the tip to elevated temperature). The electronic unit controls the laser power so that the tip temperature is held constant, irrespective from experimental conditions; in particular, the exothermic reactions that are likely to occur upon immersion of the probe tip into gaseous plasma. The flux of reactive plasma species onto the probe tip is then calculated from the difference in the laser power between plasma on and off conditions. After evacuation of the plasma reactor and before turning on the discharge, the catalytic tip was heated to a constant temperature (750 K, in this case) by the laser. When the discharge was turned on, the laser power dropped due to the extensive heterogeneous surface recombination of O atoms to parent molecules, and the electronic unit displayed the corresponding O-atom density in the vicinity of the tip. The details of the probe construction are described elsewhere [21]. The probe allows for rather precise measurements of the O-atom density, with the time resolution of less than a second and the start-up time of a few seconds. The absolute accuracy was about 20% and the relative accuracy about 3%, provided that the gas pressure does not change during the measurements.

Samples were placed onto a sample holder mounted at the bottom of the experimental system, as shown in Figure 1. The sample holder was in a form of a cup of diameter 20 cm and height 7 cm. A device for continuous mixing of granules upon plasma treatment was mounted onto the bottom of the cup. The granules filled the cup up to the level of about 1 cm from the bottom of the cup. A rather large diameter of granules (2 mm) allowed for efficient mixing, which in turn, allowed for the rather uniform treatment of the entire surface area of all the granules within a reasonable time. The treatment time was 1 min. The samples were commercially available polyethylene granules of diameter 2 mm. After evacuation of the discharge tube to the ultimate pressure, air was introduced into the discharge chamber through the flow controller upon continuous pumping so that a desired pressure was established. Different flow rates were adopted for this experiment. After the treatment, the plasma effects were studied by X-ray photoelectron spectroscopy (XPS).

We used a high-resolution instrument with monochromatized X-rays and a hemispherical electron analyzer (model TFA XPS Physical Electronics, Chanhassen, MN, USA), which employs monochromatic Al K$\alpha_{1,2}$ radiation at 1486.6 eV for excitation of the surface over an area of about 400 μm^2. Take-off angle was set at 45°. XPS survey-scan spectra were acquired at a pass-energy of 187 eV using an energy step of 0.4 eV. The high-resolution C1s spectra were acquired at a pass-energy of 23.5 eV using an energy step of 0.1 eV. The main C1s peak was fixed to a value of 284.8 eV for charge compensation. The XPS spectra were analysed using MultiPak v8.1c software (Ulvac-Phi Inc., Kanagawa, Japan, 2006) from Physical Electronics.

The morphology of granules was determined by an atomic force microscope. We used the instrument made by Solver PRO, NT-MDT, Moscow, Russia, which was operating in a tapping mode. The samples were scanned with a standard Si cantilever with a force constant of 22 N/m and at a resonance frequency of 325 kHz. The imaging was done on a 5 $\mu m \times$ 5 μm area of the samples.

3. Results and Discussion

Figure 2 represents a typical spectrum of radiation arising from air plasma. The spectrum is rich in molecular bands. In the range of short wavelengths (roughly from 300 to 400 nm), there is a band corresponding to transitions between the $C^3\Pi_u$ to $B^3\Pi_g$ electronic states of a neutral nitrogen molecule. This radiation is not suppressed by any quantum rule, so it appears in a very short time after the state has been excited due to an electron impact. The potential energy of the upper state $C^3\Pi_u$ (ground vibrational level) is about 11 eV. Theoretically, the $C^3\Pi_u$ state can be created from the ground state of the nitrogen molecule ($x^1\Sigma^+_g$). However, due to the large excitation energy of 11 eV, it is more likely that this state is created at an inelastic collision of an electron with a nitrogen molecule in a metastable state, most probably $A^3\Sigma^+_u$, which has a radiative lifetime of 0.9 s [22].

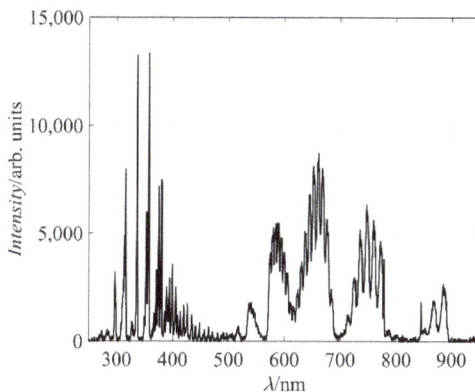

Figure 2. Optical emission spectrum of air plasma at the pressure of 80 Pa.

The spectral range from about 500 to 900 nm is rich in another set of bands, arising from the relaxation of the $B^3\Pi_g$ to $A^3\Sigma^+_u$ states. Numerous spectral features indicate rich population of vibrational states of both upper and lower electronic states. In fact, the most extensive "line" (broadened by rotational population) corresponds to the transition from the sixth vibrational level of the $B^3\Pi_g$ electronic state to the third vibrational level of the $A^3\Sigma^+_u$ state. The excitation energy of the ground vibrational level of the $B^3\Pi_g$ state is just over 7 eV: about one-third of the excitation energy of the $C^3\Pi_u$ state. Also, the excitation cross section is large [23], so the intensity of the radiation arising from the $B^3\Pi_g$ to $A^3\Sigma^+_u$ transitions is predominant in the spectrum shown in Figure 2.

Radiative transitions of atomic lines are barely visible, except from the oxygen line at 845 nm. The major oxygen line should be at 777 nm, but it overlaps with the nitrogen band in the spectrum, which was acquired by a rather low-esolution spectrometer. The excitation level of the O atoms radiating at 845 nm is about 11 eV: almost the same as the $C^3\Pi_u$ level of the nitrogen molecules. The reason for poor radiation from O atoms as compared to the $C^3\Pi_u$ to $B^3\Pi_g$ system is a smaller excitation cross section [24], rather than any difference in excitation energies, so the spectrum shown in Figure 2 only qualitatively shows the presence of neutral oxygen atoms in the plasma reactor. No radiation from excited N atoms is observed in Figure 2, although some N lines are positioned in the range of wavelengths not occupied by N_2 transitions.

The optical spectra were measured at different experimental conditions. The behavior of selected spectral features versus pressure at fully open butterfly valve (no pressure difference before or after the valve) is presented in Figure 3. The scale on the y-axis is logarithmic. The intensity of transitions within neutral nitrogen molecules (from the $C^3\Pi_u$ state at 337 nm and from the $B^3\Pi_g$ state at 679 nm) slowly decreases with decreasing pressure in the low-pressure range, while the intensity of positively charged molecular ions at 391 nm increases. This can be explained by a lower density of molecules and thus less frequent electron collisions; therefore, the electron temperature is somehow larger at lower pressures. This effect, however, is not pronounced, indicating that the plasma parameters do not change much with changing pressure. The atomic hydrogen line (H_β of the Balmer series) also increases with decreasing pressure for the same reasons. The emission from the neutral O atoms at 845 nm remains almost intact in the entire range of pressures.

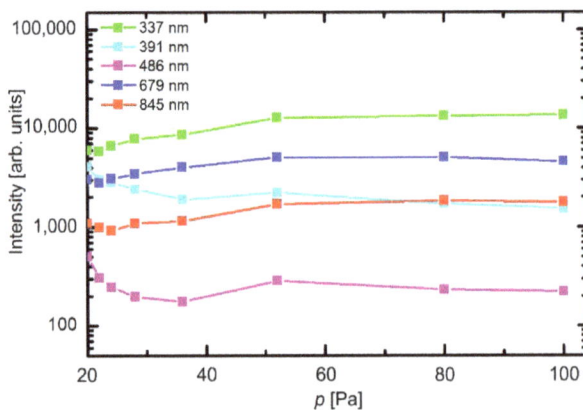

Figure 3. Intensity of selected spectral features versus pressure (p) in the discharge chamber.

Figure 4 represents the variation of the spectral intensities versus gas flow at a constant pressure of 80 Pa. These set of experiments was performed by simultaneous change of the flow rate and opening of the butterfly valve. Figure 4 clearly shows that the flow rate does not influence the plasma parameters as long as the pressure remains constant. The resident time of gas inside the discharge chamber is obviously long enough that steady parameters are obtained irrespective of the gas drift through the discharge chamber. The optical spectra therefore indicate rather small variations of plasma parameters with pressure and almost constant parameters versus the flow rate, which make the system suitable for highly repeatable treatments of any samples mounted into the discharge chamber.

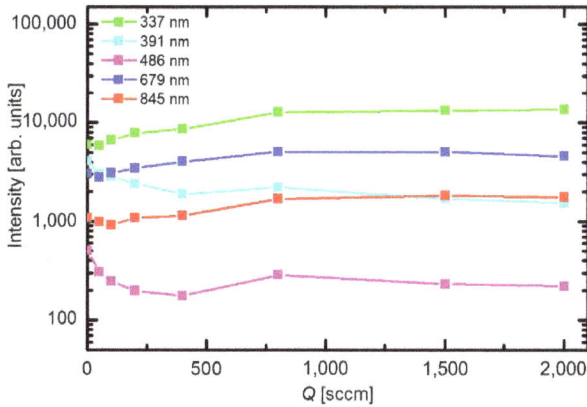

Figure 4. Intensity of selected spectral features versus air flow rate at the pressure of 80 Pa.

The absolute value of the O-atom density was measured with a catalytic probe. A typical probe signal versus time at constant discharge parameters is shown in Figure 5. As mentioned earlier, the start-up time of the commercial probe is several seconds. The probe shows rather constant O-atom density about half a minute after turning on the discharge, but a maximum is observed about 15 s after plasma ignition. The maximum is often explained by the presence of water vapor in the system; specifically, the fresh polymer granules contain some water, which is released from the surface upon plasma treatment. The water molecules are dissociated upon collisions with energetic gaseous species and contribute to the probe signal. The amount of adsorbed water is limited, and the vapor is continuously pumped away from the discharge chamber, so after about 30 s of plasma treatment, this contribution to the probe signal is marginal as compared to O atoms arising from the dissociation of oxygen molecules. A slow decrease in the O-atom density is observed for prolonged treatments and this could be due to thermal effects. It was already explained that the probability for heterogeneous surface recombination of atoms on surfaces of material facing plasma increases with increasing surface temperature [20]. The effect becomes important in the case of fibrous materials of poor thermal conductivity [25] and rather detrimental if more aggressive plasma is applied [26]. However, in our case, it remains marginal even for prolonged plasma treatment since in our case, the highly luminous (and thus reactive) plasma is concentrated to the region next to the MW sources, which are about 30 cm away from the sample holder.

The catalytic probe allowed for rather precise measurements of the O-atom density in our system. The reactor was loaded with PE granules and evacuated to ultimate pressure, and then air was introduced and the atom density was measured. This procedure was repeated six times in order to test and confirm repeatability. Figure 6 shows the density versus pressure at different duty cycles. As expected, the atom density is the highest at large duty cycles and decreases monotonously with decreasing cycles. At 10%, it is about four times smaller than at 100% or 80%. The best condition in terms of high atom density and energy efficiency is therefore at the duty cycle of about 70%.

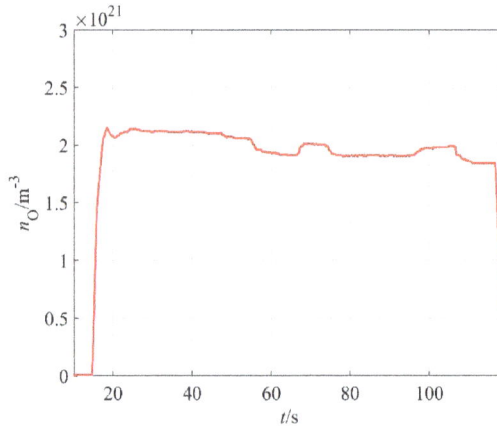

Figure 5. The behavior of O-atom density versus time (*t*) at 80 Pa and duty cycle 70%.

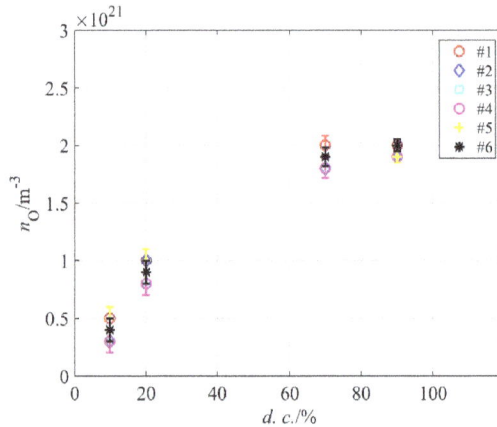

Figure 6. The O-atom density in air plasma at 80 Pa versus MW generator duty cycle.

The O-atom density versus the pressure inside the discharge chamber is shown in Figure 7. One can observe a broad maximum: the atom density hardly changes in a range of pressures between 30 and 80 Pa. The absolute value is just over 2×10^{21} m^{-3}, which is typical for plasma created in pure oxygen or a mixture of oxygen and argon [27–29]. The high density of O atoms in our weakly ionized plasma created in air can be attributed to differences in the excitation energies of oxygen and nitrogen molecules. As already mentioned, the first electronically excited level of a nitrogen molecule is $A^3\Sigma^+_u$ at the potential energy of just over 6 eV. This state is the metastable state of a long radiative lifetime [22]. The dissociation energy of the oxygen molecule is somewhat lower at 5.2 eV. The quenching of the nitrogen metastable state by dissociative collision with an oxygen molecule is therefore energetically feasible. Furthermore, oxygen molecules exhibit two metastable electronic states at the potential energies of roughly 1 and 2 eV. The first metastable state, $a^1\Delta$g, has a lifetime of almost an hour, so it is very stable at low pressure, at which the channels for relaxation are limited. Furthermore, according to Ionnin et al. [24], the cross section for excitation of the second state ($b^1\Sigma^+$g) from the metastable $a^1\Delta$g state is rather large, even for electrons of kinetic energy 1 eV, and it peaks at a few eV, which is a typical electron temperature in low-pressure plasma [30]. Therefore, there are numerous channels for the dissociation of oxygen molecules in weakly ionized air plasma, which explains a rather high density of O atoms, as shown in Figure 7.

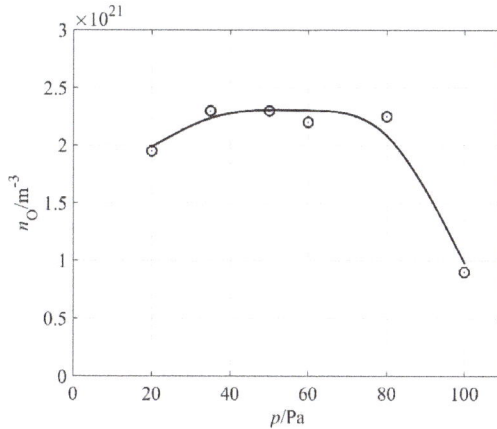

Figure 7. The O-atom density versus pressure at duty cycle 70%.

Samples of polyethylene granules were treated in air plasma and characterized by XPS soon after the treatment in order to limit any ageing effects. Specifically, the activation is not permanent, but hydrophobic recovery has been reported for numerous polymers with polar oxygen-rich functional groups on the surface [31]. Figure 8 shows survey spectra for an untreated sample and a sample treated for a minute at the pressure of 80 Pa, where the O-atom density was just over 2×10^{21} m^{-3} (see Figure 7). The survey spectrum for an untreated polyethylene shows only the carbon peak (which is sound with the polymer composition) and traces of oxygen. After the treatment, another peak appeared, indicating the incorporation of oxygen and an O-to-C ratio of 0.17.

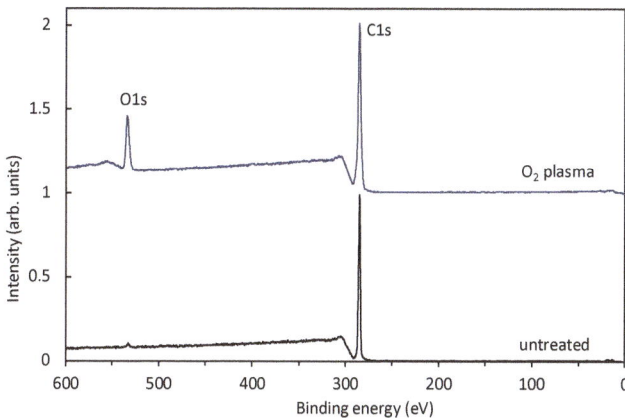

Figure 8. XPS survey spectra for an untreated and treated sample.

Interestingly enough, no nitrogen peak is observed in Figure 8. Samples were treated at different pressures, and the survey spectra were practically identical to those of Figure 8. The absence of nitrogen on the polymer surface is attributed to a higher affinity for reactions with O atoms and also to the shortage of nitrogen atoms in the plasma reactor. As revealed from Figure 2, plasma cannot be rich in

N atoms, because no peak is observed. The very low concentration of N atoms as compared to O atoms can be explained by numerous reasons. The first one is the high dissociation energy of N_2 molecules, which is about 10 eV (twice that of the value for oxygen molecules). The plasma electrons are therefore unlike to dissociate a nitrogen molecule in the ground electronic state, even if vibrationally excited. Nitrogen atoms could be created by dissociation of the metastable nitrogen molecule in the $A^3\Sigma^+_u$ electronic state. This state is the final state of the neutral molecule emission bands observed in Figure 2. Furthermore, the $A^3\Sigma^+_u$ electronic state is also well populated with vibrational states, since many radiative transitions occur to higher vibrational levels [32]. On the other hand, this electronically excited state is quenched by reactions with other plasma species as explained above, thus its density is probably much lower than the O-atom density in our plasma reactor. In fact, Ricard et al. [33] measured the density of the $A^3\Sigma^+_u$ electronic state in a similar plasma system and observed values of the order of 10^{17} m^{-3}, which is four orders of magnitude smaller than the O-atom density in our experimental setup. The density of N atoms is therefore orders of magnitude smaller than the density of O atoms, which together with other considerations, explains the absence of an N peak in Figure 8.

Both untreated and plasma treated samples were characterized by atomic force microscopy (AFM). A typical image is shown in Figure 9. Images were obtained at various spots on a granule and they differed significantly. The as-produced granules were therefore of rich morphology. After plasma treatment, no significant deviation was observed, but it should be stressed again that the granules were already rough before the treatment, so any change in topography that could be a result of plasma treatment was difficult to detect.

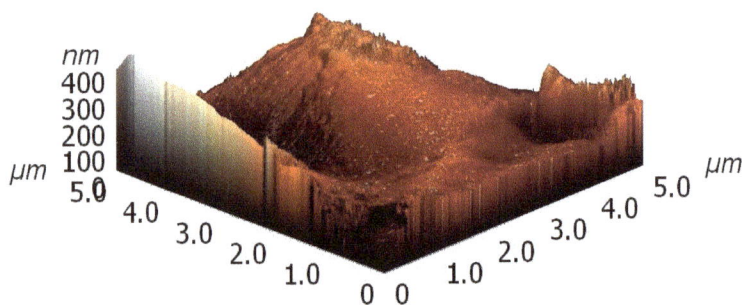

Figure 9. Three-dimensional atomic force microscopy (AFM) image of a 5 μm × 5 μm area of an untreated granule.

Figure 10 represents the high-resolution C1s peak of the photoelectron spectrum for an untreated polyethylene sample. The peak is almost perfectly symmetrical, indicating no functionalization of the polymer surface with groups that cause asymmetrical peaks. After treatment with air plasma, the peak becomes asymmetrical due to surface oxidation. The C1s peak for a treated polymer sample is shown in Figure 11. Since Figure 8 shows negligible (if any) concentration of other elements, it is feasible to deconvolute the peak in Figure 11 with oxygen-containing functional groups. The best fit was obtained using only hydroxyl, carbonyl, and carboxyl/ester groups. The subpeaks corresponding to these functional groups are shown in Figure 11. The result was quantified using the software package and revealed the following concentrations of groups: C–C 88.5%, C–O 5.4%, C=O 2.6%, and O=C–O 3.5%. The fit is almost perfect at the high-energy tail, indicating that carboxyl/ether groups are those of the highest energy shift. Some authors have also observed a carbonate peak on plasma-treated polymers, but such functional groups obviously have not appeared on polyethylene in our case [34–39].

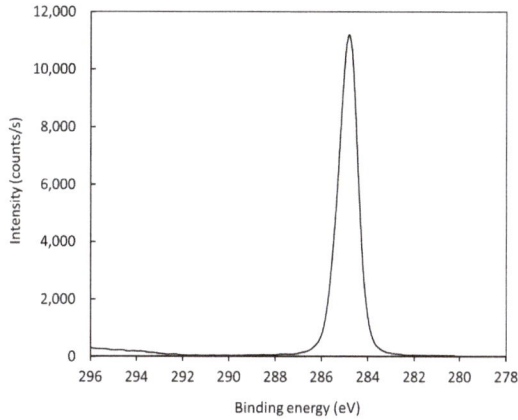

Figure 10. High-resolution C1s spectrum for an untreated sample.

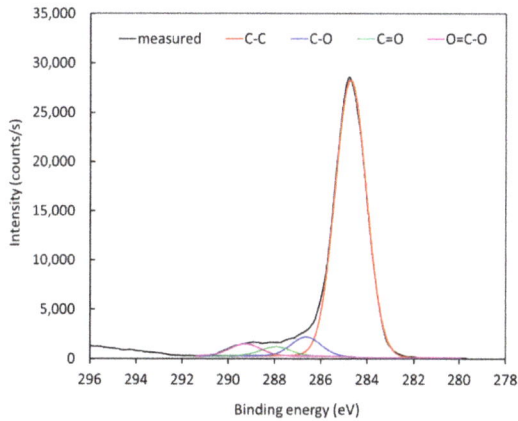

Figure 11. High-resolution C1s spectrum for a treated sample.

Our results for activation of polyethylene granules are similar to those for the same material, but of different morphology. Lopez-Santo [15] employed a remote MW source for the treatment of originally smooth PE, but using an argon–oxygen gas mixture, and observed the same O/C ratio (0.2) as we did in the case of the granules. Abou Rich et al. [40] employed Ar/O_2 and HeO_2 mixtures using a completely different discharge (plasma torches) and observed the much higher O/C ratio of 0.36, irrespective from the gas mixture. Pandiyaraj et al. [41] used a simple direct current (DC) discharge in different gases and observed the highest O/C ratio using air: it was up to twice as high as in the case of pure oxygen plasma. Bílek et al. [42] observed a much lower O/C ratio after treatment of PE in air plasma, but provided no details about the discharge system. Sanchis [43] and Vesel [44] used radiofrequency discharges to functionalize polyethylene in oxygen plasma and observed quick saturation of the PE surface with oxygen groups. The O/C ratio depended on the coupling of the radiofrequency generator: higher values were observed for electrodeless coupling using a coil than for standard capacitive coupling with electrodes inside the discharge chamber. In contrary to the above works, Borcia et al. [45] obtained a O/C ratio as high as 0.68, but they used an atmospheric pressure discharge in air, which may not be suitable for the uniform treatment of large quantities of PE granules.

4. Conclusions

Thorough characterization of the reactor employing a pulsed microwave discharge for sustaining plasma in a metallic chamber revealed interesting properties, which are extremely suitable for the functionalization of polyethylene granules. The plasma parameters do not depend much on the pressure in the discharge chamber and are practically independent from the flow of gas through the reactor. The O-atom density exhibited a broad maximum in the range of pressures between about 30 Pa and 80 Pa, so the surface functionalization with polar oxygen-rich functional groups can be accomplished in a highly reproducible manner even if the pressure slightly varies from batch to batch. The absolute value of the O-atom density is about 2×10^{21} m^{-3}, which assures for rapid functionalization of the polypropylene granules. The XPS spectra revealed a negligible concentration of nitrogen on the polymer surface, which was explained by a small density of N atoms in the plasma reactor as well as preferential interaction with O atoms. The plasma treatment resulted in functionalization with hydroxyl, carbonyl, and carboxyl/ester groups.

Author Contributions: H.Š. performed plasma treatment of polymer granules and prepared the corresponding part of the paper. G.P. performed plasma characterization and interpretation of the results. P.Š. supervised the research, suggested appropriate solutions, contributed to discussion and performed final scientific editing of the paper.

Funding: This research has been performed and funded within the program NPU1, project number LO 1207, from the Ministry of the Education, Youth, and Sport of the Czech Republic and program OPPI. Authors acknowledge financial support from the ESIF, EU Operational Programme Research, Development and Education, and from the Center of Advanced Aerospace Technology (CZ.02.1.01/0.0/0.0/16_019/0000826), Faculty of Mechanical Engineering, Czech Technical University in Prague. Authors also give thanks for the partial financial support by the Ministry of Education of the Czech Republic project SGS No. 21176/115 of the Technical University of Liberec and by the Slovenian Research Agency (research core funding no. P2-0082).

Conflicts of Interest: The authors declare no conflict of interest.

References

1. Vesel, A.; Mozetic, M. New developments in surface functionalization of polymers using controlled plasma treatments. *J. Phys. D Appl. Phys.* **2017**, *50*, 293001. [CrossRef]
2. Gorjanc, M.; Mozetič, M. *Modification of Fibrous Polymers by Gaseous Plasma: Principles, Techniques and Applications*; LAP Lambert Academic Publishing: Riga, Latvia, 2014; ISBN 978-3-659-61460-6.
3. Vukušić, T.; Vesel, A.; Holc, M.; Ščetar, M.; Jambrak, A.R.; Mozetič, M. Modification of physico-chemical properties of acryl-coated polypropylene foils for food packaging by reactive particles from oxygen plasma. *Materials* **2018**, *11*, 372. [CrossRef] [PubMed]
4. Orendáč, M.; Čižmár, E.; Kažiková, V.; Orendáčová, A.; Řezníčková, A.; Kolská, Z.; Švorčík, V. Radicals mediated magnetism in Ar plasma treated high-density polyethylene. *J. Magn. Magn. Mater.* **2018**, *454*, 185–189. [CrossRef]
5. Vartiainen, J.; Pasanen, S.; Kentt, E.; Vähä-Nissi, M. Mechanical recycling of nanocellulose containing multilayer packaging films. *J. Appl. Polym. Sci.* **2018**, *135*, 46237. [CrossRef]
6. Zhao, Z.P.; Zhang, A.S.; Wang, X.L.; Lu, P.; Ma, H.Y. Controllable modification of polymer membranes by LDDLT plasma flow: Grafting acidic ILs into PPF membrane for catalytic performance. *J. Membr. Sci.* **2018**, *553*, 99–110. [CrossRef]
7. Hu, Z.; Chng, S.; Liu, Y.; Moloney, M.G.; Parker, E.M.; Wu, L.Y.L. One-step chemical functionalization of polyethylene surfaces via diarylcarbene insertion. *Mater. Lett.* **2018**, *218*, 157–160. [CrossRef]
8. Ozaltin, K.; Lehocky, M.; Humpolicek, P.; Vesela, D.; Mozetic, M.; Novak, I.; Saha, P. Preparation of active antibacterial biomaterials based on sparfloxacin, enrofloxacin, and lomefloxacin deposited on polyethylene. *J. Appl. Polym. Sci.* **2018**, *135*, 46174. [CrossRef]
9. Muzammil, I.; Li, Y.P.; Li, X.Y.; Lei, M.K. Duty cycle dependent chemical structure and wettability of RF pulsed plasma copolymers of acrylic acid and octafluorocyclobutane. *Appl. Surf. Sci.* **2018**, *436*, 411–418. [CrossRef]

10. Van Vrekhem, S.; Vloebergh, K.; Asadian, M.; Vercruysse, C.; Declercq, H.; Van Tongel, A.; De Wilde, L.; De Geyter, N.; Morent, R. Improving the surface properties of an UHMWPE shoulder implant with an atmospheric pressure plasma jet. *Sci. Rep.* **2018**, *8*, 4720. [CrossRef] [PubMed]

11. Lindner, M.; Rodler, N.; Jesdinszki, M.; Schmid, M.; Sängerlaub, S. Surface energy of corona treated PP, PE and PET films, its alteration as function of storage time and the effect of various corona dosages on their bond strength after lamination. *J. Appl. Polym. Sci.* **2018**, *135*. [CrossRef]

12. Popelka, A.; Khanam, P.N.; Almaadeed, M.A. Surface modification of polyethylene/graphene composite using corona discharge. *J. Phys. D Appl. Phys.* **2018**, *51*, 105302. [CrossRef]

13. Popelka, A.; Novák, I.; Al-Maadeed, M.A.S.A.; Ouederni, M.; Krupa, I. Effect of corona treatment on adhesion enhancement of LLDPE. *Surf. Coat. Technol.* **2018**, *335*, 118–125. [CrossRef]

14. López-Santos, C.; Yubero, F.; Cotrino, J.; González-Elipe, A.R. Nitrogen plasma functionalization of low density polyethylene. *Surf. Coat. Technol.* **2011**, *205*, 3356–3365. [CrossRef]

15. López-Santos, C.; Yubero, F.; Cotrino, J.; Barranco, A.; Gonzlez-Elipe, A.R. Plasmas and atom beam activation of the surface of polymers. *J. Phys. D Appl. Phys.* **2008**, *41*, 225209. [CrossRef]

16. López-García, J.; Bílek, F.; Lehocký, M.; Junkar, I.; Mozetič, M.; Sowe, M. Enhanced printability of polyethylene through air plasma treatment. *Vacuum* **2013**, *95*, 43–49. [CrossRef]

17. Oberbossel, G.; Güntner, A.T.; Kündig, L.; Roth, C.; Von Rohr, P.R. Polymer powder treatment in atmospheric pressure plasma circulating fluidized bed reactor. *Plasma Process. Polym.* **2015**, *12*, 285–292. [CrossRef]

18. Inagaki, N.; Tasaka, S.; Abe, H. Surface modification of polyethylene powder using plasma reactor with fluidized bed. *J. Appl. Polym. Sci.* **1992**, *46*, 595–601. [CrossRef]

19. Park, S.H.; Kim, S.D. Plasma surface treatment of HDPE powder in a fluidized bed reactor. *Polym. Bull.* **1994**, *33*, 249–256. [CrossRef]

20. Mozetič, M.; Primc, G.; Vesel, A.; Zaplotnik, R.; Modic, M.; Junkar, I.; Recek, N.; Klanjšek-Gunde, M.; Guhy, L.; Sunkara, M.K.; et al. Application of extremely non-equilibrium plasmas in the processing of nano and biomedical materials. *Plasma Sources Sci. Technol.* **2015**, *24*, 015026. [CrossRef]

21. Primc, G.; Mozetič, M.; Cvelbar, U.; Vesel, A. Method and Device for Detection and Measuring the Density of Neutral Atoms of Hydrogen, Oxygen or Nitrogen. WIPO Patent Application WO/2015/176733, 26 November 2015.

22. Zipf, E.C. Measurement of the Diffusion Coefficient and Radiative Lifetime of Nitrogen Molecules in the $A3\Sigma_u^+$ State. *J. Chem. Phys.* **1963**, *38*, 2034–2035. [CrossRef]

23. Itikawa, Y. Cross sections for electron collisions with nitrogen molecules. *J. Phys. Chem. Ref. Data* **2006**, *35*, 31–53. [CrossRef]

24. Ionin, A.A.; Kochetov, I.V.; Napartovich, A.P.; Yuryshev, N.N. Physics and engineering of singlet delta oxygen production in low-temperature plasma. *J. Phys. D Appl. Phys.* **2007**, *40*, 25–61. [CrossRef]

25. Gorjanc, M.; Mozetič, M.; Vesel, A.; Zaplotnik, R. Natural dyeing and UV protection of plasma treated cotton. *Eur. Phys. J. D* **2018**, *72*, 41. [CrossRef]

26. Gorjanc, M.; Mozetič, M.; Primc, G.; Vesel, A.; Spasić, K.; Puač, N.; Petrović, Z.L.; Kert, M. Plasma treated polyethylene terephthalate for increased embedment of UV-responsive microcapsules. *Appl. Surf. Sci.* **2017**, *419*. [CrossRef]

27. Altaweel, A.; Imam, A.; Ghanbaja, J.; Mangin, D.; Miska, P.; Gries, T.; Belmonte, T. Fast synthesis of ultrathin ZnO nanowires by oxidation of Cu/Zn stacks in low-pressure afterglow. *Nanotechnology* **2017**, *28*, 085602. [CrossRef] [PubMed]

28. Kutasi, K.; Noël, C.; Belmonte, T.; Guerra, V. Tuning the afterglow plasma composition in $Ar/N_2/O_2$ mixtures: Characteristics of a flowing surface-wave microwave discharge system. *Plasma Sources Sci. Technol.* **2016**, *25*. [CrossRef]

29. Gueye, M.; Gries, T.; Noël, C.; Migot-Choux, S.; Bulou, S.; Lecoq, E.; Choquet, P.; Belmonte, T. Interaction of (3-Aminopropyl)triethoxysilane With Late Ar-N_2 Afterglow: Application to Nanoparticles Synthesis. *Plasma Process. Polym.* **2016**, *13*, 698–710. [CrossRef]

30. Gibson, A.R.; Foucher, M.; Marinov, D.; Chabert, P.; Gans, T.; Kushner, M.J.; Booth, J.P. The role of thermal energy accommodation and atomic recombination probabilities in low pressure oxygen plasmas. *Plasma Phys. Control. Fusion* **2017**, *59*, 024004. [CrossRef]

31. Praveen, K.M.; Thomas, S.; Grohens, Y.; Mozetič, M.; Junkar, I.; Primc, G.; Gorjanc, M. Investigations of plasma induced effects on the surface properties of lignocellulosic natural coir fibres. *Appl. Surf. Sci.* **2016**, *368*, 146–156. [CrossRef]

32. Pashov, A.; Popov, P.; Knöckel, H.; Tiemann, E. Spectroscopy of the $a^3\Sigma_u$ + state and the coupling to the $X^1\Sigma_g$ + state of K_2. *Eur. Phys. J. D* **2008**, *46*, 241–249. [CrossRef]

33. Ricard, A.; Oh, S.; Jang, J.; Kim, Y.K. Quantitative evaluation of the densities of active species of N_2 in the afterglow of Ar-embedded N_2 RF plasma. *Curr. Appl. Phys.* **2015**, *15*, 1453–1462. [CrossRef]

34. Quoc Toan Le, Q.T.; Pireaux, J.J.; Caudano, R. XPS study of the PET film surface modified by CO_2 plasma: Effects of the plasma parameters and ageing. *J. Adhes. Sci. Technol.* **1997**, *11*, 735–751. [CrossRef]

35. Arefi, F.; Andre, V.; Montazer-Rahmati, P.; Amouroux, J. Plasma polymerization and surface treatment of polymers. *Pure Appl. Chem.* **1992**, *64*, 715–723. [CrossRef]

36. Vesel, A.; Mozetic, M.; Hladnik, A.; Dolenc, J.; Zule, J.; Milosevic, S.; Krstulovic, N.; Klanjek-Gunde, M.; Hauptmann, N. Modification of ink-jet paper by oxygen-plasma treatment. *J. Phys. D Appl. Phys.* **2007**, *40*, 3689–3696. [CrossRef]

37. O'Hare, L.A.; Leadley, S.; Parbhoo, B. Surface physicochemistry of corona-discharge-treated polypropylene film. *Surf. Interface Anal.* **2002**, *33*, 335–342. [CrossRef]

38. Kregar, Z.; Bišćan, M.; Miloševiá, S.; Mozetič, M.; Vesel, A. Interaction of argon, hydrogen and oxygen plasma early afterglow with polyvinyl chloride(PVC) materials. *Plasma Process. Polym.* **2012**, *9*, 1020–1027. [CrossRef]

39. Zaldivar, R.; Nokes, J.; Patel, D.N.; Morgan, B.A.; Steckel, G.; Kim, H.I. Effect of using oxygen, carbon dioxide, and carbon monoxide as active gases in the atmospheric plasma treatment of fiber-reinforced polycyanurate composites. *J. Appl. Polym. Sci.* **2012**, *125*, 2510–2520. [CrossRef]

40. Abou Rich, S.; Dufour, T.; Leroy, P.; Reniers, F.; Nittler, L.; Pireaux, J.J. LDPE surface modifications induced by atmospheric plasma torches with linear and showerhead configurations. *Plasma Process. Polym.* **2015**, *12*, 771–785. [CrossRef]

41. Pandiyaraj, K.N.; Deshmukh, R.R.; Ruzybayev, I.; Shah, S.I.; Su, P.G.; Halleluyah, M.; Halim, A.S. Influence of non-thermal plasma forming gases on improvement of surface properties of low density polyethylene (LDPE). *Appl. Surf. Sci.* **2014**, *307*, 109–119. [CrossRef]

42. Bílek, F.; Křížová, T.; Lehocký, M. Preparation of active antibacterial LDPE surface through multistep physicochemical approach: I. Allylamine grafting, attachment of antibacterial agent and antibacterial activity assessment. *Colloids Surf. B Biointerfaces* **2011**, *88*, 440–447. [CrossRef] [PubMed]

43. Sanchis, R.; Fenollar, O.; García, D.; Sánchez, L.; Balart, R. Improved adhesion of LDPE films to polyolefin foams for automotive industry using low-pressure plasma. *Int. J. Adhes. Adhes.* **2008**, *28*, 445–451. [CrossRef]

44. Vesel, A. XPS study of surface modification of different polymer materials by oxygen plasma treatment. *Inf. MIDEM* **2008**, *38*, 257–265.

45. Borcia, G.; Anderson, C.A.; Brown, N.M.D. The surface oxidation of selected polymers using an atmospheric pressure air dielectric barrier discharge. Part I. *Appl. Surf. Sci.* **2004**, *221*, 203–214. [CrossRef]

![materials logo] *materials*

MDPI

Article

Evaluation of Surface Roughness by Image Processing of a Shot-Peened, TIG-Welded Aluminum 6061-T6 Alloy: An Experimental Case Study

Anas M. Atieh [1,*], Nathir A. Rawashdeh [2] and Abdulaziz N. AlHazaa [3,4]

[1]   Industrial Engineering Department, School of Applied Technical Sciences,
      German Jordanian University (GJU), Amman 11180, Jordan
[2]   Mechatronics Engineering Department, School of Applied Technical Sciences,
      German Jordanian University (GJU), Amman 11180, Jordan; Nathir.Rawashdeh@gju.edu.jo
[3]   Department of Physics & Astronomy, College of Science, King Saud University, Riyadh 11451, Saudi Arabia;
      alhazaa@gmail.com
[4]   King Abdullah Institute for Nanotechnology (KAIN), King Saud University, Riyadh 11451, Saudi Arabia
*   Correspondence: anas.atieh@gju.edu.jo or anas.m.attieh@gmail.com; Tel.: +962-79-830-9030

Received: 12 April 2018; Accepted: 4 May 2018; Published: 10 May 2018

Abstract: Visual inspection through image processing of welding and shot-peened surfaces is necessary to overcome equipment limitations, avoid measurement errors, and accelerate processing to gain certain surface properties such as surface roughness. Therefore, it is important to design an algorithm to quantify surface properties, which enables us to overcome the aforementioned limitations. In this study, a proposed systematic algorithm is utilized to generate and compare the surface roughness of Tungsten Inert Gas (TIG) welded aluminum 6061-T6 alloy treated by two levels of shot-peening, high-intensity and low-intensity. This project is industrial in nature, and the proposed solution was originally requested by local industry to overcome equipment capabilities and limitations. In particular, surface roughness measurements are usually only possible on flat surfaces but not on other areas treated by shot-peening after welding, as in the heat-affected zone and weld beads. Therefore, those critical areas are outside of the measurement limitations. Using the proposed technique, the surface roughness measurements were possible to obtain for weld beads, high-intensity and low-intensity shot-peened surfaces. In addition, a 3D surface topography was generated and dimple size distributions were calculated for the three tested scenarios: control sample (TIG-welded only), high-intensity shot-peened, and low-intensity shot-peened TIG-welded Al6065-T6 samples. Finally, cross-sectional hardness profiles were measured for the three scenarios; in all scenarios, lower hardness measurements were obtained compared to the base metal alloy in the heat-affected zone and in the weld beads even after shot-peening treatments.

Keywords: materials characterization; shot-peening; image processing; TIG welding; aluminum 6061-T6

---

1. Introduction

In the past two decades, owing to the good mechanical properties, i.e., high strength to weight ratio, good thermal and electrical conductivity, aluminum and its alloys have been used in versatile engineering applications such as marine vessels, automobiles, railway cars, and aircraft [1–5].

Aluminum is strengthened through precipitation hardening (for instant the 6000 aluminum series) due to the presence of silicon and magnesium alloying elements (0.3–15 wt % Si and Mg). The addition of those alloying elements resulted in further developments of Aluminum characteristics, which include good formability, corrosion resistance, and weldability [6]. Aluminum 6061 has been widely used in aerospace applications, especially with the T6 tempered solution-heat-treated and artificially aged status [7,8].

Of the different welding techniques, Tungsten Inert Gas (TIG) welding, an arc welding process that uses a non-consumable electrode, has gained more attention in welding aerospace alloys, i.e., Al6061-T6. TIG welding quality was considered to be the most effective technique not only because of the high strength but also due to the cleanliness of the resulting weld bead with minimal defects produced [9].

One of aluminum's drawbacks is the low wear resistance; furthermore, welded aluminum alloys are usually characterized by sever metallurgical changes within the heat-affected zone and grain coarsening in the adjacent base metal alloy [9–11]. Therefore, a surface treatment process is required to maintain the good mechanical characteristics of aluminum alloys. Moreover, because of the different performance and surface property requirements of aluminum and its alloys, different surface treatments could be done including surface mechanical attrition treatment [12–14], ultrasonic shot-peening [15,16], laser shot-peening [17–20], and air-blast shot-peening [19,21,22]. Among them, air-blast shot-peening has received increased attention. This process is a cold-working surface treatment, where a large amount of small spherical particles (called shots) are bombarded onto a metallic surface at high velocity [23–25].

In shot-peening, only a very thin surface layer is affected by the impact of shots, which produce tensile plastic strain, resulting in favorable compressive residual stresses. Those residual stresses are approximately equal to the process-induced stress and could be estimated and calculated analytically [25]. However, shot-peening must be optimized for specific materials and specific conditions, i.e., welded material; otherwise, it may have undesirable results such as surface contamination, reduction in strength, and surface crack formation [26–28].

Recently, many researchers have investigated the effect of different shot-peening techniques on the weldment's characteristics [16,17,29,30]. Yang et al. studied the shot-peening effects on dovetail specimens' joints of Ti-6Al-4V fretting fatigue behavior. It was reported that shot-peening altered the crack initiation mechanisms and enhanced the fretting fatigue performance for the titanium alloy [31]. Dissimilar TIG-welded joints of magnesium and titanium alloys were also subjected to high-energy shot-peening and the microstructure and mechanical properties in terms of tensile strength were evaluated by Chuan Xu et al. [32]. Surface defect elimination, strengthening by grain refinement, and strain hardening were a result of subjecting Mg/Ti welded joints to high-energy shot-peening. Furthermore, the tensile strength of Mg/Ti TIG-welded samples was increased by 24.5% to 241 MPa [32].

Different researchers have been utilizing different image processing techniques and real-time measurements. The main focus was to capture the properties and characteristics of weldments and to anticipate the quality of weld beads during live welding processes [33–38]; however, the authors are not aware of any report on the application of image processing in the detection of surface properties of welded and shot-peened alloys. The image processing techniques have been utilized to characterize different material properties [39–46]. For example, external welding defects have been widely inspected by vision-based and image processing techniques [41,43,47].

In this work, we present experimental results from a post-TIG-welded shot-peened Al6061-T6 alloy with such an analysis approach. Image processing algorithms have been utilized to extract information from the acquired images of the samples after surface processing to evaluate the surface roughness parameters. Using the proposed technique, it is expected to accelerate obtaining the characterization of the surface properties of the welded samples compared to traditional destructive testing. This may also enhance the detection reliability and overcome the traditional measurement limitations of weldment characteristics. The novel contribution of this research work is the development of an image processing technique that can measure the crater/dimple size of the post-weld shot-peened Al 6061-T6 alloy. Furthermore, we reconstruct the 3D weld surface morphology accurately and reliably for monitoring and evaluating post-process surface properties to be utilized in industrial applications.

2. Methodology

2.1. Materials, Properties, and Testing Specimens

The materials used in this study are aluminum alloy Al 6061-T6 supplied by Jordan Airmotive Company-JALCO from ALCOA Inc., Davenport, IA, USA. Raw materials were provided in sheet metal format with chemical composition as in Table 1.

Table 1. Chemical composition of aluminum alloy Al 6061-T6.

Chemical Composition (wt %)								
Si	Fe	Cu	Mg	Mn	Ti	Zn	Cr	Al
0.8	0.7	0.4	1.2	0.15	0.15	0.25	0.35	Rem.

This solution and precipitation heat-treated lightly oil coiled sheet aluminum alloy was produced according to aerospace materials specifications AMS4027N (ASTM B209-14). Rectangular specimens of 10 cm × 20 cm and 8 mm thickness were cut and then butt form welded according to AWS17.1 standard using filler wire of AMS4191. The TIG welding was carried out under 99.99% pure argon shielding gas and 72 amps AC welding current. In addition to the control sample (only TIG-welded, see Figure 1), two other sets of TIG-welded samples were prepared for shot-peening. The welded samples were subjected to two shot-peening scenarios after the welding: high-intensity and low-intensity shot-peening. The controlled shot-peening treatment was performed on the Al6061-T6 alloy by means of an E-S-1580 Pangborn-controlled shot-peening machine.

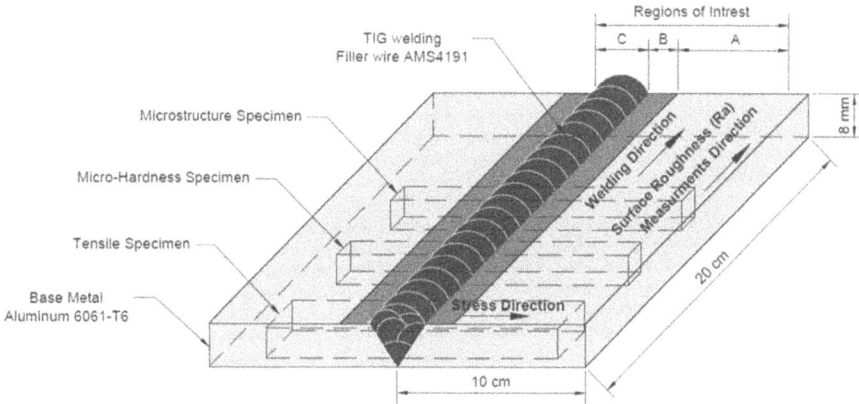

Figure 1. Geometry of TIG-welded Al6061-T6 specimen and schematic of testing specimens showing welding and surface roughness measurement direction.

The low-intensity shot-peening (6.94 N) was produced by bombarding glass shot beads. The glass shot beads were aerospace controlled level (OMAT 1/239 glass beads) with a grit number ranging from 150 to 300/RR and the size of glass beads ranged from 150 to 300 microns. This conforms to Rolls Royce CSS8 issue 5 styker orthopedics MS00097 issue 4, under a nozzle pressure of 50 Psi with a 3/8-inch nozzle size hole and a 5.5-inch nozzle distance. The low-intensity shot-peening lasted for 5.25 min, and the peening coverage was 100%, as confirmed by an optical magnifier (see Figure 2).

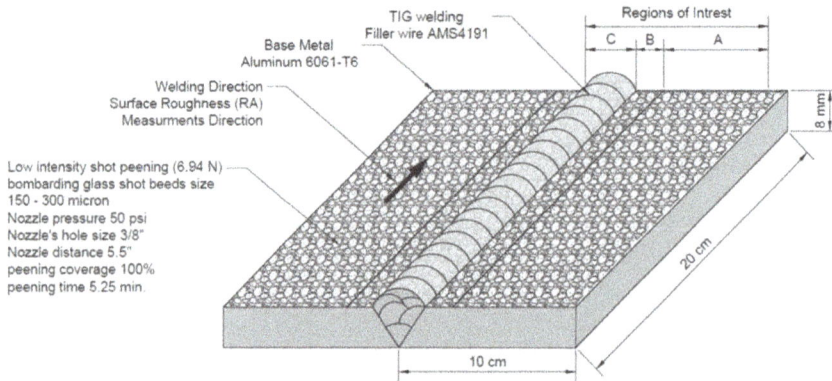

Figure 2. Geometry of Al6061-T6 low-intensity shot-peened, TIG-welded specimen.

As for the high-intensity shot-peening (5.35 A2), it was produced by bombarding steel shot beads. The steel shot beads were aerospace controlled level, spherically condition wire cut with 0.014-inch size (SCCW-0.014-inch) under nozzle pressure of 100 Psi with 3/8-inch nozzle size hole and 4-inch nozzle distance. The high-intensity shot-peening lasted for 3 min, and the peening coverage was 100%, as confirmed by an optical magnifier (see Figure 3).

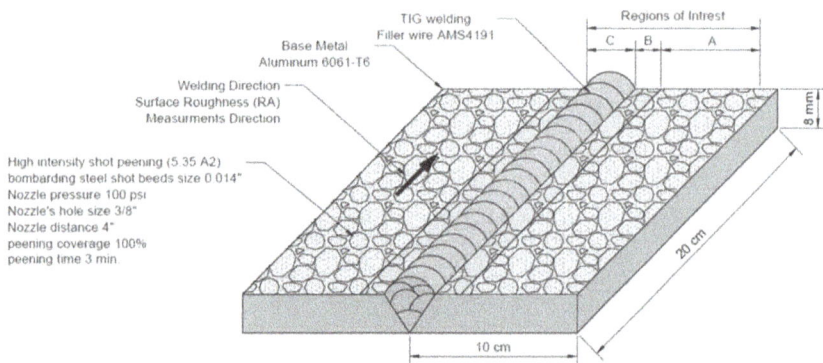

Figure 3. Geometry of Al6061-T6 high-intensity shot-peened, TIG-welded specimen.

At both intensities, the Almen gauge #2 model TSP-3 Rev. B were used with different Almen strips to achieve the required intensities by subjecting the strips to peening at different table speeds until the required saturation point (intensity) was achieved.

2.2. Characterization

Vickers hardness measurements were performed on shot-peened treated samples at both intensities along with the control samples. Measurements were performed according to ASTM E384-11$^{\varepsilon 1}$ standard on a Tru-Blue united hardness tester (Tru-Blue U/10 version F13, San Diego, CA, USA) by applying a 10 kg load for 15 s dwell time. Four measurement sets were taken on the cross section of the mounted samples, 1 mm away from the shot-peened surface (see Figure 4).

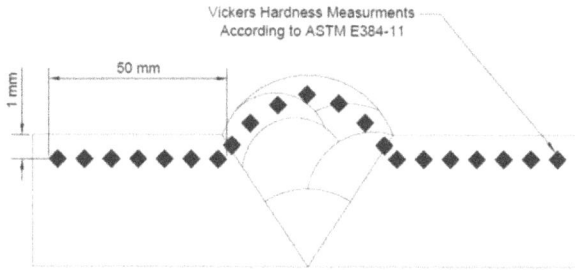

Figure 4. Micro-hardness measurement profile.

Surface roughness (Ra) SJ-210 Mitutoyo Japan tester (Mitutoyo, Kawasaki City, Japan) was used according to ISO11562 standard, where measurements were taken parallel to the weld bead's in the base metal alloy region only. Roughness on the weld-line and on the heat-affected zone was very rough and out of device measurement capabilities. The roughness readings' sampling length was 8 mm and the evaluation length was 40 mm at a constant reading speed of 0.5 mm/s (see Figure 1). The microstructure and tensile specimens obtained from those samples were kept for further analysis in stage 2 of this project.

2.3. Image Processing

An optical microscope was used to capture 8-bit intensity Tagged Image File Format (TIFF) images at a resolution of 2067 pixels per millimeter. Three regions of interest (ROI) were identified, A, B, and C, with an area of 0.8 mm². The microscope (Zeiss Discovery V20) (Carl Zeiss MicroImaging GmbH, Göttingen, Germany) was equipped with an LED ring-light around the microscope lens, and a digital camera (AxioCam ERc 5s) (Carl Zeiss MicroImaging GmbH, Göttingen, Germany) configured for 95× magnification. Then equivalent surface topography images were generated in which the light areas are shown as mountains and the dark areas as valleys. These images approximate the real topography. The goal is to compute optical metrics from the magnified surface texture image, for various material properties including hardness, roughness, and dimple size distribution. Hence, for comparative analysis, images were taken from identical regions A, B, and C as shown in Figures 1–3.

3. Results and Discussion

The image processing steps are summarized in the block diagram shown in Figure 5. The acquired color images are cropped to represent a square of 0.8 mm² for all regions of interest. The region of interest (ROI) image is then converted to an 8-bit grayscale image where an intensity value of 0 represents black, and 255 represents white. The image is then blurred to produce smoother intensity histograms, i.e., more natural intensity level distribution, where all intensities are present. From there, it is possible to produce line profiles and approximate 3D topography images for qualitative comparisons. In addition, further processing produces optical roughness and hardness metrics that correlate with the measured data produced using specialized laboratory tools such as SJ-210 Mitutoyo for surface roughness and Tru-Blue U/10 for hardness. Lastly, a matched filter process on the 2D image produces dimple size distribution plots that can be used qualitatively to compare the roughness of imaged sample regions. More details about these steps are presented in the example shown in Figure 6.

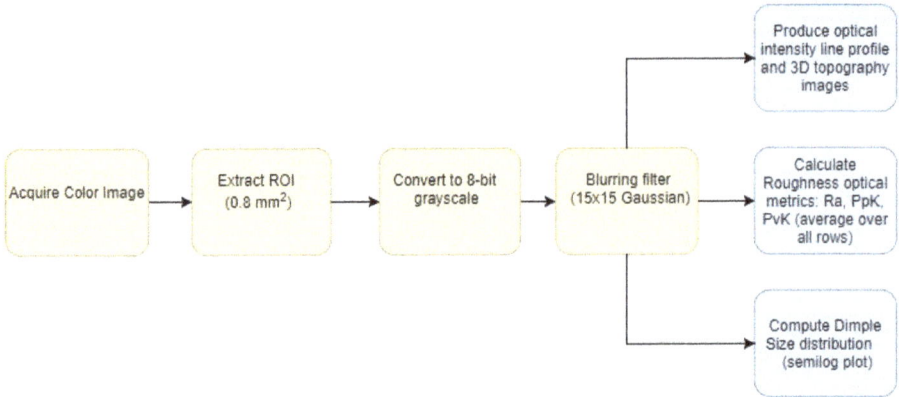

Figure 5. Image processing algorithm for analysis.

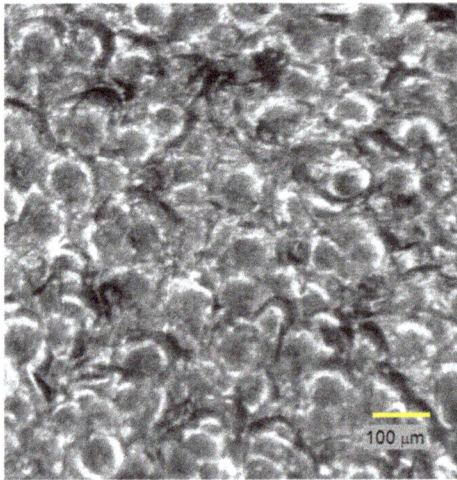

Figure 6. Example of a low-intensity shot-peened metal, of size 0.8 mm^2, i.e., 1653 × 1653 pixels.

Mostly, pixel-intensity image processing techniques are applied on the obtained microscope images to analyze various surface properties. The analysis starts by applying a slight blur to the image to enhance the intensity histogram detail. The blurring filter used a 15 × 15 Gaussian mask for convolution. The effect of blurring on the intensity histogram is shown in Figure 7. It can be seen that in the smoothed image histogram, the pre-blur histogram has empty intensity bins, which is attributed to the TIFF image format compression artifact. With the blurred image histogram, it is possible to choose intensity-level cutoffs to classify pixels as being of dark or medium intensity.

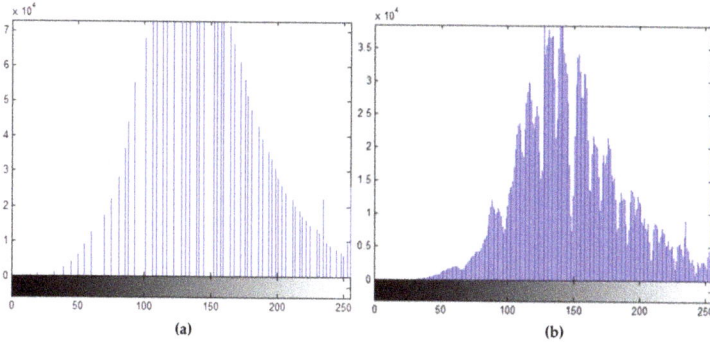

Figure 7. Intensity histograms of a low-intensity shot-peened sample: (**a**) before blurring; (**b**) after blurring.

In order to investigate the samples' surface, it is helpful to generate an optical line profile of the ROI along with the average intensity on that line. Such a profile, shown in Figure 8, approximates the line topography, where high-intensity pixels represent elevated points and low-intensity pixels represent low elevation points. This shows the pixel intensities of a horizontal line through the center of the ROI, i.e., the center row of the example image in Figure 6. Similarly, an optical approximation of the 3D topography can be generated and used for qualitative analysis. This is shown in Figure 9 for the same example in the image presented in Figure 6.

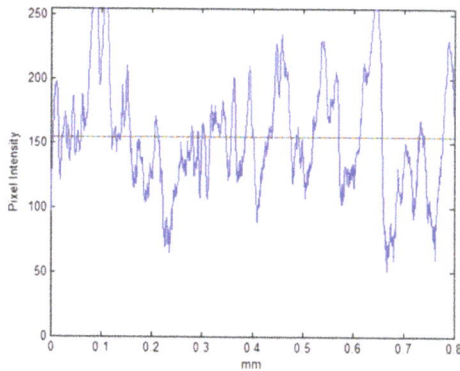

Figure 8. Optical intensity line profile of the center horizontal line in the example image. The average intensity level is depicted as a line.

Figure 9. Optical approximation of the 3D topography of the example image of size 0.8 mm^2.

The optical line profile is used to calculate roughness metrics to overcome the measurement limitations using laboratory instruments. This imaging approach has the advantage of calculating the roughness metrics as an average of all line profiles in the ROI, i.e., all horizontal lines starting at the top and ending at the bottom, of which there are J = 1653. In order to define the equations for this procedure, the first step is to describe the optical roughness profile as containing I ordered, equally spaced points along the trace, and defining y_i as the vertical distance from the mean line to the ith data point. Height is assumed to be positive in the up direction, away from the bulk material. The following equations define the roughness metrics, Roughness Average Ra, Max Peak Height P_pK, and Max Valley Depth P_vK:

$$R_a = \frac{1}{J} \sum_{j=1}^{J} \left(\frac{1}{I} \sum_{i=1}^{I} |y_{ij}| \right) \tag{1}$$

$$P_pK = \frac{1}{J} \sum_{j=1}^{J} \left(\max[y_{ij}] \right) \tag{2}$$

$$P_vK = \frac{1}{J} \sum_{j=1}^{J} \left(\min[y_{ij}] \right), \tag{3}$$

where j is the image row index, i.e., the current line profile number, and y_{ij} is the vertical distance from the mean profile intensity of the pixel at row j and column i.

The final metric values are represented as a percentage of the 8-bit intensity range from 0 to 255 as follows:

$$\text{Final } R_a = (R_a \: / \: 255) \times 100 \tag{4}$$

$$\text{Final } P_pK = (P_pK \: / \: 255) \times 100 \tag{5}$$

$$\text{Final } P_vK = (P_vK \: / \: 255) \times 100. \tag{6}$$

Figure 10 shows these metric values across ROI row number j. These intermediate plots illustrate the variability of the metrics over line profiles from top to bottom. The final metrics are averages of this data and deliver roughness estimates across the whole ROI rather than a few selected lines as in the case of laboratory equipment measurements.

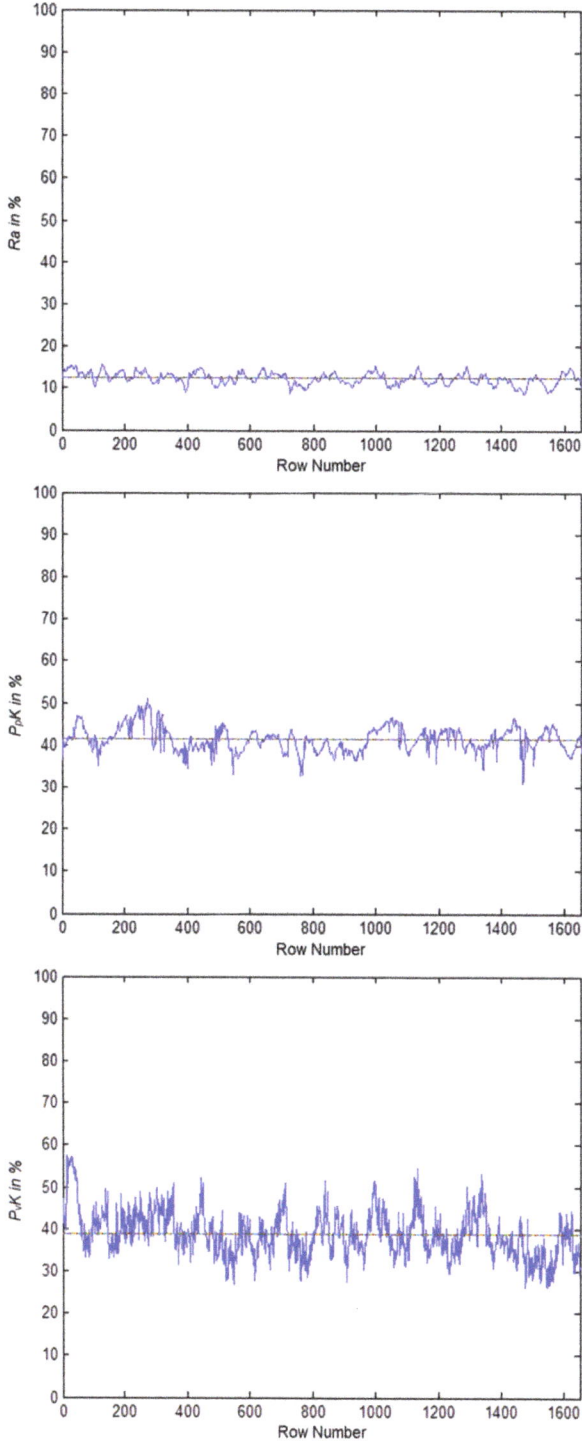

Figure 10. Optical roughness metrics versus ROI row number for the example image shown in Figure 6.

To validate the method of optically obtaining roughness measurements, three samples were chosen: A welded sample without shot-peening; a welded, low-intensity shot-peened sample; and a welded, high-intensity shot-peened sample. The ROIs for these samples are shown in Figure 11, and three regions for each sample are defined as in Figure 2. These regions are *A* (Base metal alloy), *B* (the heat affected zone), and *C* (the weld beads). Figures 12–16 are all based on these selected samples' ROIs.

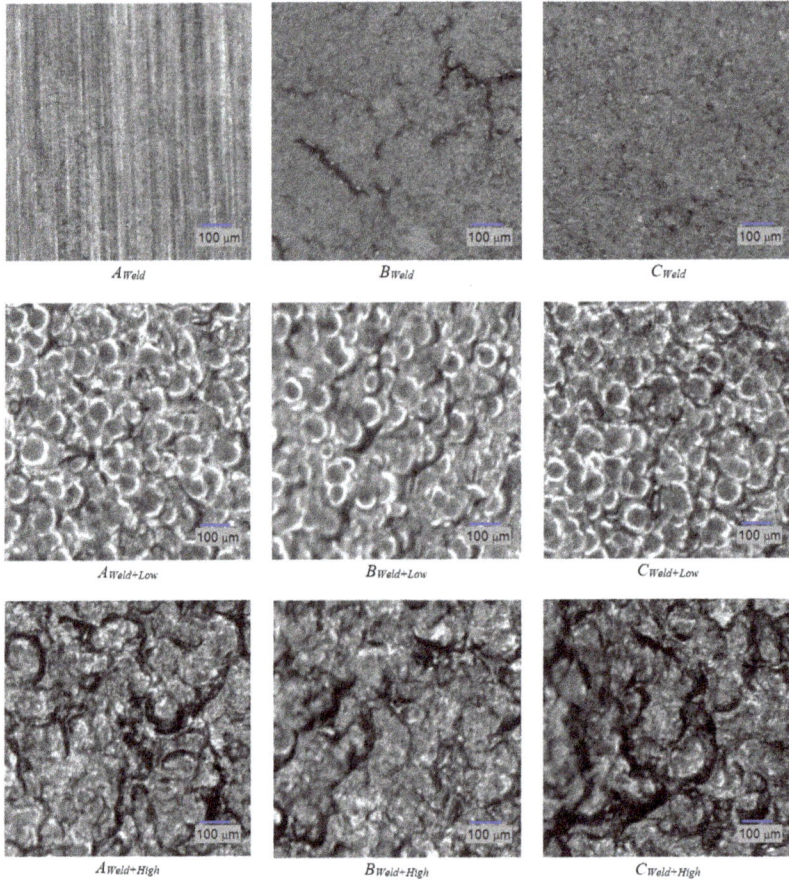

Figure 11. Grayscale images of three samples with three regions each.

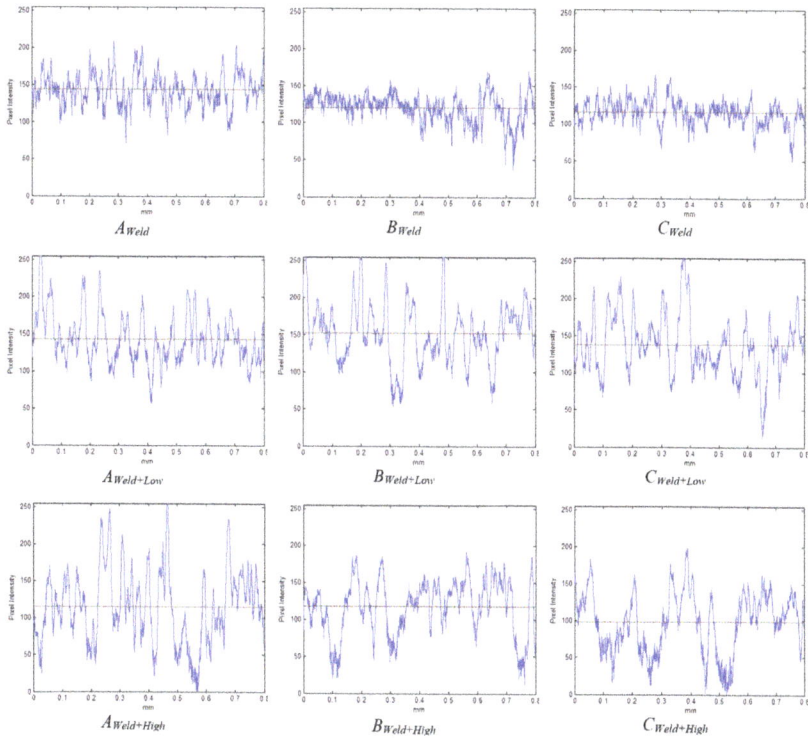

Figure 12. Line profile (center row) for three samples with three regions each. Length = 0.8 mm. Intensity is 0 for black and 255 for white pixels.

As for Figures 11 and 12, they show that the welded sample (top row) is the least rough. The heat-affected zone, B_{weld}, of this sample contains microcracks that can be attributed to the temperature gradient and difference between the high-temperature left side and low-temperature right side, and the uneven cooling after welding. Hence, a post-weld shot-peening treatment is necessary for the crack closures. It is also evident that the sample that was shot-peened with high intensity (bottom row in the figures) is the roughest. This is plausible since the high-intensity shot-peening was performed with small steel balls, while the low-intensity shot-peening utilized ground glass balls. To study the roughness across zones A, B, and C, Figures 11 and 12 must be viewed column-wise. It can be concluded that the heat-affected zone (region B) has the least roughness compared to the weld bead (region C) and the unaffected area (region A); shot-peening did not change the relative roughness. In addition, low-intensity shot-peening closes microcracks, as can be seen by comparing the top two images in the center column of Figure 11. High-intensity shot-peening not only closes the microcracks, but also introduces major surface deformations.

Figure 13 shows the optical approximations of the 3D surface topographies in the samples of Figure 11. This qualitative view may confirm that the heat-affected zones (the center column of the figures) have the least roughness relative to the other two regions. The last two rows also show the indentations (or dimples) resulting from the shot-peening process.

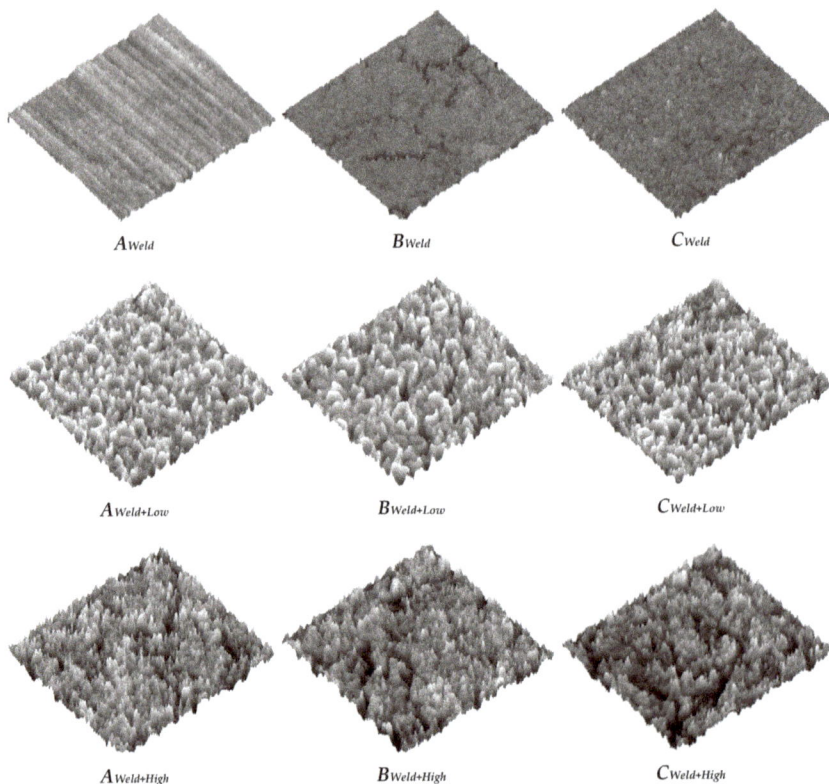

Figure 13. Surface topography images for three samples with three regions each, Image size = 0.8 mm^2.

Further image analysis was performed to find the distribution of dimple size in each of the shot-peened ROIs. These results are shown in Figure 14. The number of dimples was counted for a size range from 20 to 150 micrometers. This was achieved by running a matched filter with a variable-diameter disk as a filter mask. The analysis shows that the low-intensity shot-peened ROIs (top row) do not contain dimples larger than 100 micrometers, whereas the high-intensity shot-peened samples do. Within the sample, using row-wise inspection, it is evident that the heat-affected zones in $B_{Weld+Low}$ and $B_{Weld+High}$ have the most dimples of 100-micrometer size.

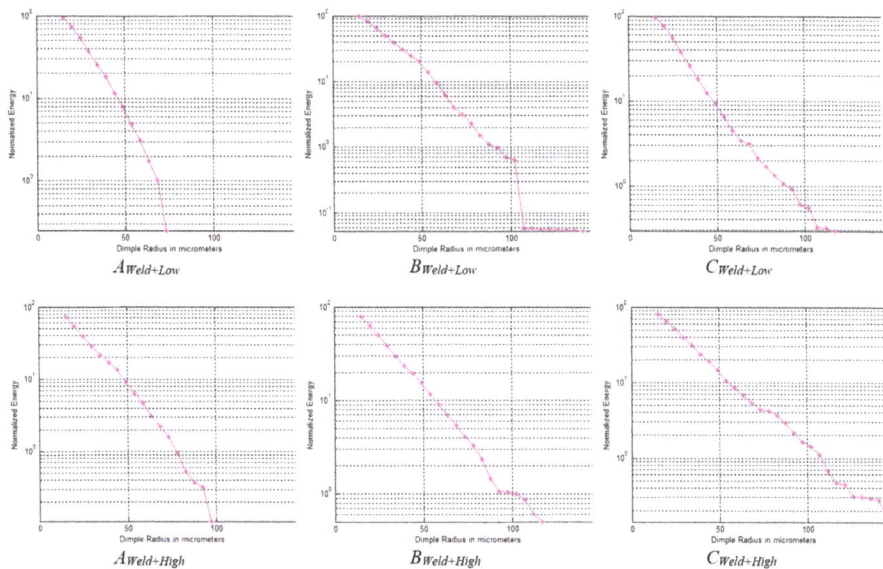

Figure 14. Dimple size distribution for two shot-peened samples with high and low intensity and three regions each.

The roughness was calculated for the samples shown in Figure 11 using the metrics defined in Equations (4) through (6), and plotted in Figure 15. It is evident that the shot-peening process increases the roughness of the surface. Our proposed algorithm was capable of capturing and quantifying surface roughness parameters R_a, P_pK and P_vK. The overlapping observed in the errors bars in Figure 15 shows that our optical metrics can detect the roughness caused by shot-peening; however, they are only marginally suitable for differentiating roughness caused between high- and low-intensity shot-peening.

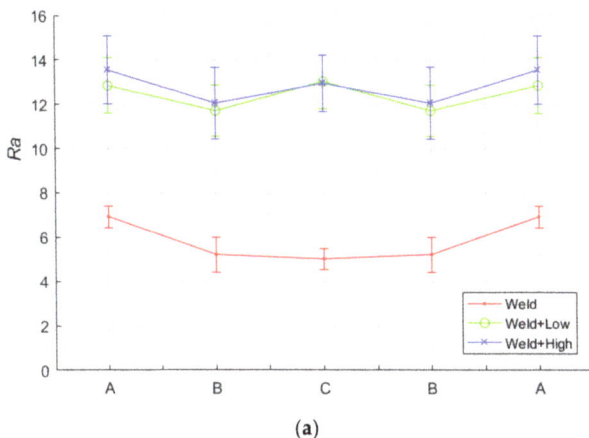

(a)

Figure 15. *Cont.*

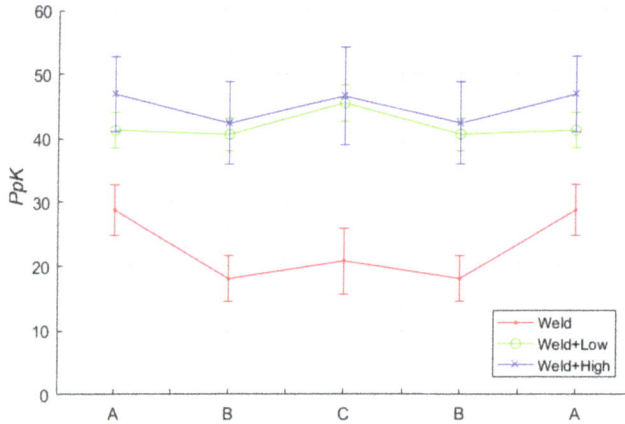

(b)

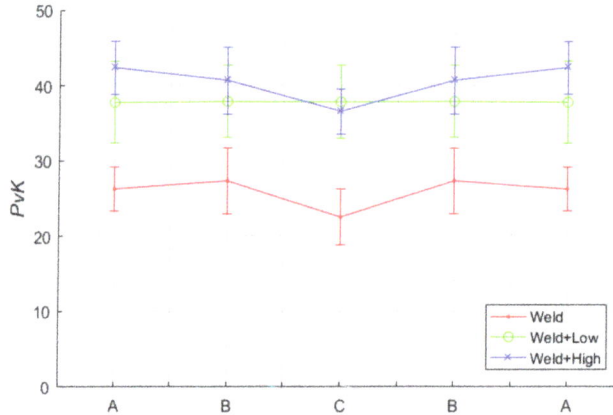

(c)

Figure 15. Optical roughness metrics (in %) for three samples in the three ROIs shown in Figure 11. (a) R_a; (b) P_pK; (c) P_vK.

The high-intensity shot-peened ROIs are the roughest, which corresponds with the qualitative observations explained previously. The results also confirm that the heat-affected zones are the least rough compared to the shot-peened weld bead and unaffected zone. It is also evident that the weld-beads zone responded similarly for both high- and low-intensity shot-peening. In other words, the surface of the weld beads does not change significantly.

This was attributed to the higher hardness of the weld beads zone (region A) compared to the heat-affected zone (region B), which made it less affected by the shot-peening intensity after a certain level. Hence, it is concluded that a measurement of sub-surface hardness is necessary, as shown in Figure 16. It was observed that the surface roughness profile somehow mimics the hardness profile for both high- and low-intensity shot-peened samples, where the W shape was observed. This was attributed to the higher hardness region, which corresponded to higher hardness values; as a result, it is less affected by shot-peening surface treatments.

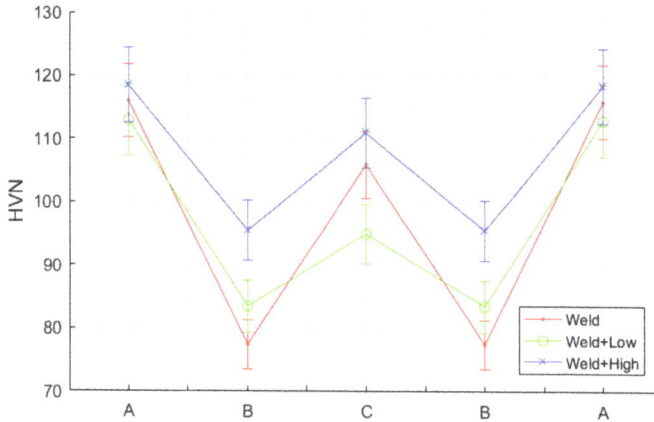

Figure 16. Hardness profile measurements for welded and shot-peened samples for the three regions of interest in Figure 11.

4. Conclusions

In this paper, we have described an image processing algorithm to measure the surface roughness in TIG-welded aluminum 6065-T6 alloy. Using the proposed technique, it was possible to measure the surface roughness of the weld beads, heat-affected zone, and base metal alloy for control samples (TIG-welded only), as well as for high- and low-intensity shot-peened, TIG-welded samples. Furthermore, 3D topography images were generated in a simple manner and used for qualitative analysis. Furthermore, the experimental results showed that dimple size and distribution measurements are possible for shot-peened samples. Optical line profiles were used to calculate roughness metrics to overcome the limitations of measurement equipment in the irregular weld bead area. The final metrics calculated were the averages of the whole region of interest (ROI) data, such that it covers all line profiles and delivers roughness estimates across the whole ROI rather than just a few selected lines, as is the case of laboratory measurements. Also, a blurring filter was used with a 15 × 15 Gaussian mask for convolution and it showed a smoothed image histogram, in which it is evident that the pre-blur histogram has empty intensity bins, which is attributed to the TIFF image format compression artifact compared to blurred ones. Hardness profiles showed that for all tested scenarios of welded only, high- and low-intensity shot-peened samples, softening was observed in the region of weld beads and heat-affected zones compared to base metal alloy. In addition, the heat-affected zone has the least hardness in general, and high-intensity shot-peening is harder than low intensity. Lastly, the presented optical roughness metrics are able to differentiate between shot-peened and untreated alloy surfaces as the metric averages do not lie with the variances of each other. However, this is not the case for high-intensity versus low-intensity shot-peening roughness measurement. Here, only qualitative conclusions are possible because the metric averages lie within the variances of each other.

Author Contributions: A.M.A. and A.N.A. conceived and designed the experiments; A.M.A. and N.A.R. performed the experiments; N.A.R. developed the software code; A.M.A., N.A.R. and A.N.A. analyzed the results; A.M.A. prepare the original draft; A.M.A., N.A.R. and A.N.A. contributed for the writing-review/editing/ and visualization.

Funding: This research was funded by the German Jordanian University, Deanship of Graduate Studies and Scientific Research/GSSR (SATS 12/2015) and Deanship of Scientific Research, King Saud University for funding through Vice Deanship of Scientific Research Chairs.

Acknowledgments: The authors gratefully acknowledge the financial support provided by the German Jordanian University, Deanship of Graduate Studies and Scientific Research/GSSR, grant number: SATS 12/2015.

Testing samples provided by Jordan Air-motive Company (JALCO) is highly appreciated. The authors are also thankful to the financial support of King Abdullah Institute for Nanotechnology, Deanship of Scientific Research, King Saud University, Riyadh, Saudi Arabia.

Conflicts of Interest: The authors declare no conflict of interest.

References

1. Shin, J.; Kim, T.; Kim, D.; Kim, D.; Kim, K. Castability and Mechanical Properties of New 7xxx Aluminum Alloys for Automotive Chassis/body Applications. *J. Alloy. Compd.* **2017**, *698*, 577–590. [CrossRef]
2. Ashraf, P.M.; Shibli, S.M.A. Development of Cerium Oxide and Nickel Oxide-Incorporated Aluminium Matrix for Marine Applications. *J. Alloy. Compd.* **2009**, *484*, 477–482. [CrossRef]
3. Cerik, B.C. Damage Assessment of Marine Grade Aluminium Alloy-Plated Structures due to Air Blast and Explosive Loads. *Thin-walled Struct.* **2017**, *110*, 123–132. [CrossRef]
4. Teimouri, R.; Amini, S.; Mohagheghian, N. Experimental Study and Empirical Analysis on Effect of Ultrasonic Vibration during Rotary Turning of Aluminum 7075 Aerospace Alloy. *J. Manuf. Process.* **2017**, *26*, 1–12. [CrossRef]
5. Heinz, A.; Haszler, A.; Keidel, C.; Moldenhauer, S.; Benedictus, R.; Miller, W.S. Recent Development in Aluminium Alloys for Aerospace Applications. *Mater. Sci. Eng. A* **2000**, *280*, 102–107. [CrossRef]
6. Milligan, J.; Brochu, M. Cladding AA7075 with a Cryomilled Al-12Si Alloy Using Spark Plasma Sintering. *Mater. Sci. Eng. A* **2013**, *578*, 323–330. [CrossRef]
7. Xing, W.Q.; Yu, X.Y.; Li, H.; Ma, L.; Zuo, W.; Dong, P.; Wang, W.X.; Ding, M. Effect of Nano Al_2O_3 Additions on the Interfacial Behavior and Mechanical Properties of Eutectic Sn-9Zn Solder on Low Temperature Wetting and Soldering of 6061 Aluminum Alloys. *J. Alloy. Compd.* **2017**, *695*, 574–582. [CrossRef]
8. Gómora, C.M.; Ambriz, R.R.; Curiel, F.F.; Jaramillo, D. Heat Distribution in Welds of a 6061-T6 Aluminum Alloy Obtained by Modified Indirect Electric Arc. *J. Mater. Process. Technol.* **2017**, *243*, 433–441. [CrossRef]
9. Liang, Y.; Hu, S.; Shen, J.; Zhang, H.; Wang, P. Geometrical and Microstructural Characteristics of the TIG-CMT Hybrid Welding in 6061 Aluminum Alloy Cladding. *J. Mater. Process. Technol.* **2017**, *239*, 18–30. [CrossRef]
10. Ahn, J.; Chen, L.; He, E.; Davies, C.M.; Dear, J.P. Effect of Filler Metal Feed Rate and Composition on Microstructure and Mechanical Properties of Fibre Laser Welded AA 2024-T3. *J. Manuf. Process.* **2017**, *25*, 26–36. [CrossRef]
11. Zalnezhad, E.; Sarhan, A.A.D.M.; Hamdi, M. Prediction of TiN Coating Adhesion Strength on Aerospace AL7075-T6 Alloy Using Fuzzy Rule Based System. *Int. J. Precis. Eng. Manuf.* **2012**, *13*, 1453–1459. [CrossRef]
12. Chang, H.W.; Kelly, P.M.; Shi, Y.N.; Zhang, M.X. Thermal Stability of Nanocrystallized Surface Produced by Surface Mechanical Attrition Treatment in Aluminum Alloys. *Surf. Coat. Technol.* **2012**, *206*, 3970–3980. [CrossRef]
13. Liu, Y.; Jin, B.; Lu, J. Mechanical Properties and Thermal Stability of Nanocrystallized Pure Aluminum Produced by Surface Mechanical Attrition Treatment. *Mater. Sci. Eng. A* **2015**, *636*, 446–451. [CrossRef]
14. Heydari Astaraee, A.; Miresmaeili, R.; Bagherifard, S.; Guagliano, M.; Aliofkhazraei, M. Incorporating the Principles of Shot Peening for a Better Understanding of Surface Mechanical Attrition Treatment (SMAT) by Simulations and Experiments. *Mater. Des.* **2017**, *116*, 365–373. [CrossRef]
15. Pandey, V.; Chattopadhyay, K.; Srinivas, N.C.S.; Singh, V. Low Cycle Fatigue Behavior of AA7075 with Surface Gradient Structure Produced by Ultrasonic Shot Peening. *Procedia Struct. Integr.* **2016**, *2*, 3288–3295. [CrossRef]
16. He, Y.Z.; Wang, D.P.; Wang, Y.; Zhang, H. Correction of Buckling Distortion by Ultrasonic Shot Peening Treatment for 5A06 Aluminum Alloy Welded Structure. *Trans. Nonferrous Met. Soc. China* **2016**, *26*, 1531–1537.
17. Hatamleh, O.; Lyons, J.; Forman, R. Laser and Shot Peening Effects on Fatigue Crack Growth in Friction Stir Welded 7075-T7351 Aluminum Alloy Joints. *Int. J. Fatigue* **2007**, *29*, 421–434. [CrossRef]
18. Sathyajith, S.; Kalainathan, S. Effect of Laser Shot Peening on Precipitation Hardened Aluminum Alloy 6061-T6 Using Low Energy Laser. *Opt. Lasers Eng.* **2012**, *50*, 345–348. [CrossRef]
19. Gao, Y.K. Improvement of Fatigue Property in 7050-T7451 Aluminum Alloy by Laser Peening and Shot Peening. *Mater. Sci. Eng. A* **2011**, *528*, 3823–3828. [CrossRef]

20. Zhang, X.Q.; Chen, L.S.; Li, S.Z.; Duan, S.W.; Zhou, Y.; Huang, Z.L.; Zhang, Y. Investigation of the Fatigue Life of Pre- and Post-Drilling Hole in Dog-Bone Specimen Subjected to Laser Shot Peening. *Mater. Des.* **2015**, *88*, 106–114. [CrossRef]

21. Luong, H.; Hill, M.R. The Effects of Laser Peening and Shot Peening on High Cycle Fatigue in 7050-T7451 Aluminum Alloy. *Mater. Sci. Eng. A* **2010**, *527*, 699–707. [CrossRef]

22. Nam, Y.S.; Jeong, Y.I.; Shin, B.C.; Byun, J.H. Enhancing Surface Layer Properties of an Aircraft Aluminum Alloy by Shot Peening Using Response Surface Methodology. *Mater. Des.* **2015**, *83*, 566–576. [CrossRef]

23. Honga, T.; Ooia, J.Y.; Shawb, B. A Numerical Simulation to Relate the Shot Peening Parameters to the Induced Residual Stresses. *Eng. Fail. Anal.* **2008**, *15*, 1097–1110. [CrossRef]

24. Han, K.; Peric, D.; Crook, A.J.L.; Owen, D.R.J. A Combined Finite/discrete Element Sim- Ulation of Shot Peening Processes-Part I: Studies on 2D Interaction Laws. *Eng. Comput.* **2000**, *17*, 593–620. [CrossRef]

25. Tu, F.; Delbergue, D.; Miao, H.; Klotz, T.; Brochu, M.; Bocher, P.; Levesque, M. A Sequential DEM-FEM Coupling Method for Shot Peening Simulation. *Surf. Coat. Technol.* **2017**, *319*, 200–212. [CrossRef]

26. Hatamleh, O. Effects of Peening on Mechanical Properties in Friction Stir Welded 2195 Aluminum Alloy Joints. *Mater. Sci. Eng. A* **2008**, *492*, 168–176. [CrossRef]

27. Hatamleh, O.; Hill, M.; Forth, S.; Garcia, D. Fatigue Crack Growth Performance of Peened Friction Stir Welded 2195 Aluminum Alloy Joints at Elevated and Cryogenic Temperatures. *Mater. Sci. Eng. A* **2009**, *519*, 61–69. [CrossRef]

28. Hatamleh, O.; Forth, S.; Reynolds, A.P. Fatigue Crack Growth of Peened Friction Stir-Welded 7075 Aluminum Alloy under Different Load Ratios. *J. Mater. Eng. Perform.* **2010**, *19*, 99–106. [CrossRef]

29. Sano, Y.; Masaki, K.; Gushi, T.; Sano, T. Improvement in Fatigue Performance of Friction Stir Welded A6061-T6 Aluminum Alloy by Laser Peening without Coating. *Mater. Des.* **2012**, *36*, 809–814. [CrossRef]

30. Atieh, A.M.; Allaf, R.M.; AlHazaa, A.N.; Barghash, M.; Mubaydin, H. Effect of Pre- and Post-Weld Shot Peening on The Mechanical & Tribological Properties of Tig Welded Al 6061-T6 Alloy. *Trans. Can. Soc. Mech. Eng.* **2017**, *41*, 197–209.

31. Yang, Q.; Zhou, W.; Gai, P.; Zhang, X.; Fu, X.; Chen, G.; Li, Z. Investigation on the Fretting Fatigue Behaviors of Ti-6Al-4V Dovetail Joint Specimens Treated with Shot-Peening. *Wear* **2017**, *372–373*, 81–90. [CrossRef]

32. Xu, C.; Sheng, G.; Wang, H.; Jiao, Y.; Yuan, X. Effect of High Energy Shot Peening on the Microstructure and Mechanical Properties of Mg/Ti Joints. *J. Alloy. Compd.* **2017**, *695*, 1383–1391. [CrossRef]

33. Zhang, G. J.; Yan, Z. H.; Wu, L. Visual Sensing of Weld Pool in Variable Polarity TIG Welding of Aluminium Alloy. *Trans. Nonferrous Met. Soc. China* **2006**, *16*, 522–526. [CrossRef]

34. Xu, Y.; Lv, N.; Fang, G.; Du, S.; Zhao, W.; Ye, Z.; Chen, S. Welding Seam Tracking in Robotic Gas Metal Arc Welding. *J. Mater. Process. Technol.* **2017**, *248*, 18–30. [CrossRef]

35. Bae, K.Y.; Lee, T.H.; Ahn, K.C. An Optical Sensing System for Seam Tracking and Weld Pool Control in Gas Metal Arc Welding of Steel Pipe. *J. Mater. Process. Technol.* **2002**, *120*, 458–465. [CrossRef]

36. Nilsen, M.; Sikström, F.; Christiansson, A.; Ancona, A. Vision and Spectroscopic Sensing for Joint Tracing in Narrow Gap Laser Butt Welding. *Opt. Laser Technol.* **2017**, *96*, 107–116. [CrossRef]

37. Gao, X.; Liu, Y.; You, D. Detection of Micro-Weld Joint by Magneto-Optical Imaging. *Opt. Laser Technol.* **2014**, *62*, 141–151. [CrossRef]

38. Gao, X.; Chen, Y.; You, D.; Xiao, Z.; Chen, X. Detection of Microgap Weld Joint by Using Magneto-Optical Imaging and Kalman Filtering Compensated with RBF Neural Network. *Mech. Syst. Signal Process.* **2016**, *84*, 570–583. [CrossRef]

39. Shankar, N.G.; Zhong, Z.W. Defect Detection on Semiconductor Wafer Surfaces. *Microelectron. Eng.* **2005**, *77*, 337–346. [CrossRef]

40. Valavanis, I.; Kosmopoulos, D. Multiclass Defect Detection and Classification in Weld Radiographic Images Using Geometric and Texture Features. *Expert Syst. Appl.* **2010**, *37*, 7606–7614. [CrossRef]

41. Chu, H.H.; Wang, Z.Y. A Vision-Based System for Post-Welding Quality Measurement and Defect Detection. *Int. J. Adv. Manuf. Technol.* **2016**, *86*, 3007–3014. [CrossRef]

42. Kafieh, R.; Lotfi, T.; Amirfattahi, R. Automatic Detection of Defects on Polyethylene Pipe Welding Using Thermal Infrared Imaging. *Infrared Phys. Technol.* **2011**, *54*, 317–325. [CrossRef]

43. Li, Y.; Li, Y.F.; Wang, Q.L.; Xu, D.; Tan, M. Measurement and Defect Detection of the Weld Bead Based on Online Vision Inspection. *IEEE Trans. Instrum. Meas.* **2010**, *59*, 1841–1849.

44. Elbehiery, H.; Hefnawy, A.; Elewa, M. Surface Defects Detection for Ceramic Tiles Using Image Processing and Morphological Techniques. *Proc. World Acad. Sci. Eng. Technol.* **2005**, *5*, 158–162.

45. Vachtsevanos, G.J.; Dar, I.M.; Newman, K.E.; Sahinci, E. Inspection System and Method for Bond Detection and Validation of Surface Mount Devices. U.S. Patent 5,963,662, 5 October 1999.

46. Bonser, G.R.; Lawson, S.W. Defect Detection in Partially Completed SAW and TIG Welds Using Online Radioscopy and Image Processing. *Process Control Sens. Manuf.* **1998**, *3399*, 231.

47. Kumar, G.S.; Natarajan, U.; Ananthan, S.S. Vision Inspection System for the Identification and Classification of Defects in MIG Welding Joints. *Int. J. Adv. Manuf. Technol.* **2012**, *61*, 923–933. [CrossRef]

materials

MDPI

Article

Gradient Nanostructured Tantalum by Thermal-Mechanical Ultrasonic Impact Energy

Jong-Min Chae [1], Keun-Oh Lee [1] and Auezhan Amanov [2,*]

[1] Department of Safety Engineering, Seoul National University of Science and Technology, Seoul 01811, Korea; jmchae72@daum.net (J.-M.C.); leeko@seoultech.ac.kr (K.-O.L.)
[2] Department of Mechanical Engineering, Sun Moon University, Asan 31460, Korea
* Correspondence: avaz2662@sunmoon.ac.kr; Tel.: +82-41-530-2892; Fax: +82-41-530-8018

Received: 8 March 2018; Accepted: 19 March 2018; Published: 20 March 2018

Abstract: Microstructural evolution and wear performance of Tantalum (Ta) treated by ultrasonic nanocrystalline surface modification (UNSM) at 25 and 1000 °C were reported. The UNSM treatment modified a surface along with subsurface layer with a thickness in the range of 20 to 150 μm, which depends on the UNSM treatment temperature, via the surface severe plastic deformation (S^2PD) method. The cross-sectional microstructure of the specimens was observed by electron backscattered diffraction (EBSD) in order to confirm the microstructural alteration in terms of effective depth and refined grain size. The surface hardness measurement results, including depth profile, revealed that the hardness of the UNSM-treated specimens at both temperatures was increased in comparison with those of the untreated ones. The increase in UNSM treatment temperature led to a further increase in hardness. Moreover, both the UNSM-treated specimens with an increased hardness resulted in a higher resistance to wear in comparison with those of the untreated ones under dry conditions. The increase in hardness and induced compressive residual stress that depend on the formation of severe plastically deformed layer with the refined nano-grains are responsible for the enhancement in wear resistance. The findings of this study may be implemented in response to various industries that are related to strength improvement and wear enhancement issues of Ta.

Keywords: tantalum; hardness; gradient nanostructured layer; grain size; residual stress; dry wear behavior

1. Introduction

Tantalum (Ta) is a rare, refractable, malleable, and lustrous metal, which is widely used in various industries, in particular, aerospace, electronic devices, and nuclear applications, owing to its high ductility at temperatures below 150 °C, good forging, and chemical and physical properties [1]. Especially, Ta exhibits a superior corrosion resistance due to a natural protective layer created by oxides of Ta on its surface [2]. The chemical inertness of Ta makes it an ideal substance for equipment and a substitute for platinum (Pt) [3]. Ta is a highly bioinert metal to manufacture biomedical components, such as hip, knee joints, and other orthopaedic implants because it is not harmed by bodily fluids and also does not irritate the flesh of the implant. The elasticity of Ta helps to avoid stress shielding of hip and knee replacements as well [4]. Ta is also a candidate material to be used in prostheses instead of Ti in the near future. In spite of the facts that Ta is a highly corrosion resistant, chemically inert, etc., but its main advantages are low strength, low wear resistance, and low fatigue strength, which may prevent its successful potential applications in a wide variety energy and fatigue ranges, starting from biomedical through chemical process equipment. The realization of components made of Ta suited for harsh and high-temperature conditions is a challenge since it is problematic to control the required mechanical properties and the fatigue strength of Ta in aerospace applications, such as gas turbines or engines where the temperature rises up to 650 °C [5]. In general, a usage of Pt is a possible option due to its high chemical inertness and high temperature stability, where Ta can be substituted for more expensive Pt, but it is not considerable from the economic and commercial point of view. Therefore, an increase in strength and wear performance, and also an extension in service life of Ta are in high demand.

One of the easy and possible ways to control the wear performance of metallic materials by increasing its strength is controlling its microstructure, in particular, producing nano-grains with grain boundaries of mostly high angle misorientation via surface severe plastic deformation (S²PD), which is a cold-forging process [6,7]. A wide range of S²PD processes, such as shot peening (SP) [8], surface mechanical attrition treatment (SMAT) [9], surface rolling treatment (SRT) [10], and ultrasonic shot peening (USP) [11], and severe plastic deformation (SPD) processes, such as equal channel angular processing (ECAP) [12] and high-pressure torsion (HPT) [13] were developed in the past. These processes develop a severe deformation and high strain, which cause the creation of gradient micro- and sub-micrometer grains, whose size is gradually increasing with the depth and finally reaches the actual initial size of matrix and coarse grains in a certain of depth [10,13]. As a consequence, sufficient S²PD and SPD processes result in the apparent modification in microstructure in terms of highly misoriented nano-sized grains. The mechanism behind the nanocrystallization and grain refinement via S²PD and SPD processes lies shear bands associated with the dynamic recrystallization (DRX) [14]. The results of S²PD and SPD processes, such as microstructural state alteration, are strongly responsible for the subsequent increase in mechanical properties of metallic materials, which in turn has a direct correlation to the wear performance of metallic materials [6]. S²PD and SPD processes are capable of increasing the hardness, yield strength, and elastic strain featuring tendency to saturation, but it is not beneficial in terms of ductility [15]. Generally, nano-sized grains have some benefits in comparison with the coarse grains, not only in terms of strength of a material, but also superplastic deep drawing.

A wide range of metallic materials such as aluminum [6], titanium [7,9,12], stainless steel [8], Inconel [11], etc. were subjected to S²PD and SPD processes to achieve a refined nano-grained and nanocrystalline structure in the past, but only a few limited studies concerning the application of those processes to Ta can be found in the literature. For example, Huang et al. studied the effects of HPT on microstructure and the hardness of pure Ta [16]. It was found that the grain size was refined significantly due to the increase in hardness, but some coarse grains appeared with increasing the numbers of rotations. Mathaudhu et al. produced a fine grained Ta via multi-pass equal channel angular extrusion (ECAE) process [17]. It was concluded that the refined nano-grains with the size of 100–400 nm was found to be beneficial for Nb₃Sn superconductor. Another study by Mathaudhu et al. on grain size refinement of Ta by ECAE process has attracted wide attention in view of a variety of findings [18]. It was found that ECAE process refined grains at $\varepsilon = 2.3$, which is more important than initial grain size to occur a recrystallization. Moreover, Zhang et al. reported the possibility of generating a nanostructured Ta with a grain size of 20 nm via sliding friction treatment (SFT) [19]. Nevertheless, there are no systematic investigations on the mechanical properties and wear performance of Ta treated by ultrasonic nanocrystalline surface modification (UNSM) technique [20]. A precision control of UNSM technique is a key advantage to produce a gradient nanostructured material over other SPD and S²PD processes. The UNSM treatment used to be applied for various metallic materials, ceramics, Si wafer, various coatings to improve the friction behavior, wear and corrosion resistance, and fatigue strength through the presence of nanostructured surface layer with the refined nano-sized grains [21–25]. Furthermore, a high-temperature UNSM treatment along with local heat treatment (LHT) was developed and recently patented, where the UNSM treatment temperature can be reached up to 1400 °C. Interestingly, so far it was discovered that the hardness and wear performance of Ti-6Al-4V alloy was improved via UNSM treatment at a high temperature [26]. However, the hardness can be continuously increased with temperature, where a softening may occur as well. For example, a softening occurred in Ni-based superalloy (Inconel 690 alloy) when the high-temperature UNSM treatment was performed at 700 °C [27]. Moreover, the highest hardness of α-Ta treated by high-temperature UNSM treatment was found at 800 °C [20]. It is therefore of interest to further increase the high-temperature UNSM treatment temperature up to 1000 °C, and to investigate its effects on the grain size refinement, hardness, and wear resistance of Ta since the increase or decrease in hardness and also grain size refinement or grain growth depend on the nature of a material. In this regard, the main purpose of the investigation is to provide systematic experimental results on the grain size refinement, hardness, and wear performance of Ta that is treated by UNSM treatment at room temperature (RT) and high temperature (HT) of 1000 °C. It is strongly believed, according to the results, that the role of a high-temperature UNSM in Ta-related applications, such as aerospace, nuclear, electrical, etc., will be significantly important.

2. Materials and Methods

2.1. Specimen Preparation

Refractory Ta is rarely used since alloying makes other metals brittle, with an exception of steel, in which case it increases the ductility and strength. In this study, the specimens with dimensions of 20×5 mm^2 was prepared from bulk Ta with a hardness of 195 ± 6 HV (after cold working). Important properties of Ta are shown in Table 1.

Table 1. Mechanical and physical properties of Tantalum (Ta).

UTS, MPa	Yield Strength, MPa	Elastic Modulus, GPa	Poisson's Ratio	Density, g/cm^3	Elongation, %
900	170	186	0.35	16.6	5

2.2. Ultrasonic Nanocrystal Surface Modification (UNSM) and Local Heat Treatment (LHT)

The UNSM parameters that are shown in Table 2 were selected to treat the specimens at RT and HT (1000 °C) temperatures. At HT, the specimens were heated up with a halogen-based high-temperature heating system, where the actual temperature was measured using a portable pyrometer. More detailed information about the UNSM technique, including high-temperature heating setting of LHT, can be found in our previous publications [26,27]. Following the UNSM treatment, both of the disk specimens were mounted in bakelite and were polished with sand papers down to 2400 grit, and then a colloidal solution was used with a powder of 1 μm in diameter to achieve a mirror-like surface. Afterwards, the disk specimens were electrolytically etched in H_2O, H_2SO_4, HF with a couple of drops of H_2O_2 solution at 5 V for 30 s using an electropolisher-etcher (ElectroMet^{TM}4, Buehler, Uzwil, Switzerland) to reveal the microstructural features, such as grains, grain boundaries, etc.

Table 2. Ultrasonic nanocrystalline surface modification (UNSM) parameters at room temperature (RT) and high temperature (HT) (1000 °C).

Frequency, kHz	Amplitude, μm	Static Load, N	Speed, mm ε	Tip Material	Tip Diameter, mm	Interval, μm
20	40	30	2000	WC	2.38	70

2.3. Wear Behavior in Dry Conditions

A commercially available ball-on-disk tribometer (Anton Paar, Graz, Austria) was used to evaluate the wear performance of Ta that came into contact with a steel ball under the test conditions, as shown in Table 3. Each test was replicated three times. The surface roughness ($R_a < 0.08$ μm) of the specimens was considered to be nearly close to escape from the influence of surface roughness on the wear performance under dry conditions.

Table 3. Wear test conditions under dry conditions.

Applied Normal Load, N	Reciprocating Speed, cm/s	Sliding Distance, m	Temperature, °C	Contact Stress, GPa
10	2.51	30	25	0.82

2.4. Characterizations

The surface hardness of the specimens was measured using a micro-Vickers hardness tester (MVK E3, Mitutoyo, Takatsui, Japan) at a load of 300 gf for dwell time of 10 s, while the nanoindentation was performed using a depth sensing tester (MTS, Nanoindenter XP, Eden Prairie, MN, USA), with a diamond Berkovich indenter at a frequency of 45 Hz, strain rate of 0.05 s^{-1} with a maximum load of 100 mN. The average surface roughness (R_a) and cross-sectional wear track profiles to quantify the wear rate were measured using a portable two-dimensional surface profilometer (SJ-210, Mitutoyo, Japan). X-ray diffraction (XRD) was performed with a CuKα radiation (k = 1.5418 Å) at a wavelength of 1.54, a tube current of 40 mA, and a voltage of 30 kV using a sin^{2}ψ method over the range of 20–130°, with a scanning speed of 1°/min using a Bruker D8 Advance X-ray diffractometer to measure the residual stress and to identify phases before and after UNSM treatment. The dimension of the specimen was 10 × 10 × 3 mm^3. The residual stress was also measured by using an indentation method that was based on the nanoindentation results. The obtained residual stress measurement data with a huge error bar represent the average of three measurements. A gradient nanostructured surface layer was observed using an electron backscatter diffraction (EBSD: Oxford Instruments HKL Nordlys Max, Abingdon, UK) installed into a scanning electron microscopy (SEM: JEOL 6610LV, Tokyo, Japan) at an accelerating voltage of 20 kV with a large beam current of 10 nA. Surface microstructure and wear tracks, and the chemical composition were characterized by SEM, along with energy-dispersive X-ray spectroscopy (EDX: EDAX/AMETEK, Mahwah, NJ, USA).

3. Results and Discussion

3.1. Microstructural Evolution by EBSD

Cross-sectional Inverse Pole Figure (IPF) maps of the specimens obtained using an electron backscattered diffraction (EBSD) are shown Figure 1. The untreated specimen, as shown in Figure 1a, presents a homogenous microstructure with equiaxed grains at the topmost surface in the range of 10–30 μm in diameter. Obviously, it was observed that the UNSM treatment at RT generated a gradient nanostructured surface layer, as shown in Figure 1b. In addition, an effective depth of UNSM treatment at RT was found to be about 35–40 μm, while some refined grains are also randomly visible in a depth of about 100 μm between the elongated coarse grains with further increasing the depth beyond the actual effective depth of UNSM treatment at RT. This irregular deformation may be attributed to the inhomogeneous of a large amount of plastic deformation that was introduced during UNSM treatment. Figure 1c shows the IPF color map of the specimen LHT at 1000 °C without UNSM treatment. It can be seen that some elongated coarse grains at the top surface were refined remarkably, which may be a result of machining and mechanical polishing. Figure 1d shows the IPF color map of the specimen LHT at 1000 °C with UNSM treatment. The modification in microstructure with depth can be further extended by performing an LHT with UNSM treatment at 1000 °C in comparison with the UNSM-treated specimen at RT. The effective depth was found to be deeper than ~150 μm, which is about five times deeper than that of the UNSM at RT specimen. Interestingly, elongated coarse grained was refined due to the results of heat treatment, but both grain size refinement and extension in effective depth occurred by performing an LHT with UNSM treatment at 1000 °C (see Figure 1c). Moreover, it was observed that a lamellar structure with low-angle grain boundaries (LAGB) is visible (see Figure 1b,d)). It is worth mentioning here that no banded structures, such as deformation and shear, were observed in the untreated and LHT at 1000 °C without UNSM treatment specimens.

Microbands are visible feature in the both UNSM-treated at RT and HT specimens, as shown by the dashed lines in Figure 1b,d. These banded structure produced by UNSM treatment is an important factor in high stacking fault energy (SFE) metallic materials subjected to deformation [28]. In general, α-Ta has needle-shape morphology, while β-Ta has a spherical-shaped one [1]. As shown in Figure 1, the elongated shape of needle-shape grains of α-Ta was changed into the spherical-shaped grains, owing to ultrasonic-based strikes at a frequency of 20 kHz.

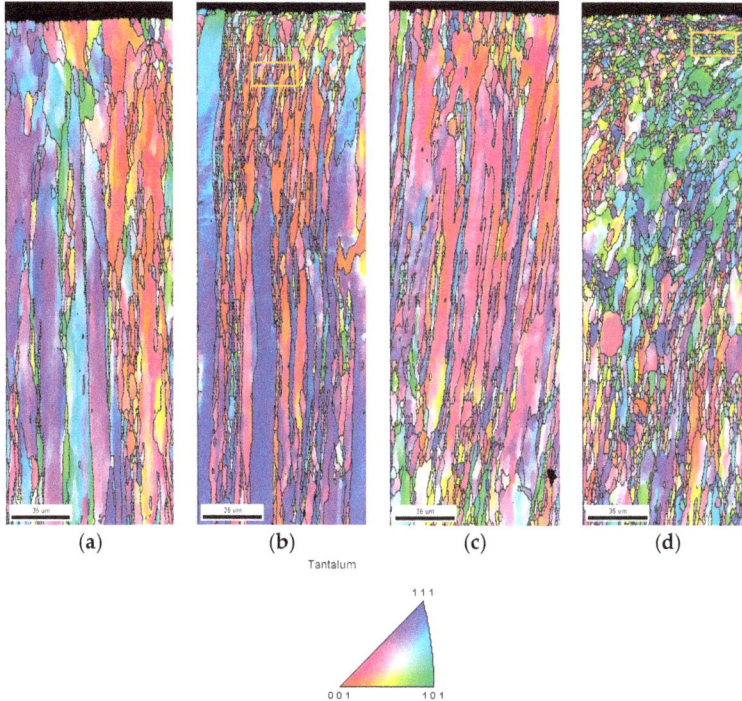

Figure 1. Cross-sectional electron backscatter diffraction (EBSD) Inverse Pole Figure (IPF) maps of the untreated (**a**), UNSM-treated at RT (**b**) and local heat treatment (LHT) without (**c**) and with (**d**) UNSM treatment at 1000 °C specimens.

The grain size distributions with area fraction of the specimens are presented in Figure 2. The grain size distribution of the untreated specimen (see Figure 2a) shows that the grains in the range of 5–60 μm are distributed uniformly, while the UNSM treatment at RT was capable of producing a gradient nanostructure layer with a high fraction (~18.8%) of (sub) grains with a size less than 0.1 μm at the topmost surface, which is deliberately increased, as shown in Figure 2b. The fraction of grains of the specimen treated by solely LHT treatment (without UNSM treatment) was found to be higher in comparison with the untreated specimen, as presented in Figure 2c. The grain fraction was increased significantly by high-temperature UNSM treatment at 1000 °C, as shown in Figure 2d and also the slope (angle) of a gradient nanostructure was higher in comparison with the UNSM-treated specimen at RT. Gradient nanostructured materials have a number of advantages over the homogeneous nanostructured materials in terms of mechanical properties, especially ductility. For example, Lu has discovered the possibility of making a balance between the strength and ductility of materials by producing a gradient nanostructured material [29]. Kang et al. have also demonstrated the features and importance of a gradient nanostructured material that is produced by high pressure torsion (HPT) [13]. It was concluded that the gradient nanostructured material had a significant higher strength with no loss in ductility in comparison with the nanostructured material. Yin et al. have also produced a gradient nanostructure surface layer on Cu by SMAT, pointing out that a gradient nanostructure can exhibit superior strength-ductility synergy [30]. Recently, Wang et al. reported the possibility of generating a gradient nanostructured surface layer in Cu with a grain size of 85 nm via rotationally accelerated shot peening (RASP) [31]. This newly developed RASP technology may apply much higher impact energy in comparison with the conventional SP process. Yang et al. investigated the role of volume fraction of gradient nanostructures Cu that is produced by SMAT at cryogenic temperature [32]. It was found that the gradient nanostructure exhibited a great correlation between strength and ductility. Wu et al. combined gradient nanostructure with transformation-induced plasticity that is produced by SMAT to synthesized gradient nanostructure in austenitic 304 stainless steel [33]. As a result, a gradient nanostructured layer provided a good correlation between strength and ductility. Therefore, it is now desirable to produce a gradient nanostructured material with a thick layer as much as possible as the homogeneous nanostructured material due to the lack of ductility. Moreover, AlMangour and Yang improved the mechanical properties of 17-4 steel that was fabricated by direct metal laser sintering (DMLS), which is a type of additive manufacturing (AM), through grain size refinement by means of SP [34]. It was reported that severe plastically deformed layer along with grain size refinement by SP led to an increase in mechanical properties.

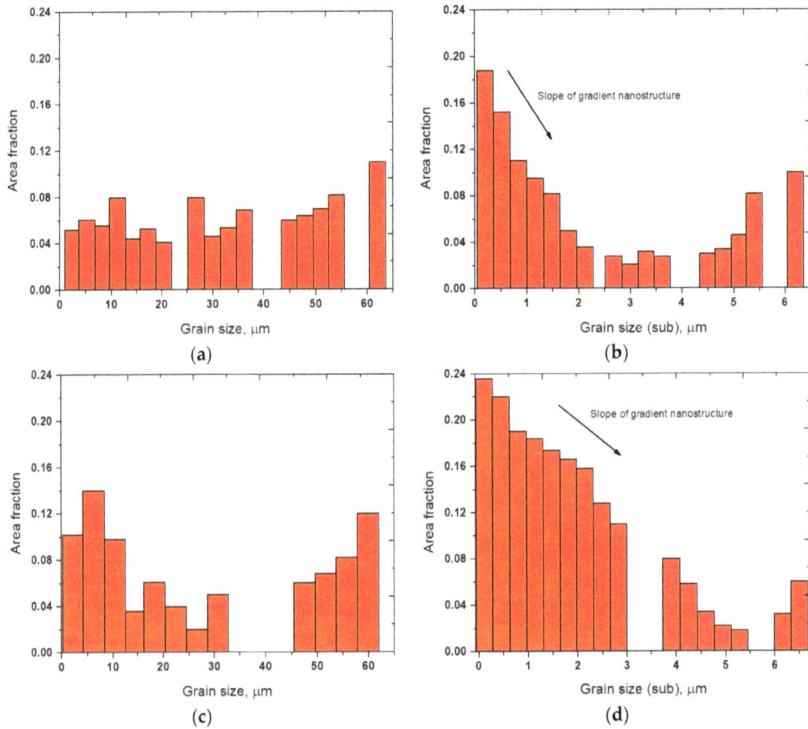

Figure 2. Histogram of the grain size distribution of the untreated (**a**), UNSM-treated at RT (**b**) and LHT without (**c**) and with (**d**) UNSM treatment at 1000 °C specimens representing the presence of gradient nanostructured surface layer.

3.2. Residual Stress and XRD Pattern

A comparison in residual stress of the specimens measured at $\phi0°$ and $\phi90°$ is presented in Figure 3a. It is apparent that the untreated and heat up without UNSM treatment specimens exhibited a tensile residual stress, while the UNSM treatment at both RT and HT induced a great compressive residual stress. The value of the compressive residual stress of the UNSM-treated at RT specimens that were measured both perpendicular and along orthogonal directions was about −600 MPa, which was reached a greater −1200 and −1375 MPa by using an LHT with UNSM treatment at 1000 °C, respectively. Consequently, the compressive residual stress of both the UNSM-treated at RT and HT specimens that were measured along orthogonal direction of $\phi90°$ had a higher compressive residual stress in comparison with the compressive residual stress measured along orthogonal direction of $\phi0°$. This difference is due to the different number of strikes. It is well established that the induced compressive residual stress is the most crucial property, which determines the strength and fatigue lifespan of a material [35,36]. It was also reported earlier that SP and USP processes induces high compressive residual stress in the surface layer thanks to the severe plastically deformed layer [11,34].

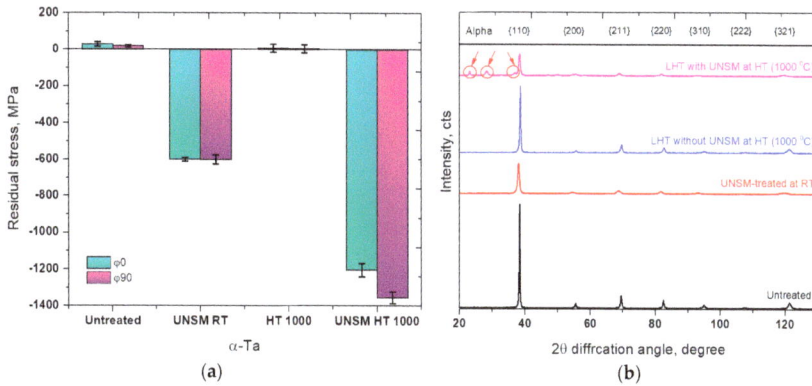

Figure 3. Variation in surface residual stress (**a**) and X-ray diffraction (XRD) pattern (**b**) of the untreated, UNSM-treated at RT and LHT without and with UNSM treatment at 1000 °C specimens.

XRD patterns of the specimens are presented in Figure 3b. The relative intensities of all the diffraction peaks of the untreated specimen were reduced significantly after UNSM treatment at RT. By treating the specimen by LHT without UNSM treatment at HT led to a negligible reduction in intensity in comparison with the untreated specimen, but a substantial reduction in intensity was found for the specimen by LHT with UNSM treatment at HT, which is lower in comparison with the UNSM-treated specimen at RT. Post-polishing process of the specimens treated by LHT with and without UNSM treatment at HT was responsible for the absence of any diffraction peaks of the oxide particles on the surface. The high-temperature UNSM treatment, regardless the treatment temperature, may result in the formation of thick oxide layer, which is responsible for the deteriorated mechanical properties [26]. In addition, new diffraction β diffraction peaks {101}, {400} and {410} are detected after LHT with UNSM treatment at HT (see the inset in Figure 3b), which means that the phase transformation occurred from α → β. Beta Ta is a metastable phase that is transformed from α Ta when it heated up to 900 °C [37]. The α phase is tend to have excellent corrosion, thermal ductility properties, while the β Ta provides additional hardenability, therefore the newly appeared β phases are expected to beneficially affect the strength, wear performance, and fatigue strength.

A comparison in relative intensity, full width at half maximum (FWHM) and d spacing with respect to diffraction angle of the specimens is presented in Figure 4. It is clear that the relative intensity of the primary alpha peak {110} diffracted at an angle of 38.7 was about 25,000, 5000, 16,700, and 4900 for the UNSM-treated at RT, and LHT with and without UNSM treatment at HT specimens, respectively. Other secondary {200}, {211}, {220}, {310}, {222}, and {321} peaks also reduced remarkably by UNSM treatment both at RT and HT, where they were not visible in Figure 4, while those peaks were reduced as well after LHT without UNSM treatment. In addition, it is noticeable from Figure 4 that the FWHM of the UNSM-treated specimens at both RT and HT got broadening in comparison with the untreated specimen, where the FWHM was increased with increasing the diffraction angle. The highest FWHM of the UNSM-treated specimens at both RT and HT was found to be about 2.5° at a diffraction angle of 108°, as shown in Figure 4b,d. From this, it is clear that the FWHM depends on the relative intensity of the peaks, where the lower relative intensity the higher FWHM. As shown in Figure 4, no significant difference in d spacing of the specimens was found, which is related to the change of diffraction peak position through Bragg's law [38], where it was gradually reduced with increasing the diffraction angle because when the UNSM treatment applied load is removed, the d spacing returned to normal position unless a compressive residual stress induced by UNSM treatment controls the original strain. It is worth mentioning here that the changes, such as strain, work hardening, etc. of the UNSM treatment at both RT and HT specimens can be estimated by quantitatively analyzing the broadening in FWHM and reduction in relative intensity of diffraction peaks [39]. Top surface grain size of the specimens quantified based on the Scherer equation was in consistent with the cross-sectional EBSD IPF maps, where the refinement of coarse grains into (sub) grains is clearly seen in Figure 1b,d, where the refined (sub) grain size by UNSM treatment at RT was further refined with increasing the temperature, leading to the highest fraction (~23.8%) of sub (grains) at the top surface. Apparently, it is expected to achieve more grain size refinement towards ultrafine grain (UFG) scale with a high-angle grain boundaries (HAGB) by increasing the temperature of UNSM treatment, which can be reached up to 1400 °C so far. In order to confirm the presence of UFG scale at the top surface since the EBSD method does not allow due to the resolution, some advanced quantitative surface analysis by transmission electron microscopy (TEM) is in need to determine the exact nano-sized grain.

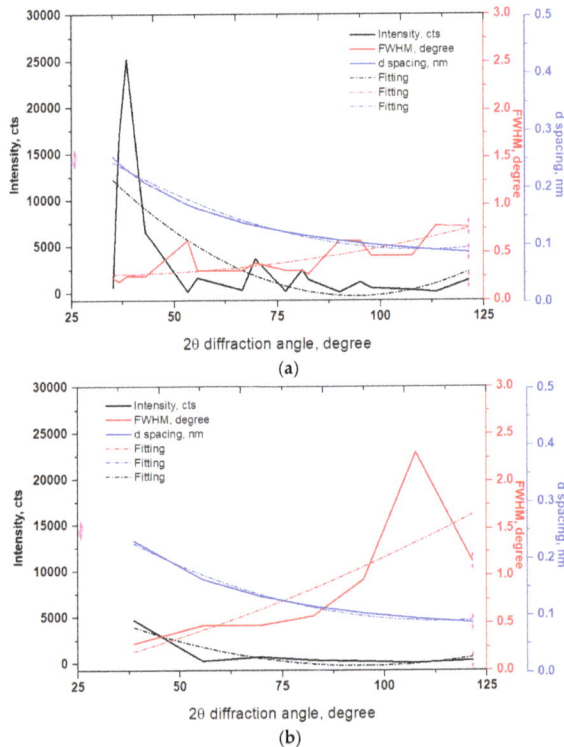

(a)

(b)

Figure 4. *Cont.*

(c)

(d)

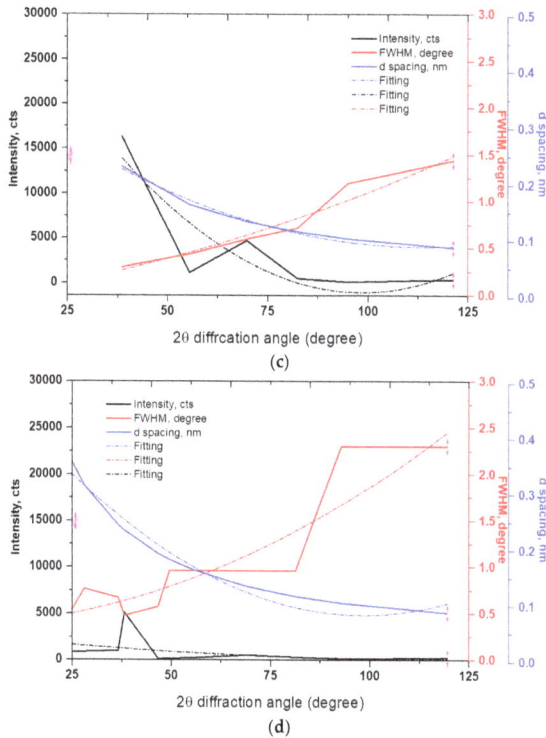

Figure 4. Variation in intensity, full width at half maximum (FWHM) and spacing of the untreated (**a**), UNSM-treated at RT (**b**) and LHT without (**c**) and with (**d**) UNSM treatment at 1000 °C specimens.

Furthermore, the diffraction peaks of the UNSM-treated specimens at both RT and HT shifted to a lower diffraction angle (see Figure 5), which is an indicator of the induced compressive residual stress [40], while on the contrary, the diffraction peak of the LHT without UNSM treatment at 1000 °C specimen shifted to a higher diffraction angle. The presence of uniform compressive strain that was derived during grain size refinement process is responsible for the diffraction peak shift to a lower angle [41], while the diffraction peak shift to a higher angle is responsible for the tensile stress [42]. It is well established that the internal stresses, planar faults (stacking faults or twinning) are responsible for the changes in relative diffraction peaks (intensity, FWHM, and shift) of the metallic materials that are subjected to both S^2PD and SPD processes [43,44].

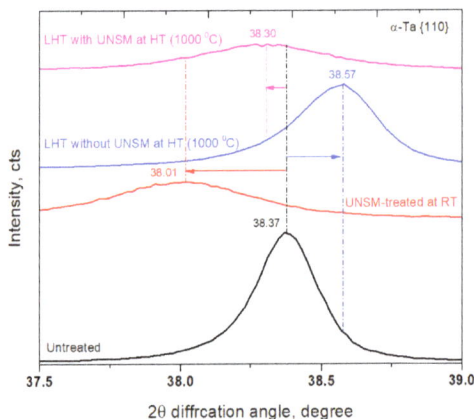

Figure 5. Comparison in intensity peak {110} shift to lower and higher angles of the untreated, UNSM-treated at RT, LHT without and with UNSM treatment specimens.

3.3. Microhardness and Nanoindentation

A comparison in surface hardness of the specimens is shown in Figure 6. The UNSM treatment at RT led to an increase in hardness by about 20% in comparison with the untreated one. In turn, the hardness was further increased from 193 to 511 HV by heating up the specimen up to 1000 °C, which is corresponding to a 62% increase in comparison with the untreated one. The untreated specimen that was treated by the combination of an LHT with UNSM treatment was able to further increase the surface hardness by about 16%, 58%, and 70% in comparison with the LHT specimen without UNSM, UNSM-treated at RT, and untreated ones, respectively. The increase in surface hardness by UNSM treatment at both RT and HT is associated with the grain size refinement, which may be explained well by the Hall-Petch relationship, where the grain size is a key factor, in other words, the smallest grain size the highest hardness [45], while the increase in hardness by LHT is related to the movement of atoms from their original position [46]. However, Chokshi et al. have reported a negative slope, where the hardness or strength of a material start dropping with reducing the grain size less than 10 nm due to the grain boundary sliding [47]. Therefore, it is always desirable to refine the grain size bigger than that critical value.

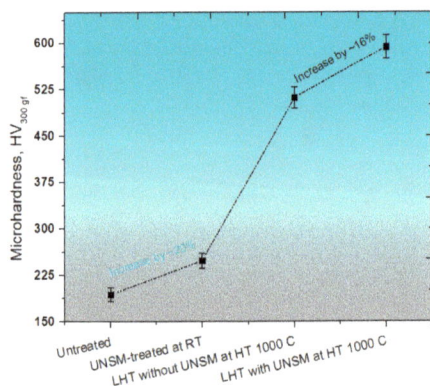

Figure 6. Comparison in surface hardness of the untreated, UNSM-treated at RT and LHT without and with UNSM treatment at 1000 °C specimens.

Load and depth of penetration curve of the specimens that was obtained by nanoindentation method is depicted in Figure 7. The UNSM-treated at RT specimen had a shallower depth of penetration in comparison with the untreated specimen under the same load of 100 mN. The depth of penetration occurred on the surface of the LHT without UNSM specimen at HT got shallowed at the same load in comparison with both the untreated and UNSM-treated at RT specimens due to the higher surface hardness, as shown in Figure 6. It was noticed that the UNSM-treated at HT specimen exhibited two times shallower penetration depth in comparison with the untreated one. The increase in hardness by UNSM treatment at both RT and HT thanks to the generation of a gradient nanostructured surface layer and a relatively high dislocation density [13,19,30,48]. Estimation of residual stress is made by various measurement methods, such as XRD, ultrasonic, neutron diffraction, strain curve, magnetic, hole drilling, and Raman spectroscopy as well [49], but these methods have some drawbacks in terms of spatial resolution, data accuracy, and reliability. Zhu et al. discovered a new method to measure the residual stress by the nanoindentation method [50]. The inset in Figure 7 shows the magnified loading curves of the specimen at the onset of the indentation. According to the conclusion of the study [51], the UNSM-treated at RT and LHT with UNSM treatment at HT specimens exhibited a compressive residual stress since both of these specimens required a higher load to be indented in comparison with the untreated and LHT without UNSM treatment at HT specimens, as shown in Figure 7. In turn, the LHT with UNSM treatment at HT specimen required larger load in comparison with the UNSM-treated at RT because of the higher induced compressive residual stress, as shown in Figure 3. In order to validate the results of residual stress that was measured by XRD and nanoindentation methods, a newly proposed nanoindentation method [51,52], which was adopted based on the difference between the contact areas of the specimens, was used using the following equations [51]:
for tensile residual stress

$$\sigma_r = H(1 - A_0/A) \tag{1}$$

for compressive residual stress

$$\sigma_r = H\left(1 - \frac{A_0}{A}\right)/f \tag{2}$$

where A and A_0 are the indentation contact area of the specimens with tensile and compressive residual stress (σ_r), respectively. H is the material hardness. $f = \sin \alpha$ is a geometric factor, where α is related to the indentation angle of the indenter. For a Berkovich indenter, $\alpha = 24.7°$ and $f = 0.418$.

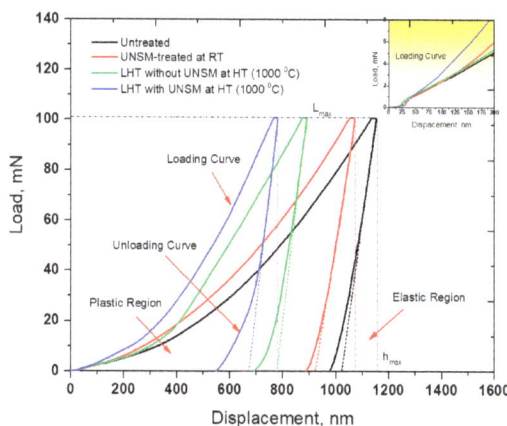

Figure 7. Comparison in load-displacement curve of the untreated, UNSM-treated at RT and LHT without and with UNSM treatment at 1000 °C specimens.

After obtaining a contact area of the specimens with tensile and compressive residual stresses, the tensile residual stress of the untreated and LHT without UNSM treatment at HT, and the UNSM-treated at RT and HT specimens was calculated by Equations (1) and (2), respectively. It was found that the calculated tensile residual stress of the untreated and LHT without UNSM treatment at HT was 24.6 ± 9 and 9.4 ± 3 MPa, while the tensile residual stress that was measured by the XRD method was 20.47 ± 7 and 6.54 ± 2 MPa, respectively. The calculated compressive residual stress of the UNSM-treated at RT and HT specimens was −636.26±86 and −1284.71 ± 114 MPa, while the compressive residual stress that was measured by XRD method was −607.44 ± 82 and −1228 ± 59 MPa, respectively. It is apparent from the residual stress results that were obtained by the nanoindentation method are in good consistence with the residual stress results obtained by XRD method ensuring a standard deviation in the range of about 10–14%. It is worth mentioning here that the residual stress of Au/TiW bilayer was estimated by deflection of double clamped beams, where the beam deflection was corrected by the indent penetration [53]. This method can also be adopted to Ta as well. As a consequence, the residual stress measurement results of the specimens by nanoindentation method are applicable to predict the induced compressive residual stress by UNSM treatment. Additionally, it is important to mention here that this method can be adopted only to materials that create a pile-up around the indent after nanoindentation. The cross-sectional profile of the residual indent on the surface of the UNSM-treated at RT specimen is shown in Figure 8, where the pile up around the indent after nanoindentation is clearly observed. Hence, the calculated tensile and compressive residual stress results by nanoindentation method in this study were found to be absolutely accurate and reliable.

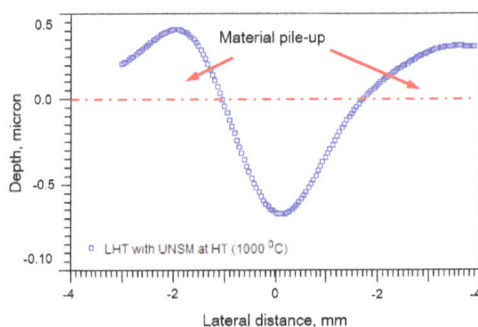

Figure 8. Cross-sectional profile of the residual indent on the surface of the UNSM-treated at RT specimen.

3.4. Friction and Wear Performance

Friction of metallic materials is usually relatively high under dry conditions due to the frictional mating contact inducing plastic deformation in relative motion, leading to a mating surface roughening and progressive wear, delamination, or even fatigue. In this regard, the friction of metallic materials is a crucial property to improve the performance, reliability, and efficiency of metallic materials contacts in various industries. Figure 9 shows the variation in friction coefficient of the specimens as a function of sliding distance. The friction coefficient of the specimens was increased drastically at the beginning of the test, but the friction coefficient of the LHT with UNSM treatment at HT specimen increased gradually. All of the specimens demonstrated running-in, transition, and steady-state periods, as partially shown each periods in Figure 9. Obviously, the untreated specimen had the highest friction coefficient among other specimens with a friction coefficient of 0.72 during running-in period, which reduced gradually till the friction coefficient of 0.58 in transition period, and then approached a stabilization in friction coefficient of 0.44 in the steady-state period. The friction coefficient of the UNSM-treated at RT specimen was also increased drastically at the beginning of the test, and then continued to increase to a friction coefficient of 0.62 during the running-in period and then reduced rapidly till the friction coefficient of 0.52 in transition period and then finally approached a stabilization in friction coefficient of 0.44 in steady-state period. The UNSM treatment was found to be beneficial in running-in and transition periods, but not in steady-state one, which is attributed to the lack of change in initial surface integrity of the UNSM-treated at RT specimen under dry conditions, where the effectiveness of UNSM treatment can be easily lost under severe plastic deformation. It has been reported earlier that the UNSM treatment reduced the friction coefficient of metallic materials under both oil-lubricated and dry conditions, due to the features of UNSM treatment, such as improvement in surface integrity, grain size refinement, the presence of micro-dimples on the surface, etc. [21,22,51]. On the contrary, Chen et al. have pointed out that refining the grain size of metallic materials cannot reduce the friction coefficient under dry conditions even though its' hardness may be increased significantly [52]. Fortunately, UNSM treatment not only increases the hardness of metallic materials, but it also reduces the surface roughness and creates a bunch of dimples (dint) that can behave as traps for wear particles under both oil-lubricated and dry conditions [51,54]. In case of the LHT without UNSM at HT specimen, the friction coefficient exhibited absolutely the same friction coefficient trend with the untreated and UNSM-treated at RT specimens, where the friction coefficient was about 0.51 in steady-state period, which was increased continuously throughout the friction test with relatively high fluctuation (see Figure 9). The continuous rise in friction coefficient is due to the generated wear particles or debris derived from roughened surface, due to repeated sliding under dry conditions. Interestingly, in the case of the LHT with UNSM at HT specimen, the friction coefficient slowly increased first to the highest friction coefficient of 0.59, and then reduced again to a value of 0.42, and finally approached stabilization in a friction coefficient of 0.39, which reduced slightly throughout the friction test. In addition, a shift in transition periods can be observed where the running-in and transition periods were found to be shortened by LHT with UNSM at HT, while a steady-state period achieved faster than other specimens, as shown in Figure 9.

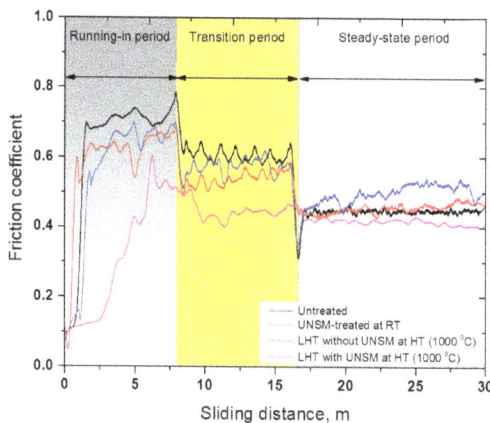

Figure 9. Comparison in friction coefficient with respect to sliding distance of the untreated, UNSM-treated at RT and LHT without and with UNSM treatment at 1000 °C specimens.

The wear resistance of the specimens showing the rate how fast or slow wear occurred is shown in Figure 10, where the wear rate of the UNSM-treated at RT specimen was increased remarkably by about 18–20% in comparison with the untreated specimen. The corresponding wear resistance of the LHT without UNSM at HT specimen increased substantially in comparison with those specimens due to the increase in hardness, as shown in Figure 6. In turn, the corresponding wear resistance of the LHT with UNSM at HT specimen was further increased by an additional 36% in comparison with the LHT without UNSM at HT specimen. Overall, the corresponding wear rate of untreated specimen can be decreased by over 90% by the application of thermal-mechanical UNSM treatment at HT of 1000 °C. Wear resistance of a material and its hardness, compressive residual stress, and grain size have a linear correlation. The enhancement in wear resistance of the UNSM-treated specimen was attributed to the presence of gradient nanostructured surface layer, along with refined coarse grains into nano-sized grains, induced compressive residual stress, increased surface, and subsurface hardness. Furthermore, the wear resistance of the LHT with UNSM at HT specimen was further enhanced by increasing the UNSM treatment temperature up to 1000 °C in comparison with that of the UNSM-treated specimen. It was found that the combination of LHT with UNSM at HT of 1000 °C was able to produce a thicker gradient nanostructured surface layer, higher and deeper compressive residual stress, and also higher surface hardness and deeper subsurface hardness in comparison with that of the UNSM-treated specimen at RT. The development of LHT with UNSM treatment successfully demonstrated the possibility of further improvement in wear resistance of Ti-6Al-4V alloy by increasing the hardness, compressive residual stress, refining grain size in comparison with the UNSM-treated specimen at RT [26]. Moreover, it has been reported in our previous study that a gradient nanostructured surface layer with a thickness of about 60 μm was produced in Ti-6Al-4V by UNSM treatment at RT, while LHT with UNSM at HT of 800 °C was able to increase the thickness of nanostructured surface layer by up to about 100 μm [39].

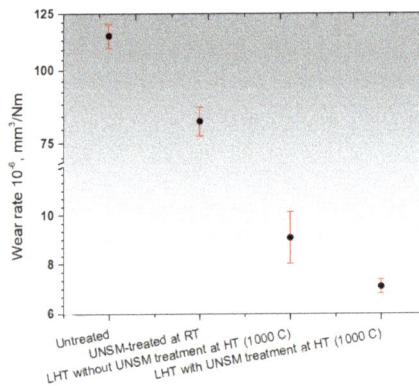

Figure 10. Comparison in wear rate of the untreated, UNSM-treated at RT and LHT without and with UNSM treatment at 1000 °C specimens.

The presence of gradient nanostructured surface layer was found to be responsible not only for the increase in wear resistance but also improvement in frictional behavior of the LHT with UNSM at HT specimen. In addition, in order to shed light on the friction and wear behavior, the surface morphology and the roughening after sliding distance of 30 m under dry conditions of the UNSM-treated at RT and LHT with UNSM at HT specimens are investigated as shown in Figure 11. It can be seen that the worn out deep scars and damages parallel to sliding direction were formed on the surface of the UNSM-treated at RT specimen that roughened the contact interface significantly without any cracks in comparison with the LHT with UNSM at HT specimen due to its high hardness. The average surface roughness inside the wear track of the UNSM-treated at RT and LHT with UNSM at HT specimens increased to 2.4 and 1.8 μm from its initial surface roughness of 0.08 μm. It means that the resistance against the sliding-induced surface roughening of the LHT with UNSM at HT specimen remained greater in comparison with the UNSM-treated at RT one, which is also owing to the presence of gradient nanostructured surface layer [54]. Moreover, the change in chemistry after wear test of the UNSM-treated at RT and LHT with UNSM at HT specimens is shown in Figure 12. It was found that the oxidative wear also played a crucial role in controlling the frictional behavior, where the amount of formed oxide-rich tribolayer was much more on the surface of the LHT with UNSM at HT specimen, as presented in the inset of Figure 12. Consequently, a newly developed thermo-mechanical UNSM treatment gives an opportunity to produce a gradient nanostructured surface layer with a thickness of several hundreds in microns, where the refined grain size is increased gradually at an incremental angle of about 30°, as shown in Figure 2d. Many researchers mentioned the advantages of gradient nanostructured surface layer over the nanostructured surface layer in terms of mechanical and thermal stabilities, and also strain localization [13,29,54].

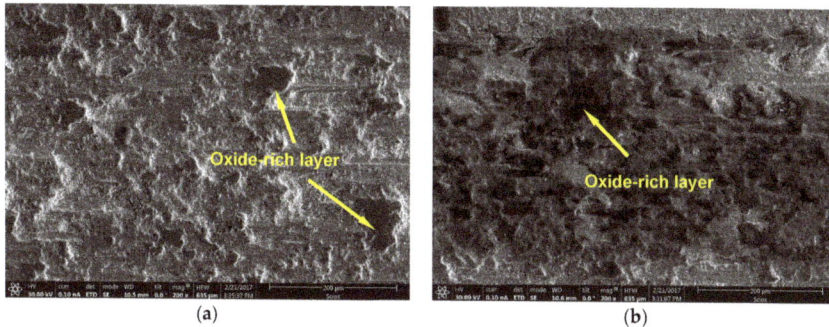

Figure 11. Scanning electron microscopy (SEM) images of the wear track generated on the surface of the UNSM-treated at RT (**a**) and LHT with UNSM treatment at HT of 1000 °C (**b**) specimens.

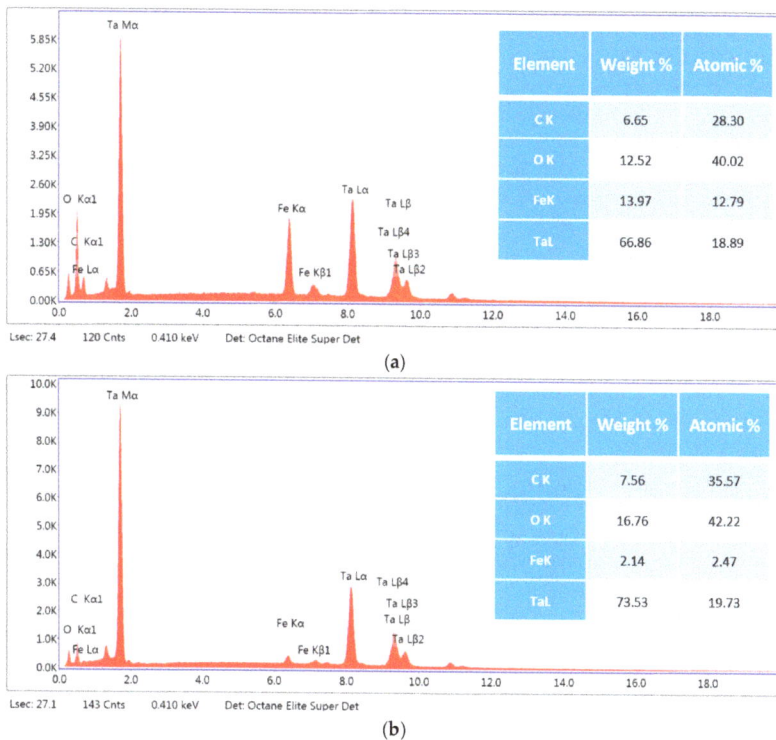

Element	Weight %	Atomic %
C K	6.65	28.30
O K	12.52	40.02
Fe K	13.97	12.79
Ta L	66.86	18.89

Lsec: 27.4   120 Cnts   0.410 keV   Det: Octane Elite Super Det

(**a**)

Element	Weight %	Atomic %
C K	7.56	35.57
O K	16.76	42.22
Fe K	2.14	2.47
Ta L	73.53	19.73

Lsec: 27.1   143 Cnts   0.410 keV   Det: Octane Elite Super Det

(**b**)

Figure 12. EDX (energy-dispersive X-ray spectroscopy) of the wear track generated on the surface of the UNSM-treated at RT (**a**) and LHT with UNSM treatment at HT of 1000 °C; (**b**) specimens.

4. Conclusions

In the current investigation, the effects of UNSM treatment with and without LHT at RT and HT (1000 °C) on the microstructure, hardness, and wear resistance of Ta were systematically investigated. The hardness of the UNSM-treated specimens at RT and HT (1000 °C) increased by about 20 and 62% in comparison with the untreated one. The LHT with UNSM at HT (1000 °C) was capable of inducing a greater compressive residual stress (~1400 MPa) at the surface layer in comparison with the UNSM-treated at RT and LHT without UNSM at HT. It was confirmed by cross-sectional EBSD observations that the combination of LHT with UNSM treatment at HT (1000 °C) produced a stable gradient nanostructured surface layer, with a thickness of several tens of microns, which led to an increase in wear resistance and a reduction in friction behavior of Ta. In general, the possibility of producing a stable gradient nanostructured layer by controlling the UNSM treatment temperature is significant and would find any potential applications of Ta in various industries.

Acknowledgments: This study was supported by the Start-Up Research Project through the Ministry of Science, ICT and Future Planning of Korea (NRF-2017R1C1B5017434). This study was also partially supported by the Korea Technology and Information Promotion Agency (TIPA) for Small and Medium Enterprises. Project (No. S2544322).

Author Contributions: Jong-Min Chae and Auezhan Amanov conceived and designed the experiments; Jong-Min Chae and Auezhan Amanov performed the experiments; Jong-Min Chae analyzed the data; Keun-Oh Lee contributed materials and XRD analysis; Auezhan Amanov wrote the paper.

Conflicts of Interest: The authors declare no conflict of interest.

References

1. Navid, A.A.; Hodge, A.M. Nanostructured alpha and beta tantalum formation—Relationship between plasma parameters and microstructure. *Mater. Sci. Eng. A* **2012**, *536*, 49–56. [CrossRef]
2. Robin, A.; Rosa, J.L. Corrosion behaviour of niobium, tantalum and their alloys in boiling sulfuric acid solutions. *Int. J. Refract. Metals Hard Mater.* **1997**, *15*, 317–323. [CrossRef]
3. Chamelot, P.; Palau, P.; Massot, L.; Savall, A.; Taxil, P. Electrodeposition process of tantalum (V) species in molten fluorides containing oxide ions. *Electrochim. Acta* **2002**, *47*, 3423–3429. [CrossRef]
4. Levine, B.R.; Sporer, S.; Poggie, R.A.; Della Valle, C.J.; Jacobs, J.J. Experimental and clinical performance of porous tantalum in orthopaedic surgery. *Biomaterials* **2006**, *27*, 4671–4681. [CrossRef] [PubMed]
5. Sandim, H.R.Z.; Padiha, A.F.; Randle, V.; Blum, W. Grain subdivision and recrystallization in oligocrystalline tantalum during cold swaging and subsequent annealing. *Int. J. Refract. Metals Hard Mater.* **1999**, *17*, 431–435. [CrossRef]
6. Abdulstaar, M.; Mhaede, M.; Wollmann, M.; Wagner, L. Investigating the effects of bulk and surface severe plastic deformation on the fatigue, corrosion behaviour and corrosion fatigue of AA5083. *Surf. Coat. Technol.* **2014**, *254*, 244–251. [CrossRef]
7. Serra, G.; Moralis, L.; Elias, C.N.; Semenova, I.P.; Valiev, R.; Salimgareeva, G.; Pithon, M.; Lacerda, R. Nanostructured severe plastic deformation processed titanium for orthodontic mini-implants. *Mater. Sci. Eng. C* **2013**, *33*, 4197–4202. [CrossRef] [PubMed]
8. Unal, O.; Varol, R. Surface severe plastic deformation of AISI 304 via conventional shot peening, severe shot peening and repeening. *Appl. Surf. Sci.* **2015**, *351*, 289–295. [CrossRef]
9. Fu, T.; Zhan, Z.; Zhang, L.; Yang, Y.; Liu, Z.; Liu, J.; Li, L.; Yu, X. Effect of surface mechanical attrition treatment on corrosion resistance of commercial pure titanium. *Surf. Coat. Technol.* **2015**, *280*, 129–135. [CrossRef]
10. Wang, Q.; Yin, Y.; Sun, Q.; Sun, L.; Sun, J. Gradient nano microstructure and its formation mechanism in pure titanium produced by surface rolling treatment. *J. Mater. Res.* **2014**, *29*, 569–577. [CrossRef]
11. Chaise, T.; Li, J.; Nelias, D.; Kubler, R.; Taheri, S.; Douchet, G.; Robin, V.; Gilles, P. Modeling of multiple impacts for the prediction of distortions and residual stresses induced by ultrasonic shot peening. *J. Mater. Process. Technol.* **2012**, *212*, 2080–2090. [CrossRef]
12. Park, J.W.; Kim, Y.J.; Park, C.H.; Lee, D.H.; Ko, Y.G.; Jang, J.H.; Lee, C.S. Enhanced osteoblast response to an equal channel angular pressing-processed pure titanium substrate with microrough surface topography. *Acta Biomater.* **2009**, *5*, 3272–3280. [CrossRef] [PubMed]

13. Kang, J.K.; Kim, J.G.; Park, H.W.; Kim, H.S. Multiscale architectured materials with composition and grain size gradients manufactured using high-pressure torsion. *Sci. Rep.* **2016**, *6*, 26590. [CrossRef] [PubMed]

14. Sakai, T.; Belyakov, A.; Kaibyshev, R.; Miura, H.; Jonas, J.J. Dynamic and post-dynamic recrystallization under hot, cold and severe plastic deformation conditions. *Prog. Mater. Sci.* **2017**, *60*, 130–207. [CrossRef]

15. Valiev, R.Z. Strength and ductility of nanostructured SPD metals. *Metall. Mater. High Struct. Effic.* **2004**, *146*, 79–90. [CrossRef]

16. Huang, Y.; Maury, N.; Zhang, N.X.; Langdon, T.G. Microstructure and mechanical properties of pure tantalum processed by high-pressure torsion. *IOP Conf. Ser. Mater. Sci. Eng.* **2014**, *63*, 012100. [CrossRef]

17. Mathaudhu, S.N.; Hartwig, K.T.; Barber, R.E. Fine grained tantalum for composite Nb$_3$Sn superconductor diffusion barrier sheet. *IEEE Trans. Appl. Supercond.* **2007**, *17*, 2660–2663. [CrossRef]

18. Mathaudhu, S.N.; Hartwig, K.T. Grain refinement and recrystallization of heavily worked tantalum. *Mater. Sci. Eng. A* **2006**, *426*, 128–142. [CrossRef]

19. Zhang, Y.S.; Wei, Q.M.; Niu, H.Z.; Li, Y.S.; Chen, C.; Yu, Z.T.; Bai, X.F.; Zhang, P.X. Formation of nanocrystalline structure in tantalum by sliding friction treatment. *Int. J. Refract. Metals Hard Mater.* **2014**, *45*, 71–75. [CrossRef]

20. Amanov, A.; Pyun, Y.S.; Vasudevan, V.K. High strength and wear resistance of tantalum by ultrasonic nanocrystalline surface modification technique at high temperatures. *IOP Conf. Ser. Mater. Sci. Eng.* **2017**, *63*, 012100. [CrossRef]

21. Amanov, A.; Lee, S.W.; Pyun, Y.S. Low friction and high strength of 316L stainless steel tubing for biomedical applications. *Mater. Sci. Eng. C* **2017**, *71*, 176–185. [CrossRef] [PubMed]

22. Amanov, A.; Penkov, O.V.; Pyun, Y.S.; Kim, D.E. Effects of ultrasonic nanocrystalline surface modification on the tribological propertied of AZ91D magnesium alloy. *Tribol. Int.* **2012**, *54*, 106–113. [CrossRef]

23. Amanov, A.; Pyun, Y.S.; Kim, J.H.; Sasaki, S. The usability and preliminary effectiveness of ultrasonic nanocrystalline surface modification technique on surface properties of silicon carbide. *Appl. Surf. Sci.* **2014**, *311*, 448–460. [CrossRef]

24. Amanov, A.; Kwon, H.G.; Pyun, Y.S. The possibility of reducing the reflectance and improving the tribological properties of Si wafer by UNSM technique. *Tribol. Int.* **2017**, *105*, 175–784. [CrossRef]

25. Amanov, A. Wear resistance and adhesive failure of thermal spray coatings deposited onto graphite in response to ultrasonic nanocrystal surface modification technique. *Appl. Surf. Sci.* **2018**, in press. [CrossRef]

26. Amanov, A.; Pyun, Y.S. Local heat treatment with and without ultrasonic nanocrystal surface modification of Ti-6Al-4V alloy: Mechanical and tribological properties. *Surf. Coat. Technol.* **2017**, *326*, 343–354. [CrossRef]

27. Amanov, A.; Umarov, R. The effects of ultrasonic nanocrystal surface modification temperature on the mechanical and fretting wear resistance of Inconel 690 alloy. *Appl. Surf. Sci.* **2018**, *441*, 515–529. [CrossRef]

28. An, X.; Lin, Q.; Qu, S. Influence of stacking-fault energy on the accommodation of severe shear in Cu-Al alloys during equal-channel angular pressing. *J. Mater. Res.* **2009**, *24*, 3636–3646. [CrossRef]

29. Lu, K. Making strong nanomaterials ductile with gradients. *Science* **2014**, *345*, 1455–1456. [CrossRef] [PubMed]

30. Yin, Z.; Yang, X.; Ma, X.; Moering, J.; Yang, J.; Gong, Y.; Zhu, Y.; Zhu, X. Strength and ductility of gradient structured copper obtained by surface mechanical attrition treatment. *Mater. Des.* **2016**, *105*, 89–95. [CrossRef]

31. Wang, X.; Li, Y.S.; Zhang, Q.; Zhao, Y.H.; Zhu, Y.T. Gradient structured copper by rotationally accelerated shot peening. *J. Mater. Sci. Technol.* **2017**, *33*, 758–761. [CrossRef]

32. Yang, X.; Ma, X.; Moering, J.; Zhou, H.; Wang, W.; Gong, Y.; Tao, J.; Zhu, Y.; Zhu, X. Influence of gradient structure volume fraction on the mechanical properties of pure copper. *Mater. Sci. Eng. A* **2015**, *645*, 280–285.

33. Wu, X.L.; Yang, M.X.; Yuan, F.P.; Chen, L.; Zhu, Y.T. Combining gradient structure and TRIP effect to produce austenite stainless steel with high strength and ductility. *Acta Mater.* **2016**, *112*, 337–346. [CrossRef]

34. AlMangour, B.; Yang, J.M. Improving the surface quality and mechanical properties by shot-peening of 17-4 stainless steel fabricated by additive manufacturing. *Mater. Des.* **2016**, *110*, 914–924. [CrossRef]

35. Amanov, A.; Pyun, Y.S.; Kim, J.H.; Suh, C.M.; Cho, I.S.; Kim, H.D.; Wang, Q.; Khan, M.K. Ultrasonic fatigue performance of high temperature structural material Inconel 718 alloys at high temperature after UNSM treatment. *Fatigue Fract. Eng. Mater. Struct.* **2015**, *38*, 1266–1273. [CrossRef]

36. Khan, M.K.; Fitzpatrick, M.E.; Wang, O.Y.; Pyun, Y.S.; Amanov, A. Effect of ultrasonic nanocrystal surface modification on residual stress and fatigue cracking in engineering alloys. *Fatigue Fract. Eng. Mater. Struct.* **2017**, *39*, 1–12. [CrossRef]

37. Lee, Y.J.; Lee, T.H.; Kim, D.Y.; Nersisyan, H.H.; Han, M.H.; Kang, K.S.; Bae, K.K.; Shin, Y.J.; Lee, J.H. Microstructural and corrosion characteristics of tantalum coatings prepared by molten salt electrodeposition. *Surf. Coat. Technol.* **2013**, *235*, 819–826. [CrossRef]

38. Ortiz, A.L.; Shaw, L. X-ray diffraction analysis of a severely plastically deformed aluminium alloy. *Acta Mater.* **2004**, *52*, 2185–2197. [CrossRef]

39. Amanov, A.; Urmanov, B.; Amanov, T.; Pyun, Y.S. Strengthening of Ti-6Al-4V alloy by high temperature ultrasonic nanocrystal surface modification technique. *Mater. Lett.* **2017**, *196*, 198–201. [CrossRef]

40. Islamgaliev, R.K.; Kuzel, R.; Mikov, S.N.; Igo, A.V.; Burianek, J.; Chmelik, F.; Valiev, R.Z. Structure of silicon processed by severe plastic deformation. *Mater. Sci. Eng. A* **1999**, *266*, 205–210. [CrossRef]

41. Wang, M.; Vo, N.Q.; Campion, M.; Nguyen, T.D.; Setman, D.; Dillon, S.; Bellon, P.; Averback, R.S. Forced atomic mixing during severe plastic deformation: Chemical interactions and kinetically driven segregation. *Acta Mater.* **2014**, *66*, 1–11. [CrossRef]

42. Tian, J.W.; Dai, K.; Villegas, J.C.; Shaw, L.; Liaw, P.K.; Klastrom, D.L.; Ortiz, A.L. Tensile properties of a nickel-base alloy subjected to surface severe plastic deformation. *Mater. Sci. Eng. A* **2008**, *493*, 176–183. [CrossRef]

43. Satheesh Kumar, S.S.; Raghu, T. Mechanical behaviour and microstructural evolution of constrained groove pressed nickel sheets. *J. Mater. Process. Technol.* **2013**, *213*, 214–220. [CrossRef]

44. Ungar, T. Microstructural parameters from X-ray diffraction peak broadening. *Scr. Mater.* **2004**, *21*, 777–781. [CrossRef]

45. Estrin, Y.; Vinogradov, A. Extreme grain refinement by severe plastic deformation: A wealth of challenging science. *Acta Mater.* **2013**, *61*, 782–817. [CrossRef]

46. Rajan, T.V.; Sharma, C.P.; Sharma, A. *Heat Treatment: Principles and Techniques*, 2nd ed.; PHI: New Delhi, India, 2011.

47. Chokshi, A.H.; Rosen, A.; Karch, J.; Gleiter, H. On the validity of the Hall-Petch relationship in nanocrystalline materials. *Scr. Metall.* **1989**, *23*, 1679–1683. [CrossRef]

48. Wang, S.; Feng, S.K.; Chen, C.; Jia, Y.L.; Wang, M.P.; Li, Z.; Zhong, Z.H.; Lu, P.; Li, P.; Cao, S.F.; et al. A twin orientation relationship between {001} <210> and {111} <110> obtained in Ta-2.5W alloy during heavily cold rolling. *Mater. Character.* **2017**, *125*, 108–113. [CrossRef]

49. Suresh, S.; Giannakopoulos, A.E. A new method for estimating residual stress by instrumented sharp indentation. *Acta Mater.* **1998**, *46*, 5755–5767. [CrossRef]

50. Zhu, L.N.; Xu, B.S.; Wang, H.D.; Wang, C.B. Measurement of residual stress in quenched 1045 steel by nanoindentation method. *Mater. Character.* **2010**, *61*, 1359–1362. [CrossRef]

51. Amanov, A.; Cho, I.S.; Pyoun, Y.S.; Lee, C.S.; Park, I.G. Micro-dimpled surface by ultrasonic nanocrystal surface modification and its tribological effects. *Wear* **2012**, *286–287*, 136–144. [CrossRef]

52. Chen, X.; Han, Z.; Li, X.Y.; Lu, K. Lowering friction coefficients in Cu alloys with stable gradient nanostructures. *Sci. Adv.* **2016**, *2*, e1601942. [CrossRef] [PubMed]

53. Ghidelli, M.; Sebastiani, M.; Collet, C.; Guillemet, R. Determination of the elastic moduli and residual stresses of freestanding Au-TiW bilayer thin films by nanoindentation. *Mater. Des.* **2016**, *106*, 436–445. [CrossRef]

54. Amanov, A.; Pyun, Y.S.; Zhang, B.; Nohava, J. Preliminary study on the effect of micro-scale dimple size on friction and wear under oil-lubricated sliding contact. *Tribol. Online* **2011**, *6*, 284–290. [CrossRef]

![materials logo] *materials*

MDPI

Article

Modification of Physico-Chemical Properties of Acryl-Coated Polypropylene Foils for Food Packaging by Reactive Particles from Oxygen Plasma

Tomislava Vukušić [1], Alenka Vesel [2], Matej Holc [3], Mario Ščetar [1], Anet Režek Jambrak [1] and Miran Mozetič [2],*

[1]   Department of Food Engineering, University of Zagreb, Pierottijeva 6, 10000 Zagreb, Croatia;
      tvukusic@pbf.hr (T.V.); mscetar@pbf.hr (M.Š.); arezek@pbf.hr (A.R.J.)
[2]   Department of Surface Engineering, Jozef Stefan Institute, Jamova cesta 39, 1000 Ljubljana, Slovenia;
      alenka.vesel@guest.arnes.si
[3]   Jozef Stefan International Postgraduate School, Jamova cesta 39, 1000 Ljubljana, Slovenia; matej.holz@ijs.si
*   Correspondence: miran.mozetic@guest.arnes.si; Tel.: +386-1-477-3405

Received: 30 January 2018; Accepted: 2 March 2018; Published: 3 March 2018

Abstract: This investigation was focused on the influence of long-living neutral reactive oxygen species on the physico-chemical properties of acryl-coated polypropylene foils for food packaging. Reactive species were formed by passing molecular oxygen through a microwave discharge and leaking it to a processing chamber of a volume of 30 L, which was pumped by a rotary pump. The density of neutral O-atoms in the chamber was tuned by adjustment of both the effective pumping speed and the oxygen leak rate. The O-atom density was measured with a catalytic probe and was between 3×10^{18} and 5×10^{19} m^{-3}. Commercial foils of biaxially oriented polypropylene (BOPP) coated with acrylic/ poly(vinylidene chloride) (AcPVDC) were mounted in the chamber and treated at room temperature by O atoms at various conditions, with the fluence between 1×10^{21} and 3×10^{24} m^{-2}. The evolution of the surface wettability versus the fluence was determined by water contact angle (WCA) measurements, the formation of functional groups by X-ray photoelectron spectroscopy (XPS), and the morphology by atomic force microscopy (AFM). The WCA dropped from the initial 75° to approximately 40° after the fluence of a few 10^{22} m^{-2} and remained unchanged thereafter, except for fluences above 10^{24} m^{-2}, where the WCA dropped to approximately 30°. XPS and AFM results allowed for drawing correlations between the wettability, surface composition, and morphology.

Keywords: plasma surface modification; polymer polypropylene; neutral oxygen atom density; initial surface functionalization; food packaging; wettability

1. Introduction

Today, food technology is constantly evolving in response to different challenges. The changes in consumer demands and the necessity for the production of safe and high-quality foods are responsible for the innovation and improvement of already established food processes. In this sense, the introduction of new technologies could lead to a reduction of the processing time or an improvement in operating conditions, thereby decreasing both environmental and financial costs. Plasma treatments cause several chemical and physical changes on the plasma-polymer interface, which improve the surface properties [1–6]. Plasma-induced effects on the polymer surface are nowadays exploited in surface functionalization of the packaging polymers for promoting adhesion or sometimes anti-adhesion [7], enhanced printability [8], sealability [9], assuring anti-mist properties, improving the polymer's resistance to mechanical failure [1], and adhesion of antibacterial coatings [10–14].

Polypropylene (PP) is an important commercial polymer which is often used for producing package films [15], because of its low cost and good thermal stability. Extruded PP film is amorphous, while the crystallization can be achieved by two-way stretching (monoaxially or biaxially oriented films) at elevated temperatures. Biaxial orientation (BO) slightly improves the silky structure of the film and significantly reduces turbidity, enhancing the barrier properties and flexural toughness at a low temperature. Biaxially oriented polypropylene (BOPP) film is often coated with an additional polymer layer to improve its mechanical properties or barrier properties against gases and moisture. If BOPP is used for food packaging applications, it is often coated with an acrylic layer or poly(vinylidene chloride) (PVDC). Acrylic coating (Ac) is durable, flexible, and resistant to degradation caused by ultraviolet rays [16]. If the PP foil is coated with a PVDC layer, this topcoat enhances PP barrier properties against water vapor and gases. Excellent protective properties of this layer make PP foil suitable for the packaging of confectionery products which require barrier protection from moisture [17].

As mentioned above, plasma can improve surface properties of polymers such as wettability and surface functionalization, and consequently also adhesion properties. This may be important for coating the food-packaging foils with antibacterial layers [18]. Many authors have investigated treatment of a pure PP foil rather than industrial-grade foils, which are often covered with an ultra-thin film of a protective coating. A reason for this might be unknown details regarding the composition and structure of the coating, let alone the method applied for deposition of the coating. Pandirayaj et al. [19] used a low pressure weakly ionized plasma created by a DC glow discharge to improve the wettability of the PP foil. The water contact angle dropped from the original value of 98° down to 58° upon treatment for 10 min. Similar results were reported by Choi et al. [20], who obtained 60° in low-pressure oxygen DC plasma. Additionally, Morent et al. [21] obtained a water contact angle of 60°, although he used an extremely weak plasma at the discharge power of solely 1.4 W and moderate pressure of 5 kPa. A dielectric barrier discharge (DBD) was applied. Leroux et al. [22] obtained the contact angle of 64° using plasma created in air at atmospheric pressure by classical DBD and a treatment speed of 2 m/min. Lower water contact angles were observed by some other authors. Aguiar et al. [23] achieved a water contact angle of 40° (initially 110°) on PP surface treated in oxygen plasma at 700 W and 6.7 Pa. Vishnuvarthanan et al. [24] observed that the contact angle depended on the discharge power (7.2–29.6 W) and treatment time (0–300 s). The lowest water contact angle was ~44°; however, the initial angle was just 74.5°, which indicates that the initial surface was probably already contaminated with surface impurities. Mirabedini et al. [25] obtained a minimal contact angle of 34.4° in RF oxygen plasma at 50 W and 0.35×10^5 Pa. However, Wanke et al. [26] managed to achieve only 24° (initially 98°) at 15 min of treatment. Unlike other authors, who observed a decreasing water contact angle with increasing treatment time until reaching saturation, he observed that at long treatment times (after 15 min of treatment), the contact angle increased to 53°. The reason that some authors obtained such low contact angles can be associated with polymer overtreatment leading to the formation of low-molecular weight fragments (LMWOM) because of polymer degradation [27]. The main papers and key results are summarized in Table 1.

Although oxygen plasma treatment causes beneficial effects such as improved wettability, it also causes other modifications of the surface and sub-surface layer which may not be tolerated. Oxygen plasma is rich in different reactive species and represents a source of ultraviolet (UV) radiation. The reactive gaseous species that interact with a polymer sample include positively charged molecular and atomic ions, neutral atoms in the ground and metastable states, and neutral molecules in both "a" and "b" metastable states and ozone. The major UV radiation occurs at the wavelength of 130 nm due to the transition from a highly excited $2s^2 2p^3 (^4S°) 3s^3S°$ state to the ground state ($2s^2 2p^4\,^3P$). The photon energy for this transition is 9.52 eV. The penetration depth of such UV radiation in a polymer material is around a micrometer [28]. The energetic photons cause bond scission and thus modification of the polymer properties well below its surface. Furthermore, there is always some water vapor in a low-pressure plasma reactor. The vapor is the major constitute of the residual atmosphere and is also formed due to chemical etching of the polymer upon oxygen plasma treatment. The water molecules dissociate under plasma conditions and the resulting OH and H radicals are excited upon inelastic collisions with energetic electrons. The excited states de-excite to the corresponding ground states by ration in the UV range: Lyman hydrogen series in the vacuum UV range and OH band of bandhead at 309 nm. All this radiation causes bond scission in the polymer film of a thickness of the order of several μm. The reactive species interact with dangling bonds on the polymer surface, causing the formation of LMWOM that are often volatile. Therefore, rather extensive etching is observed upon the treatment of a polymer material with oxygen plasma [29]. In fact, precise measurements of the oxidation rate for the same polymer exposed to oxygen plasma and only neutral O-atoms at the same O-atom flux on the sample surface showed a two orders of magnitude higher etching rate for the case where synergistic effects of radiation and reactive species were effective [30,31]. Such synergies should therefore be avoided if functionalization of the polymer surface with oxygen functional groups is the goal.

The aim of this research was to examine the effect of surface oxidation of commercial PP foils used for food packaging. Such foils are covered with a very thin acrylic coating. Unlike other authors, neutral reactive particles from late afterglow were used instead of gaseous plasma, because glowing plasma always causes the etching of polymers and the acrylic coating could have been removed by direct exposure to oxygen plasma [30]. Furthermore, in afterglow, a density of oxygen species interacting with the polymer can also be precisely determined. This allowed determination of the minimal oxygen atom fluence necessary for saturation of the surface with polar functional groups and thus optimal wettability at a minimal treatment time.

Table 1. Treatment conditions and results obtained by other authors.

Plasma Treatment	Discharge Parameters	Wettability	Surface Analysis	Reference
Low-pressure oxygen plasma	– RF source 8–12 MHz – Power: 7.2, 10.2, 29.6 W – Pressure: 26.7–80 Pa – Flow: 5–10 scfh – Exposure time: up to 5 min	WCA [1] was decreasing with increasing power and treatment time. From 74.5° it decreased to approx. 44° at the highest power of 29.6 W and at the longest treatment time 300 s. SFE [2] increased from 56.5 to max. 94.1 mJ/m².	AFM [3]: surface roughness RMS [4] increased from 1.5 to 7.3 nm. XRD [5]: higher degree of crystallinity observed after oxygen plasma treatment. Mechanical properties: tensile strength decreased after plasma treatment. Barrier properties: water vapor transmission was increasing with increasing power and time.	[24]
Low-pressure oxygen and argon plasma	– commercial plasma reactor from Diener Co. – LF plasma 40 kHz – Power: 50 W – Pressure: 0.2 × 10⁵ Pa – Flow: 50 cm³/min – Exposure time: up to 5 min	Increase of the surface energy of oxygen plasma-treated sample was higher than for the one treated in Ar. SFE was 70 and over 50 mJ/m² for O₂ and Ar plasma, respectively.	ATR-FTIR [6]: carbonyl groups observed as well as C=C which could be a sign of crosslinking. AFM: O₂ plasma caused higher roughness than Ar. Adhesion: High improvement of surface adhesion strength, especially for O₂ plasma.	[32]
Low-pressure oxygen plasma	– RF source 13.56 MHz, capacitive sample placed on the grounded electrode – Power: 70 W – Pressure: 6.7 Pa – Flow: 49 sccm – Exposure time: up to 5 min	WCA decreased from 110° to 40°.	AFM: surface roughness first decreased with treatment time. At longer treatment times, a significant increase is observed. FTIR: C=O and –OH peaks observed for plasma-treated samples.	[23]
Low-pressure air plasma	– DBD plasma sample placed on the grounded electrode – AC power source 10 kHz – Power: 1.4 W – Pressure: 5 × 10³ Pa – Flow: 20 sccm – Exposure time: 0.2–30 s – Energy load: up to 3.34 J/cm²	WCA decreased from 94.9° to 60°. Saturation reached after 10 s.	XPS [7]: Oxygen concentration increased from 4.3 to 13.7 at.%. Nitrogen (0.8 at.%) was also found, the rest being carbon. 8.2% C–O, 2.7% C=O and O–C=O and 86.4% C–C, C–H groups observed on plasma treated sample. ATR-FTIR: peaks attributed to OH and C=O in ketones, aldehydes and carboxylic acids.	[21]

Table 1. *Cont.*

Plasma Treatment	Discharge Parameters	Wettability	Surface Analysis	Reference
Atmospheric pressure air plasma	— DBD plasma — "Coating Star" device from Ahlbrandt System — sample placed on the grounded electrode — 30 kHz, 15 kV — Power: 300–1000 W — Pressure: atmospheric — Flow: 20 sccm — Treatment speed: 2–10 m/min — Energy load: up to 60 kJ/m^2	WCA decreased from 104° to 64° even at 1.2 kJ/m^2. SFE increased from 33.7 to almost 50 mN/m.	XPS: O/C ratio increased over 0.16. Nitrogen (2 at %) was also found. After one month, O/C decreased to 0.12. Groups like C–O (22.5%), C=O or O–C–O (8.4%) and O=C–O (5.3%) were found. Maximum concentration was obtained at the lowest treatment speed. AFM: Ra [8] increased from 5.8 to 12.9 nm. Bumps were observed on the surface. The height and width were increasing with treatment power and reached 60 and 500 nm, respectively.	[22]
Low-pressure oxygen and argon plasma	— RF source 13.56 MHz — Commercial K 1050X Plasma Asher Model from Emitech Co. — sample placed in the middle of the chamber on the glass substrate — Power: 10, 30, 50 W — Pressure: 0.35 × 10^5 Pa — Flow: 15 mL/min — Exposure time: up to 5 min	WCA was decreasing with increasing power and treatment time. The lowest WCA was 34.4° for O$_2$ and 38.2° for Ar plasma (initially 98.3°). SFE increased to ~45 mN/m.	SEM [9] and AFM: topology and roughness changed significantly, especially for Ar plasma (nodules observed on the surface). RMS roughness increased from 3.6 to 6.9 and 6.1 nm for O$_2$ and Ar plasma, respectively. ATR-FTIR: C=O stretching bond and C=C vibration observed. Some peaks attributed also to carboxylic/ester, aldehydes and ketone groups.	[25]
Low-pressure oxygen plasma	— RF power source — sample was on the tray in the middle of the chamber — Power: 500 W — Pressure: 13.3 Pa — Flow: 49 sccm — Exposure time: up to 40 min	WCA decreased from 121.5° to 84° on PP nonwoven mats. Ageing for 90 days did not have significant effect on WCA. SFE increased from 13.7 to 29.2 mN/m.	SEM: etching of PP fibers observed. XRD: no significant effect on the crystallinity of the treated fibers.	[33]

Table 1. *Cont.*

Plasma Treatment	Discharge Parameters	Wettability	Surface Analysis	Reference
Low-pressure oxygen plasma	– RF source 13.56 MHz, capacitive – Power: 0–150 W – Pressure: 0–120 Pa – Exposure time: 30 s–3 min – Ageing: 30 days	The lowest WCA—bellow 10° was observed at 150 W, 3.33 Pa and 60 s. After 30 days of ageing WCA increased to ~50°.	Ageing and crystallinity: Two polymers with different initial crystallinity were used. More crystalline PP was ageing slower—WCA after 30 days was for ~5° lower than for less crystalline one. Degree of crosslinking was increased after the treatment for both polymers. XPS: ~25 at % of oxygen was found on less-crystalline polymer. O concentration on more crystalline polymer was few at % lower. However, after ageing the O concentration changed in favor of more crystalline one.	[34]
Low-pressure oxygen plasma	– RF power source commercial reactor Inverse Sputter Etcher ISE 90 model 2001 (Von Ardenne Anlagentechnik GmBh) – Power: 50 W – Pressure: 5.1 Pa – Exposure time: up to 40 min	WCA decreased from 98° to 24°. At long treatment times, it increased to 53°.	AFM: roughness RMS increased from ~12 nm to ~44 nm. ATR-FTIR: OH, C=O and CO–C=O peaks observed for plasma treated samples.	[26]
Low-pressure oxygen plasma	– DC plasma (20 mA, 2 kV) sample was placed on glass walls of discharge chamber positioned between the electrodes separated 42 cm – Pressure: 30 Pa – Exposure time: up to 200 s	WCA decreased from 83° to 60°. SFE increased from 25.7 to 43 mJ/m².	ATR-FTIR: OH, C=O groups in ester, ketone and carboxyl groups, C=O groups in unsaturated ketones and aldehydes.	[20]
Low-pressure oxygen plasma	– Capacitor plate plasma commercial K1050 X Plasma Asher from Emitech Ltd. sample positioned on the holder – Power: 0–100 W – Pressure: 60 Pa – Flow: 15 mL/min – Exposure time: up to 10 min – Ageing: 90 days in air or water	WCA was decreasing with the increasing power and treatment time. Minimal achievable WCA was 55.6° (initially 103°). Ageing in water was faster than in air. After 90 days, the WCA was 81.7° (in water) and 71.2° (in air).	AFM: Roughness RMS increased after treatment from 2.1 nm to ~10 nm (in air) and ~5 nm (in water). Lower roughness of samples stored in water was explained by removing of water-soluble short-chain species.	[35]

[1] Water contact angle (WCA); [2] Total surface free energy (SFE); [3] Atomic-force microscopy (AFM); [4] Root mean squared (RMS) roughness; [5] X-ray diffraction (XRD); [6] Attenuated total reflection Fourier transform-infrared spectroscopy (ATR-FTIR); [7] X-ray photoelectron spectroscopy (XPS); [8] Average roughness (Ra); [9] Scanning electron microscope (SEM).

2. Materials and Methods

2.1. Materials

Biaxially oriented polypropylene (PP) films (Bicor 32MB777, ExxonMobil, Antwerp, Belgium) were used in the experiments. One side of the film had an acrylic acid coating and the other side was coated with a thin film of poly(vinylidene chloride) (PVDC), which means that plasma interacted with the coating and not with the PP substrate. Only the acrylic side was treated with plasma. The thickness of the foil was 32 μm.

2.2. Plasma Afterglow Treatment

The polymer foil was cut into pieces of 2×2 cm^2 and treated with reactive neutral oxygen species created in the center of the processing chamber. Oxygen species which were created in the surfatron plasma were passed through the narrow glass tube to the processing chamber. The experimental system is shown schematically in Figure 1. The processing chamber was a pyrex cylinder with a diameter of 33 cm and a length of 40 cm. The chamber was pumped with a two-stage oil rotary pump of a nominal pumping speed of 40 m$^3 \cdot$h^{-1} and ultimate pressure well below 1 Pa. A zeolite trap was used to prevent back-diffusion of the oil vapor. The pump was mounted on the flange at the bottom of the processing chamber via bellows of a large conductivity at the pressure of 20 Pa and above, and a manually adjustable shutter valve which allowed for suppressing the effective pumping speed in a gradual manner from the maximal speed (40 m$^3 \cdot$h^{-1}) down to zero. The upper flange of the Pyrex tube was equipped with a pressure gauge, a discharge tube, and a movable catalytic probe which was used for O-atom density measurements [36]. Oxygen of commercial purity 99.99% was leaked continuously in the discharge tube through a manually adjusted leak valve. A standard quartz tube with an inner diameter of 6 mm was used. The pressure was measured with an absolute gauge (baratron) calibrated for the pressure range 0.1–100 Pa. A microwave cavity of approximately 5 cm in length was mounted onto the discharge tube and connected to the microwave power supply. The configuration allowed for sustaining the gaseous plasma in the surfatron mode inside the discharge tube. The microwave power was set to 200 W. Continuous leakage of oxygen on one side and pumping of the processing chamber on the other side allowed for a drift of gas through the discharge into the processing chamber. Molecular oxygen from the flask partially ionized, dissociated, and excited in the plasma within the microwave cavity. Charged particles quickly neutralized and excited species relaxed on the way between the gaseous plasma and the processing chamber. Therefore, the only highly reactive oxygen species left for treatment of the polymer samples was neutral O atoms. The density of O atoms above the surface of the polymer samples was measured with a calibrated catalytic probe. The probe consists of a catalytic tip which is heated in the plasma because of the recombination of O atoms to O$_2$ molecules on the surface of the catalyst [36]. The temperature of the catalyst is measured by a thermocouple. The heating rate of the probe is proportional to the flux of oxygen atoms. The O-atom density (n) was calculated from the probe temperature derivate using the following equation [37]:

$$n = \frac{8 \cdot m \cdot c_p}{v \cdot W_D \cdot \gamma \cdot A} \cdot \left(\frac{\mathrm{d}T}{\mathrm{d}t} \right) \tag{1}$$

where m is the mass of the probe tip, c_p is its specific heat capacity, W_D is the dissociation energy of an oxygen molecule, γ is the recombination coefficient for O atoms on the catalyst surface, A is the area of the catalyst, and $\mathrm{d}T/\mathrm{d}t$ is the time derivative of the probe temperature just after turning off the discharge. More details regarding the O-atom density calculation are explained in the works [36,37]. We have used cobalt as the catalyst. This material is particularly suitable for the detection of atomic oxygen at a low density. The lower detection limit of the probe was approximately 2×10^{18} at a pressure above 10 Pa, whereas the upper at 10^{22} m^{-3}.

The experiments presented here were performed at the pressure of 20 Pa. At these conditions, the O-atom density in the system was 5.3×10^{19} m^{-3} when the shutter valve was fully open (the effective pumping speed was equal to the nominal pumping speed of the vacuum pump). By adjusting the shutter and leak valves simultaneously, it was possible to keep the pressure in the processing chamber constant but the O-atom density variable: less opened valves caused a lower atom density because the drift velocity of the gas through the discharge chamber was suppressed by closing valves. A detailed description of this effect was reported elsewhere [38]. Four adjustments of the O-atom density in the vicinity of the samples were chosen: 5.3×10^{19}, 2.9×10^{19}, 1.0×10^{19}, 8.7×10^{18}, and 3×10^{18} m^{-3}. The corresponding fluxes of O-atoms onto the sample surface were calculated as:

$$j = \frac{1}{4} nv \qquad (2)$$

where n is the measured density of oxygen atoms and v is an average thermal velocity of O atoms at room temperature ($v = 630$ m·s^{-1}). The fluence of O atoms to the surface of the sample was calculated as $j \times t$, where j is the flux of oxygen atoms to the surface and t is the treatment time. Various treatment times were used for modification of the sample's surface. Such an experimental setup allowed for the treatment of samples in a broad range of fluences from 5×10^{21} to 3×10^{24} m^{-2}—almost three orders of magnitude.

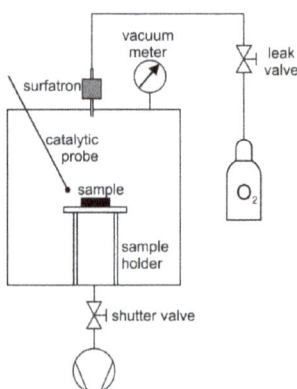

Figure 1. Experimental plasma system used for treating polymer samples.

2.3. X-ray Photoelectron Spectroscopy (XPS) Characterization

Chemical composition of the samples was determined with an XPS instrument model TFA XPS (Physical Electronics, Ismaning, Germany) from Physical Electronics. Analyses were performed 15 min after the plasma treatment. Monochromatic Al K$\alpha_{1,2}$ radiation at 1486.6 eV was used for sample excitation. Photoelectrons were detected at an angle of 45° with respect to the normal of the sample surface. XPS survey spectra were measured at a pass-energy of 187 eV using an energy step of 0.4 eV. High-resolution spectra of carbon C1s were measured at a pass-energy of 23.5 eV using an energy step of 0.1 eV. Because the samples are insulators, an electron gun was used for the additional charge compensation. The spectra were analyzed using MultiPak v8.1c software (Ulvac-Phi Inc., Kanagawa, Japan, 2006) from Physical Electronics.

2.4. Atomic Force Microscopy (AFM) Measurements

The surface morphology of the samples was analyzed with an AFM (Solver PRO, NT-MDT, Moscow, Russia). Images were recorded in a tapping mode using ATEC-NC-20 tips (Nano And More GmbH, Germany). A resonance frequency of the tip and the force constant were 210–490 kHz and 12–110 Nm^{-1}, respectively. An average surface roughness of the samples (Ra) was determined by using the program Spip 5.1.3 (Image Metrology A/S). The average surface roughness was calculated from the images taken over an area of 5×5 μm^2.

2.5. Contact Angle Measurements

Changes of the surface wettability of the plasma-treated samples were determined immediately after the plasma treatment. An instrument by See System (Advex Instruments, Brno, Czech Republic) was used. A demineralized water droplet of a volume of 3 µL was applied to the surface. The measured contact angles were analyzed by the software supplied by the producer. For each sample, three measurements were taken to minimize the statistical error.

3. Results and Discussion

Figure 2 illustrates the variation of the water contact angle of the acrylic coating versus the fluence of oxygen atoms. As mentioned earlier, the treatment was performed at several different densities of O atoms in the vicinity of the sample and at various treatment times. It seems that the water contact angle only depends on the fluence and not on the O-atom density because all measured points in Figure 2 follow the same curve. The contact angle at first decreases rapidly with the increasing fluence, but later the decrease becomes less and less rapid until the water contact angle becomes constant at approximately 40°. The particular measured points in Figure 2 are somehow scattered; however, the trend is obvious: no knee is observed in the curve which is only plotted for eye guidance. The contact angle becomes constant (approximately 40°) after the fluence of a few 10^{22} m^{-2} is used. Further exposure to O-atoms does not influence the wettability of this particular material. The exemptions are both measured points at very large fluences where the contact angles are approximately 30°. A feasible explanation for this effect will be presented and discussed later in this paper.

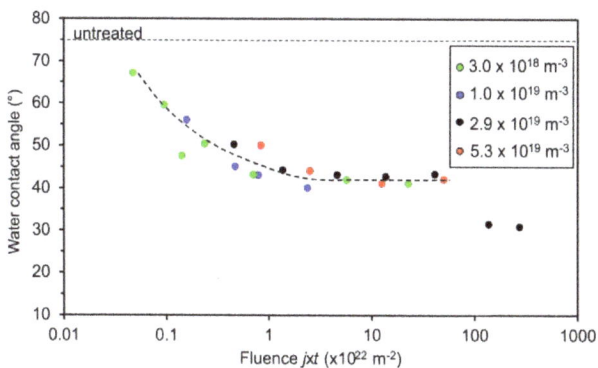

Figure 2. Variation of the water contact angle of the plasma-treated acryl-coated PP with the fluence of oxygen atoms. The different colors represent experiments with different O-atom densities.

Figure 3 represents the required treatment time for the fluences of 1×10^{22} and 1×10^{23} m^{-2}. From this figure, one can conclude that the required treatment time for receiving the fluence of 1×10^{22} m^{-2} is only 6 ms at the atom density of 1×10^{22} m^{-3}, which is typical for the extremely reactive oxygen plasma [39]. Such a short treatment time is achievable only when using pulsed discharges. Unfortunately, this experimental setup does not allow for verification of the calculated values presented in Figure 3. Furthermore, in practice, such small treatment times are not very suitable, because the treated surface may be contaminated with impurities. This means that at such a short treatment time, plasma radicals interact with the contaminants rather than with a pure polymer surface.

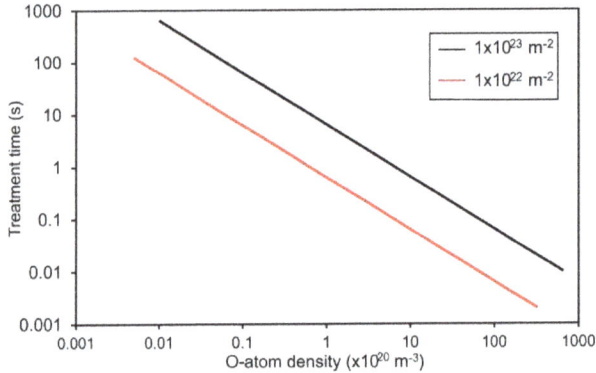

Figure 3. Recommended treatment times for achieving good wettability (~40°) of the acryl-coated polypropylene foils at two O-atom fluences.

Figure 4 shows the variation of oxygen concentration and the O/C ratio on the polymer surface as determined by XPS. The oxygen concentration of the untreated sample was approximately 18 at %. The rather high concentration of oxygen in the surface film as detected by XPS (several nm thick) arises from the acrylic coating. After the treatment, the oxygen concentration on the surface increased. The increase is at first rapid but then less pronounced; however, the x-axis in Figure 4 is plotted in the logarithmic scale and therefore the measured points appear in a line. The oxygen concentration thus increases as a logarithm of the fluence. It is interesting that the oxygen concentration keeps increasing after the fluence that corresponds to the saturation of the wettability. Numerous explanations can be stated for this observation. A trivial one is that already approximately 30 at % of oxygen is enough for the optimal wettability. The second possibility is that the surface (which influences the wettability) is already saturated with the polar functional groups at a moderate fluence and oxidation of the sub-surface layers occurs at higher fluences. Yet another explanation could be the formation of oxides on the surface—this effect will be discussed later.

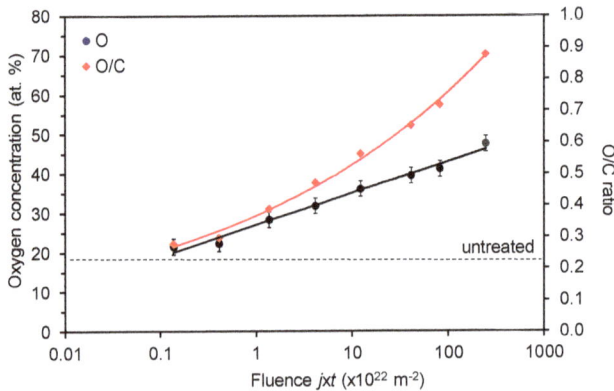

Figure 4. Variation of the oxygen concentration and the O/C ratio on the acryl-coated PP polymer surface with the O-atom fluence.

The high-resolution spectra of the carbon C1s peak for selected samples are presented in Figure 5. The spectra are normalized to the height of the main peak at 285 eV. The deconvolution of selected spectra is presented in Figure 6. The untreated sample (Figure 6a) contains three peaks: the main one at 285 eV corresponding to C–C, C–H bonds, and two small peaks at 286.5 and 289 eV corresponding to C–O and O=C–O groups, respectively. The spectrum in Figure 6a supports the information that the original sample has the acrylic coating. Figure 6b,c show an example of deconvolution of the sample treated at short (low oxygen fluence) and long (high oxygen fluence) treatment times. It can be observed that the intensity of C–O and O=C–O groups increased, especially for longer treatment times. It is difficult to judge about the formation of additional peaks corresponding to functional groups like C=O; however, if such groups develop upon treatment with the O atoms, their concentration on the polymer surface is much lower than the concentration of C–O and O=C–O groups. Figure 5 shows a gradual increase of the polar functional groups versus the fluence of the O-atoms, thus it is in good agreement with Figure 4. The increase is not equal for C–O and O=C–O groups, though. This can be seen from Figure 7, which shows the concentration of the functional groups versus the O-atom fluence. The highly polar O=C–O group increases somehow more intensively than the C–O group and actually prevails at the highest fluence. Interesting enough, this observation is not sound with the wettability presented in Figure 2. Namely, on the basis of the results presented in Figure 7, one would expect a monotonous decrease of the water contact angle with the increasing O-atom fluence. As mentioned above, this phenomenon could be related to surface saturation with the polar functional groups already at moderate fluences, and to oxidation of the sub-surface layers at higher fluences, or to the formation of Si oxides (discussed later).

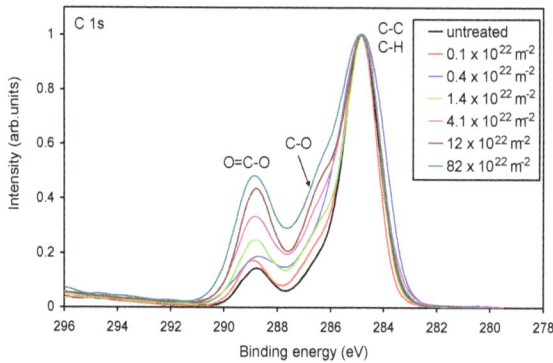

Figure 5. Comparison of high-resolution XPS carbon C 1s spectra of the acryl-coated PP polymer. The parameter is the O-atom fluence.

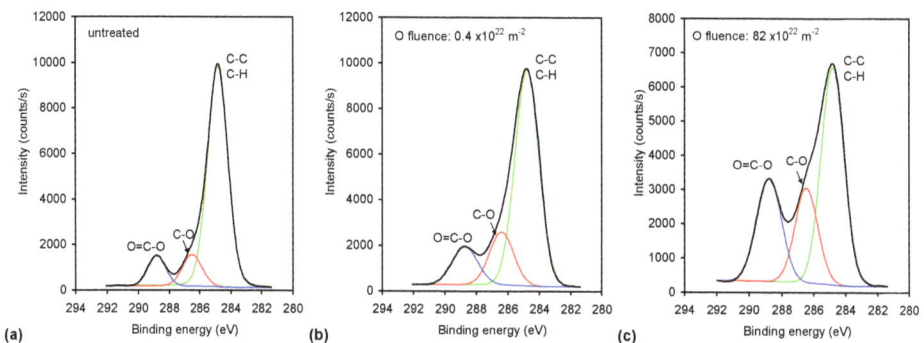

Figure 6. An example of fitting of XPS spectra: (**a**) untreated sample; (**b**) sample treated with a low O-atom fluence; and (**c**) sample treated with a high O-atom fluence.

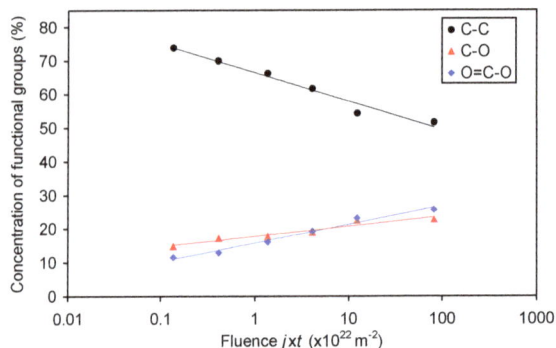

Figure 7. Variation of the concentration of various oxygen functional groups versus oxygen fluence. Concentrations were determined by fitting C 1s XPS spectra.

Another observation about the surface composition is worth stressing and discussing. Figure 8 represents survey XPS spectra for selected samples. Apart from carbon and oxygen, one can observe tiny peaks at binding energies of approximately 102 and 153 eV. The peaks correspond to silicon levels of Si 2p and Si 2s, respectively. The peaks are easily overlooked for the untreated sample (lowest curve in Figure 8), but become more pronounced after the sample has received a large fluence (upper curve). Detailed spectrum in the range 88–188 eV is shown in the insert of Figure 8. Doubtlessly, silicon is presented in the as-received sample and its concentration as detected by XPS increases with the increasing O-atom fluence. Figure 9 represents the concentration of Si in the surface of selected samples. Although the initial concentration is at the limit of this experimental technique, the trend is well justified. The origin of Si in the untreated sample is known to polymer scientists: i.e., silicon is often added to polymers as an anti-block or slipping agent in order to improve their performance. When the polymers are exposed to oxygen atoms, etching occurs. The effect has been elaborated elsewhere [40]. The oxygen atoms at first cause surface functionalization, but as the polymer surface becomes saturated with the O-rich functional groups, they form unstable molecular fragments which desorb from the surface. The polymer is thus slowly etched, leaving on the surface compounds that do not form volatile oxides. The effect is sometimes called plasma ashing [41]. Here, the acryl coating is slowly degraded and thus etched, leaving oxidized silicon nanoparticles on the surface. This effect explains the increase of Si concentration versus the O-atom fluence presented in Figure 9. It may or may not be a coincidence that the Si concentration (Figure 9) starts rising as the sample wettability becomes stable (Figure 2).

The virtual discrepancy between Figures 2 and 4 can be attributed to the appearance of silicon on the polymer surface. As explained above, the wettability (Figure 2) assumes a rather constant value after the fluence of about 3×10^{22} m^{-2}, but the concentration of oxygen on the polymer surface still increases (Figure 4). Taking into account the measured values of Si (Figure 9) and assuming that silicon is in the form of oxide (SiO_2), one can replot Figure 4 by considering that a part of oxygen is bonded to silicon, i.e., subtracting $2 \times$ [Si] oxygen from the curves. The new plot of O concentration and the O/C ratio by considering this effect is plotted in Figure 10. The behavior of the curve for oxygen in Figure 10 is now almost sound with the observations presented in Figure 2. Namely, the oxygen concentration as determined from XPS results also approaches a constant value for large fluences. Unfortunately, the saturation in Figure 10 does not appear at the same fluence as in Figure 2.

The role of silicon dioxide on the sample wettability is worth discussing. Figure 2 represents numerous measured data that fit the curve well, but the two points at the highest fluences definitely do not fit the general behavior. The decrease of the WCA for the highest fluences could be explained by oxidized silica nanoparticles on the sample surface, because well activated silicon oxide (treated by oxygen plasma) is hydrophilic [42]. The hydrophilicity is, however, lost soon after the plasma treatment because of the adsorption of organic impurities. That is one of the reasons why wettability tests were performed just after the treatment of samples with the O-atoms; however, hydrophobic recovery cannot be excluded completely.

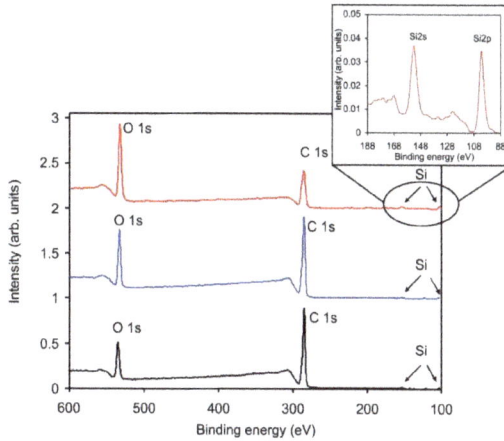

Figure 8. Selected XPS survey spectra of the untreated (**lowest curve**) and treated polymer at a low fluence of 0.4×10^{22} m^{-2} (**middle**) and at a high fluence of 82×10^{22} m^{-2} (**upper curve**).

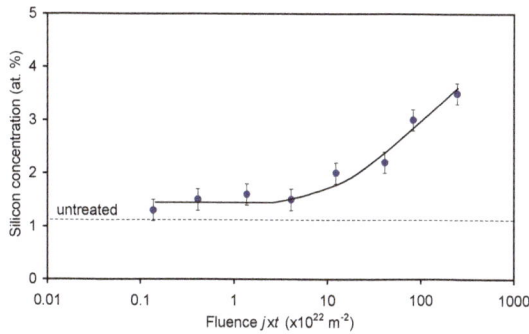

Figure 9. Silicon concentration versus O-atom fluence.

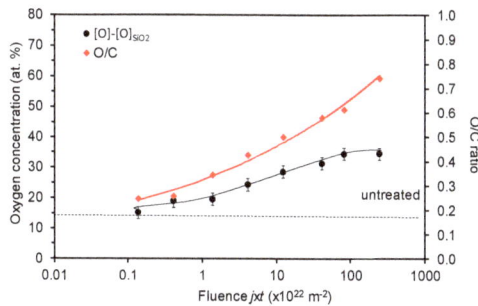

Figure 10. Variation of the oxygen concentration and the O/C ratio on the acryl-coated PP polymer surface with the O-atom fluence for the case when oxygen bonded to silicon is subtracted.

In view of the upper discussion, let us also discuss the AFM images of selected samples. The images are shown in Figure 11. The images were taken over the area of $5 \times 5\ \mu m^2$. The untreated sample (Figure 11a) exhibits small un-evenly distributed particles of virtually the same lateral size protruding from the surface. The typical lateral dimension of the particles is almost 100 nm and the height as determined by AFM is several 10 nm. The origin of these particles is probably polymer additives containing silicon. According to the XPS results (Figure 9), the density of the particles fits the concentration of silicon on the surface of the untreated sample. Figure 11b is the image of the sample after receiving a small O-atom fluence. According to the upper results and discussion, the fluence received by this sample was too small to cause any detectable polymer etching. The image actually does not differ significantly from Figure 11a. Also, the surface roughness of the sample shown in Figure 11b did not change much (from the initial 5.8 nm it increased to 5.9 nm). One can qualitatively conclude that the concentration of the particles protruding from the sample surface is similar in Figure 11a,b, which is sound with the observations presented in Figure 9.

The AFM images in Figure 11c,d vary significantly from Figure 11a,b. The particles protruding from the surface are now much denser, which could be a consequence of the polymer etching. Moreover, the surface roughness increased to 6.8 nm. From Figure 11, one can therefore assume that the surface is enriched with silica nanoparticles, which has been proposed on the basis of the XPS results presented in Figure 9.

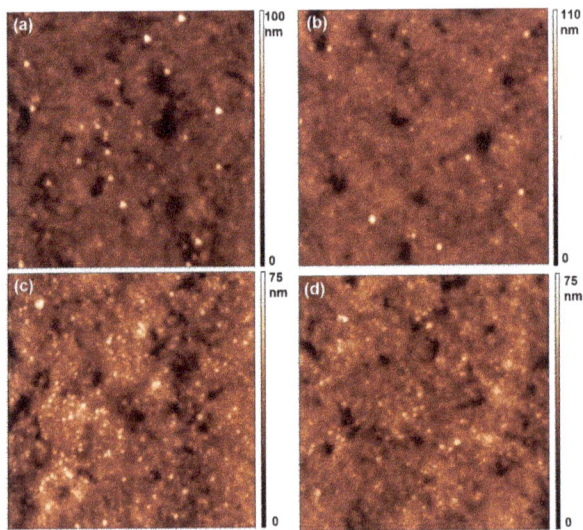

Figure 11. AFM images ($5 \times 5\ \mu m^2$) of selected samples: untreated (**a**) and treated at various fluences: (**b**) $0.1 \times 10^{22}\ m^{-2}$; (**c**) $82 \times 10^{22}\ m^{-2}$; and (**d**) $247 \times 10^{22}\ m^{-2}$.

In Figure 12, AFM topographic and phase images of the untreated sample and of one the selected treated sample recorded at a higher magnification of $2 \times 2\ \mu m^2$ are shown. The phase signal depends on the viscoelastic properties of the materials; therefore, the signal variation between the soft polymer surface and stiff silica particles can be observed. Figure 12 clearly shows a big difference in the variation of the phase signal for the treated sample in comparison to the untreated one. Many black spots with a big phase shift are observed on the treated sample, which confirms our conclusions about the presence of silica particles.

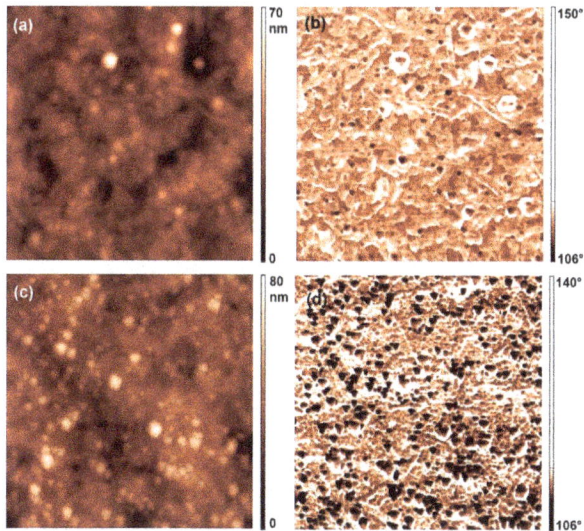

Figure 12. AFM topography (2×2 µm^2) and phase images of selected samples: (**a**) topography of the untreated sample; (**b**) phase image of the untreated sample; (**c**) topography of the sample treated with a fluence of 82×10^{22} m^{-2}; and (**d**) phase image of the sample treated with a fluence of 82×10^{22} m^{-2}.

4. Conclusions

An early stage of activation of commercial acrylic coated polypropylene foils for food packaging has been elaborated. The results clearly show that the maximum achievable surface wettability is already obtained at a rather low fluence of O-atoms of the order of a few 10^{22} m^{-3}. This information is particularly useful for users who want to activate the material without losing the acrylic surface film. Namely, larger fluences (in practice it means prolonged treatment time) has little or no effect on the surface wettability but causes etching of the thin acrylic film and thus loss of the functional properties of such foils. As stated in the introduction to this paper, the acrylic coating protects the polypropylene foil from external influences, and should therefore remain on the PP foil after accomplishing the activation procedure.

Acknowledgments: This research was co-financed by the European Union from the European Regional Development Fund and by the Ministry of Education, Science and Sport (F4F "Food for future").

Author Contributions: M.H. and T.V. conceived and designed the experiments under the supervision of M.M., M.S. and A.R.J.; M.H. and T.V. performed the experiments; M.H. performed AFM measurements; A.V. measured and analyzed the XPS results; M.S. contributed materials; M.M. wrote the paper; M.S. and A.R.J. participated in the discussion.

Conflicts of Interest: The authors declare no conflict of interest. The founding sponsors had no role in the design of the study; in the collection, analyses, or interpretation of data; in the writing of the manuscript, and in the decision to publish the results.

References

1. Pankaj, S.K.; Bueno-Ferrer, C.; Misra, N.N.; Milosavljević, V.; O'Donnell, C.P.; Bourke, P.; Keener, K.M.; Cullen, P.J. Applications of cold plasma technology in food packaging. *Trends Food Sci. Technol.* **2014**, *35*, 5–17. [CrossRef]
2. Gorjanc, M.; Savic, A.; Topalic-Trivunovic, L.; Mozetic, M.; Zaplotnik, R.; Vesel, A.; Grujic, D. Dyeing of plasma treated cotton and bamboo rayon with fallopia japonica extract. *Cellulose* **2016**, *23*, 2221–2228. [CrossRef]

3. Gorjanc, M.; Jazbec, K.; Sala, M.; Zaplotnik, R.; Vesel, A.; Mozetic, M. Creating cellulose fibres with excellent uv protective properties using moist cf_4 plasma and zno nanoparticles. *Cellulose* **2014**, *21*, 3007–3021. [CrossRef]

4. Lopez-Garcia, J.; Primc, G.; Junkar, I.; Lehocky, M.; Mozetic, M. On the hydrophilicity and water resistance effect of styrene-acrylonitrile copolymer treated by cf_4 and o_2 plasmas. *Plasma Process. Polym.* **2015**, *12*, 1075–1084. [CrossRef]

5. Lopez-Garcia, J.; Lehocky, M.; Humpolicek, P.; Novak, I. On the correlation of surface charge and energy in non-thermal plasma-treated polyethylene. *Surf. Interface Anal.* **2014**, *46*, 625–629. [CrossRef]

6. Lehocky, M.; St'ahel, P.; Koutny, M.; Cech, J.; Institoris, J.; Mracek, A. Adhesion of rhodococcus sp. S3E2 and rhodococcus sp. S3E3 to plasma prepared teflon-like and organosilicon surfaces. *J. Mater. Process. Technol.* **2009**, *209*, 2871–2875. [CrossRef]

7. Poncin-Epaillard, F.; Brosse, J.C.; Falher, T. Reactivity of surface groups formed onto a plasma treated poly(propylene) film. *Macromol. Chem. Phys.* **1999**, *200*, 989–996. [CrossRef]

8. Izdebska, J.; Sabu, T. *Printing on Polymers: Fundamentals and Applications*; Elsevier Science Publishing Co Inc.: Waltham, MA, USA, 2016.

9. Gorjanc, M.; Mozetic, M. *Modification of Fibrous Polymers by Gaseous Plasma: Principles, Techniques and Applications*; Lambert Academic Publishing: Saarbrücken, Germany, 2014.

10. Primc, G.; Tomsic, B.; Vesel, A.; Mozetic, M.; Razic, S.E.; Gorjanc, M. Biodegradability of oxygen-plasma treated cellulose textile functionalized with zno nanoparticles as antibacterial treatment. *J. Phys. D Appl. Phys.* **2016**, *49*, 324002. [CrossRef]

11. Gorjanc, M.; Sala, M. Durable antibacterial and uv protective properties of cellulose fabric functionalized with ag/tio_2 nanocomposite during dyeing with reactive dyes. *Cellulose* **2016**, *23*, 2199–2209. [CrossRef]

12. Popelka, A.; Novak, I.; Lehocky, M.; Bilek, F.; Kleinova, A.; Mozetic, M.; Spirkova, M.; Chodak, I. Antibacterial treatment of ldpe with halogen derivatives via cold plasma. *Express Polym. Lett.* **2015**, *9*, 402–411. [CrossRef]

13. Asadinezhad, A.; Novak, I.; Lehocky, M.; Sedlarik, V.; Vesel, A.; Junkar, I.; Saha, P.; Chodak, I. A physicochemical approach to render antibacterial surfaces on plasma-treated medical-grade pvc: Irgasan coating. *Plasma Process. Polym.* **2010**, *7*, 504–514. [CrossRef]

14. Sulovska, K.; Lehocky, M. Characterization of plasma treated surfaces for food safety by terahertz spectroscopy. *Proc. SPIE* **2014**, *9252*, 925209.

15. Lin, Y.J.; Dias, P.; Chen, H.Y.; Chum, S.; Hiltner, A.; Baer, E. Oxygen permeability of biaxially oriented polypropylene films. *Polym. Eng. Sci.* **2008**, *48*, 642–648. [CrossRef]

16. Decker, C.; Biry, S. Light stabilisation of polymers by radiation-cured acrylic coatings. *Prog. Org. Coat.* **1996**, *29*, 81–87. [CrossRef]

17. Vujković, I.; Galić, K.; Vereš, M. *Ambalaža za Pakiranje Namirnica*; Tectus: Zagreb, Croatia, 2007.

18. Tkavc, T.; Petrinič, I.; Luxbacher, T.; Vesel, A.; Ristić, T.; Zemljič, L.F. Influence of O_2 and CO_2 plasma treatment on the deposition of chitosan onto polyethylene terephthalate (pet) surfaces. *Int. J. Adhes. Adhes.* **2014**, *48*, 168–176. [CrossRef]

19. Pandiyaraj Navaneetha, K.; Selvarajan, V.; Deshmukh, R.R.; Gao, C. Adhesive properties of polypropylene (pp) and polyethylene terephthalate (pet) film surfaces treated by dc glow discharge plasma. *Vacuum* **2008**, *83*, 332–339. [CrossRef]

20. Choi, H.S.; Rybkin, V.V.; Titov, V.A.; Shikova, T.G.; Ageeva, T.A. Comparative actions of a low pressure oxygen plasma and an atmospheric pressure glow discharge on the surface modification of polypropylene. *Surf. Coat. Technol.* **2006**, *200*, 4479–4488. [CrossRef]

21. Morent, R.; De Geyter, N.; Leys, C.; Gengernbre, L.; Payen, E. Comparison between xps- and ftir-analysis of plasma-treated polypropylene film surfaces. *Surf. Interface Anal.* **2008**, *40*, 597–600. [CrossRef]

22. Leroux, F.; Campagne, C.; Perwuelz, A.; Gengembre, L. Polypropylene film chemical and physical modifications by dielectric barrier discharge plasma treatment at atmospheric pressure. *J. Colloid Interface Sci.* **2008**, *328*, 412–420. [CrossRef] [PubMed]

23. Aguiar, P.H.L.; Oliveira, E.C.; Cruz, S.A. Modification of clarified polypropylene by oxygen plasma to improve the adhesion of thin amorphous hydrogenated carbon films deposited by plasma enhanced chemical vapor deposition. *Polym. Eng. Sci.* **2013**, *53*, 1065–1072. [CrossRef]

24. Vishnuvarthanan, M.; Rajeswari, N. Effect of mechanical, barrier and adhesion properties on oxygen plasma surface modified pp. *Innov. Food Sci. Emerg. Technol.* **2015**, *30*, 119–126. [CrossRef]

25. Mirabedini, S.M.; Arabi, H.; Salem, A.; Asiaban, S. Effect of low-pressure o$_2$ and ar plasma treatments on the wettability and morphology of biaxial-oriented polypropylene (bopp) film. *Prog. Org. Coat.* **2007**, *60*, 105–111. [CrossRef]

26. Wanke, C.H.; Feijo, J.L.; Barbosa, L.G.; Campo, L.F.; de Oliveira, R.V.B.; Horowitz, F. Tuning of polypropylene wettability by plasma and polyhedral oligomeric silsesquioxane modifications. *Polymer* **2011**, *52*, 1797–1802. [CrossRef]

27. Strobel, M.; Jones, V.; Lyons, C.S.; Ulsh, M.; Kushner, M.J.; Dorai, R.; Branch, M.C. A comparison of corona-treated and flame-treated polypropylene films. *Plasmas Polym.* **2003**, *8*, 61–95. [CrossRef]

28. Beke, S.; Anjum, F.; Tsushima, H.; Ceseracciu, L.; Chieregatti, E.; Diaspro, A.; Athanassiou, A.; Brandi, F. Towards excimer-laser-based stereolithography: A rapid process to fabricate rigid biodegradable photopolymer scaffolds. *J. R. Soc. Interface* **2012**, *9*, 3017–3026. [CrossRef] [PubMed]

29. Vesel, A.; Mozetic, M. New developments in surface functionalization of polymers using controlled plasma treatments. *J. Phys. D Appl. Phys.* **2017**, *50*, 293001. [CrossRef]

30. Doliska, A.; Vesel, A.; Kolar, M.; Stana-Kleinschek, K.; Mozetic, M. Interaction between model poly(ethylene terephthalate) thin films and weakly ionised oxygen plasma. *Surf. Interface Anal.* **2012**, *44*, 56–61. [CrossRef]

31. Vesel, A.; Kolar, M.; Doliska, A.; Stana-Kleinschek, K.; Mozetic, M. Etching of polyethylene terephthalate thin films by neutral oxygen atoms in the late flowing afterglow of oxygen plasma. *Surf. Interface Anal.* **2012**, *44*, 1565–1571. [CrossRef]

32. Chashmejahanbin, M.R.; Salimi, A.; Ershad Langroudi, A. The study of the coating adhesion on pp surface modified in different plasma/acrylic acid solution. *Int. J. Adhes. Adhes.* **2014**, *49*, 44–50. [CrossRef]

33. Masaeli, E.; Morshed, M.; Tavanai, H. Study of the wettability properties of polypropylene nonwoven mats by low-pressure oxygen plasma treatment. *Surf. Interface Anal.* **2007**, *39*, 770–774. [CrossRef]

34. Yun, Y.I.; Kim, K.S.; Uhm, S.-J.; Khatua, B.B.; Cho, K.; Kim, J.K.; Park, C.E. Aging behavior of oxygen plasma-treated polypropylene with different crystallinities. *J. Adhes. Sci. Technol.* **2004**, *18*, 1279–1291. [CrossRef]

35. Chen, R.; Bayon, Y.; Hunt, J.A. Preliminary study on the effects of ageing cold oxygen plasma treated pet/pp with respect to protein adsorption. *Colloid. Surface B* **2012**, *96*, 62–68. [CrossRef] [PubMed]

36. Zaplotnik, R.; Vesel, A.; Mozetic, M. A fiber optic catalytic sensor for neutral atom measurements in oxygen plasma. *Sensors* **2012**, *12*, 3857–3867. [CrossRef] [PubMed]

37. Šorli, I.; Ročak, R. Determination of atomic oxygen density with a nickel catalytic probe. *J. Vac. Sci. Technol. A* **2000**, *18*, 338–342. [CrossRef]

38. Primc, G.; Vesel, A.; Dolanc, G.; Vrančić, D.; Mozetič, M. Recombination of oxygen atoms along a glass tube loaded with a copper sample. *Vacuum* **2017**, *138*, 224–229. [CrossRef]

39. Mozetic, M.; Primc, G.; Vesel, A.; Zaplotnik, R.; Modic, M.; Junkar, I.; Recek, N.; Klanjsek-Gunde, M.; Guhy, L.; Sunkara, M.K.; et al. Application of extremely non-equilibrium plasmas in the processing of nano and biomedical materials. *Plasma Sour. Sci. Technol.* **2015**, *24*, 015026. [CrossRef]

40. Junkar, I.; Cvelbar, U.; Vesel, A.; Hauptman, N.; Mozetič, M. The role of crystallinity on polymer interaction with oxygen plasma. *Plasma Process. Polym.* **2009**, *6*, 667–675. [CrossRef]

41. Pike, S.; Dewison, M.G.; Spears, D.A. Sources of error in low temperature plasma ashing procedures for quantitative mineral analysis of coal ash. *Fuel* **1989**, *68*, 664–668. [CrossRef]

42. Alam, A.U.; Howlader, M.M.R.; Deen, M.J. The effects of oxygen plasma and humidity on surface roughness, water contact angle and hardness of silicon, silicon dioxide and glass. *J. Micromech. Microeng.* **2014**, *24*, 035010. [CrossRef]

Article

Tribological and Wear Performance of Nanocomposite PVD Hard Coatings Deposited on Aluminum Die Casting Tool

Jose Mario Paiva [1,2,*], German Fox-Rabinovich [1], Edinei Locks Junior [1], Pietro Stolf [1], Yassmin Seid Ahmed [1], Marcelo Matos Martins [2], Carlos Bork [1,3] and Stephen Veldhuis [1]

[1]   McMaster Manufacturing Research Institute (MMRI), Department of Mechanical Engineering, McMaster University, 1280 Main Street West, Hamilton, ON L8S4L7, Canada; gfox@mcmaster.ca (G.F.-R.); lockse@mcmaster.ca (E.L.J.); stolfp@mcmaster.ca (P.S.); seidahmy@mcmaster.ca (Y.S.A.); carlosbork@gmail.com (C.B.); veldhu@mcmaster.ca (S.V.)

[2]   Department of Mechanical and Materials Science, Catholic University of Santa Catarina, Rua Visconde de Taunay, 427-Centro, Joinville, SC 89203-005, Brazil; marcelo.martins@catolicasc.org.br

[3]   IFSul—Federal Institute Sul-rio-grandense—Campus Sapucaia do Sul, Av Copacabana, 100, Sapucaia do Sul, RS 93216-120, Brazil

*   Correspondence: paivajj@mcmaster.ca; Tel.: +1-905-525-9140 (ext. 27800)

Received: 1 February 2018; Accepted: 24 February 2018; Published: 28 February 2018

Abstract: In the aluminum die casting process, erosion, corrosion, soldering, and die sticking have a significant influence on tool life and product quality. A number of coatings such as TiN, CrN, and (Cr,Al)N deposited by physical vapor deposition (PVD) have been employed to act as protective coatings due to their high hardness and chemical stability. In this study, the wear performance of two nanocomposite AlTiN and AlCrN coatings with different structures were evaluated. These coatings were deposited on aluminum die casting mold tool substrates (AISI H13 hot work steel) by PVD using pulsed cathodic arc evaporation, equipped with three lateral arc-rotating cathodes (LARC) and one central rotating cathode (CERC). The research was performed in two stages: in the first stage, the outlined coatings were characterized regarding their chemical composition, morphology, and structure using glow discharge optical emission spectroscopy (GDOES), scanning electron microscopy (SEM), and X-ray diffraction (XRD), respectively. Surface morphology and mechanical properties were evaluated by atomic force microscopy (AFM) and nanoindentation. The coating adhesion was studied using Mersedes test and scratch testing. During the second stage, industrial tests were carried out for coated die casting molds. In parallel, tribological tests were also performed in order to determine if a correlation between laboratory and industrial tests can be drawn. All of the results were compared with a benchmark monolayer AlCrN coating. The data obtained show that the best performance was achieved for the AlCrN/Si$_3$N$_4$ nanocomposite coating that displays an optimum combination of hardness, adhesion, soldering behavior, oxidation resistance, and stress state. These characteristics are essential for improving the die mold service life. Therefore, this coating emerges as a novelty to be used to protect aluminum die casting molds.

Keywords: PVD nanocomposite coatings; aluminum die casting; tool life; tribological performance

1. Introduction

Aluminum is a widely used material in the automotive industry. The use of special aluminum alloys as materials to manufacture components and automotive parts allows for the construction of light-weight components that lead to an overall weight reduction, and thus, to reduced fuel consumption [1]. These aluminum alloys are commonly cast using the high pressure die casting (HPDC) process, which is one of the most efficient methods for the production of complex shape castings in today's manufacturing industry [2]. For that reason, recent research in this field has focused to improve tool life of aluminum die casting molds.

During the casting of aluminum alloys, the cyclic process leads to thermocyclic loads on the tool surface from T = 90 °C at cooling up to T = 600 °C at die casting. In this way, the tools are exposed to erosion, corrosion, soldering, or die sticking due to the frequent contact between the tool surface and the casting alloy. Any one of these phenomena can result in damage to the die and poor surface quality of the casting, as well as a notable decrease in the productivity and efficiency of the casting operation [3].

In order to protect die casting molds, hard coatings deposited by physical vapor deposition (PVD), such as TiN, CrN, (Cr,Al)N, Ti(C,N), Ti(B,N), or (Ti,Al)(C,N) [4–6] have been employed to act as a physical barrier to the die casting molding to prevent the erosion and soldering of aluminum and improve the resistance against thermal cracking [7]. Previous studies show that TiN presented good corrosion and erosion wear resistance [4–6]. However, this coating is not a good solution for HPDC due to its low oxidation temperature [4]. TiAlN and TiSiN coatings exhibit good mechanical properties and also exhibit a better oxidation resistance up to 700 °C, however, their adhesion under substrate is reduced, which leads to sudden failures [8] during service. Thus, novel solutions need to be studied in order to reduce the intensity of frictional interaction between the tool surface and the aluminum casting alloy and reduce wear of the molds.

The coatings previously studied do not show properties good enough to solve the problems outlined above. Therefore, the development of advanced PVD hard coatings, such as nanocomposites, which are designed to resist under severe mechanical and thermal stress conditions emerge as possible solutions to drastically reduce outlined problems. This family of coatings has a nano-crystalline structure that results in high mechanical and functional properties. The coatings of such structure are able to maintain a low friction coefficient (self-lubricating coatings) in extreme environments, combined with high hardness to improve the mechanical resistance [9,10]. Furthermore, the adhesion strength for nanocomposite coating at the interface between the coating and substrate is improved in comparison to regular coatings, and it can guarantee desirable surface properties, as well as long durability of the tool [11].

When considering the different characteristics of the nanocomposite coatings during service, the goal of this paper is to present the nanocomposite (AlTiN/Si$_3$N$_4$ and AlCrN/Si$_3$N$_4$) coatings that have been deposited by PVD as novel coatings for the aluminum die casting process. To demonstrate suitability, studies of the mechanical properties and tribological performance have been investigated in both laboratory and industrial tests. The results that were obtained in terms of tribological and wear performance were compared with a monolayer AlCrN, which is the coating currently used in industry.

2. Materials and Methods

Coatings were deposited onto (AISI H13) hot worked tool steel substrate (48 ± 1 HRC) using a physical vapor deposition (PVD) process. The chemical composition of AISI H13 was provided by the supplier, and it is presented in Table 1.

Table 1. Chemical composition of AISI H13 used as die cast mold substrate.

Element/Amount	C	Cr	Mo	Mn	Si	V	Fe
wt %	0.49	4.99	1.3	0.4	0.97	0.93	Balance

One set of samples (die casting mold, core pins, and blocks) were used for each coating. Figure 1 shows the samples that were used for coating characterization and tool life tests.

The PVD nanocomposites (AlTiN/Si$_3$N$_4$ and AlCrN/Si$_3$N$_4$), and AlCrN coatings were deposited on the samples by means of cathodic arc evaporation in a Platit π411 industrial deposition unit, equipped with three lateral arc-rotating cathodes (LARC) and one central rotating cathode (CERC). Plansee, AG—Germany produced all of the targets that were employed in this work. The target that was used for AlCrN coating contained Al/Cr (Al60:Cr40 at %). The deposition process was carried out at a temperature of 500 °C in a 99.999% pure nitrogen atmosphere at a pressure of 4 Pa and a bias voltage of −40 V. To produce nanocomposite coatings (AlTiN/Si$_3$N$_4$ and AlCrN/Si$_3$N$_4$), targets containing pure metals (Cr, Ti) and the AlSi (88:12 at %) were employed. The coatings were deposited at a temperature of 480 °C, and two different gases, argon and nitrogen (both 99.999% pure), were used. The pressure of the process was about 3.5 to 4 Pa and the bias voltage of −40 V. During the process, evaporated metals and metal alloys enter the plasma state to combine with the ionized process gas (nitrogen) and eventually condense on the substrate surface, as part of ceramic compounds. Amorphous and micro-nanocrystalline structures and layers are developed with optimized thermodynamic and kinetic conditions. Spinoidal decomposition allows for building TiN and CrN nanocrystalline structures dispersed in a Si$_3$N$_4$ amorphous matrix, with a typical crystallite size of about 10 nm [12].

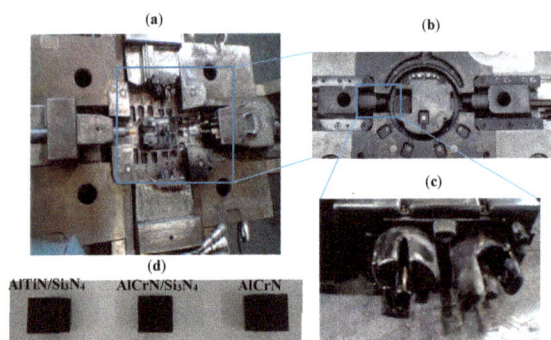

Figure 1. Set of coated samples (**a**) General view from the coated mold; (**b**) cavity details; (**c**) core pins; and (**d**) H13 blocks used for coating characterization studies.

The roughness of the substrate and coating surfaces were determined by atomic force microscopy (AFM) equipped with a scanning probe (Shimadzu SPM 9500 J3, Kyoto, Japan). A single crystal silicon with a long rectangular cantilever was used as a scanning probe. The tips are pyramidal shaped, with a nominal radius of 10 nm with a spring constant of 0.5 N/m. The scanning mode was configured as contact, with a scanning rate of 1 Hz and high resolution. A region with the size of 30 µm × 30 µm was selected for the characterization of the coated surface samples. The images that were produced were processed to remove background signals, and to extract results such as surface roughness (Ra) and topographic profiles.

The cross-sections of the coatings were inspected with a field emission gun microscope Supra 55VP by Zeiss (Oberkochen, Germany), equipped with energy-dispersive X-ray spectroscopy (EDS) Quantax XFlash 6/30 by Bruker (Berlin, Germany). The elemental compositions of the nanocomposite coatings were determined by glow discharge optical emission spectroscopy (GDOES) GDS-850A by Leco (Saint Joseph, MI, USA). Five points were analyzed for each sample that was coated, and the average result was taken. The phases of the coatings and residual stress were characterized using X-ray diffraction XDR 7000 by Shimadzu (Kyoto, Japan) with 0.5° grazing angle, scanning range from 30° to 70°, the angle of incidence of 0.02°, and scanning speed of 1°/min.

Hardness (H) and elastic modulus (E) of the coatings were measured using a G200 XP-MTS Nano Indenter System (Agilent, Boston, MA, USA) equipped with a Berkovich Indenter. To determine the Hardness (H) and elastic modulus (E), the load of 400 mN was applied through 25 indentations that were arranged in a matrix of 5 × 5. To obtain a reliable mean value and standard deviation, at least six points were tested for each sample. Table 2 shows mechanical properties, residual stress, and the thickness of the coatings that were investigated in this work.

Table 2. Mechanical properties, residual stress, and thickness of the coatings.

Coatings	H—Hardness (GPa)	E—Reduced Elastic Modulus (GPa)	H/E Ratio	H3/E2 Ratio	Residual Stress (GPa)	Thickness (μm)
AlCrN	27 ± 3	360 ± 25	0.0750	0.151	4	3.2
AlTiN/Si$_3$N$_4$	39 ± 3	400 ± 25	0.0975	0.370	4.8	2.8
AlCrN/Si$_3$N$_4$	44 ± 3	430 ± 25	0.0102	0.460	5.2	2.3

Indentation adhesion tests evaluated the quality of the adhesion between the coating and substrate. These tests were carried out on the Wilson instruments Rockwell hardness tester (Buehler, Norwood, UK) at an indentation load of 1471 N (150 kgf) was performed to assess the quality of the coatings. The test procedure followed the VDI 3198 (1991) standard [13]. In addition, a CSM Instruments Revetest Scratch Tester (Anton-Paar, Buch, Switzerland) equipped with a diamond cone (radius of 200 μm, cone angle of 120°) was employed to determine the adhesion of the coatings on the H13 blocks. A progressive normal load, ranging from 1 to 200 N, was applied over a length of 3 mm. An average of three scratch tests was carried out for each coating.

Tribological tests were performed on a pin-on-disk CSM Tribometer (Anton-Paar, Buch, Switzerland) to determine the friction behavior of the coatings at different ranges of temperature. The tests were performed at room and elevated temperature (200, 400 and 600 °C) with a constant load of 20 N on the pin (WC ball was used as a pin), the sliding speed of 0.2 m/s and slide distance of 300 m. The pin-on-disk experiments were repeated at least three times for each temperature to check the reproducibility of the test results. Based on that, the scatter of the friction coefficient values was found to be approximately 5%. The three-dimensional surface profile and depth of the friction zone was assessed using a non-contact Optical Surface Profilometer Zygo-6000 Series (ZYGO, Middlefield, OH, USA).

To evaluate the performance of the coatings, coated mold and core pins were tested during the industrial process using an industrial pressure die casting of the Aluminum alloy AlSi$_{12}$Cu$_3$. The casting conditions and the spraying of the lubricating compound were not changed from the standard procedures. The melt temperature in the holding furnace was maintained at 680 °C throughout the experiments. The die gate velocity was 60 m/s, and a final intensification pressure of approximately 65 MPa was applied. The tool life results were measured in terms of parts and hours of production. After the operation, the cores were examined using SEM microscopy (ZEISS, Oberkochen, Germany).

3. Results and Discussion

3.1. Coatings Characterizations

The morphology of the surfaces was evaluated by Atomic Force Microscopy (AFM). The substrate and coatings' roughness values obtained in three-dimensional (3D) and two-dimensional (2D) models are presented in Figure 2.

As it is shown, coatings present higher roughness values compared to the substrate, which can be confirmed through of surface topographies analyzed by AFM (Figure 2a). As a final result after the coating deposition process, the lowest surface roughness Ra was observed in AlCrN/Si$_3$N$_4$ coating (Table 3), and then, confirmed by AFM images (Figure 2b). A summary of the average and standard deviation values of the surface roughness are presented in Table 3.

Table 3. The surface roughness of the coatings.

Coatings	Surface Roughness—Ra (μm)
Uncoated	0.88 ± 0.02
AlCrN	1.59 ± 0.02
AlTiN/Si$_3$N$_4$	1.43 ± 0.02
AlCrN/Si$_3$N$_4$	1.38 ± 0.02

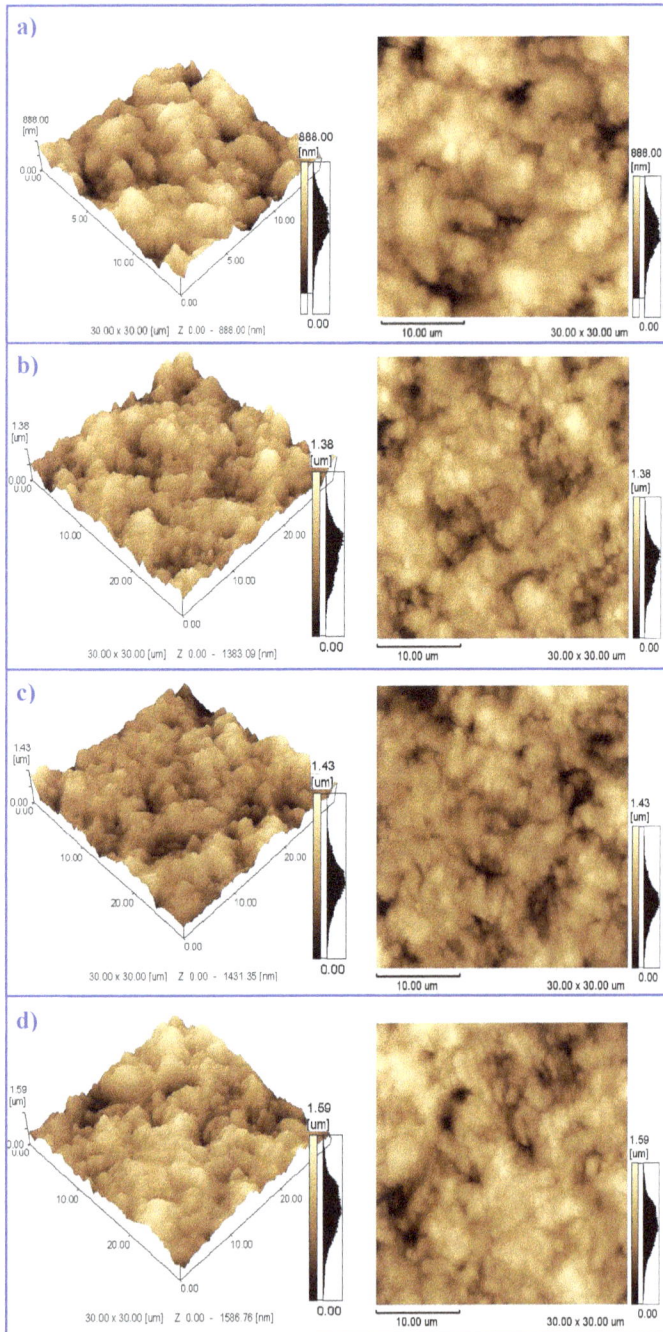

Figure 2. Atomic force microscope images of surface topography. (**a**) uncoated sample; (**b**) AlCrN/Si₃N₄ physical vapor deposition (PVD) nanocomposite coating; (**c**) AlTiN/Si₃N₄ PVD nanocomposite coating and (**d**) AlCrN PVD Coating.

From the data presented in Figure 2 and Table 3, we can confirm an increase in surface roughness compared to the uncoated state. Nanocomposite hard coatings with increased AlSi deposited by PVD using the Lateral Rotating Cathode (LARC) system allows for obtaining an increase of Ra surface roughness if compared to the substrate. This is due to the lower melting point of the evaporated material during cathodic arc deposition [14]. Increasing cathode current leads to the emission of a higher quantity of particles with a bigger volume and size. As a final result, it is expected that the amount of cathode material would rise with a lower melting point, which would then lead to an increase in the surface roughness [14]. Even though a quick increasing in Ra has been seen, the cathodic arc evaporation process using LARC technology is able to produce very smooth surface roughness when compared to conventional cathodic arc evaporation process [15]. However, in this way, one realizes that nanocomposite coatings present low roughness than AlCrN coating.

Figure 3 shows SEM cross-section images for H13 blocks coated with AlCrN, AlTiN/Si_3N_4, and AlCrN/Si_3N_4. The coatings present a compact structure, without any visible delaminations or defects.

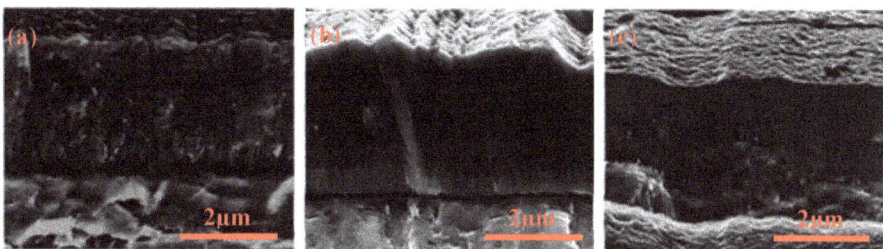

Coating	(a)—AlCrN	(b)—AlTiN/Si_3N_4	(c)—AlCrN/Si_3N_4
Element	Atomic %	Atomic %	Atomic %
Al K	31.6	20.5	27.3
Cr K	18.1	-	22.7
Ti K	-	16	-
Si K	-	3.5	3.9
N K	50.3	60	51.1

Figure 3. Scanning electron microscopy (SEM) images of coatings cross-section and glow discharge optical emission spectroscopy (GDOES) results for coating chemical composition, (**a**) AlCrN PVD Coating; (**b**) AlTiN/Si_3N_4 PVD nanocomposite; and (**c**) AlCrN/Si_3N_4 PVD nanocomposite coating.

A dense and uniform structure characterizes the morphology of the fracture of coatings studied (Figure 3). AlCrN and AlTiN/Si_3N_4 coatings present a columnar structure, while AlCrN/Si_3N_4 coating, a randomized, the nearly flawless structure can be observed. Furthermore, the interface of AlCrN/Si_3N_4 and the H13 substrate shows no irregularities. However, for AlCrN/Si_3N_4 coating, it is apparent that, especially in the lower part of the PVD layer, a slight orientation in the growth direction is existent. Furthermore, the coating gets finer with a decreasing orientation and becoming an amorphous structure as dropless. This observation is supported by the presence of higher content of Silicon at the structure. The GDOES results for the chemical composition of the coatings are shown in Figure 3, as well. The results indicate that the amount of Al content (at %) for AlCrN/Si_3N_4 nanocomposite coatings is higher than for AlTiN/Si_3N_4. This higher Al content can increase the hardness and the temperature oxidation resistance [16], and may partly explain the better performance for coated die casting.

Figure 4 shows the XRD patterns of the AlCrN and nanocomposite (AlCrN/Si_3N_4, AlTiN/Si_3N_4) coatings that are deposited over AISI H13 block samples. The patterns clearly show the characteristic peaks of Face Center Cubic (FCC) structure with (111), (200), (220) for all of the coatings.

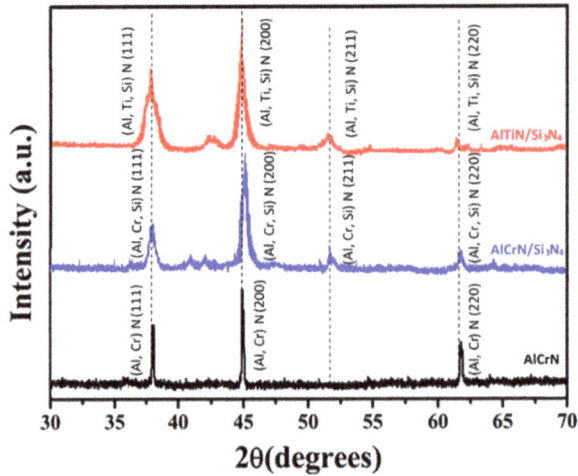

Figure 4. X-ray diffraction (XRD) patterns for coatings deposited under H13.

This FCC metastable solid solution can be obtained in the PVD coating under non-equilibrium conditions of the coating synthesis [17]. In contrast, with the presence of Si for nanocomposite coatings, the diffractograms show that the coatings exhibit a structure with multiple orientations of crystal planes, corresponding to (111), (200), and (220). This result indicates the presence of an amorphous phase of Si_3N_4 that is formed during the PVD process deposition on either AlCrSiN or AlTiSiN coatings [17]. Kao et al. [18] using the XRD technique, also provided evidence for the same structure for AlCrSiN and AlTiSiN coating system. According to them, with the presence of Si content, the peaks exhibited a broadening and weakening trend as a whole, which indicated the formation of a fine-grained structure and the decrease of crystalline size; this was due to the incorporation of amorphous Si_3N_4 in the coatings. At the same time, the development of the crystal phase was disturbed by the amorphous Si_3N_4, causing the nitride grains to grow discontinuously and forming a general nitride mix of aluminum and chromium, or aluminum and titanium, which could effectively affect the property of coating [19,20].

3.2. Coatings Mechanical Properties

Table 2 is showing the mechanical properties of the coatings studied in this work. The hardness (H) and reduced elastic modulus (E) were determined by nanoindentation. The results show that, with the presence of Si content, both hardness and reduced elastic modulus of coating increase slightly. The highest values for hardness and reduced elastic modulus were reached by AlCrN/Si_3N_4 nanocomposite coating. This improvement was also observed for AlTiN/Si_3N_4 nanocomposite coatings. The hardness difference of the nanocomposite coatings is related to the different Si contents (Figure 3), which may result from different hardening mechanisms. The presence of Si can lead to the formation of a solid solution hardening (Figure 4), created by the dissolving Si and Al atoms in CrN or TiN (in the case of AlCrN/Si_3N_4–AlTiN/Si_3N_4) [21,22]. This process will result in a lattice distortion due to having a different atom radius. The amorphous Si_3N_4 is thin and envelops CrN or TiN grains, and the interfaces between different phases can hinder dislocation formation or movement, which will lead to a super hard effect. However, it does not mean that a large volume of Si is beneficial for the coatings. For instance, residual stress for AlCrN/Si_3N_4 is higher than other coatings that are studied in this work. If the amount of amorphous Si_3N_4 matrix increases too much, then the percent of interface area exceeds a certain optimum value, and the phase grain separation and the blocking effect of grain boundaries are limited, which results in high residual stresses [23].

Also, from the H and E values, the toughness of the coatings was evaluated in terms of the relation between H/E and H3/E2 ratios. The values that were calculated are listed in Table 2. This relation results in the ability of a material to resist crack initiation and propagation; the toughness is reflected in the resistance against elastic strain to failure (H/E) and the resistance against plastic deformation (H3/E2). In this case, the H/E and H3/E2 ratios of $AlCrN/Si_3N_4$ were approximately 0.0102 and 0.460 GPa, respectively. This indicates that $AlCrN/Si_3N_4$ coating exhibited the best elastic strain to failure parameter and presents the higher resistance to plastic deformation. This could be confirmed throughout Rockwell adhesion tests done at load of 150 kgf. According to the results presented Figure 5, for all of coatings studied, radial cracks were found around the indentation margin without the presence of peeling.

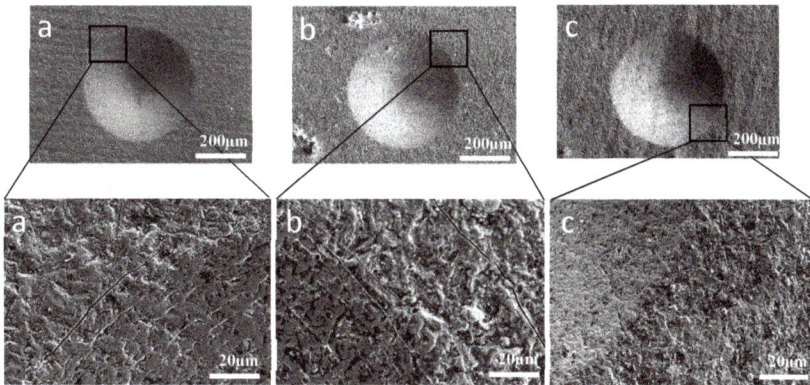

Figure 5. SEM micrographs after Rockwell C indentations with a load of 150 kgf on different coated surfaces (**a**) AlCrN; (**b**) $AlTiN/Si_3N_4$; and (**c**) $AlCrN/Si_3N_4$.

Based on VDI guideline 3198 the compound adhesion can be evaluated to adhesion class HF 1 for all samples. This indicates a good adhesion strength of the coatings. However, $AlCrN/Si_3N_4$ presents only very small (almost invisible) cracks around the indentation margin in comparison with other coatings (Figure 5c). This shows that this coating has a better adhesion strength. As it is known, the adhesion strength between the coating and substrate is a critical property in wear-resistance for coatings. If a failure of the coating occurs during operation, then the coating's capabilities can be greatly reduced, and it can cause severe abrasive wear on the friction system. Therefore, in order to measure this noticeable improvement for the adhesion strength, a scratch test was employed for each sample.

Figure 6 presents the SEM images of scratch tracks for the different coatings that were studied in this paper. From these results, it is possible to see only minor delamination that is found on the scratch track (Figure 6b), demonstrating that $AlCrN/Si_3N_4$ coating has high adhesion strength with the substrate. This can be related to the elastic recovery behavior that takes place in front of the indenter path, which is caused by the compressive stresses that are generated by the indenter and the inability of the coating to deform plastically. This failure mode was observed in LC2 for $AlCrN/Si_3N_4$ with the highest load (125 N), among the three coating system, indicating, therefore, the highest adhesive strength. Obviously, delamination is also formed in the scratch track for AlCrN and $AlTiN/Si_3N_4$ coatings. For these coatings, the load measured in LC2 was 100 N and 92 N, respectively, which proves that the coatings have weaker adhesion strength. Therefore, $AlCrN/Si_3N_4$ is a coating with a higher level of adhesion force, shows a combination of excellent toughness and adhesion strength. These characteristics are essential for improving the die mold service life.

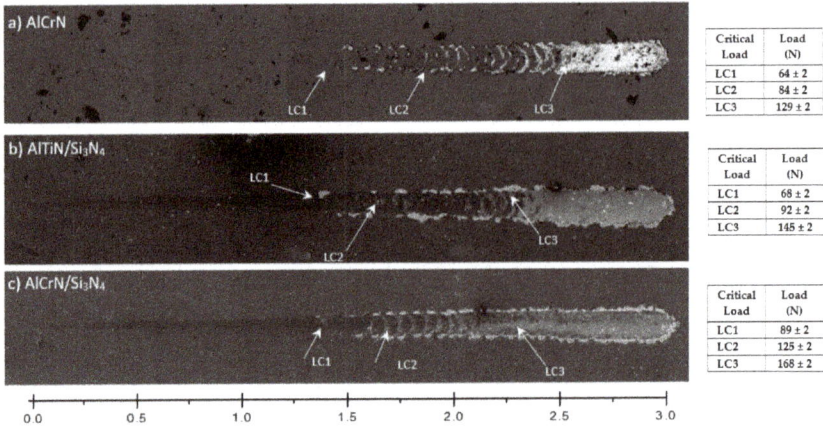

Figure 6. SEM micrographs after scratch test of the coated studied and the Critical load measured at the three stages LC1—First Critical Load (Cohesive Failure); LC2—Second Critical Load (Adhesion Failure) and LC3—Third Critical Load (Substrate Exposure): (**a**) AlCrN; (**b**) AlTiN/Si$_3$N$_4$; and, (**c**) AlCrN/Si$_3$N$_4$.

3.3. Tribological Properties

The coefficient of friction (COF) vs. temperature data for the coatings in contact with WC ball pin is shown in Figure 7a.

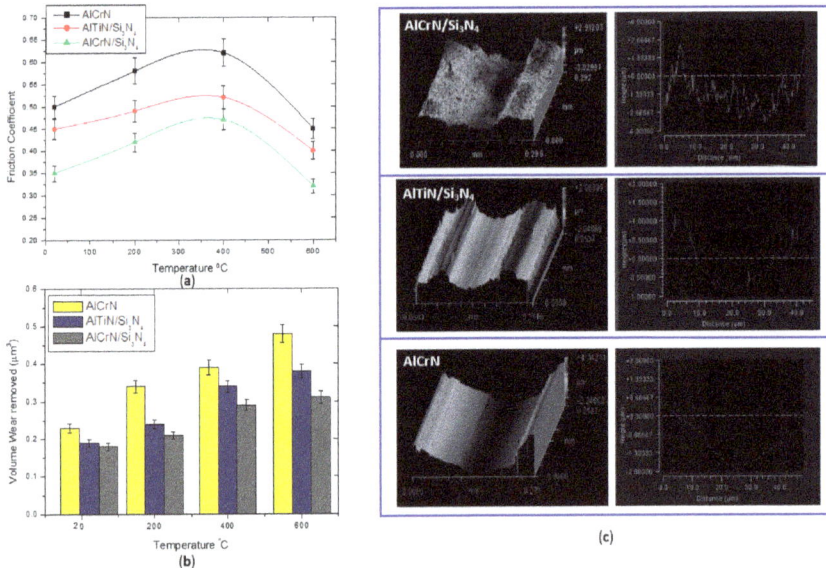

Figure 7. (**a**) Friction coefficient of the coatings at different temperatures; (**b**) Volume wear measured on the top of the surface after pin on disc test and (**c**) three-dimensional (3D) and two-dimensional (2D) profilometer images of the wear track profiles measured on the samples tested at high temperature (600 °C).

The coefficient of friction of the AlCrN/Si$_3$N$_4$ nanocomposite coating is noticeably lower when compared to the other coatings, especially at elevated temperatures within the range of 600 °C. All of the coatings present the characteristics to improve lubricity under elevated temperatures, which can be explained by the presence of oxides that are generated at high temperatures, which tend to reduce the friction conditions at the surface [24]. Also, the COF of both nanocomposite coatings gradually decreases in all stages studied. This behavior is related to the presence of the Si content in the coatings (Figure 3). Similar observation of a decrease in the COF as the Si content increased was also widely reported in other Al-Ti-Si-N based nanocomposite systems [25]. Lower friction might result in longer tool life. To support this hypothesis, soon after the pin-on-disk tests, the overall volume wear that was removed for the coatings was calculated and the results confirm that less friction, in the case of AlCrN/Si$_3$N$_4$, resulted in much less intensive wear when compared to the other coatings that were studied (Figure 7b). The analysis of the wear volume also revealed a higher wear volume of the AlCrN coating than the nanocomposite coatings for all the different temperatures. Figure 7c shows the corresponding wear tracks at 600 °C for all coatings. The surfaces presented grooves along the sliding direction, indicating that adhesive wear is the prevailing wear mechanism.

A way to explain the improvement at frictions conditions for AlCrN/Si$_3$N$_4$ coating is evaluating the presence of oxides in EDS analysis. This analysis has been done on the wear tracks for AlCrN/Si$_3$N$_4$ coating. The results that are presented in Figure 8 (a and b) revealed the presence of elements such as Si, Cr, Al, and O, showing a possible formation of the Al$_2$O$_3$ phase. This oxide is considered a thermal barrier and lubricant layer, and may improve the friction conditions at the work surface. This is a result of the high temperature that is inherent to the process, which causes the appearance of a thin film amorphous structure of the top layer. As shown in Figure 7a, the friction conditions of nanocomposite AlCrN/Si$_3$N$_4$ tend to reduce at high temperature. Therefore, these results show that, due to the lubricant characteristic and elevated hardness, AlCrN/Si$_3$N$_4$ nanocomposite coating has excellent wear performance and can contribute to the better wear life of the mold.

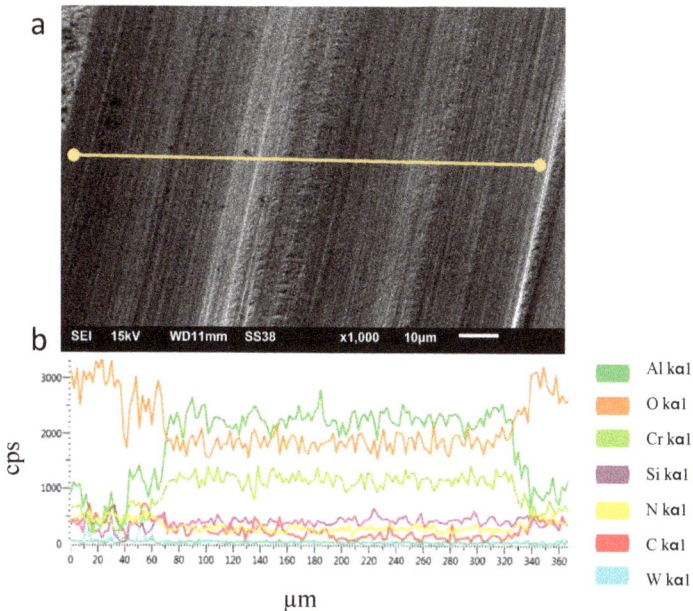

Figure 8. (**a**) EDS spectrum of the drawn line over the track for AlCrN/Si$_3$N$_4$ coated sample tested at 600 °C and (**b**) spectrum lines.

3.4. Mold's Tool Life

The behavior of PVD coatings was studied during industrial applications. The tests were done under usual production conditions. For that, the end of the life was determined by obtaining pieces that were free of defects, in other words, the period of time that the machinery could run without stopping for maintenance of the die cast mold. The results of these tests showed that nanocomposite coatings could be used as a novelty in this segment, and that results in an increase of the lifetime significantly in comparison with AlCrN coating (Figure 9). Consequently, the molds coated with AlCrN/Si$_3$N$_4$ represent a tool life increase of approximately 92% in comparison to the AlCrN coating. This confirms that tool life progress is caused by an improvement of mechanical properties of these coatings (Table 2) within a range of operating temperatures.

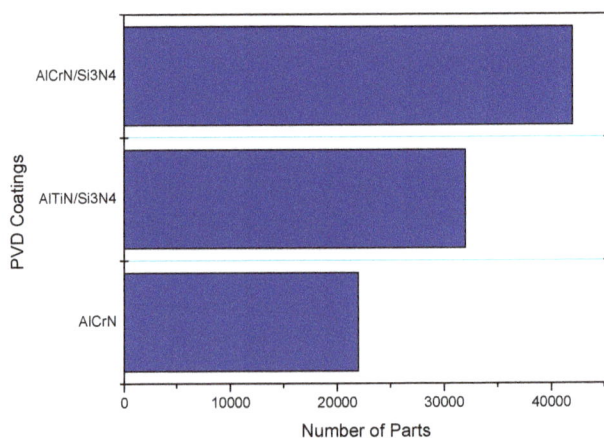

Figure 9. Number of parts produced by high pressure die casting (HPDC) until quality insufficient be achieved.

Figure 10 shows SEM images of the top surface of the coated molds and the cross-section of the core pins after the tests. The pins coated with AlCrN suffered severe adhesion and cracking (Figure 10a,b). The microcrack, especially, is caused by the thermal fatigue due to the alternate heating and cooling over the die surface during die casting. Therefore, the die surface tends to be in compression during heating and tension during cooling of the die. This results in thermal cracking on the die surface, which degrades part of the surface finish and ultimately leads to die failure. For the nanocomposite coatings, the main wear problems during continuous die-casting were not microcracks, but erosion and sticking of the molten aluminum due to the regular flow during the casting process. For AlTiN/Si$_3$N$_4$, areas of the coating were detached by erosion (Figure 10c,d). Adhesion and erosion phenomena were particularly reduced for the AlCrN/Si$_3$N$_4$ (Figure 10e,f).

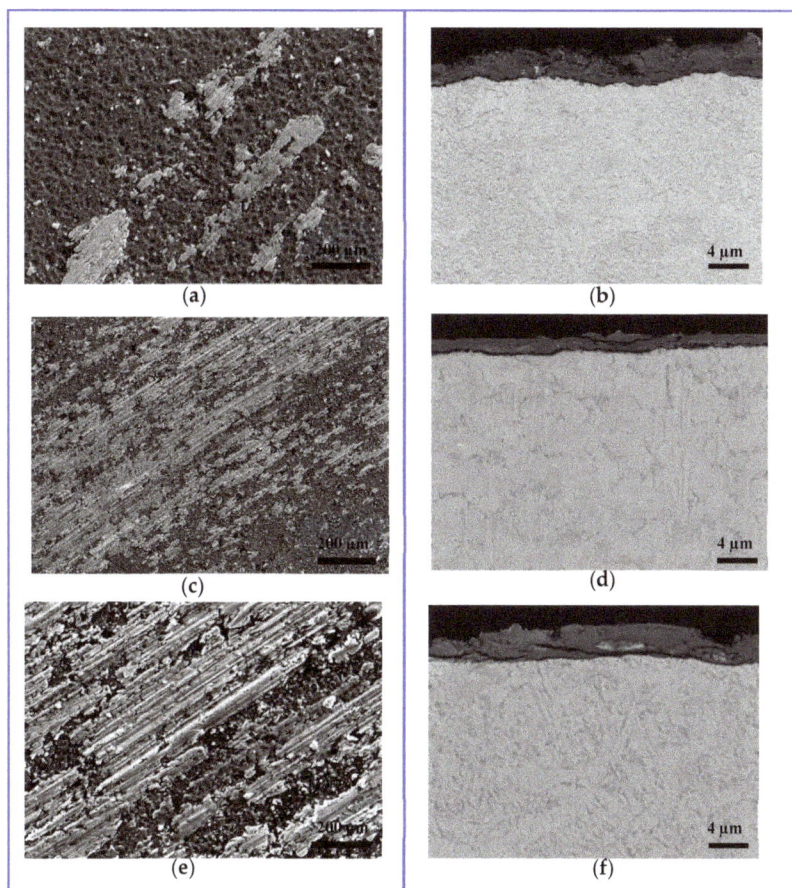

Figure 10. SEM Images of the mold's surface topography (**a,c,d**) and core pins cross section (**b,d,f**) after the HPDC tests. AlCrN/Si$_3$N$_4$ (**a,b**); AlTiN/Si$_3$N$_4$ (**c,d**) and AlCrN (**e,f**).

3.5. Results Outline

The use of conventional coatings, such as TiN, CrN, (Cr,Al)N, employed in aluminum die cast molds has been showed a limited performance during operation at temperatures range of 400 °C to 600 °C. Requirements of higher operating temperatures up to 1200 °C have led to the development of more complex nanocomposite nitride coatings such as AlTiN and AlCrSiN. Nanocomposite coatings consist of a hard crystalline phase (e.g., AlTiN, AlCrN grains) that is embedded in an amorphous matrix (a-Si$_3$N$_4$, a-C) with Si being present either in the form of a solid solution or as a separate a-Si$_3$N$_4$ matrix/phase in which nanocrystalline AlTiN or AlCrN phase is embedded, forming consequently nc-AlTiN/a-Si$_3$N$_4$ or nc-AlCrN/a-Si$_3$N$_4$. Due to their unique nanostructural design, nanocomposite coatings exhibit increased hardness levels of >40 GPa and strongly improved thermal stability at temperatures of up to 1200 °C [26,27]. Therefore, the wear resistance during their operation is known to be governed by the composition of some functional and mechanical properties for instance adhesion, hardness, hot hardness, residual stress level, and other mechanical properties.

In this way, it was shown experimentally that nano-composite coatings, in particular, AlCrN/Si$_3$N$_4$ coating demonstrated the best performance. This is an optimum combination of hardness, stress state adhesion, oxidation resistance and soldering behavior. The higher adhesion resistance of AlCrN/Si$_3$N$_4$ is a result of the combination of the microstructure of the CrAlSiN coatings, which was composed of (Cr,Al)N crystallites combined with the amorphous Si$_3$N$_4$ phase [28]. It leads to the better performance during the die casting process.

This coating also exhibited low friction characteristics at 600 °C in relation the other coatings studied. The high hardness of nanocomposite coatings at elevated temperature leads to a reduction of seizure intensity at the tool/workpiece interface that is caused by plastic deformation of the surface layers. The application of AlCrN/Si$_3$N$_4$ also leads to an increase in the lifetime of aluminum die-casting dies through reduction of erosion, corrosion, and soldering processes, and due to an increase of the thermal fatigue limit. Moreover, the molds are generally subjected to a series of operational stresses during die casting. These are mainly thermal stresses, resulting from thermal cycling of the material each time the mold is filled. Based on the results that were obtained in this work, we propose that the nanocomposite coatings that are deposited by LERC technology are very well suited for large and heavy dies with a complex shape, providing a uniform coating thickness and homogeneity.

4. Conclusions

AlCrN/Si$_3$N$_4$ nanocomposite PVD coating showed an optimum combination of hardness, adhesion, soldering behavior, oxidation resistance, and stress state. These characteristics are essential for improving the die mold service life.

Due to the presence of nanocrystals of AlCrN dispersed in an amorphous matrix of Si$_3$N$_4$, this coating showed the ability to resist severe mechanical and thermal stress conditions. It provides an efficient stress barrier, preventing crack propagation that is caused by dynamic pressing. As a final result, the application of nano-composite coating increases the lifespan by approximately three times as compared to the benchmark AlCrN coating. The surface damage due to wear for this coating was minimal.

Therefore, according to the goal of this paper and considering the experimental results, nano-composite AlCrN/Si$_3$N$_4$ PVD coating deposited by pulsed cathodic arc evaporation equipped with lateral and central arc-rotating cathodes has been proposed as a suitable novel solution for surface engineering of aluminum die casting molds. The results showed that this coating exhibited advantages in terms of wear and tribological performance.

Acknowledgments: The authors acknowledge the financial support from the Natural Sciences and Engineering Research Council of Canada (NSERC) and the Canadian Network for Research and Innovation in Machining Technology (CANRIMT). The authors also acknowledge the MMRI for the use of its facilities.

Author Contributions: Jose Mario Paiva conceived, designed the experiments and co-wrote the paper; German Fox-Rabinovich co-wrote and revision the paper; Edinei Locks Junior prepared the samples and perform the HPDC experiments; Pietro Stolf performed SEM characterizations; Yassmin Seid Ahmed performed the COF experiments and XRD analysis; Marcelo Matos Martins performed AFM characterizations and co-wrote the paper; Carlos Bork performed hardness tests for all coatings and co-wrote the paper; and Stephen Veldhuis directed the experiments and analyzed the data.

Conflicts of Interest: The authors declare no conflict of interest.

References

1. Hirsch, J.; Al-Samman, T. Superior light metals by texture engineering: Optimized aluminum and magnesium alloys for automotive applications. *Acta Mater.* **2013**, *61*, 818–843. [CrossRef]
2. Wang, F.; Ma, Q.; Meng, W.; Han, Z. Experimental study on the heat transfer behavior and contact pressure at the casting-mold interface in squeeze casting of aluminum alloy. *Int. J. Heat Mass Transf.* **2017**, *112*, 1032–1043. [CrossRef]
3. Klobčar, D.; Kosec, L.; Kosec, B.; Tušek, J. Thermo fatigue cracking of die casting dies. *Eng. Fail. Anal.* **2012**, *20*, 43–53. [CrossRef]

4. Wang, B.; Bourne, G.R.; Korenyi-Both, A.L.; Monroe, A.K.; Midson, S.P.; Kaufman, M.J. Method to evaluate the adhesion behavior of aluminum-based alloys on various materials and coatings for lube-free die casting. *J. Mater. Process. Technol.* **2016**, *237*, 386–393. [CrossRef]

5. Birol, Y.; İsler, D. Thermal cycling of AlTiN- and AlTiON-coated hot work tool steels at elevated temperatures. *Mater. Sci. Eng. A* **2011**, *528*, 4703–4709. [CrossRef]

6. Tentardini, E.K.; Kunrath, A.O.; Aguzzoli, C.; Castro, M.; Moore, J.J.; Baumvol, I.J.R. Soldering mechanisms in materials and coatings for aluminum die casting. *Surf. Coat. Technol.* **2008**, *202*, 3764–3771. [CrossRef]

7. Bobzin, K.; Brögelmann, T.; Hartmann, U.; Kruppe, N.C. Analysis of CrN/AlN/AlO and two industrially used coatings deposited on die casting cores after application in an aluminum die casting machine. *Surf. Coat. Technol.* **2016**, *308*, 374–382. [CrossRef]

8. Bouzakis, K.-D.; Skordaris, G.; Gerardis, S.; Katirtzoglou, G.; Makrimallakis, S.; Pappa, M.; LilI, E.; M'Saoubi, R. Ambient and elevated temperature properties of TiN, TiAlN and TiSiN PVD films and their impact on the cutting performance of coated carbide tools. *Surf. Coat. Technol.* **2009**, *204*, 1061–1065. [CrossRef]

9. Yu, D.; Wang, C.; Cheng, X.; Zhang, F. Microstructure and properties of TiAlSiN coatings prepared by hybrid PVD technology. *Thin Solid Films* **2009**, *517*, 4950–4955. [CrossRef]

10. Miletić, A.; Panjan, P.; Škorić, B.; Čekada, M.; Dražič, G.; Kovač, J. Microstructure and mechanical properties of nanostructured Ti-Al-Si-N coatings deposited by magnetron sputtering. *Surf. Coat. Technol.* **2014**, *241*, 105–111. [CrossRef]

11. Ma, Q.; Li, L.; Xu, Y.; Ma, X.; Xu, Y.; Liu, H. Effect of Ti content on the microstructure and mechanical properties of TiAlSiN nanocomposite coatings. *Int. J. Refract. Met. Hard Mater.* **2016**, *59*, 114–120. [CrossRef]

12. Settineri, L.; Faga, M.G.; Gautier, G.; Perucca, M. Evaluation of wear resistance of AlSiTiN and AlSiCrN nanocomposite coatings for cutting tools. *CIRP Ann. Manuf. Technol.* **2008**, *57*, 575–578. [CrossRef]

13. *Verein Deutscher Ingenieure Normen*; VDI 3198; VDI-Verlag: Dusseldorf, Germany, 1991.

14. Haršáni, M.; Sahul, M.; Zacková, P.; Čaplovič, Ľ. Study of cathode current effect on the properties of CrAlSiN coatings prepared by LARC. *Vacuum* **2017**, *139*, 1–8. [CrossRef]

15. Antonov, M.; Hussainova, I.; Sergejev, F.; Kulu, P.; Gregor, A. Assessment of gradient and nanogradient PVD coatings behaviour under erosive, abrasive and impact wear conditions. *Wear* **2009**, *267*, 898–906. [CrossRef]

16. Chen, L.; Du, Y.; Wang, A.J.; Wang, S.Q.; Zhou, S.Z. Effect of Al content on microstructure and mechanical properties of Ti–Al–Si–N nanocomposite coatings. *Int. J. Refract. Met. Hard Mater.* **2009**, *27*, 718–721. [CrossRef]

17. Soldán, J.; Neidhardt, J.; Sartory, B.; Kaindl, R.; Čerstvý, R.; Mayrhofer, P.H.; Tessadri, R.; Polcik, P.; Lechthaler, M.; Mitterer, C. Structure–property relations of arc-evaporated Al–Cr–Si–N coatings. *Surf. Coat. Technol.* **2008**, *202*, 3555–3562. [CrossRef]

18. Kao, C.M.; Lee, J.W.; Chen, H.W.; Chan, Y.C.; Duh, J.G.; Chen, S.P. Microstructures and mechanical properties evaluation of TiAlN/CrSiN multilayered thin films with different bilayer periods. *Surf. Coat. Technol.* **2010**, *205*, 1438–1443. [CrossRef]

19. Vladescu, A.; Braic, V.; Braic, M.; Balaceanu, M. Arc plasma deposition of TiSiN/Ni nanoscale multilayered coatings. *Mater. Chem. Phys.* **2013**, *138*, 500–506. [CrossRef]

20. Polcar, T.; Cavaleiro, A. High-temperature tribological properties of CrAlN, CrAlSiN and AlCrSiN coatings. *Surf. Coat. Technol.* **2011**, *206*, 1244–1251. [CrossRef]

21. Xie, Z.; Wang, L.; Wang, X.; Huang, L.; Lu, Y.; Yan, J. Influence of Si content on structure and mechanical properties of TiAlSiN coatings deposited by multi-plasma immersion ion implantation and deposition. *Trans. Nonferrous Met. Soc. China* **2011**, *21*, s476–s482. [CrossRef]

22. Sun, S.Q.; Ye, Y.W.; Wang, Y.X.; Liu, M.Q.; Liu, X.; Li, J.L.; Wang, L.P. Structure and tribological performances of CrAlSiN coatings with different Si percentages in seawater. *Tribol. Int.* **2017**, *115*, 591–599. [CrossRef]

23. Voevodin, A.A.; Muratore, C.; Aouadi, S.M. Hard coatings with high temperature adaptive lubrication and contact thermal management: Review. *Surf. Coat. Technol.* **2014**, *257*, 247–265. [CrossRef]

24. Feng, Y.P.; Zhang, L.; Ke, R.X.; Wan, Q.L.; Wang, Z.; Lu, Z.H. Thermal stability and oxidation behavior of AlTiN, AlCrN and AlCrSiWN coatings. *Int. J. Refract. Met. Hard Mater.* **2014**, *43*, 241–249. [CrossRef]

25. Barshilia, H.C.; Ghosh, M.; Ramakrishna, R.; Rajam, K.S. Deposition and characterization of TiAlSiN nanocomposite coatings prepared by reactive pulsed direct current unbalanced magnetron sputtering. *Appl. Surf. Sci.* **2010**, *256*, 6420–6426. [CrossRef]

26. Uhlmann, E.; Fuentes, J.A.O.; Gerstenberger, R.; Frank, H. nc-AlTiN/a-Si$_3$N$_4$ and nc-AlCrN/a-Si$_3$N$_4$ nanocomposite coatings as protection layer for PCBN tools in hard machining. *Surf. Coat. Technol.* **2013**, *237*, 142–148. [CrossRef]

27. Veprek, S.; Reiprich, S. A concept for the design of novel superhard coatings. *Thin Solid Films* **1995**, *268*, 64–71. [CrossRef]

28. Ravi, N.; Markandeya, R.; Joshi, S.V. Effect of nitrogen pressure on mechanical properties of nc-TiAlN/a-Si$_3$N$_4$ nanocomposite coatings deposited by cathodic arc PVD process. *Mater. Today Proc.* **2016**, *3*, 3002–3011. [CrossRef]

materials

MDPI

Article

Comparison of SF$_6$ and CF$_4$ Plasma Treatment for Surface Hydrophobization of PET Polymer

Matic Resnik [1,2], Rok Zaplotnik [2], Miran Mozetic [2] and Alenka Vesel [2,*]

[1]   Jozef Stefan International Postgraduate School, Jamova 39, Ljubljana 1000, Slovenia; matic.resnik@ijs.si
[2]   Jozef Stefan Institute, Jamova 39, Ljubljana 1000, Slovenia; rok.zaplotnik@ijs.si (R.Z.);
      miran.mozetic@ijs.si (M.M.)
*   Correspondence: alenka.vesel@guest.arnes.si; Tel.: +386-1-477-3502

Received: 20 January 2018; Accepted: 18 February 2018; Published: 21 February 2018

Abstract: The fluorination of the polymer polyethylene terephthalate in plasma created from SF$_6$ or CF$_4$ gas at various pressures was investigated. The surface was analysed by X-ray photoelectron spectroscopy and water contact angle measurements, whereas the plasma was characterized by optical emission spectroscopy. The extent of the polymer surface fluorination was dependent on the pressure. Up to a threshold pressure, the amount of fluorine on the polymer surface and the surface hydrophobicity were similar, which was explained by the full dissociation of the SF$_6$ and CF$_4$ gases, leading to high concentrations of fluorine radicals in the plasma and thus causing the saturation of the polymer surface with fluorine functional groups. Above the threshold pressure, the amount of fluorine on the polymer surface significantly decreased, whereas the oxygen concentration increased, leading to the formation of the hydrophilic surface. This effect, which was more pronounced for the SF$_6$ plasma, was explained by the electronegativity of both gases.

Keywords: sulphur hexafluoride (SF$_6$) plasma; tetrafluoromethane (CF$_4$) plasma; polymer polyethylene terephthalate (PET); surface modification; functionalization and wettability; optical emission spectroscopy (OES); electronegativity

1. Introduction

Fluorine-containing plasmas are often used for the surface hydrophobization of polymer materials [1–8] and for dry-etching in the semiconducting industry [9–12]. In the latter, the addition of oxygen is used to enhance the etching rate [11]. If no oxygen is added, etching of the surface could be done by using substrate biasing. However, high ion energies can cause sample graphitization [10]. When fluorine plasmas are used to enhance the surface hydrophobicity, two effects can be obtained, namely, functionalization or deposition (polymerization of fluorocarbons), depending on the F/C ratio [13,14]. If the F/C ratio is high (F/C > 3), there is no polymerization, whereas if the F/C ratio is low (F/C ≤ 2), fluorocarbons will polymerize on the surface. Thus, gases such as CF$_4$, SF$_6$ and C$_2$F$_6$ do not cause polymerization [14,15] unless CH$_4$ is added to change the F/C ratio [16]. CF$_4$ is therefore often used for polymer surface modification to introduce nonpolar functional groups. SF$_6$ is rarely used, and therefore, literature is scarce. SF$_6$ plasma has been used to treat polyethylene terephthalate PET (fabric [1], fibres [7] or film [5]), cotton fibres [7], polypropylene (PP) [3,4], polyethylene (PE) [5], polyvinyl chloride (PVC) [5] and polymethyl methacrylate [2]. The authors reported increased hydrophobicity; however, different authors reported different stabilities of the hydrophobic surface. Selli et al. found that repeated SF$_6$ treatment caused more stable hydrophobicity [7]. Walton et al. found a negligible ageing effect after one year for the sample treated for the longest treatment time of 60 s, but this was not the case for the samples treated for shorter times [6]. Mrad et al. observed the ageing of PET, whereas PVC was stable even 210 days after treatment [5]. Polyethylene was quite stable as well, because the contact angle did not change in the first 40 days of ageing, whereas later it slightly decreased. Here, it is worth mentioning that all authors observed fluorine at the surface treated in SF$_6$ plasma; however, few authors observed sulphur because of grafting of SF$_x$ species on the treated surface, which were very sensitive to ablation [8,17].

There have also been some reports in the literature on using O$_2$ plasma followed by SF$_6$. Mangindaan et al. prepared gradient PP surfaces with wettability between 20° and 135° by applying an O$_2$ pretreatment followed by SF$_6$ plasma treatment under a specially designed mask with an open end and a closed end, which allowed the diffusion of reactive fluorine species [3,4]. The highest fluorine content of 44 at % was found at the open end, and only 3 at % was found at the close end of the mask. In contrast, the oxygen concentration was approximately 11 and 30 at % at the open and closed ends, respectively. The sulphur content was very small at approximately 0.4 at %. The authors also studied the adhesion of fibroblast cells and found that the number of cells decreased from the hydrophobic surface at the open end to the hydrophilic surface at the closed end. Consecutive O$_2$ and SF$_6$ plasma treatments were also applied by Bi et al. for treatment of Parylene-C to obtain a superhydrophobic surface [18]. Oxygen plasma treatment time was varied, whereas the treatment time in the SF$_6$ plasma was kept constant. The hydrophobicity increased with increasing pretreatment time in the O$_2$ plasma until saturation was achieved with a contact angle of 169°. The obtained superhydrophobic surface was a result of the increased surface nanoroughness induced by O$_2$ plasma treatment, followed by surface fluorination with SF$_6$ plasma treatment.

In this paper, we investigated and compared the SF$_6$ and CF$_4$ plasma created at various pressures on the surface modification of PET polymer films.

2. Materials and Methods

2.1. Plasma Treatment

A semi-crystalline PET polymer with a thickness of 0.250 mm was obtained from Goodfellow (Goodfellow Cambridge Ltd., Huntingdon, England). It was cut into small samples of 1×1 cm^2. The samples were treated in a plasma system, as shown in Figure 1. The plasma was created in a Pyrex discharge tube with a length of 80 cm and a diameter of 4 cm. A coil with 6 turns was placed in the centre of the tube. The coil was connected to a radiofrequency RF generator (13.56 MHz) via a matching network. The generator nominal power was fixed to 200 W. The discharge chamber was pumped with a rotary pump with a nominal pumping speed of 80 m$^3 \cdot$h^{-1}. The base pressure was 1 Pa. Sulphur hexafluoride (SF$_6$) or tetrafluoromethane (CF$_4$) gas (supplied by Messer, Messer Group GmbH, Bad Soden, Germany) was leaked into the plasma chamber, and the gas purity was 99.998 and 99.995, respectively. Samples of the PET polymer were placed in the middle of the coil and treated by plasma at various gas pressures. The lowest pressure was set at 10 Pa, whereas the maximum pressure was determined as a pressure at which it was still possible to ignite the plasma. For the SF$_6$ plasma, the highest pressure was 200 Pa, whereas for the CF$_4$ plasma, it was 500 Pa. The treatment time was kept constant at 40 s.

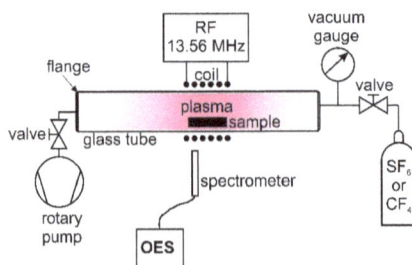

Figure 1. Schematic diagram of the plasma system for sample treatment.

2.2. Plasma Characterization

The plasma was characterized using optical emission spectroscopy (OES). OES measurements were performed in a quartz tube with a 16-bit Avantes AvaSpec 3648 fibre optic spectrometer (Avantes Inc., Louisville, CO, USA). A nominal spectral resolution was 0.8 nm, and the spectra were recorded in the range from 200 to 1100 nm. A combined deuterium-tungsten reference light source was used to determine the spectral response of the spectrometer. The measured OES spectra were calibrated with this spectral response.

2.3. Surface Characterization

Approximately 20 min after plasma treatment, the surface composition of the samples was analysed by X-ray photoelectron spectroscopy (XPS). An XPS instrument model TFA XPS from Physical Electronics (Munich, Germany) was used. The samples were excited using monochromatic Al K$\alpha_{1,2}$ radiation at 1486.6 eV. Photoelectrons were detected at an angle of 45° with respect to the normal of the sample surface. XPS survey spectra were measured at a pass-energy of 187 eV using an energy step of 0.4 eV. High-resolution C 1s spectra were measured at a pass-energy of 23.5 eV using an energy step of 0.1 eV. An additional electron gun was used for the surface charge compensation. All spectra were referenced to the main C 1s peak with a position set to 284.8 eV. The measured spectra were evaluated using MultiPak v8.1c software (Ulvac-Phi, Inc., Kanagawa, Japan, 2006) from Physical Electronics.

The surface wettability was measured 5 min after plasma treatment by a See System (Advex Instruments, Brno, Czech Republic). Contact angles (WCA) were determined with a demineralized water droplet of a volume of 3 µL. Three measurements were taken to minimize the statistical error.

The surface roughness and morphology were analysed by atomic force microscopy (AFM) using a Solver PRO (NT-MDT, Moscow, Russia) in tapping mode. The surface roughness, R_a, was measured over an area of 5 µm × 5 µm.

3. Results and Discussion

Figure 2a shows the variation of the XPS surface composition of the PET polymer treated in SF$_6$ plasma versus pressure. The values for the atomic concentration at a pressure of 0 Pa correspond to an untreated sample. The measured values for the untreated sample, i.e., 25 at % oxygen and 75 at % carbon, are close to the theoretical values for pure polyethylene terephthalate. These values are altered upon plasma treatment, as demonstrated in Figure 2a.

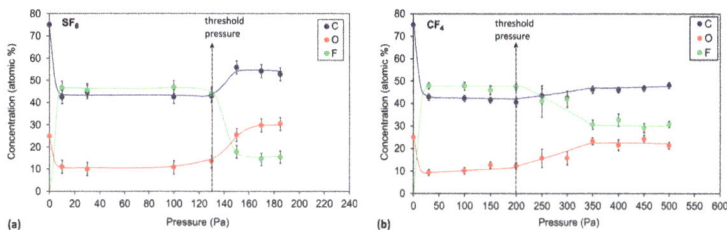

Figure 2. Surface composition of the PET samples treated at various pressures, as determined by XPS: (a) treated in SF$_6$ plasma and (b) treated in CF$_4$ plasma. Two different regions regarding the surface composition are observed at low/high pressures. The values of the atom concentration at a pressure of 0 Pa correspond to the untreated sample.

As expected, fluorine appeared on the surface, and its concentration was dependent upon the pressure in the discharge chamber during the plasma treatment of the polymer sample. Two regions can be distinguished. The first one appeared at pressures up to approximately 130 Pa, where the surface composition was relatively constant and was independent of the pressure. However, at pressures higher than 130 Pa, a drastic (and rather abrupt) modification of the surface composition occurred. Hereinafter, a pressure of 130 Pa is considered the threshold pressure. In the first region, below the threshold pressure, plasma treatment resulted in intensive fluorination of the polymer because a high fluorine content of approximately 46 at % was found. Furthermore, the oxygen concentration decreased from the initial 26 to 10 at %. In the second region, above the threshold pressure, the fluorine concentration on the polymer surface dropped to only ~16 at %, whereas the oxygen concentration increased to almost 30 at %. Another important difference in both regions was the presence of a minor concentration of other elements. In addition to carbon and oxygen, which were already present in the original polymer, only fluorine was found in the first region. However, in the second region, a minor concentration of sulphur from SF_x radicals was found as well (<1 at %). A detailed reason for this transition will be explained later in the text. It is correlated with the concentration of F atoms in the plasma, which was lower after the threshold pressure; therefore, surface fluorination was less efficient. Furthermore, all vacuum systems contain water vapour, which dissociates to O and OH radicals that compete with F atoms and cause oxidation. For this reason, a higher oxygen concentration was found above the threshold pressure. This phenomenon, whereby oxidation may occur when treating materials in F-containing plasmas, has been observed before and was published in [19].

The significant change in the surface concentration of F and O before and after the threshold pressure is also observed in the high-resolution carbon C 1s spectra shown in Figure 3. The samples treated at pressures below the threshold pressure were rich in CF_3 and CF_2 as well as CF functional groups (see Figure 4 also), whereas the sample treated at pressures higher than the threshold pressure had only some CF groups and an insignificant number of CF_2 functional groups in the surface film probed by the photoelectrons. This result is shown in more detail in Figure 4, where an example of a detailed curve deconvolution of the C 1s spectra showing peak assignment is presented. The C 1s peak was fitted with five components positioned at the binding energies of 284.8 eV assigned to C–C, 286.5 eV assigned to C–O and C̲–CF, 289 eV assigned to O–C=O and CF, 291.3 eV to CF_2 and 293 eV to CF_3 [20,21].

When treating the polymer in CF_4 plasma (Figure 2b), a similar behaviour was observed as when treated in SF_6 plasma. However, the transition between the regions of high and low fluorine content was not very sharp (it appeared approximately at a threshold pressure of 200 Pa) and was less intense (the fluorine concentration dropped to only 30 at %, and oxygen increased to almost 23 at %). In the first region below the threshold pressure, there was no significant difference in the fluorine concentration on the sample treated in CF_4 or SF_6 plasma, (Figure 2a,b, respectively). Figure 5 shows a comparison of selected carbon peaks for the samples treated in CF_4 and SF_6 plasma at 100 Pa (low-pressure region). We see only minor differences in the intensity of the various fluorine functional groups of CF, CF_2, and CF_3 and the presence of OCF_3 at ~295 eV for the sample treated in CF_4 plasma.

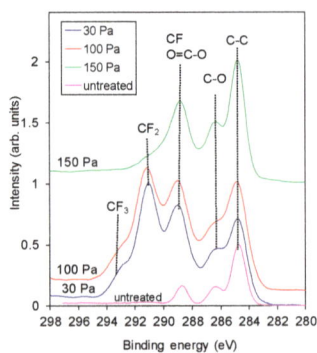

Figure 3. Comparison of high-resolution spectra C 1s of samples treated at various SF_6 pressures. A spectrum for the untreated PET is added for comparison.

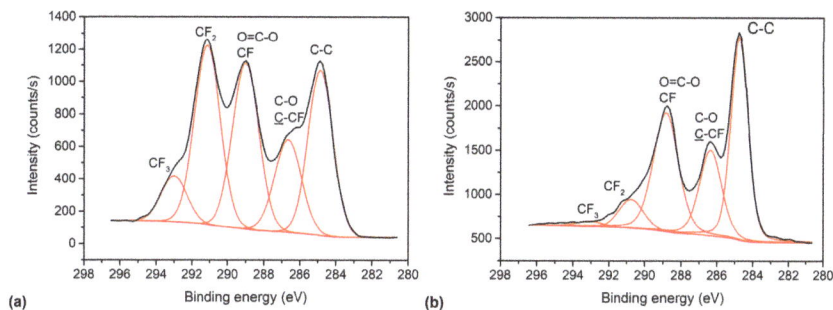

Figure 4. Deconvolution of C 1s spectra: (**a**) below and (**b**) above the threshold pressure.

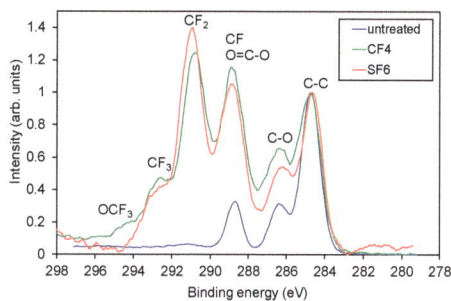

Figure 5. Comparison of selected high-resolution C 1s spectra of the untreated PET sample and the plasma-treated samples in the first region, below the threshold pressure.

Plasma treatment changed the surface hydrophobicity of the samples. The water contact angle increased from the initial 76° to approximately 106° regardless of using the CF_4 or SF_6 plasma as long as the pressure was low enough (below the threshold). The value of 106° is typical for hydrophobic materials with a smooth surface [22]. The variation of the contact angle with pressure is interesting, as shown in Figure 6a. We can see that after the threshold pressure, when a decrease in fluorine and an increase in the oxygen concentration were observed (Figure 2a), the water contact angle significantly decreased to approximately 35°. The surface lost its hydrophobic character and became hydrophilic because of a lack of nonpolar fluorine functional groups and the presence of more polar oxygen groups. Interestingly, the water contact angle for PET treated at high pressure in SF_6 plasma was much lower than that for the untreated polymer, which was 76°. Measurements of the surface roughness by AFM showed only a slight increase in the roughness from 1.2 nm measured for the untreated sample to 2.3 and 2.6 nm measured for the samples treated in SF_6 and CF_4 plasma, respectively. Therefore, only a minor influence of the surface roughness on the contact angles was observed, and the major reason for modified wettability is thus chemical modification of the surface. Figure 6b also shows the results for the CF_4 plasma. Similar to the results obtained for SF_6 plasma treatment, we observed that, after the threshold pressure, the contact angle decreased. However, the decrease was less pronounced, which is correlated with a lower oxygen content in comparison to the sample treated in SF_6 plasma (Figure 2).

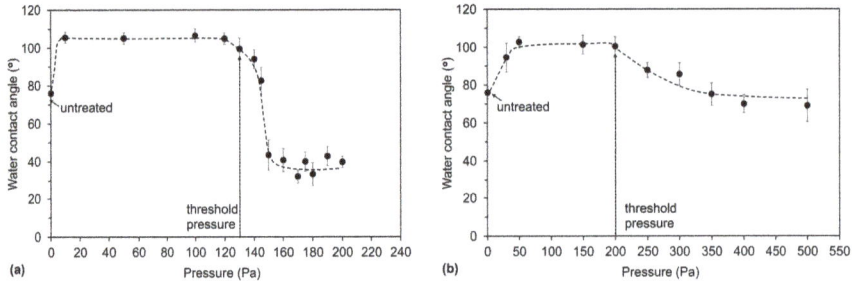

Figure 6. Water contact angles for the samples treated in: (**a**) SF$_6$ plasma and (**b**) CF$_4$ plasma at various pressures.

To explain such unusual behaviour of the surface composition and surface wettability with the pressure, we performed OES characterization of the plasma. OES spectra are shown in Figure 7. Figure 7a shows the spectra measured at low pressures (before the threshold pressure), while Figure 7b shows the spectra measured above the threshold pressure.

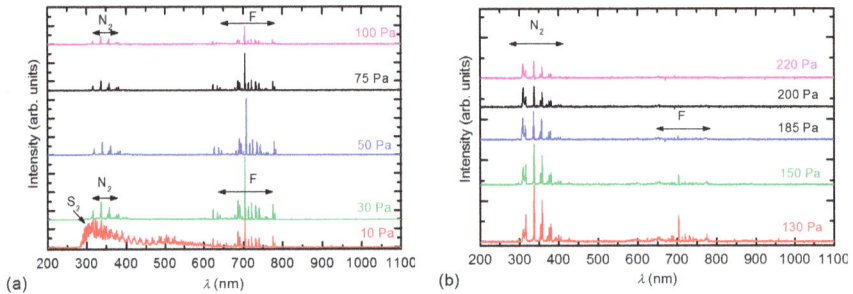

Figure 7. OES spectra of SF$_6$ plasma at various pressures: (**a**) below the threshold pressure and (**b**) above the threshold pressure.

At low pressures, intensive atomic F lines in addition to bands corresponding to the N$_2$ molecule are observed (Figure 7a). One exception is the spectrum measured at the lowest pressure 10 Pa, where bands corresponding to the S$_2$ molecule are observed as well. The appearance of the S$_2$ molecules can only be explained by the almost full dissociation of SF$_6$ and the subsequent recombination to sulphur dimers. The presence of nitrogen, which is known to be a strong emitter, was explained as an impurity present in the original gas according to the manufacturer's specifications. At high pressures (Figure 7b), the situation was different because the intensity of the F lines decreased. The variation in the F emission intensity with pressure is plotted in Figure 8. Figure 8 is in excellent agreement with Figure 2; at low pressures, where the emission intensity of F is high, the concentration of fluorine on the polymer surface is high. Whereas at higher pressures, when the OES intensity of F decreased, the XPS concentration of fluorine decreased.

This phenomenon deserves further discussion. In SF$_6$ plasma, SF$_x$ dissociates according to [23,24]:

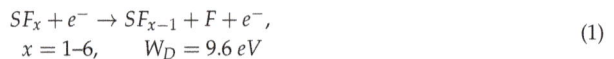

$$SF_x + e^- \rightarrow SF_{x-1} + F + e^-,$$
$$x = 1\text{-}6, \qquad W_D = 9.6 \, eV \tag{1}$$

The extent of dissociation and thus the concentration of radicals such as SF$_5$, SF$_4$, SF$_3$, SF$_2$, SF, S and F depends on the electron density and temperature, which in turn depends on the pressure. According to Kokkoris et al., a loss of SF$_x$ and F species on the reactor walls is also important for the production and consumption of neutral plasma species [23]. The F atoms tend to associate with F$_2$ molecules either by heterogeneous surface recombination or in the gas-phase—the probability of gas-phase loss increases as a square of the pressure, because three-body collisions are necessary.

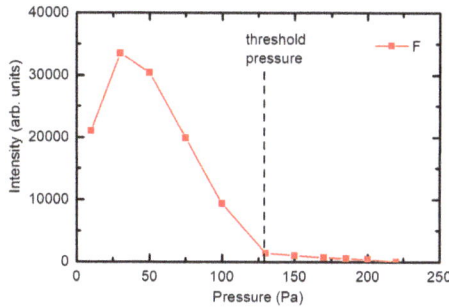

Figure 8. OES intensity of the F emission line at 703 nm versus SF$_6$ pressure.

SF$_6$ gas is also known to be a highly electronegative gas, which means that it has a strong tendency to acquire free electrons, thus forming negative ions: $e^- + SF_6 \rightarrow SF_6^-$ [24,25]. At low pressures, the electron temperature is high, thus causing a strong dissociation of SF$_6$ and, thus, the occurrence of a high density of F atoms in the plasma. The F atoms diffuse and eventually reach the polymer surface where they interact chemically and cause at least partial substitution of oxygen in the surface film of the PET polymer. The exact interaction mechanism is still unknown, but a high fluence of F atoms onto the surface of the polymer will guarantee the substitution of almost all oxygen in the PET surface by fluorine. The curves in Figure 3 obtained at 30 and 100 Pa confirm this simplified explanation.

At high pressures, however, the electron density and temperature decrease; therefore, the dissociation of SF$_6$, which has a relatively high dissociation energy of W_D = 9.6 eV, is less effective [24]. Furthermore, the loss of F atoms due to the gas phase reactions becomes important. Electrons are also lost by the attachment to SF$_6$ molecules. The lack of electrons capable of SF$_6$ dissociation caused the density of the fluorine atoms in the plasma to decrease significantly at elevated pressure.

Similar conclusions can be drawn for SF$_6$ plasma as for CF$_4$ plasma. However, CF$_4$ is as strongly electronegative as SF$_6$; therefore, this effect is not very pronounced. For the CF$_4$ plasma, F atomic lines are observed in the OES spectra at low pressures up to the threshold pressure (Figure 9a). The CF$_3$ continuum is not observed below the threshold pressure (200 Pa), where it is barely noticeable.

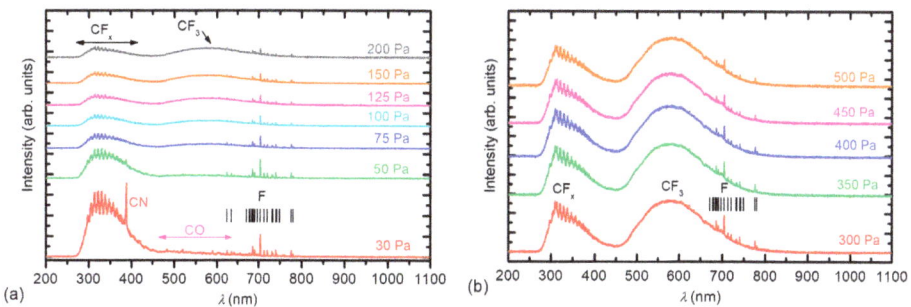

Figure 9. OES spectra of CF$_4$ plasma at various pressures: (**a**) below the threshold pressure and (**b**) above the threshold pressure.

Therefore, we can expect good dissociation of CF_4 at low pressures according to [26]:

$$CF_x + e^- \rightarrow CF_{x-1} + F + e^-,$$
$$x = 2\text{--}4, \qquad W_D = 12.5\ eV \tag{2}$$

The intensity of the F line decreased with increasing pressure. Furthermore, a continuum corresponding to CF_3 appeared at high pressures (Figure 9b). The appearance of the CF_3 continuum coincided with a decrease in the F intensity. Figure 10 represents the radiation intensity arising from the F atoms and the CF_3 radicals. At elevated pressure, the radiation from the F-atoms became marginal, indicating qualitatively that the dissociation of multiple CF_4 molecules was scarce. Comparing Figures 2 and 10, we can again conclude that at high pressures, the electron temperature and density was so low that it was insufficient to cause substantial dissociation of CF_4, and thus, the substitution of oxygen with fluorine on the PET polymer upon plasma treatment was poor.

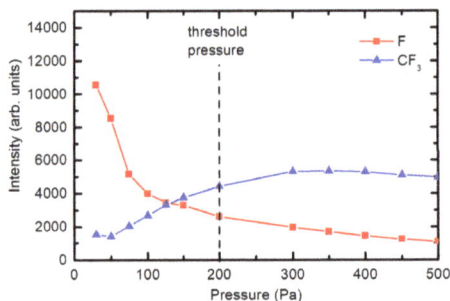

Figure 10. OES intensity of the F emission line at 703 nm and the CF_3 band at 580 nm versus CF_4 pressure.

4. Conclusions

Fluorination of the polymer surface in SF_6 and CF_4 plasma was investigated. Plasma was created at various pressures. It was observed that at low pressures up to the threshold pressure, the XPS concentration of fluorine on the polymer surface was high (~46 at %) regardless of the gas used. After the threshold pressure, a sudden decrease in the fluorine concentration was observed, which was more pronounced for the SF_6 plasma. Simultaneously, the concentration of oxygen increased. Therefore, the surface changed from hydrophobic to hydrophilic. The threshold pressure for the SF_6 plasma was ~130 Pa, whereas for the CF_4 plasma it was slightly higher at ~200 Pa. This effect was explained by the electronegativity of both gases, especially SF_6. At low pressures up to the threshold pressure, electrons can cause the full dissociation of gas molecules in plasma giving rise to a high concentration of fluorine radicals, which are responsible for surface fluorination. Because the density of fluorine in the plasma was high, the surface was fully saturated with the fluorine functional groups. At high pressures, the electron density and temperature decreased. Furthermore, they were also lost by electron attachment; therefore, the gas dissociation was weak, thus causing poor surface reactions.

Acknowledgments: The authors acknowledge the financial support from the Slovenian Research Agency (research core funding No. P2-0082) as well as the Young researcher grant.

Author Contributions: Matic Resnik designed the experiments and performed the plasma treatment under the supervision of Alenka Vesel and Miran Mozetic; Alenka Vesel performed the surface characterization of the samples; Rok Zaplotnik performed the plasma characterization; Alenka Vesel and Miran Mozetic analysed the data and wrote the paper.

Conflicts of Interest: The authors declare no conflicts of interest.

References

1. Barni, R.; Riccardi, C.; Selli, E.; Massafra, M.R.; Marcandalli, B.; Orsini, F.; Poletti, G.; Meda, L. Wettability and dyeability modulation of poly(ethylene terephthalate) fibers through cold SF_6 plasma treatment. *Plasma Process. Polym.* **2005**, *2*, 64–72. [CrossRef]

2. Korotkov, R.Y.; Goff, T.; Ricou, P. Fluorination of polymethylmethacrylate with SF_6 and hexafluoropropylene using dielectric barrier discharge system at atmospheric pressure. *Surf. Coat. Technol.* **2007**, *201*, 7207–7215. [CrossRef]

3. Mangindaan, D.; Kuo, C.-C.; Lin, S.-Y.; Wang, M.-J. The diffusion-reaction model on the wettability gradient created by SF_6 plasma. *Plasma Process. Polym.* **2012**, *9*, 808–819. [CrossRef]

4. Mangindaan, D.; Kuo, W.-H.; Wang, Y.-L.; Wang, M.-J. Experimental and numerical modeling of the controllable wettability gradient on poly(propylene) created by SF_6 plasma. *Plasma Process. Polym.* **2010**, *7*, 754–765. [CrossRef]

5. Mrad, O.; Saloum, S.; Al-Mariri, A. Effect of a new low pressure SF_6 plasma sterilization system on polymeric devices. *Vacuum* **2013**, *88*, 11–16. [CrossRef]

6. Walton, S.G.; Lock, E.H.; Ni, A.; Baraket, M.; Fernsler, R.F.; Pappas, D.D.; Strawhecker, K.E.; Bujanda, A.A. Study of plasma-polyethylene interactions using electron beam-generated plasmas produced in Ar/SF_6 mixtures. *J. Appl. Polym. Sci.* **2010**, *117*, 3515–3523. [CrossRef]

7. Selli, E.; Mazzone, G.; Oliva, C.; Martini, F.; Riccardi, C.; Barni, R.; Marcandalli, B.; Massafra, M.R. Characterisation of poly(ethylene terephthalate) and cotton fibres after cold SF_6 plasma treatment. *J. Mater. Chem.* **2001**, *11*, 1985–1991. [CrossRef]

8. Barni, R.; Zanini, S.; Beretta, D.; Riccardi, C. Experimental study of hydrophobic/hydrophilic transition in SF_6 plasma interaction with polymer surfaces. *Eur. Phys. J.-Appl. Phys.* **2007**, *38*, 263–268. [CrossRef]

9. Chen, Q.W.; Wang, Z.Y.; Tan, Z.M.; Liu, L.T. Characterization of reactive ion etching of benzocyclobutene in SF_6/O_2 plasmas. *Microelectron. Eng.* **2010**, *87*, 1945–1950. [CrossRef]

10. Joubert, O.; Pelletier, J.; Fiori, C.; Tan, T.A.N. Surface mechanisms in O_2 and SF6 microwave plasma-etching of polymers. *J. Appl. Phys.* **1990**, *67*, 4291–4296. [CrossRef]

11. Kim, S.H.; Woo, S.G.; Ahn, J.H. Effects of SF_6 addition to O_2 plasma on polyimide etching. *Jpn. J. Appl. Phys.* **2000**, *39*, 7011–7014. [CrossRef]

12. Kim, G.S.; Steinbruchel, C. Plasma etching of benzocyclobutene in CF_4/O_2 and SF_6/O_2 plasmas. *J. Vac. Sci. Technol. A* **2006**, *24*, 424–430. [CrossRef]

13. Gilliam, M.A. A Plasma Polymerization Investigation and Low Temperature Cascade Arc Plasma for Polymeric Surface Modification. Ph.D. Thesis, University of Missouri, Columbia, MO, USA, August 2006.

14. Iriyama, Y.; Yasuda, H. Fundamental aspect and behavior of saturated fluorocarbons in glow-discharge in absence of potential source of hydrogen. *J. Polym. Sci. Pol. Chem.* **1992**, *30*, 1731–1739. [CrossRef]

15. Strobel, M.; Corn, S.; Lyons, C.S.; Korba, G.A. Surface modification of polypropylene with CF_4, CF_3H, CF_3Cl, and CF_3Br plasmas. *J. Polym. Sci. Pol. Chem.* **1985**, *23*, 1125–1135. [CrossRef]

16. Inagaki, N.; Katsuura, K. Glow-discharge polymerization of CF_4/CH_4 mixture investigated by infrared-spectroscopy and esca. *J. Macromol. Sci. Chem.* **1982**, *A18*, 661–672. [CrossRef]

17. Leonard, D.; Bertrand, P.; Khairallahabdelnour, Y.; Arefikhonsari, F.; Amouroux, J. Time-of-flight secondary-ion mass-spectrometry (ToF-SIMS) study of SF_6 and SF_6–CF_4 plasma-treated low-density polyethylene films. *Surf. Interface Anal.* **1995**, *23*, 467–476. [CrossRef]

18. Bi, X.P.; Crum, B.P.; Li, W. Super hydrophobic parylene-C produced by consecutive O_2 and SF_6 plasma treatment. *J. Microelectromech. Syst.* **2014**, *23*, 628–635. [CrossRef]

19. Gorjanc, M.; Jazbec, K.; Šala, M.; Zaplotnik, R.; Vesel, A.; Mozetic, M. Creating cellulose fibres with excellent UV protective properties using moist CF_4 plasma and ZnO nanoparticles. *Cellulose* **2014**, *21*, 3007–3021. [CrossRef]

20. Vandencasteele, N.; Fairbrother, H.; Reniers, F. Selected effect of the ions and the neutrals in the plasma treatment of PTFE surfaces: An OES-AFM-contact angle and XPS study. *Plasma Process. Polym.* **2005**, *2*, 493–500. [CrossRef]

21. Kim, Y.; Lee, Y.; Han, S.; Kim, K.-J. Improvement of hydrophobic properties of polymer surfaces by plasma source ion implantation. *Surf. Coat. Technol.* **2006**, *200*, 4763–4769. [CrossRef]

22. Test, A.D. Critical Surface Tension and Contact Angle with Water for Various Polymers. Available online: https://www.accudynetest.com/polytable_03.html?sortby=contact_angle (accessed on 20 December 2017).

23. Kokkoris, G.; Panagiotopoulos, A.; Goodyear, A.; Cooke, M.; Gogolides, E. A global model for SF_6 plasmas coupling reaction kinetics in the gas phase and on the surface of the reactor walls. *J. Phys. D Appl. Phys.* **2009**, *42*. [CrossRef]

24. Lallement, L.; Rhallabi, A.; Cardinaud, C.; Peignon-Fernandez, M.C.; Alves, L.L. Global model and diagnostic of a low-pressure SF_6/Ar inductively coupled plasma. *Plasma Sources Sci. Technol.* **2009**, *18*. [CrossRef]

25. Christophorou, L.G.; Olthoff, J.K. Electron interactions with SF_6. *J. Phys. Chem. Ref. Data* **2000**, *29*, 267–330. [CrossRef]

26. Hiroshi, F.; Isao, F.; Yoshinori, T.; Koji, E.; Kouichi, O. Plasma chemical behaviour of reactants and reaction products during inductively coupled CF_4 plasma etching of SiO_2. *Plasma Sources Sci. Technol.* **2009**, *18*. [CrossRef]

materials

MDPI

Review

Biocompatibility of Plasma-Treated Polymeric Implants

Nina Recek

Department of Surface Engineering and Optoelectronics, Jožef Stefan Institute, Jamova cesta 39, 1000 Ljubljana, Slovenia; nina.recek@ijs.si; Tel.: +386-1-477-36-72

Received: 30 November 2018; Accepted: 2 January 2019; Published: 12 January 2019

Abstract: Cardiovascular diseases are one of the main causes of mortality in the modern world. Scientist all around the world are trying to improve medical treatment, but the success of the treatment significantly depends on the stage of disease progression. In the last phase of disease, the treatment is possible only by implantation of artificial graft. Most commonly used materials for artificial grafts are polymer materials. Despite different industrial procedures for graft fabrication, their properties are still not optimal. Grafts with small diameters (<6 mm) are the most problematic, because the platelets are more likely to re-adhere. This causes thrombus formation. Recent findings indicate that platelet adhesion is primarily influenced by blood plasma proteins that adsorb to the surface immediately after contact of a synthetic material with blood. Fibrinogen is a key blood protein responsible for the mechanisms of activation, adhesion and aggregation of platelets. Plasma treatment is considered as one of the promising methods for improving hemocompatibility of synthetic materials. Another method is endothelialization of materials with Human Umbilical Vein Endothelial cells, thus forming a uniform layer of endothelial cells on the surface. Extensive literature review led to the conclusion that in this area, despite numerous studies there are no available standardized methods for testing the hemocompatibility of biomaterials. In this review paper, the most promising methods to gain biocompatibility of synthetic materials are reported; several hypotheses to explain the improvement in hemocompatibility of plasma treated polymer surfaces are proposed.

Keywords: biomaterial; polymer; plasma; functionalization; surface properties; thrombosis; hemocompatibility; endothealization; vascular graft; biocompatibility; endothelial cells

1. Introduction

In the developed world, cardiovascular diseases are the most frequent cause of morbidity and mortality of the population, and represent one of the greatest health problems. In Europe alone, the cost of treating patients with these diseases is over 200 billion euros a year. In the first place is atherosclerosis, which causes the internal walls of the vessels to constrict, which means the blood can no longer run freely through the veins, and therefore its flow slows down. Treatment of such diseases is possible with a vascular stent, or by replacing a damaged vessel with a synthetic vascular implant. Approximately 500 surgeries per year are performed per million inhabitants, in which the damaged vein is replaced by a vascular implant (artificial blood vessel). This number is still growing every year. Both treatment options are commonly used, but in the long term, the recovery of patients with vascular stent and, in particular, artificial blood vessel, is still unsatisfactory. About 10% of patients with artificial vessels experience post-operative complications, mainly due to inflammatory reactions, infections and aneurysms. In such cases, it is necessary to replace the artificial vessel with an autologous vein, which further increases the cost of treatment [1].

Therefore, for the treatment of highly calcified vascular constrictions, a surgical procedure is necessary, where by inserting a synthetic vascular implant, a bypass to restore the blood flow is made. The materials used in cardiovascular applications for prosthetic heart valves, catheters, heart assist devices, hemodialysers, synthetic vascular implants and stents have to meet the requirements for biocompatibility/hemocompatibility and should also have appropriate mechanical properties, in particular the flexibility and ease of surgical implantation [2,3]. Today, the following polymers are used for this purpose; polyamids, polyolefin, polyesters, polyuretans, polyethylene terephthalate and polytetrafluoroethylene [4]. All these materials have been used for synthetic vascular prosthesis for many years, but, unfortunately, they do not offer sufficient hemocompatibility, especially when used for replacement of veins of smaller diameters (<6 mm). The main reason for this is that the probability of thrombosis occurrence is even greater in the narrower part of the veins. On the wall of the artificial vein, there is a non-specific binding of plasma proteins, which also affects the platelet binding and is one of the main causes of thrombosis [5]. Lack of endothealization is another main cause of thrombosis.

Biological response to biomaterials is very complex and still poorly known. Since the surface of the biomaterials is the one that enables the first interaction with the body, the properties of the surface of the biomaterials are of key importance for an appropriate biological response. For years, the most suitable materials were inert materials that do not react with the body and do not allow the integration of biomaterials with the body. Today, the opinion is that biocompatible materials that are in contact with blood should enable interaction with the body and prevent infections, inflammatory reactions, blood clotting and other related reactions. For hemocompatible materials, it is particularly important that the surface has anti-thrombogenic properties that prevent the occurrence of thrombosis. Thrombosis begins with the binding of plasma proteins to the surface of the biomaterial and is strongly dependent on the physical and chemical properties of the surface of the biomaterial. Clinical studies showed that poly-L-lactic acid (PLLA) stent, which was the first absorbable stent implanted in humans, had low complication rates for thrombosis and thus very high hemocompatibilty. Another clinical study also showed that metallic base stent, coated with poly-D,L-lactide, used to carry the antiproliferative drug everolimus, lacked stent thrombosis and even ensured total vascular function restoration [6,7]. However, stents used in elder patients resulted in significantly higher target vessel failure rates compared with younger patients. Moreover, with increasing age, revascularization rates were also higher. In addition, there was no difference in stent thrombosis [8]. In the past two decades, the experimental and clinical studies have grown significantly, but there is a still need to develop materials that mimic the properties of natural cardiac tissues, i.e., composite materials. Furthermore, novel surface modifications should also be evolved to develop better biocompatible cardiac biomaterials.

In order to improve the properties of materials that are in contact with blood, various methods of surface treatment are used. In general, these methods are divided into mechanical and chemical. Mechanical methods of treatment are not particularly interesting because they often cause damage and changes in the other properties of the material that we want to avoid. Chemical processing methods are further divided into wet chemical methods, which include treatments with various chemical reagents in aqueous or other liquid media, and gaseous treatments, including plasma treatments, ion beam treatments, electron jets, photon jets (lasers), X-ray and other energy rays. To improve biocompatible properties, antithrombotic deposits are often used, such as heparin, albumin, or chitosan. In addition to this kind of application, the pre-treatment of artificial vessels with attachment of endothelial cells is also used to improve the properties. Covering the prosthetic implants in vitro with endothelial cells was first suggested by Heering et al. [9], although many polymer surfaces are not optimal for cell adhesion as such, unless modified. Nevertheless, all these methods have a limited degree of success [10–13]. One of the methods for improving the biocompatibility of materials is the use of gaseous plasma, which has many advantages over other methods. Before processing, samples do not need any special pre-treatment; procedures are quick and environmentally friendly. Plasma can also modify surface charge, roughness and polymer crystallinity, which has an important influence on cell adhesion [14,15]. For surface modifications, many different types of plasma can be applied, depending on requirements and the particular application. Depending on the gas used (e.g., oxygen, nitrogen, CF₄), specific functional groups are formed on the surface and it can be either hydrophobic or hydrophilic [16–19].

2. Biomaterials

Biomaterials are materials that are either natural or synthetic and are used to regulate, supplement or replace the function of a tissue in the human body [20]. Their role is to replace or restore the function of an injured or degenerate tissue or organ. They are helpful to treat, improve performance or correct abnormalities, which all improve the quality of the patient's life. Biomaterials used in the manufacture of medical devices are metals, ceramics, composites and polymers. Metals are relatively strong, flexible and fairly resistant to wear. Their disadvantage is poor biocompatibility, corrosion and excessive strength compared to tissue and elimination of metal ions, which can eventually lead to allergic reactions. Ceramics are more compatible than metals and are more resistant to corrosion. Poor properties are fragility, demanding production and low mechanical durability and flexibility. Polymers are the most useful materials, since they are easy to manufacture and are available in various compositions and forms, such as, for example, gels, fibers, films, and solids. Polymers are widely used in many industries such as electronics, the automotive industry, the food industry, and nowadays their usefulness has gained great importance in the field of medical science. Polymers most commonly used for these purposes are polyurethane, silicone, polytetrafluoroethylene (PTFE), polyethylene (PE), polymethyl methacrylate (PMMA), polyethylene terephthalate, etc. The production of these materials is usually relatively simple, quick and cost effective. In addition, their physical and chemical properties are usually good, however, they are often too flexible and too weak to meet mechanical requirements for certain applications [21,22].

2.1. Artificial Vascular Grafts

In parallel with advances in vascular surgery, development and production of vascular implants were also carried out. Various substitutes are available for the replacement of a damaged or unusable blood vessel, which may either be biological or synthetic. Natural blood vessels inside the human body are arterial or venous and can be categorized as autologous, allograft and xenographic. Becouse of these differences in the size and anatomy of blood vessels it is not certain they will match between the donor and recipient host. Synthetic vascular implants are used as arterial supplements. The industry strives to produce materials with properties that are identical or, to the extent possible, similar to the real veins. In the synthesis, it is therefore necessary to consider certain criteria. The artificial vessels produced must be sterile and must not contain toxic substances. In addition, production processes must comply with strict regulations in this field. Production must be cost-effective. The artificial vessels must be flexible and elastic, but with time they must not lose their flexibility. During the post-implantation period, the expansion of the vessel must not exceed 15% of the internal diameter. Within five years of transplantation, a frequency of 2% for anastomotic aneurysms should be acceptable. In the same period, the infections were reported to occur in only 3% of cases. Depending on the size of the artificial vein, they are divided into veins with a large diameter (d > 6 mm), with a middle diameter (d = 4–6 mm) and veins with small diameters (d < 4 mm) [23,24].

The most commonly used materials for the synthetic artificial artifacts are PET-Dacron and PTFE. Dacron is a multifilament polymer, which is formed into artificial vessels with various knitting methods. In the first mode, the fibers are wrapped in a simple pattern, in which they are grouped together and arranged to each other. In the second mode, fiber sets are closely interconnected in the whales. Such structures are relatively strong and almost unmanageable, strongly reducing the likelihood extension and stretching after integration. The tightness between the fibers and, consequently, the porosity of the material can vary during the production process [25]. Due to high permeability, plaited artificial vessels are impregnated with albumin or collagen [26,27]. The treated surfaces of the veins were further improved by chemical treatment with glutaraldehyde, formaldehyde, polyethylene glycol or heparin [28]. Despite the various properties, the artificial vessels are still not optimal. Dacron and polytetrafluoroethylene cores have many positive properties, but smaller diameters are still problematic. High hydrophobicity of the surface limits the endothelization of the surface. After integration in the organism, infections continue to occur. The greatest problem is the thrombogenicity of the surface, as in many cases patients experience thrombosis, which continues to occur [29,30].

2.2. Biocompatibility of Synthetic Materials

The use of synthetic materials in medicine has been growing steadily since the 1940s when they were actually applied in practice. Millions of such products are used each year. Despite all the advances and more than 50 years of research in this field, we still did not create a material that would meet all the requirements and, after application in the human body, be completely without any negative response—in this case materials would be completely biocompatible. Biocompatibility is defined as the ability of a material to induce an appropriate response in a specific application in host [13]. Hemolytic, toxicological and immune responses in the case of materials that come into contact with blood are no longer as problematic. For these materials, the main problem is thrombogenic reactions and the possibility of bleeding after implantation. There are many examples of clinical complications of cardiovascular devices. Thus, complete blockage of stents is reported already within a few weeks after implantation, acute thromboses in vessels of middle-diameter, embolisms in catheters and heart valves, complications of coronary artery bypass, etc. [4]. These problems occur despite therapy with drugs that prevent blood coagulation and the formation of clots. There are more hypotheses as to why despite all the knowledge and the long-term effort, we still do not have a fully compatible surface. One says that it is not possible to produce the industrial surface, which has the same characteristics as the natural one. Natural blood vessels have a layer of endothelial cells which is constantly renewed and thus produces antithrombotic substances, such as, for example, prostacyclin. It is produced in response to the conditions in the blood and is constantly changing. Another antithrombotic substance produced is glycocalyx molecules, which are in contact with the vascular endothelial cells of the blood vessels and due to their composition represent the antithrombotic surface. There are many attempts to imitate the natural blood vessel condition as closely as possible, but for now none are successful. Another hypothesis concerns the knowledge of blood and the processes that are associated with platelet activation and coagulation, but this is also very complex and still not fully understood. Platelets' membranes contain over 100 different oligosaccharide and protein receptors, which are important for transmitting signals between environmental factors and platelets. It is known that when the synthetic material is in contact with blood, it first comes in contact with the protein and the formation of a protein film [31]. Recently, a lot of attention is paid to this subject, but there are controversial opinions in the literature about whether platelet activation depends more on the amount of blood plasma protein adsorbed on the surface or on the final layer of the protein conformation [32]. Another problem is that there are no standardized methods that could determine the biocompatibility of materials. Research is mostly performed on individual blood components, under different conditions, which are difficult to compare with each other. For all the ongoing research in this field, there are many materials that are potentially better than those used up to now, but there is still a long way to their application, as they have to be tested and obtain the necessary documentation, which is consequently connected with high costs [33–36].

2.3. Factors That Influence the Biocompatibility of Biomaterials

The biocompatibility of the material is largely influenced by the surface properties of the material. The first few atomic layers of the material surface present the biointerface between the cells and biomaterials. Surface characteristics also trigger biological response after contact with the tissue and, ultimately, the success of the transplant or a medical device made of such material depends on it [37]. There are numerous conditions, which overlap and determine the biocompatibility. These are not only the mechanical and chemical characteristics of the material, but also place of application, individual host reaction, immune system as well as physical condition of the patient. According to Ikada, chemical and physical characteristics of the surface, which are responsible for biological response at the interface, are of the greatest importance. In the literature, there are various opinions about which surface properties are crucial for optimal biological response [38–41]. The most frequently investigated properties are the chemical composition of the surface, the topography and the wettability of the surface. The known effects on hemocompatibility as well as on cell response are presented below.

2.3.1. Impact of Plasma Treatment on the Hemo- and Biocompatibility of Synthetic Materials

Plasma modifies the surface morphology and increases surface roughness of PET. It was shown that such surface modification has a significant effect on platelet adhesion and activation. Even a short exposure of PET surface to highly non-equilibrium plasma reduced adhesion and activation of platelets mainly through oxygen surface functionalization. However effects of plasma treatment diminish with time and many oxygen functional groups are lost from the surface within 3 h of aging [42]. Plasma treatment also has an influence on the biological response, as all plasma treated surfaces exhibit improved proliferation of fibroblast and endothelia cells. The number of adherent platelets practically did not change after nitrogen plasma treatment, however, a much lower number of adherent platelets was observed on oxygen plasma treated surfaces [43]. Cvelbar et al. [44] studied the fabrication of micro- and nanostructure poly(ethylene terephthalate) (PET) polymer surfaces used for synthetic vascular grafts and their hemocompatible response to plasma-treated surfaces. The surface modification of PET polymer was performed using radio frequency (RF) weakly ionized and highly dissociated oxygen or nitrogen plasma to enable the improved proliferation of endothelial cells. Results indicate that surface treatment with both oxygen and nitrogen plasma improved the proliferation of endothelial cells, which increased with treatment time by 15 to 30%. This phenomenon was explained by the creation of new functional groups and the modification of surface morphology, which promotes the adhesion of endothelial cells. Numerous studies have proved that plasma treatment significantly improves biocompatible properties of polymer materials [17,45–50]. In their study, Jaganjac et al. [49] proved that oxygen rich coating after plasma treatment promotes binding of proteins and endothelialization of polyethylene terephthalate polymer. In another study it was shown that cells prefer to adhere on moderate hydrophobic polymer surfaces, rather than on hydrophilic or super-hydrophilic ones. Recek et al. [47] showed improved proliferation on oxygen plasma treated polystyrene. On the other hand, Garcia et al. [40] described greatly improved cell proliferation of HaCaT keratinocytes on collagen films modified by argon plasma treatment. There are countless papers in the literature describing the improvement of hemo- and biocompatibility of synthetic polymer materials using plasma treatment. Only a few were presented in this paragraph, proving that plasma really is a good method to improve polymer properties for biomedical applications.

2.3.2. The Effect of the Surface Chemical Composition on the Hemocompatibility of Biomaterials and on Cell Response

The chemical composition of the surface is one of the key characteristics when designing the hemocompatible materials from which medical devices are made. There are various ways in which we can control the surface with specific chemistry and functional groups, such as, for example, hydroxyl, methyl, sulphate, carboxyl, amino group etc. [51,52]. Their purpose is to improve the immobilization of various biomolecules such as proteins, enzymes and so on to improve the cellular response. Cell interaction with the surface of biomaterial is never direct, because the surface is previously covered with water molecules and proteins absorbed from biological fluids (see Figure 1). Initially cells respond to this adsorbed protein layer, rather than to the surface itself [53]. Cell adhesion is conducted in several phases: an early phase where short-term events take place, like physico-chemical linkage between cells and material, and a later signal transduction phase, involving biomolecules like extracellular matrix (ECM) proteins, cell membrane and cell skeleton proteins, regulating the gene expression [54]. These phases are illustrated in Figure 1. Firstly, when the biomaterial is in contact with cells in vitro or when they come in contact with an implant surface in vivo, the proteins either from culture medium or biological fluids adsorb and form the protein layer on the surface. After that, cells attach on the surface covered by proteins, spread and express cytoskeleton proteins and integrins, which help them firmly adhere to the surface. Thirdly, the proteins connect, and cytoskeleton reorganizes to adapt the surface morphology and actively spread on the substrate. Finally, at the interface with the material, cells synthesize ECM proteins, securing their shape stability and cell-matrix-substrate interfaces [55,56]. Cell adhesion differs on cell phenotype, that is why mechanisms of adhesion of blood cells are different from mechanism of cells from connective tissues, like fibroblast, osteoblasts, or cells originated from endothelia and epithelia, like endothelial vascular cell or keratinocytes. Cells from connective tissues use mostly integrins in cell-ECM interactions, whereas epithelial and endothelial cells can adhere with both adhesion molecules.

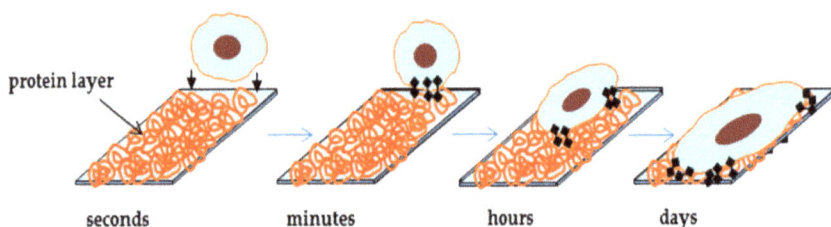

Figure 1. Kinetics and phases of cell adhesion.

The results of many studies have not yet led to the solution of what an ideal surface should be. Grunkemeier et al. [12] reported that increased oxygen groups reduced the activation of coagulation. Likewise, a higher proportion of these groups should also affect the reduction in the amount of bound fibrinogen and, according to their results, also the reduction in the number of bound platelets [57]. The same authors also reported that coagulation was reduced when methyl groups on the surface increased. Tengvall and colleagues [58] came to the same conclusions. One way to introduce new functional groups on the surface is plasma treatment, where different surface functionalities can be achieved with different types of plasma. Wang et al. [59] treated the PET polymer surface with acetylene plasma, thereby increasing the carbon content of the surface. Such films at different atomic percentages acted inhibitory at adhesion and activation of platelets. A significant reduction in contact activation of platelets was observed in the treatment of polyurethane with nitric plasma [60]. On the other hand, improvement was not observed in hemocompatibility after treatment with oxygen and argon plasma. Better cell adhesion and proliferation was observed on oxygen plasma treated PET and polystyrene (PS) samples, while samples treated in CF_4 or nitrogen plasma did not show significant improvement [45–48]. Jaganjac et al. [49] found that oxygen functional groups on PET treated by plasma, stimulated endothelial cell growth and proliferation by 25%, compared to control, plasma untreated samples, suggesting the possible use of oxygen plasma treatment to enhance endothelialization of synthetic vascular grafts.

2.3.3. The Influence of the Topography on the Hemocompatibility of Biomaterials and on Cell Response

In the production of biomaterials, it is important to take into account the impact of the structure of the material on the biological response. Natural vessels have of course the best biocompatibility, so it is necessary to know their structure, which consists of several layers. The first layer inside the vessel consists of endothelial cells that connect with the basal lamina. Its main building blocks are collagen, proteoglycans and glycoproteins such as fibronectin and laminin. They follow the layers of elastic fibers and smooth muscle cells that shrink under the control of a sympathetic nervous system. The back or the outer layer builds connective tissue. The surface of the inner side of the vessel is not smooth, but it is made up of micrometer corrugated grooves running in the direction of the blood stream. At the top of the individual protrusions, there are nano projections. Significance of the influence of the topography on biological response has been brought to the attention of many researchers, all seeking to create structures that will achieve the highest degree of hemocompatibility. There are many physical and chemical methods or combinations of both to achieve nanostructure, either by deposition of the material or by etching it. The most commonly used methods are: photolithography, colloidal lithography, laser etching, metal oxidation, nanophase ceramics production, supramolecular aggregation, surface coating with carbon nanotubes, nanowires, nanocomposites and, last but not least, plasma techniques such as plasma chemical infiltration, ion implantation, plasma sputtering, etc. [61–63]. Fan and co-workers [64] created structures with grooves of about 500 nm in width and about 100 nm in height, and 100 nm × 4 nm in size on the surface of polydimethylsiloxane (PDMS) using self-assembled layers and lithography. For comparison, unmodified, smooth PDMS surface, surface with ditches and surface with only nano-extensions were used as control. Platelet adhesion analyses have shown that only on the surface containing both grooves and protrusions was the number of bound platelets significantly reduced. Until recently, it was established that the increase in roughness of the surface due to the greater surface area available for platelet binding also increases the level of thrombogenicity. In the literature, the most commonly used term for characterization of the surface is roughness, but it is important to be aware of the fact that it does not tell us much about the actual topography, but only gives the average roughness of the surface. To explain the observed hemostatic response, we proposed few hypotheses in this review. One of the hypotheses says that hemostatic response is based on significant reduction of contact area between polymer and platelets due to high roughness of plasma treated polymer samples. According to Chen and colleagues [62], the roughness values that are crucial for platelet binding are roughly divided into three groups. The first group includes areas with a roughness of more than 2 μm, which is about the size of the platelets. The second group are surfaces with a roughness less than 2 μm, where the correct design of the structures can reduce the contact area of platelets and, consequently, the platelet adhesion, since they can be fixed only at the top of the structures as shown in Figure 2.

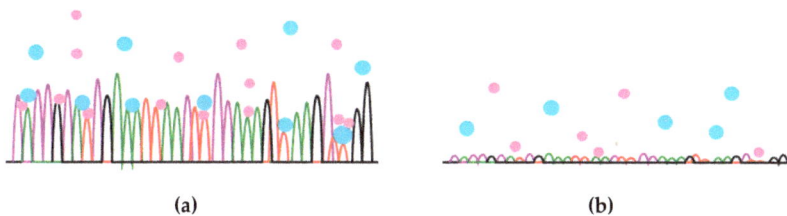

(a) (b)

Figure 2. Nanostructured topography of the surface, that attracts (**a**) and repels (**b**) adhesion of platelets.

The third group includes roughness greater than 50 nm, where the surface structures are much smaller than pseudopods. These are smooth surfaces that do not play a role in platelets adhesion. In such cases, other factors are likely to be involved in reducing thrombogenicity, among which are binding and conformation of plasma proteins, in particular fibrinogen and albumin. According to other authors, when performing structuring of the surface is important to take into account the entire dimension of the surface (height, width and distances between individual structures) [65–67]. It was shown in many studies that cells respond to surface topography and align themselves along defined surface features, e.g. ridges or grooves [68]. However, cell behavior on the nano-topography is still unknown so far. In the review of Curtis and Wilkinson on topological control of cells, they correlated the topological parameters with biological parameters, such as short- and long term adhesion and proliferation [69]. In several studies [69–71] it was demonstrated that the best adhesion of human bone cells was on less organized rough surfaces. Dalby et al. [71] developed surfaces with 120 nm diameter nanopits and demonstrated stimulation of human mesenchymal stem cells on such surfaces. In 1997, Curtis and Wilkinskon [68] described that cells reacted to discontinuities on the surface, with a radius of a certain length. It is related to the cell mechanism of mechano-sensitivity which is related to integrin mediated cell-matrix adhesion [72]. There have been different hypotheses on how cells sense the morphological discontinuities on the surface. One hypothesis is about the thermodynamics and extra-cellular matrix protein adsorption [73], another is about discontinuities acting as the energy barrier, where the size of energy barrier depends on both the geometry and surface chemistry [74]. Stevenson and Donald [75] have been investigating the attachment of cells on the different micro-meter scale substrates. They observed that the attachment of cells is dependent on the ridge spacing. At the ridge spacing between ~10–20 μm, the cells were able to attach and bridge between the neighbor ridges. At the moderate spacing, from ~30–50 μm cells attach to a single ridge or groove and at the largest spacing ≥50 μm cells connect between a ridge and a groove. From these results they proposed both a critical length and a critical slope angle of the ridge-groove surface morphology. Cells adjust their shape according to morphology, which causes reorganization of attachment and cytoskeleton structures (see Figure 3). Similarly, Berry et al. [76] described that cells were sensitive to the changed morphology, especially in the radius of curvature of pits. Thery et al. [77] have noticed that cells can memorize and recognize the adhesive substrates and in this way reorganize attachment and cytoskeleton structures. Hallab at al. [78] demonstrated that for cellular adhesion and proliferation, even more important factor is surface free energy of polymers. Other groups demonstrated that short term cells adhesion on metal substrates coated with gold-palladium is more dependent on surface chemistry, whereas the long-term adhesion is more dependent on surface roughness [70,79–82]. Ponsonnet at al. [83,84] also observed high impact of surface energy of titanium and titanium alloys on cell proliferation.

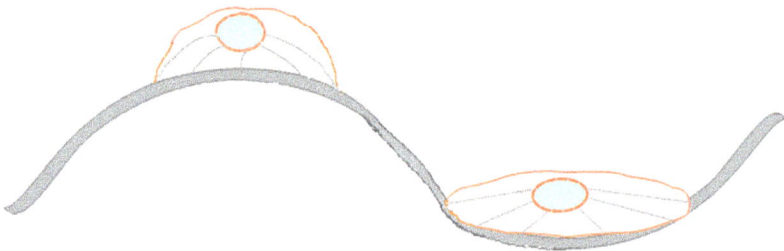

Figure 3. Illustration of the reorganization of cell actin skeleton structures, shape and attachment according to surface morphology.

2.3.4. The Effect of Wettability on the Hemocompatibility of Biomaterials and on Cell Response

Wettability is one of the important properties of the surface and has a major impact on the biological response. The second hypothesis is explained below and is based on preferential adhesion of water molecules from blood to the polar functional groups on the polymer surface. When we talk about the wettability of the surface, this is most often associated with the adsorption of proteins. In general, hydrophobic surfaces are considered to be much more susceptible to protein binding than hydrophilic, due to the strong hydrophobic reactions resulting from the contact of the protein with the surface, which results in reflective forces due to strongly bound water molecules. In addition to the amount of bound proteins, wettability also affects the conformation of bound proteins. Because contact of artificial material with blood leads to immediate contact with proteins, this is consequently important for the binding and activation of platelets and hence for the hemocompatibility of the material. Blood cells adhere with membrane adhesion proteins cadherins and selectines, which are involved in cell-cell interactions, and integrins which are involved in cell-material interaction [85–87].

Xu and Siedlecki [88] treated polyethylene with gaseous plasma and created different wettability of the surfaces. The influence of such surfaces on the binding of proteins was checked by binding of three blood plasma proteins. For all three proteins, fibrinogen, bovine serum albumin, and Human Factor XII, critical values were found to be at the water drop between 60° and 65°, where adhesion increased at these values. By measuring the force with AFM tip, it was also found that the binding forces change over time, which suggests that after binding of proteins to the surface, their conformational changes occur. Similar results were also obtained on the polyurethane polymer, where the increased binding of fibrinogen was also observed at angles greater than 65°. The conformation was monitored by the binding of monoclonal antibodies, and it was found that it varies with different surface wettability, depending on the ability of binding antibodies to different binding sites on the fibrinogen molecule [89]. The binding of molecules to surfaces should be a time-dependent process, which is supposed to be carried out on several levels. The first molecules which come into contact with the surface are water molecules, which also react with the surface according to its properties. Water molecules create a layer on the surface from which the binding of other molecules depends and diffuse later because of their size to the surface. If there is a mixture of different proteins in the solution, their binding depends on both size and their properties. Over time, their exchange can occur, as dynamic confocal changes and reorientation can affect the binding power and consequently the activity of the protein [66,88]. Due to the redistribution of amino acids, the availability of receptor binding sites may also change, which could also affect platelet adhesion. The influence of surface wettability on platelet binding is also the subject of numerous studies. In 2002, Spijker and colleagues [90] studied adhesion and platelet activation on polyethylene, in which gaseous-plasma produced a gradient of hydrophobicity and concluded that binding was greater on hydrophilic surfaces, and their activation was greater on more hydrophobic surfaces. Vogler and colleagues [91] in 1995 came to the same conclusion. Rodrigues and co-workers [57], Lee et al. [92] and Sperling and colleagues [52] came to the exact opposite conclusion. They listed the largest number of both bounded and activated forms of platelets on hydrophilic surfaces. Cell attachment was investigated by Yanagisawa and Wakamatsu [93]. They observed that cell attachment rate and cell spreading were higher on substrates with a water contact angle below 60° and that attachment decreased dramatically for more hydrophobic surfaces, whatever the time after inoculation. For the oblast cells, no correlation between wettability of the material and cell attachment and proliferation was found [94]. On the other hand, Lee et al. demonstrated that endothelial cells [95] or neural cells [96] adhesion was more increased on moderately hydrophilic surfaces, than on the superhydrophilic or hydrophobic surfaces. In 2004, it was demonstrated by Lime et al. [97] that hydrophilic substrates are better for human fetal osteoblast adhesion and proliferation than hydrophobic ones. Interesting, surface energy had no effect on cell differentiation.

Despite the fact that the wettability of the surface is likely to play an important role in the hemocompatibility of materials, it is difficult to derive clear conclusions from the contradictory results, which could explain the role of wettability. One of the reasons is definitely the complexity of the processes that take place in the blood. Individual impacts cannot be considered separately. Such a multivariable system should be taken as a whole and take into account the interaction between the individual impacts.

3. Biomaterial-Blood Interactions

When biomaterial comes in contact with the biological system, activation of the intrinsic pathway at the blood/biomaterial interface starts. There are many studies examining blood biocompatibility and the most important parameters for characterization are the number of adhered platelets and their activation [98]. Platelets are the smallest blood fragments with a diameter of 1 µm to 3 µm, without nucleous. In the blood of an adult, they are 2–3×10^8 per mL. In the bloodstream, they are present in inactive form. In the event of endothelial vein damage or when activating the coagulation cascade, the platelet shape is activated and changed. Initially, it was established that platelets are important only in stopping bleeding, but now it is known that in addition to these very important functions, they also play an important role in other physiological and pathological processes of hemostasis, inflammatory reactions, tumor metastases, and defense mechanisms [99].

Figure 4 represents the artificial PET polymer material in contact with blood. The surface of original material (as manufactured at the factory) is fully covered with platelets (Figure 4a). However, when PET is treated with oxygen plasma, there are only few platelets adhered on the surface of material (Figure 4b). When the body is in a contact with artificial biomaterial, platelets tend to adhere similar like in the case of an external injury. This is the reason why materials, which show strong platelet adhesion or provoke an increase in platelet adhesion, are considered as thrombogenic [100]. If the blood leucocytes decrease at the same time, this is a sign of a "cellular immunoresponse" of the body towards the biomaterial. Material is considered as clinically biocompatible when it does not provoke any damage of blood cells or any structural change of plasma proteins when in contact with the blood [101]. Human blood plasma contains over 300 different proteins that differ in structure and function: proteins involved in coagulation and fibrinolysis, complementary system proteins, immune system proteins, enzymes, inhibitors, lipoproteins, hormones, cytokines and growth factors, proteins that are important for transport and others [102].

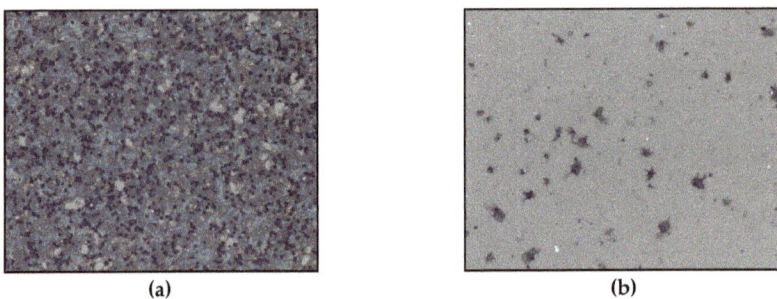

(a)   (b)

Figure 4. Platelet adhesion on untreated (**a**) and oxygen plasma treated (**b**) poly(ethylene terephthalate (PET) polymer. Incubation was performed with shaking at 250 RPM.

To explain hemostatic response, third hypothesis is based on different conformations and orientations of adsorbed plasma proteins. If proteins adsorb on the surface and blood cells adhere on the surface of material, the contact of the biomaterial with blood leads to clot formation [103–105]. Activation of the coagulation system at the blood-biomaterial interface drives sequence of reactions. Proteins compete to adhere to the biomaterial surface and this determines the pathway and adhesion of platelets. Having the exact knowledge of the material surface and the conformation of the adsorbed proteins, prediction about the interactions between the biomaterial surface and the absorbed proteins can be made. These interactions are determined both by the nature of the polymer surface and by the nature of protein parts in contact with the surface (hydrophilic/hydrophobic, charged/uncharged, polar/non-polar etc.) [106–108]. It is commonly accepted that a decrease in surface roughness increases the compatibility of material [109]. Surface tension of a material is one of the most important factors on protein adsorption. Andrade et al. [110] suggest that smaller interfacial energies between blood and polymer surface results in better blood compatibility. Contrary, Bair et al. [111] claims that higher surface tension, between 20–25 mN/m gives better hemocompatibility. On the other hand, Ratner et al. [112] prove good blood compatibility on the surfaces with the moderate relationship between their hydrophilic/hydrophobic properties. Carboxylate, sulfate or sulfonate groups on the surface may act as antithrombotic agents, as a result of repulsive electric forces between plasma proteins and platelets [113]. Norde has shown that protein adsorption increases if concentration of ionic groups in the protein and in the polymer surface decreases [114]. The relation between the electrical conductivity of biomaterials and blood biocompatibility is described by Bruck [115]. In addition, there are studies on the influence of the streaming potential on blood coagulation [114,116].

Hemostasis is the body's response to vascular damage and bleeding. It involves a complex set of events and biochemical reactions that lead to the formation of a blood clot which consequently prevents bleeding. At the beginning of the 20th century, Morawitz combined all the insights into his classic coagulation theory, which he divided into two groups. In the first, in the presence of calcium ions and thrombokinase, the conversion of prothrombin into thrombin occurs. In the second step, the resulting thrombin converts fibrinogen into fibrin. His theory touched the basics of coagulation, but it had several drawbacks. One of them was that it did not take into account the specific function of platelets, which was later described by Bürker. Fibrinogen is a large, complex bar shaped glycoprotein. It consists of three pairs of Aα, Bβ and γ polypeptide chains, which are interconnected with 29 disulfide bonds. At both ends, globular domains are interconnected with α-helices and bind calcium ions, which are important for maintaining the structure and function of fibrinogen. In blood plasma, it is usually present at a concentration of about 2.5 g/L. Fibrinogen is important for the preservation of hemostasis and platelet aggregation [57,102]. This protein also plays an essential role in binding to synthetic materials and thus has an important impact on the hemocompatibility of the material. In addition, similar to protein fibrinogen, albumin is also important in binding to the surfaces of synthetic materials, helping to maintain the surface antithrombotic [102]. Human serum albumin is the most abundant protein in human blood plasma. It is synthesized in the liver and is present in all body fluids. It consists of a single chain containing three interconnected domains. It has binding sites for various molecules like water, ions, fatty acids, hormones, bilirubin, synthetic medicines and many others. It is present in blood plasma at concentrations of 35–50 g/L. Because of its abundance and high binding capacity, albumin is the main transport protein that regulates and maintains osmotic pressure in the blood. Competitive adsorption of the protein albumin and fibrinogen is very complex and has been widely investigated. Albumin inhibits and fibrinogen activates the adhesion of platelets; in the case of hydrophobic surfaces, fibrinogen is mostly absorbed, while in the case of hydrogels, absorption of albumin is dominant [117,118]. The stationary state, which corresponds to an irreversible protein adsorption, is reached after longer contact time. The adsorbed protein films show time-dependent conformational changes, like desorption or protein exchange and are described by the Langmuir isotherms [119–121].

Research in this field has led to many different theories and numerous terminologies. Much progress was made when the International Commission introduced a common name for coagulation factors, which have since been designated with Roman numerals [122]. Understanding the processes of coagulation, which is established at the present time, is a result of years of research, but still there are numerous questions waiting for answers [123,124].

4. Methods for Improving the Biocompatibility of Synthetic Materials

Nevertheless, their chemical structure, hydrophilicity, roughness, crystallinity and conductivity are not suitable for certain applications and need to be modified [11]. A number of methods are available to improve biocompatibility of biomaterials. The most promising method is coverage of synthetic surfaces with a monolayer of human endothelial cells, since this closely imitates biological conditions. In natural blood vessel, a monocellular film of endothelial cells covers the interior of a vessel, which is in contact with blood and has an important function in blood compatibility [125]. Another common method is chemical surface modification, by including specific functional groups on the surface. These methods are relatively invasive and may also result in harmful chemical products that may lead to irregular surface etching on one hand and may be harmful to the environment on the other. Modification of a material with surface functionalization can also be achieved by ozone oxidation or gamma radiation and UV radiation, but these methods do not achieve a lasting effect, and there is also a high probability of polymer degradation [126]. Most of these disadvantages can be replaced by plasma treatment, which proved to be a very promising method for optimization of surface properties of synthetic materials [127–129].

4.1. Plasma Treatment of Polymers

Plasma treatment is an environmentally friendly method that enables easy and fast modification of the surface of polymers, whereas the polymer bulk properties remain unchanged. Plasma treatment causes formation of new functional groups on the surface, increase of surface energy, increase or decrease of hydrophobicity and hydrophilicity, change of morphology and roughness, and increase or decrease of polymer crystallinity. It also removes poorly bound layers and impurities. The reactions occurring during plasma treatment can be divided into several groups. Surface reactions as a result of plasma changes create functional groups between atoms present in gas and surface atoms and molecules. Such reactions can be achieved with oxygen, nitrogen and NO_2 plasma. With plasma, thin films from organic monomers, such as CH_4, C_2H_6, C_2F_4 to C_3F_6, can also be formed. Such polymerizations involve reactions between atoms in gas and on the surface of polymer and reactions between surface molecules. Plasma can produce volatile products from the surface of polymers by chemical reactions or by physical etching, thus removing unwanted material from the surface. Oxygen plasma is used to remove organic impurities such as oligomers, antioxidants, by-products released from molds and other microorganisms. Oxygen and fluorine plasmas are commonly used for etching of polymers [128,130,131]. Oxygen and mixtures of oxygen plasma are also widely used for treating materials, which are used in biomedical applications [47,49,67,129].

The main products generated during treatment in non-thermal plasma are electrons, ions, excited particles, radicals, as well as UV radiation. These products are mainly free radicals, unsaturated organic components, cross-links between polymer macromolecules, degradation products of polymer chains and gas products. The effects of electrons and UV radiation cause the R–H and C–C bonds to break, which can be represented by the following reactions [128]:

$$RH \rightarrow R^\bullet + H, \; RH \rightarrow R_1^\bullet + R_2^\bullet \tag{1}$$

The direct formation of unsaturated organic compounds with double bonds on the surface of polymers describes the following reaction:

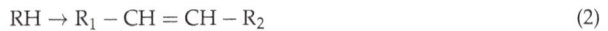

$$RH \rightarrow R_1 - CH = CH - R_2 \tag{2}$$

In the secondary reactions of atomic hydrogen through various mechanisms, molecular hydrogen is usually formed, including recombination and transfer of hydrogen to polymeric molecules. These reactions describe the following equation:

$$H + H \rightarrow H_2, \; H + RH \rightarrow R^\bullet + H_2 \tag{3}$$

In addition to recombination, in organic material atomic hydrogen may also form a double bond with an organic radical:

$$H + R^\bullet \rightarrow R_1 - CH = CH - R_2 \tag{4}$$

During the treatment of polymers with non-thermal oxygen plasma, free organic radicals form on the surface and react with molecular oxygen in the gas phase form, resulting in the formation of active peroxide radicals [128,132,133]. This process describes the following equation:

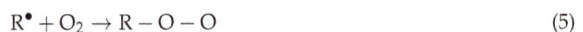

$$R^\bullet + O_2 \rightarrow R - O - O \tag{5}$$

These RO_2 peroxide radicals can trigger various other chemical reactions. The simplest processes involving RO_2 radicals are reactions where various peroxide components are formed on the surface of the polymer and can be simplified by Equations (6) and (7). Due to the low energy of electrons and ions in plasma and the high excitation coefficient of UV radiation, these peroxides on the surface are formed in a thin layer [134,135].

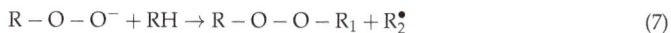

$$R - O - O^- + RH \rightarrow R - O - O - H + R^\bullet \tag{6}$$

$$R - O - O^- + RH \rightarrow R - O - O - R_1 + R_2^\bullet \tag{7}$$

In addition to the formation of new functional groups on the surface of polymers, the plasma treatment also produces an effect called etching. Etching can be explained by two mechanisms: chemical etching and physical etching, which occurs due to ion bombardment. Chemical etching results in surface reactions, which makes the surface part of the polymers to fumigate. The major molecules that usually participate in these reactions are oxygen atoms, ozone, fluorine atoms and electronically excited oxygen molecules. Both processes take place during plasma treatment, so it is difficult to separate them from each other. Nevertheless, by changing the conditions and characteristics of plasma, we can regulate the relationship between the two processes. In addition to the processing conditions, the polymer etching rate depends also on the type of gas used. In the case of PET polymer treated with different plasmas at different power levels (25, 50 and 100 W) and frequency of 13.56 MHz, it was found that the etching was linearly dependent on the power [136]. The highest degree of etching occurred in oxygen plasma. In addition, etching also depends on the type of polymer, its chemical composition and the crystallinity of the material [135].

4.1.1. Aging of Plasma-Treated Materials

The stability of the plasma treated surface is a very important feature, especially if materials are not used immediately after plasma treatment. Plasma treated biomaterials, used in medicine, where materials come in contact with the living tissue, stability is very important. After plasma treatment, the surface of polymers tends to return to its original state, what is called ageing. Many researchers studied the so-called aging of various materials that were treated with different types of plasmas. Aging depends on both the type of plasma and the treated material. Experiments on polydimethylsiloxane, which was treated in nitrogen, oxygen, argon and NH_3 plasmas and aged on air and in the buffer solution confirmed, that the surface returned back to initial, hydrophobic state after one month [137]. Wilson and colleagues [60] treated the surface of PTFE polymers with the same types of plasma and were aged under the same conditions. From their studies it was concluded that aging was present under both storage conditions and that the effect was more noticeable in aging in the buffer. For both samples, aging after one month stabilized, but the condition did not return to the initial state. On the other hand, the hydrophobicity of the polysulphonic membranes remained unchanged even after three months after treatment with CO_2 plasma [138]. Modic et al. [42] studied aging of PET polymer treated in oxygen plasma and exposed to different environmental conditions (see Figure 5). After plasma treatment, the contact angle dropped from original 73° to 10°. The first set of treated samples was left at room temperature; the second set was stored in the refrigerator at 4 °C and the third set was put in phosphate buffer solution (PBS). Aging of all samples were monitored for two weeks. Results showed that samples stored at room temperature and those stored in the refrigerator had the same relative slow aging; the contact angle changed from original 21° to 30°; again ageing in PBS turned out to be much faster. Already after 3 h contact angle increased from 10° after plasma treatment, to ~30°. The effect of ageing in all environmental conditions was observed for the first three days; later on the contact angle does not change significantly (Figure 5).

There are four proposed mechanisms of ageing:

- Reorientation and relocation of polar groups from the surface of the polymer into the bulk of the material due to thermodynamic relaxation,
- Diffusion of low molecular weight oligomers from the interior to the surface and products that are formed during plasma treatment on the surface of polymers,

- Reactions of free radicals and other active species and groups formed during treatment, with each other and with the environment in which the polymer is located.

The aging of hydrocarbon materials treated in oxygen plasma is mainly due to the reorientation and transfer of polar peroxide groups into the interior of the polymer. If the same polymers are treated with nitrogen plasma, aging results from reactions of nitrogenous surface groups with the environment after plasma treatment [135].

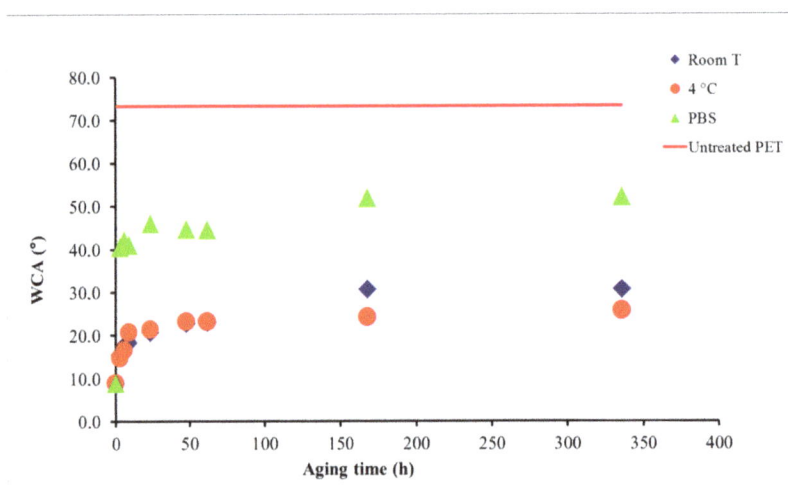

Figure 5. Effect of different aging conditions on the wettability of surface of PET polymer treated in oxygen plasma glow for 30 s.

5. Conclusions and Future Perspectives

Plasma treatment is one of the most favorable methods for treatment of synthetic materials and it greatly improves the hemocompatible properties of polymers. Systematic measurements on whole human blood of healthy volunteers have confirmed the hypothesis that the rate of hemocompatibility monotonically increases with increasing hydrophilicity and surface roughness. Biocompatibility depends on the success rate of surface endothealization, which is strongly correlated with surface properties, i.e., surface wettability, topography and chemistry. According to studies, cells prefer to adhere to moderate hydrophilic surfaces at micrometer scale. Furthermore, oxygen functional groups on the surface proved to stimulate cell adhesion and proliferation.

Based on extensive experimental results, three possible hypotheses to explain the observed hemostatic response were proposed in this review. The first hypothesis is based on preferential adhesion of water molecules from blood to the polar functional groups on the polymer surface. The second one is based on different conformations and orientations of adsorbed plasma proteins, and the third hypothesis is based on the significant reduction of contact area between polymer and platelets due to high roughness of plasma treated polymer samples. Because of extreme complexity of interactions between whole blood and polymer surface, it is not possible to declare which hypothesis is the most suitable. Results indicate that a combination of different physical and chemical processes can lead to a biological response of material used for fabrication of artificial grafts and other cardiovascular implants.

Funding: This research received no external funding.

Acknowledgments: The author gratefully acknowledges Alenka Vesel for the insightful comments and guidance in preparing the manuscript.

Conflicts of Interest: The authors declare no conflict of interest.

References

1. Allender, S.; Foster, C.; Hutchinson, L.; Arambepola, C. Quantification of urbanization in relation to chronic diseases in developing countries: A systematic review. *J. Urban Health* **2008**, *85*, 938–951. [CrossRef] [PubMed]

2. Helmus, M.N.; Hubbell, J.A. Materials selection. *Cardiovasc. Pathol.* **1993**, *2*, 53–71. [CrossRef]

3. Tsuruta, T.; Hayashi, T.; Kataoka, K.; Ishihara, K.; Kimura, Y. *Biomedical Applications of Polymeric Materials*; CRC Press: Boca Raton, FL, USA, 1993; Volume 340.

4. Jaganathan, S.K.; Supriyanto, E.; Murugesan, S.; Balaji, A.; Asokan, M.K. Biomaterials in cardiovascular research: applications and clinical implications. *BioMed Res. Int.* **2014**, *2014*. [CrossRef] [PubMed]

5. Roald, H.; Barstad, R.; Bakken, I.; Roald, B.; Lyberg, T.; Sakariassen, K. Initial interactions of platelets and plasma proteins in flowing non-anticoagulated human blood with the artificial surfaces Dacron and PTFE. *Blood Coagul. Fibrinol. Int. J. Haemost. Thromb.* **1994**, *5*, 355–363.

6. Ormiston, J.A.; Serruys, P.W.; Regar, E.; Dudek, D.; Thuesen, L.; Webster, M.W.; Onuma, Y.; Garcia-Garcia, H.M.; McGreevy, R.; Veldhof, S. A bioabsorbable everolimus-eluting coronary stent system for patients with single de-novo coronary artery lesions (ABSORB): A prospective open-label trial. *Lancet* **2008**, *371*, 899–907. [CrossRef]

7. Serruys, P.W.; Ormiston, J.A.; Onuma, Y.; Regar, E.; Gonzalo, N.; Garcia-Garcia, H.M.; Nieman, K.; Bruining, N.; Dorange, C.; Miquel-Hébert, K. A bioabsorbable everolimus-eluting coronary stent system (ABSORB): 2-year outcomes and results from multiple imaging methods. *Lancet* **2009**, *373*, 897–910. [CrossRef]

8. Damman, P.; Iñiguez, A.; Klomp, M.; Beijk, M.; Woudstra, P.; Silber, S.; Ribeiro, E.E.; Suryapranata, H.; Sim, K.H.; Tijssen, J.G. Coronary Stenting With the GenousTM Bio-Engineered R StentTM in Elderly Patients. *Circ. J.* **2011**, *75*, 2590–2597. [CrossRef]

9. Herring, M.; Gardner, A.; Glover, J. A single-staged technique for seeding vascular grafts with autogenous endothelium. *Surgery* **1978**, *84*, 498–504.

10. Bruck, S.D. Interactions of synthetic and natural surfaces with blood in the physiological environment. *J. Biomed. Mater. Res.* **1977**, *11*, 1–21. [CrossRef]

11. Chan, C.-M.; Ko, T.-M.; Hiraoka, H. Polymer surface modification by plasmas and photons. *Surf. Sci. rep.* **1996**, *24*, 1–54. [CrossRef]

12. Grunkemeier, J.; Tsai, W.; Horbett, T. Hemocompatibility of treated polystyrene substrates: Contact activation, platelet adhesion, and procoagulant activity of adherent platelets. *J. Biomed. Mater. Res. Off. J. Soc. Biomater. Jpn. Soc. Biomater. Aust. Soc. Biomater.* **1998**, *41*, 657–670. [CrossRef]

13. Ratner, B.D.; Hoffman, A.S.; Schoen, F.J.; Lemons, J.E. *Biomaterials Science: An Introduction to Materials in Medicine*; Elsevier: Amsterdam, The Netherlands, 2004.

14. Junkar, I.; Cvelbar, U.; Vesel, A.; Hauptman, N.; Mozetič, M. The Role of Crystallinity on Polymer Interaction with Oxygen Plasma. *Plasma Process. Polym.* **2009**, *6*, 667–675. [CrossRef]

15. Cui, H.; Sinko, P. The role of crystallinity on differential attachment/proliferation of osteoblasts and fibroblasts on poly (caprolactone-co-glycolide) polymeric surfaces. *Front. Mater. Sci.* **2012**, *6*, 47–59. [CrossRef]

16. Vesel, A. Hydrophobization of polymer polystirene in fluorine plasma. *Mater. Tehnol.* **2011**, *45*, 217–220.

17. Vesel, A.; Junkar, I.; Cvelbar, U.; Kovac, J.; Mozetic, M. Surface modification of polyester by oxygen- and nitrogen-plasma treatment. *Surf. Interface Anal.* **2008**, *40*, 1444–1453. [CrossRef]

18. Di Mundo, R.; Palumbo, F.; d'Agostino, R. Influence of chemistry on wetting dynamics of nanotextured hydrophobic surfaces. *Langmuir* **2010**, *26*, 5196–5201. [CrossRef] [PubMed]

19. Wang, H.; Ji, J.; Zhang, W.; Wang, W.; Zhang, Y.; Wu, Z.; Zhang, Y.; Chu, P.K. Rat calvaria osteoblast behavior and antibacterial properties of O_2 and N_2 plasma-implanted biodegradable poly(butylene succinate). *Acta Biomater.* **2010**, *6*, 154–159. [CrossRef]

20. John, M.J.; Thomas, S. Biofibres and biocomposites. *Carbohydr. Polym.* **2008**, *71*, 343–364. [CrossRef]

21. Dolenc, A.; Homar, M.; Gašperlin, M.; Kristl, J. Z nanooblaganjem do izboljšanja biokompatibilnosti vsadkov. *Medicinski Razgledi* **2006**, *4*, 411–420.

22. Lemons, J.; Ratner, B.; Hoffman, A.; Schoen, F. *Biomaterials Science: An Introduction to Materials in Medicine*; Academic Press: San Diego, CA, USA, 1996.

23. Kerdjoudj, H.; Berthelemy, N.; Rinckenbach, S.; Kearney-Schwartz, A.; Montagne, K.; Schaaf, P.; Lacolley, P.; Stoltz, J.-F.; Voegel, J.-C.; Menu, P. Small vessel replacement by human umbilical arteries with polyelectrolyte film-treated arteries: In vivo behavior. *J. Am. Coll. Cardiol.* **2008**, *52*, 1589–1597. [CrossRef]

24. Moris, D.; Sigala, F.; Georgopoulos, S.; Bramis, I. The choice of the appropriate graft in the treatment of vascular diseases. *Hellenic J. Surg.* **2010**, *82*, 274–283. [CrossRef]

25. Nunn, D.B.; Freeman, M.H.; Hudgins, P.C. Postoperative alterations in size of Dacron aortic grafts: An ultrasonic evaluation. *Ann. Surg.* **1979**, *189*, 741. [CrossRef] [PubMed]

26. Drury, J.K.; Ashton, T.R.; Cunningham, J.D.; Maini, R.; Pollock, J.G. Experimental and clinical experience with a gelatin impregnated Dacron prosthesis. *Ann. Vasc. Surg.* **1987**, *1*, 542–547. [CrossRef]

27. Guidoin, R.; Snyder, R.; Martin, L.; Botzko, K.; Marois, M.; Awad, J.; King, M.; Domurado, D.; Bedros, M.; Gosselin, C. Albumin Coating of a Knitted Polyester Arterial Prosthesis: An Alternative to Preclotting. *Ann. Thorac. Surg.* **1984**, *37*, 457–465. [CrossRef]

28. Kito, H.; Matsuda, T. Biocompatible coatings for luminal and outer surfaces of small-caliber artificial grafts. *J. Biomed. Mater. Res.* **1996**, *30*, 321–330. [CrossRef]

29. Callow, A.D. Problems in the Construction of A Small Diameter Graft. *Int. Angiol.* **1988**, *7*, 246–253.

30. Charpentier, P.A.; Maguire, A.; Wan, W.K. Surface modification of polyester to produce a bacterial cellulose-based vascular prosthetic device. *Appl. Surf. Sci.* **2006**, *252*, 6360–6367. [CrossRef]

31. Kolar, M.; Mozetič, M.; Stana-Kleinschek, K.; Fröhlich, M.; Turk, B.; Vesel, A. Covalent binding of heparin to functionalized PET materials for improved haemocompatibility. *Materials* **2015**, *8*, 1526–1544. [CrossRef]

32. Doliška, A.; Ribitsch, V.; Stana Kleinschek, K.; Strnad, S. Viscoelastic properties of fibrinogen adsorbed onto poly(ethylene terephthalate) surfaces by QCM-D. *Carbohydr. Polym.* **2013**, *93*, 246–255. [CrossRef]

33. Gorbet, M.B.; Sefton, M.V. Biomaterial-associated thrombosis: Roles of coagulation factors, complement, platelets and leukocytes. *Biomaterials* **2004**, *25*, 5681–5703. [CrossRef]

34. Otto, M.; Franzen, A.; Hansen, T.; Kirkpatrick, C.J. Modification of human platelet adhesion on biomaterial surfaces by protein preadsorption under static and flow conditions. *J. Mater. Sci. Mater. Med.* **2004**, *15*, 35–42. [CrossRef] [PubMed]

35. Ratner, B.D. The catastrophe revisited: Blood compatibility in the 21st century. *Biomaterials* **2007**, *28*, 5144–5147. [CrossRef] [PubMed]

36. Rihova, B. Biocompatibility of biomaterials: Hemocompatibility, immunocompatibility and biocompatibility of solid polymeric materials and soluble targetable polymeric carriers. *Adv. Drug Deliv. Rev.* **1996**, *2*, 157–176. [CrossRef]

37. Ratner, B.D. The blood compatibility catastrophe. *J. Biomed. Mater. Res.* **1993**, *27*, 283–287. [CrossRef] [PubMed]

38. Ninan, N.; Muthiah, M.; Park, I.-K.; Elain, A.; Wong, T.W.; Thomas, S.; Grohens, Y. Faujasites Incorporated Tissue Engineering Scaffolds for Wound Healing: In Vitro and In Vivo Analysis. *ACS Appl. Mater. Interfaces* **2013**, *5*, 11194–11206. [CrossRef] [PubMed]

39. Ghosal, K.; Latha, M.S.; Thomas, S. Poly(ester amides) (PEAs)—Scaffold for tissue engineering applications. *Eur. Polym. J.* **2014**, *60*, 58–68. [CrossRef]

40. Garcia, J.L.; Asadinezhad, A.; Pachernik, J.; Lehocky, M.; Junkar, I.; Humpolicek, P.; Saha, P.; Valasek, P. Cell Proliferation of HaCaT Keratinocytes on Collagen Films Modified by Argon Plasma Treatment. *Molecules* **2010**, *15*, 2845–2856. [CrossRef] [PubMed]

41. Asadinezhad, A.; Novák, I.; Lehocký, M.; Bílek, F.; Vesel, A.; Junkar, I.; Sáha, P.; Popelka, A. Polysaccharides Coatings on Medical-Grade PVC: A Probe into Surface Characteristics and the Extent of Bacterial Adhesion. *Molecules* **2010**, *15*, 1007–1027. [CrossRef] [PubMed]

42. Modic, M.; Junkar, I.; Vesel, A.; Mozetic, M. Aging of plasma treated surfaces and their effects on platelet adhesion and activation. *Surface Coat. Technol.* **2012**, *213*, 98–104. [CrossRef]

43. Junkar, I.; Cvelbar, U.; Lehocky, M. Plasma treatment of biomedical materials. *Mater. Tehnol.* **2011**, *45*, 221–226.

44. Uroš, C.; Ita, J.; Martina, M. Hemocompatible Poly(ethylene terephthalate) Polymer Modified via Reactive Plasma Treatment. *Jpn. J. Appl. Phys.* **2011**, *50*, 08JF02.

45. Vesel, A.; Mozetic, M.; Jaganjac, M.; Milkovic, L.; Cipak, A.; Zarkovic, N. Biocompatibility of oxygen-plasma-treated polystyrene substrates. *Eur. Phys. J. Appl. Phys.* **2011**, *56*, 24024. [CrossRef]
46. Jaganjac, M.; Milkovic, L.; Cipak, A.; Mozetic, M.; Recek, N.; Zarkovic, N.; Vesel, A. Cell adhesion on hydrophobic polymer surfaces. *Mater. Tehnol.* **2012**, *46*, 53–56.
47. Recek, N.; Mozetič, M.; Jaganjac, M.; Milkovič, L.; Žarkovic, N.; Vesel, A. Improved proliferation of human osteosarcoma cells on oxygen plasma treated polystyrene. *Vacuum* **2013**, *98*, 116–121. [CrossRef]
48. Recek, N.; Mozetic, M.; Jaganjac, M.; Milkovic, L.; Zarkovic, N.; Vesel, A. Adsorption of Proteins and Cell Adhesion to Plasma Treated Polymer Substrates. *Int. J. Polym. Mater. Polym. Biomater.* **2014**, *63*, 685–691. [CrossRef]
49. Jaganjac, M.; Vesel, A.; Milkovic, L.; Recek, N.; Kolar, M.; Zarkovic, N.; Latiff, A.; Kleinschek, K.-S.; Mozetic, M. Oxygen-rich coating promotes binding of proteins and endothelialization of polyethylene terephthalate polymers. *J. Biomed. Mater. Res. Part A* **2014**, *102*, 2305–2314. [CrossRef] [PubMed]
50. Recek, N.; Vesel, A.; Mozetic, M.; Jaganjac, M.; Milkovic, L.; Žarković, N. Influence of polymer surface on cell proliferation and cell oxidation momeostatis. In Proceedings of the ICPM5, 5th International Conference on Plasma Medicine, Nara, Japan, 18–23 May 2014.
51. Seifert, B.; Mihanetzis, G.; Groth, T.; Albrecht, W.; Richau, K.; Missirlis, Y.; Paul, D.; von Sengbusch, G. Polyetherimide: A new membrane-forming polymer for biomedical applications. *Artif. Organs* **2002**, *26*, 189–199. [CrossRef]
52. Sperling, C.; Schweiss, R.B.; Streller, U.; Werner, C. In vitro hemocompatibility of self-assembled monolayers displaying various functional groups. *Biomaterials* **2005**, *26*, 6547–6557. [CrossRef]
53. Wilson, C.J.; Clegg, R.E.; Leavesley, D.I.; Pearcy, M.J. Mediation of Biomaterial–Cell Interactions by Adsorbed Proteins: A Review. *Tissue Eng.* **2005**, *11*, 1–18. [CrossRef]
54. Anselme, K. Osteoblast adhesion on biomaterials. *Biomaterials* **2000**, *21*, 667–681. [CrossRef]
55. Ingber, D.E. Cellular mechanotransduction: Putting all the pieces together again. *FASEB J. Off. Publ. Fed. Am. Soc. Exp. Biol.* **2006**, *20*, 811–827. [CrossRef] [PubMed]
56. Trepat, X.; Lenormand, G.; Fredberg, J.J. Universality in cell mechanics. *Soft Matter* **2008**, *4*, 1750–1759. [CrossRef]
57. Rodrigues, S.N.; Goncalves, I.C.; Martins, M.C.; Barbosa, M.A.; Ratner, B.D. Fibrinogen adsorption, platelet adhesion and activation on mixed hydroxyl-/methyl-terminated self-assembled monolayers. *Biomaterials* **2006**, *27*, 5357–5367. [CrossRef] [PubMed]
58. Tengvall, P.; Askendal, A.; Lundstrom, I.; Elwing, H. Studies of surface activated coagulation: Antisera binding onto methyl gradients on silicon incubated in human plasma in vitro. *Biomaterials* **1992**, *13*, 367–374. [CrossRef]
59. Wang, J.; Chen, J.Y.; Yang, P.; Leng, Y.X.; Wan, G.J.; Sun, H.; Zhao, A.S.; Huang, N.; Chu, P.K. In vitro platelet adhesion and activation of polyethylene terephthalate modified by acetylene plasma immersion ion implantation and deposition. *Nucl. Instrum. Methods Phys. Res. Sect. B Beam Interact. Mater. Atoms.* **2006**, *242*, 12–14. [CrossRef]
60. Wilson, D.J.; Williams, R.L.; Pond, R.C. Plasma modification of PTFE surfaces. Part II: Plasma-treated surfaces following storage in air or PBS. *Surf. Interface Anal.* **2001**, *31*, 397–408. [CrossRef]
61. Anselme, K.; Davidson, P.; Popa, A.M.; Giazzon, M.; Liley, M.; Ploux, L. The interaction of cells and bacteria with surfaces structured at the nanometre scale. *Acta Biomater.* **2010**, *6*, 3824–3846. [CrossRef] [PubMed]
62. Chen, L.; Han, D.; Jiang, L. On improving blood compatibility: From bioinspired to synthetic design and fabrication of biointerfacial topography at micro/nano scales. *Colloids Surf. B Biointerfaces* **2011**, *85*, 2–7. [CrossRef] [PubMed]
63. Liu, X.; Chu, P.K.; Ding, C. Surface nano-functionalization of biomaterials. *Mater. Sci. Eng. R Rep.* **2010**, *70*, 275–302. [CrossRef]
64. Fan, H.; Chen, P.; Qi, R.; Zhai, J.; Wang, J.; Chen, L.; Chen, L.; Sun, Q.; Song, Y.; Han, D.; et al. Greatly Improved Blood Compatibility by Microscopic Multiscale Design of Surface Architectures. *Small* **2009**, *5*, 2144–2148. [CrossRef]
65. Ferraz, N.; Carlsson, J.; Hong, J.; Ott, M.K. Influence of nanoporesize on platelet adhesion and activation. *J. Mater. Sci. Mater. Med.* **2008**, *19*, 3115–3121. [CrossRef] [PubMed]
66. Sivaraman, B.; Latour, R.A. The relationship between platelet adhesion on surfaces and the structure versus the amount of adsorbed fibrinogen. *Biomaterials* **2010**, *31*, 832–839. [CrossRef] [PubMed]

67. Mozetič, M.; Primc, G.; Vesel, A.; Modic, M.; Junkar, I.; Recek, N.; Klanjšek-Gunde, M.; Guhy, L.; Sunkara, M.K.; Assensio, M.C.; et al. Application of extremely non-equilibrium plasmas in the processing of nano and biomedical materials. *Plasma Sources Sci. Technol.* **2015**, *24*, 015026. [CrossRef]
68. Curtis, A.; Wilkinson, C. Topographical control of cells. *Biomaterials* **1997**, *18*, 1573–1583. [CrossRef]
69. Anselme, K.; Bigerelle, M. Topography effects of pure titanium substrates on human osteoblast long-term adhesion. *Acta Biomater.* **2005**, *1*, 211–222. [CrossRef] [PubMed]
70. Bigerelle, M.; Anselme, K.; Noel, B.; Ruderman, I.; Hardouin, P.; Iost, A. Improvement in the morphology of Ti-based surfaces: A new process to increase in vitro human osteoblast response. *Biomaterials* **2002**, *23*, 1563–1577. [CrossRef]
71. Dalby, M.J.; Gadegaard, N.; Tare, R.; Andar, A.; Riehle, M.O.; Herzyk, P.; Wilkinson, C.D.W.; Oreffo, R.O.C. The control of human mesenchymal cell differentiation using nanoscale symmetry and disorder. *Nat. Mater.* **2007**, *6*, 997. [CrossRef]
72. Bershadsky, A.; Kozlov, M.; Geiger, B. Adhesion-mediated mechanosensitivity: A time to experiment, and a time to theorize. *Curr. Opin. Cell Biol.* **2006**, *18*, 472–481. [CrossRef]
73. Chen, Q.; Espey, M.G.; Sun, A.Y.; Lee, J.H.; Krishna, M.C.; Shacter, E.; Choyke, P.L.; Pooput, C.; Kirk, K.L.; Buettner, G.R.; et al. Ascorbate in pharmacologic concentrations selectively generates ascorbate radical and hydrogen peroxide in extracellular fluid in vivo. *Proc. Natl. Acad. Sci. USA* **2007**, *104*, 8749–8754. [CrossRef]
74. Feinberg, A.W.; Wilkerson, W.R.; Seegert, C.A.; Gibson, A.L.; Hoipkemeier-Wilson, L.; Brennan, A.B. Systematic variation of microtopography, surface chemistry and elastic modulus and the state dependent effect on endothelial cell alignment. *J. Biomed. Mater. Res. Part A* **2008**, *86A*, 522–534. [CrossRef]
75. Stevenson, P.M.; Donald, A.M. Identification of Three Regimes of Behavior for Cell Attachment on Topographically Patterned Substrates. *Langmuir* **2009**, *25*, 367–376. [CrossRef] [PubMed]
76. Berry, C.C.; Campbell, G.; Spadiccino, A.; Robertson, M.; Curtis, A.S.G. The influence of microscale topography on fibroblast attachment and motility. *Biomaterials* **2004**, *25*, 5781–5788. [CrossRef] [PubMed]
77. Théry, M.; Pépin, A.; Dressaire, E.; Chen, Y.; Bornens, M. Cell distribution of stress fibres in response to the geometry of the adhesive environment. *Cell Motil.* **2006**, *63*, 341–355. [CrossRef] [PubMed]
78. Hallab, N.J.; Bundy, K.J.; O'Connor, K.; Moses, R.L.; Jacobs, J.J. Evaluation of Metallic and Polymeric Biomaterial Surface Energy and Surface Roughness Characteristics for Directed Cell Adhesion. *Tissue Eng.* **2001**, *7*, 55–71. [CrossRef] [PubMed]
79. Anselme, K.; Bigerelle, M.; Noel, B.; Dufresne, E.; Judas, D.; Iost, A.; Hardouin, P. Qualitative and quantitative study of human osteoblast adhesion on materials with various surface roughnesses. *J. Biomed. Mater. Res.* **2000**, *49*, 155–166. [CrossRef]
80. Anselme, K.; Bigerelle, M. Modelling approach in cell/material interactions studies. *Biomaterials* **2006**, *27*, 1187–1199. [CrossRef]
81. Anselme, K.; Ploux, L.; Ponche, A. Cell/Material Interfaces: Influence of Surface Chemistry and Surface Topography on Cell Adhesion. *J. Adhes. Sci. Technol.* **2010**, *24*, 831–852. [CrossRef]
82. Bigerelle, M.; Anselme, K. Statistical correlation between cell adhesion and proliferation on biocompatible metallic materials. *J. Biomed. Mater. Res. Part A* **2005**, *72A*, 36–46. [CrossRef]
83. Ponsonnet, L.; Comte, V.; Othmane, A.; Lagneau, C.; Charbonnier, M.; Lissac, M.; Jaffrezic, N. Effect of surface topography and chemistry on adhesion, orientation and growth of fibroblasts on nickel–titanium substrates. *Mater. Sci. Eng. C* **2002**, *21*, 157–165. [CrossRef]
84. Ponsonnet, L.; Reybier, K.; Jaffrezic, N.; Comte, V.; Lagneau, C.; Lissac, M.; Martelet, C. Relationship between surface properties (roughness, wettability) of titanium and titanium alloys and cell behaviour. *Mater. Sci. Eng. C* **2003**, *23*, 551–560. [CrossRef]
85. Agnihotri, A.; Soman, P.; Siedlecki, C.A. AFM measurements of interactions between the platelet integrin receptor GPIIbIIIa and fibrinogen. *Colloids Surf. B Biointerfaces* **2009**, *71*, 138–147. [CrossRef] [PubMed]
86. Savage, B.; Bottini, E.; Ruggeri, Z.M. Interaction of Integrin αIIbβ3 with Multiple Fibrinogen Domains during Platelet Adhesion. *J. Biol. Chem.* **1995**, *270*, 28812–28817. [CrossRef] [PubMed]
87. Shaw, D.E.; Maragakis, P.; Lindorff-Larsen, K.; Piana, S.; Dror, R.O.; Eastwood, M.P.; Bank, J.A.; Jumper, J.M.; Salmon, J.K.; Shan, Y.; et al. Atomic-level characterization of the structural dynamics of proteins. *Science* **2010**, *330*, 341–346. [CrossRef]
88. Xu, L.C.; Siedlecki, C.A. Effects of surface wettability and contact time on protein adhesion to biomaterial surfaces. *Biomaterials* **2007**, *28*, 3273–3283. [CrossRef]

89. Xu, L.C.; Vadillo-Rodriguez, V.; Logan, B.E. Residence time, loading force, pH, and ionic strength affect adhesion forces between colloids and biopolymer-coated surfaces. *Langmuir* **2005**, *21*, 7491–7500. [CrossRef]

90. Spijker, H.T.; Bos, R.; Busscher, H.J.; van Kooten, T.; van Oeveren, W. Platelet adhesion and activation on a shielded plasma gradient prepared on polyethylene. *Biomaterials* **2002**, *23*, 757–766. [CrossRef]

91. Vogler, E.A.; Graper, J.C.; Harper, G.R.; Sugg, H.W.; Lander, L.M.; Brittain, W.J. Contact activation of the plasma coagulation cascade. I. Procoagulant surface chemistry and energy. *J. Biomed. Mater. Res.* **1995**, *29*, 1005–1016. [CrossRef] [PubMed]

92. Lee, J.H.; Lee, H.B. Platelet adhesion onto wettability gradient surfaces in the absence and presence of plasma proteins. *J. Biomed. Mater. Res.* **1998**, *41*, 304–311. [CrossRef]

93. Yanagisawa, I.; Sakuma, H.; Shimura, M.; Wakamatsu, Y.; Yanagisawa, S.; Sairenji, E. Effects of "wettability" of biomaterials on culture cells. *J. Oral Implantol.* **1989**, *15*, 168–177.

94. Moller, K.; Meyer, U.; Szulczewski, D.; Heide, H.; Priessnitz, B.; Jones, D. The influence of zeta-potential and interfacial-tension on osteoblast-like cells. *Cells Mater.* **1994**, *4*, 263–274.

95. Lee, J.H.; Lee, S.J.; Khang, G.; Lee, H.B. The Effect of Fluid Shear Stress on Endothelial Cell Adhesiveness to Polymer Surfaces with Wettability Gradient. *J. Colloid Interface Sci.* **2000**, *230*, 84–90. [CrossRef] [PubMed]

96. Lee, S.J.; Khang, G.; Lee, Y.M.; Lee, H.B. The effect of surface wettability on induction and growth of neurites from the PC-12 cell on a polymer surface. *J. Colloid Interface Sci.* **2003**, *259*, 228–235. [CrossRef]

97. Lim, J.Y.; Liu, X.; Vogler, E.A.; Donahue, H.J. Systematic variation in osteoblast adhesion and phenotype with substratum surface characteristics. *J. Biomed. Mater. Res. Part A* **2004**, *68A*, 504–512. [CrossRef] [PubMed]

98. George, J.; Sreekala, M.S.; Thomas, S. A review on interface modification and characterization of natural fiber reinforced plastic composites. *Polym. Eng. Sci.* **2001**, *41*, 1471–1485. [CrossRef]

99. Jurk, K.; Kehrel, B.E. Platelets: Physiology and biochemistry. *Semin. Thromb. Hemost.* **2005**, *31*, 381–392. [CrossRef]

100. Anderson, J.M.; Kottke-Marchant, K. *Platelet Interactions With Biomaterials and Artificial Devices*; CRC Press Inc: Boca Raton, FL, USA, 1987.

101. Klinkmann, H.; Falkenhagen, D.; Courtney, J.M. Clinical Relevance of Biocompatibility—The Material Cannot Be Divorced from the Device. In *Uremia Therapy*; Gurland, H., Ed.; Springer: Berlin/Heidelberg, Germany; New York, NY, USA, 1987; pp. 125–140. [CrossRef]

102. Schaller, J.; Gerber, S.; Kaempfer, U.; Lejon, S.; Trachsel, C. *Human Blood Plasma Proteins: Structure and Function*; John Wiley & Sons: New York, NY, USA, 2008.

103. Courtney, J.; Lamba, N.; Sundaram, S.; Forbes, C. Biomaterials for blood-contacting applications. *Biomaterials* **1994**, *15*, 737–744. [CrossRef]

104. Salzman, E.W. Interaction of the blood with natural and artificial surfaces. In *Blood Material Interaction*; Dekker Inc.: Ney York, NY, USA, 1986; p. 39.

105. Meyer, J.G. *Blutgerinnung und Fibrinolyse*; Deutsche Ärzte Verlag: Köln, Germany, 1986.

106. Brash, J.L. Mechanism of adsorption of proteins to solid surfaces and its relationship to blood compatibility. In *Biocompatible Polymers, Metals and Composites*; Technomic: Lancaster, UK, 1983; p. 35.

107. Baszkin, A. The effect of polymer surface composition and structure on adsorption of plasma proteins. In *Blood Compatible Materials and Their Testing*; Dawids, S., Bantjes, A., Eds.; Martins Nijhoff Publishers: Dortrecht, The Netherlands, 1986; p. 39.

108. Kottke-Marchant, K.; Anderson, J.M.; Umemura, Y.; Marchant, R.E. Effect of albumin coating on the in vitro blood compatibility of Dacron arterial prostheses. *Biomaterials* **1989**, *10*, 147–155. [CrossRef]

109. Ikada, Y. *Blood-Compatible Polymers*; Springer: Berlin/Heidelberg, Germany, 1984; pp. 103–140.

110. Andrade, J.D. Interfacial phenomena and biomaterials. *Med. Instrum.* **1973**, *7*, 110–119.

111. Baier, R.E. The role of surface energy in thrombogenesis. *Bull. N. Y. Acad. Med.* **1972**, *48*, 257–272.

112. Ratner, B.D.; Hoffman, A.S.; Hanson, S.R.; Harker, L.A.; Whiffen, J.D. Blood-compatibility-water-content relationships for radiation-grafted hydrogels. *J. Polym. Sci. Polym. Symp.* **1979**, *66*, 363–375. [CrossRef]

113. Bantjes, A. Clotting Phenomena at the Blood-Polymer Interface and Development of Blood Compatible Polymeric Surfaces. *Br. Polym. J.* **1978**, *10*, 267–274. [CrossRef]

114. Norde, W.; Lyklema, J. Proportion titration and electrokinetic studies of adsorbed protein layers. In *Surface and Interfacial Aspects of Biomedical Polymers*; Andrade, J.D., Ed.; Plenum Press: New York, NY, USA, 1985; p. 241.

115. Bruck, S.D. Physicochemical aspects of the blood compatibility of polymeric surfaces. *J. Polym. Sci. Polym. Symp.* **1979**, *66*, 283–312. [CrossRef]

116. Sawyer, P.N.; Srinivasan, S. Studies on the biophysics of intravascular thrombosis. *Am. J. Surg.* **1967**, *114*, 42–60. [CrossRef]

117. Szycher, M. (Ed.) *Biocompatible Polymers, Metals and Composites*; Technomic: Lanchester, UK, 1983.

118. Yu, J.; Sundaram, S.; Weng, D.; Courtney, J.M.; Moran, C.R.; Graham, N.B. Blood interactions with novel polyurethaneurea hydrogels. *Biomaterials* **1991**, *12*, 119–120. [CrossRef]

119. Beissinger, R.L.; Leonard, E.F. Plasma protein adsorption and desorption rates on quartz: Approach to multi-component systems. *Trans.Am. Soc. Artif. Intern. Organs* **1981**, *27*, 225–230. [PubMed]

120. Soderquist, M.E.; Walton, A.G. Structural changes in proteins adsorbed on polymer surfaces. *J. Colloid Interface Sci.* **1980**, *75*, 386–397. [CrossRef]

121. Lundström, I.; Elwing, H. Simple kinetic models for protein exchange reactions on solid surfaces. *J. Colloid Interface Sci.* **1990**, *136*, 68–84. [CrossRef]

122. Giangrande, P.L.F. Six Characters in Search of An Author: The History of the Nomenclature of Coagulation Factors. *Br. J. Haematol.* **2003**, *121*, 703–712. [CrossRef]

123. Brummel, K.E.; Butenas, S.; Mann, K.G. An Integrated Study of Fibrinogen during Blood Coagulation. *J. Biol. Chem.* **1999**, *274*, 22862–22870. [CrossRef]

124. Macfarlane, R.G. A. The blood clotting mechanism The development of a theory of blood coagulation. *Proc. R. Soc. Lond. Ser. B Biol. Sci.* **1969**, *173*, 261–268. [CrossRef]

125. Van Wachem, P.; Beugeling, T.; Feijen, J.; Bantjes, A.; Detmers, J.; Van Aken, W. Interaction of cultured human endothelial cells with polymeric surfaces of different wettabilities. *Biomaterials* **1985**, *6*, 403–408. [CrossRef]

126. Joseph, K.; Thomas, S.; Pavithran, C. Effect of chemical treatment on the tensile properties of short sisal fibre-reinforced polyethylene composites. *Polymer* **1996**, *37*, 5139–5149. [CrossRef]

127. France, R.M.; Short, R.D. Plasma treatment of polymers Effects of energy transfer from an argon plasma on the surface chemistry of poly(styrene), low density poly(ethylene), poly(propylene) and poly(ethylene terephthalate). *J. Chem. Soc. Faraday Trans.* **1997**, *93*, 3173–3178. [CrossRef]

128. Morent, R.; De Geyter, N.; Desmet, T.; Dubruel, P.; Leys, C. Plasma surface modification of biodegradable polymers: a review. *Plasma Process. Polym.* **2011**, *8*, 171–190. [CrossRef]

129. Vesel, A.; Mozetic, M. New developments in surface functionalization of polymers using controlled plasma treatments. *J. Phys. D Appl. Phys.* **2017**, *50*, 293001. [CrossRef]

130. Plasma—Materials Interactions. *Plasma Deposition, Treatment, and Etching of Polymers*; d'Agostino, R., Ed.; Academic Press: San Diego, CA, USA, 1990; p. ii. [CrossRef]

131. Yasuda, H.K.; Yeh, Y.S.; Fusselman, S. A growth mechanism for the vacuum deposition of polymeric materials. *Pure nd Appl. Chem.* **1990**, *62*, 1689. [CrossRef]

132. Kirkpatrick, M.J.; Locke, B.R. Hydrogen, oxygen, and hydrogen peroxide formation in aqueous phase pulsed corona electrical discharge. *Ind. Eng. Chem. Res.* **2005**, *44*, 4243–4248. [CrossRef]

133. Locke, B.R.; Shih, K.-Y. Review of the methods to form hydrogen peroxide in electrical discharge plasma with liquid water. *Plasma Sources Sci. Technol.* **2011**, *20*, 034006. [CrossRef]

134. Rajesh, D.; Mark, J.K. A model for plasma modification of polypropylene using atmospheric pressure discharges. *J. Phys. D Appl. Phys.* **2003**, *36*, 666.

135. Fridman, A. *Plasma Chemistry*; Cambridge University Press: Cambridge, UK, 2008; pp. 848–857.

136. Inagaki, N.; Narushim, K.; Tuchida, N.; Miyazaki, K. Surface characterization of plasma-modified poly(ethylene terephthalate) film surfaces. *J. Polym. Sci. Part B Polym. Phys.* **2004**, *42*, 3727–3740. [CrossRef]

137. Williams, R.L.; Wilson, D.J.; Rhodes, N.P. Stability of plasma-treated silicone rubber and its influence on the interfacial aspects of blood compatibility. *Biomaterials* **2004**, *25*, 4659–4673. [CrossRef] [PubMed]

138. Wavhal, D.S.; Fisher, E.R. Modification of polysulfone ultrafiltration membranes by CO_2 plasma treatment. *Desalination* **2005**, *172*, 189–205. [CrossRef]

materials

MDPI

Review

A Mini Review: Recent Advances in Surface Modification of Porous Silicon

Seo Hyeon Lee [1], Jae Seung Kang [2,3,*] and Dokyoung Kim [1,4,5,6,*]

[1] Department of Biomedical Science, Graduate School, Kyung Hee University, Seoul 02447, Korea; lee19911230@gmail.com
[2] Laboratory of Vitamin C and Anti-Oxidant Immunology, Department of Anatomy and Cell Biology, College of Medicine, Seoul National University, Seoul 03080, Korea
[3] Institute of Allergy and Clinical Immunology, Medical Research Center, Seoul National University, Seoul 03080, Korea
[4] Department of Anatomy and Neurobiology, College of Medicine, Kyung Hee University, Seoul 02447, Korea
[5] Center for Converging Humanities, Kyung Hee University, Seoul 02447, Korea
[6] Biomedical Science Institute, Kyung Hee University, Seoul 02447, Korea
* Correspondence: genius29@snu.ac.kr (J.S.K.), dkim@khu.ac.kr (D.K.); Tel.: +82-02-961-0297 (D.K.)

Received: 26 November 2018; Accepted: 13 December 2018; Published: 15 December 2018

Abstract: Porous silicon has been utilized within a wide spectrum of industries, as well as being used in basic research for engineering and biomedical fields. Recently, surface modification methods have been constantly coming under the spotlight, mostly in regard to maximizing its purpose of use. Within this review, we will introduce porous silicon, the experimentation preparatory methods, the properties of the surface of porous silicon, and both more conventional as well as newly developed surface modification methods that have assisted in attempting to overcome the many drawbacks we see in the existing methods. The main aim of this review is to highlight and give useful insight into improving the properties of porous silicon, and create a focused description of the surface modification methods.

Keywords: surface modification; porous silicon; silicon surface; carbonization; oxidation

1. Introduction

Porous silicon (abbreviated as pSi) is a silicon formulation that has introduced nanopores in its microstructure. Porous silicon was discovered in the mid-1950s, and its unique physical, chemical, optical, and biological properties allowed us to develop new disciplines [1].

Porous silicon can be generated by electrochemical etching of crystalline silicon in hydrofluoric acid (HF) containing aqueous or non-aqueous electrolytes [2]. Silicon (Si) elements in Si wafer can be dissolved out to a hexafluorosilane (SiF_6^{2-}) ion in the electrochemical etching stage, with each wafer generating a different pore diameter; p-type Si wafer (micropores, <2 nm), $p^+/p^{++}/n^+$-type Si wafer (mesopores, 2–50 nm), and n^+-type Si wafer (macropores, >50 nm) [1]. So far, various types of pSi materials have been reported, including pSi chip, pSi film, and pSi micro- and nano-particles (Figure 1). The porosity, pore size, pore pattern, and particle size can be controlled by the fabrication parameters; HF concentration, current density, electrolyte composition, and wafer (dopant type, dopant density, crystallographic orientation) [3]. As compared to anodization and sonication processes, recently a new concept in electroless etching of Si powder has been reported that is an easily scalable process for the generation of pSi particles [4].

The pSi materials have been widely used in various industries and basic science. Due to a significant amount of research and the discovery of quantum confinement effects, photoluminescence, and photonic crystal properties of pSi [5–9], the focus has mostly been on creating optoelectronic materials [10,11], displays [12,13], sensors [14–16], and bio-imaging materials [17,18]. Recently, the pSi micro- and nano-particles have been applied to drug delivery systems and controlled release systems, by using the biodegradation property of pSi [18–22]. One such approach, "drug loading in the pore and surface functionalization of pSi materials with disease targeting moiety", was one of the biggest leaps in the field of drug delivery systems. In addition, the nanostructured pSi material is a promising anode material for high-performance lithium-ion batteries [23,24]. With proper surface modification, pSi materials have shown the suppression of pulverization, low volume expansion, and a long-term cycling stability in the lithiation and delithiation stages as next-generation lithium-ion batteries.

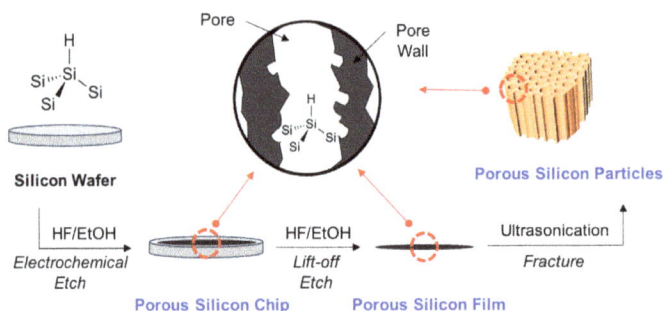

Figure 1. Schematic illustration for the preparation of porous silicon by electrochemical etching and ultrasonication. A porous silicon layer is generated on the surface of the silicon wafer using electro chemical etching, and the layer can be separated from the wafer using lift-off etch. Porous silicon micro- and nano-particles can be prepared using ultrasonication fracturing. The surface of the resulting porous silicon is covered mainly with silicon-hydrogen (Si–H) and partly with silicon-oxygen (Si–OH, Si–O–Si).

As we described above, the surface modification of pSi materials is imperative to improve the properties of pSi and its usage. Essentially, freshly etched pSi has silicon hydrides (Si–H) on the surface and residual oxides or fluorides are removed by the HF electrolyte (Figure 1). The reactive silicon hydrides on the large surface of pSi is susceptible to slow oxidation in humid air [25,26]. In the field of optoelectronics or battery application, the oxidized pSi could degrade performance of materials. However, the oxidized pSi is necessary in the development of sensors, photoluminescent bio-imaging materials, and drug-delivery systems. Accordingly, surface modification is the most important component in terms of the use of pSi materials.

In this review, we have summarized the conventional surface modification methods, newly reported surface modification methods, and our perspectives.

2. Conventional Surface Modification Methods

We categorized the well-known and typical surface modification methods into three categories: (i) hydrosilylation & carbonization, (ii) oxidation, and (iii) hydrolytic condensation.

2.1. Hydrosilylation & Carbonization

Over the last several decades, various attempts have been made at stabilizing the surface of porous silicon in order to improve its suitability for various applications [27–31]. Among them, the silicon-carbon (Si–C) bond formation yielded a very stable surface of pSi due to the low electronegativity of carbon, which possesses greater kinetic stability in comparison to silicon-oxygen (Si–O) [32]. The most ubiquitous reaction to make the Si–C bond from hydrogen-terminated pSi (Si–H) is hydrosilylation. Silicon hydride (Si–H) moiety can react with unsaturated carbon (alkene or alkyne) and bonds to one end of the hydrocarbon reagent (Figure 2) [33].

In the reported methods, the hydrosilylation reaction is achieved through thermal (heat), photon (light), catalyst, and microwave energy (Figure 2a). The thermal and light-induced hydrosilylation provides a means to place a wide variety of functional groups on a pSi surface that allow further surface functionalization, including carboxylic acid. The first hydrosilylation methods were demonstrated by Buriak and co-workers in early 1999, and elaborated by Boukherroub, Chazalviel, Lockwood, and many others afterwards [1,32–35].

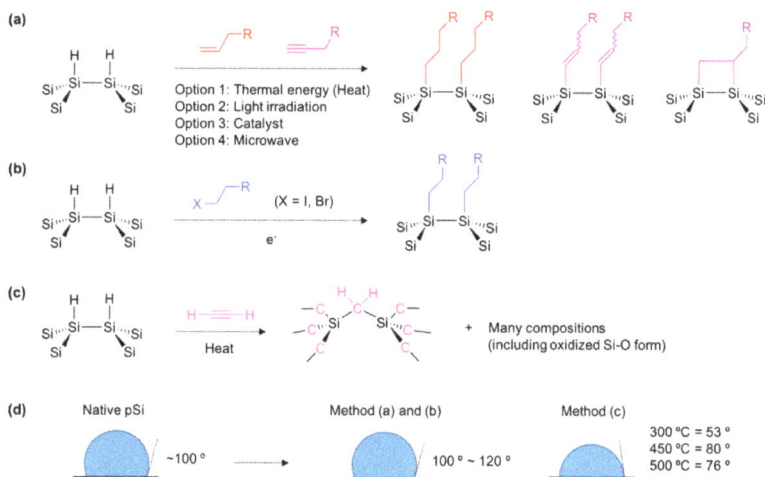

Figure 2. Surface chemistry for native porous silicon (pSi, Si–H). (**a**) Hydrosilylation with unsaturated carbon containing reagents. (**b**) Electrochemical grafting with halide containing reagents. (**c**) Carbonization with acetylene. (**d**) Water contact angle of native pSi and its surface chemistry product. R = simple aliphatic chain, functional group containing aliphatic chain.

At a similar time, Sailor and Salonen reported electrochemical reduction of organohalides and thermally-induced carbonization approaches using acetylene for the surface grafting of native pSi surfaces (Si–H) (Figure 2b,c) [36–38]. The electrochemical reduction method allows surface grafting with a methyl group, which is impossible to achieve through typical thermally induced hydrosilylation. The reaction of pSi with gas phase acetylene generates many derivatives of Si–C bonds on the surface of pSi that is stable in aqueous media. In the water contact angle measurement, typical hydrosilylation with simple aliphatic chains and an electrochemical reaction with methylhalide gives a contact angle of around 100–120° and a hydrophobic surface, but the carbonization reaction gives lower contact angle values at around 53–80°, which might vary depending on the reaction temperature, due to the undesired Si–O bond formation (Figure 2d).

The important points of hydrosilylation and carbonization reactions on the native silicon hydride terminal of the pSi surface can be summarized as: (i) maintaining porous structure, (ii) enhancing the stability of the pSi surface, (iii) making further functionalization possible, and (iv) expanding the scope of application.

2.2. Oxidation

The oxide-layer formation on the surface of pSi is also important to functionalize the materials. The hydrophobic property of native pSi became hydrophilic after the surface oxidation. This formulation can be used for bio-related applications such as biosensors, drug delivery systems, and photoluminescence bio-imaging. Generally, silicon-hydride (Si–H) and silicon-silicon (Si–Si) bonds on the surface of silicon materials can be broken with the presence of an oxidant and generates the oxide layer; hydrated silicon oxide (Si–OH) and silicon oxide (Si–O–Si) [39].

As an easy and facile method, gas-phase oxidants such as air (including oxygen) and ozone (O_3) have been widely used for the oxidation of pSi (Figure 3a,b) [25,40]. At room temperature (25 °C), a fairly thin oxide layer (Si–O–Si) grows over the course of several months in the presence of oxygen. At temperatures below 200 °C, pSi generates more hydroxy species silicon oxide layers on the surface. Oxidation at 900 °C is sufficient to completely convert the pSi to silicon oxide, which has no crystallinity of silicon, although the transformation depends on the type of pSi. Interestingly, ozone oxidation at a room temperature of 25 °C gives a more hydrated silicon oxide than the thermal oxidation procedure below 200 °C.

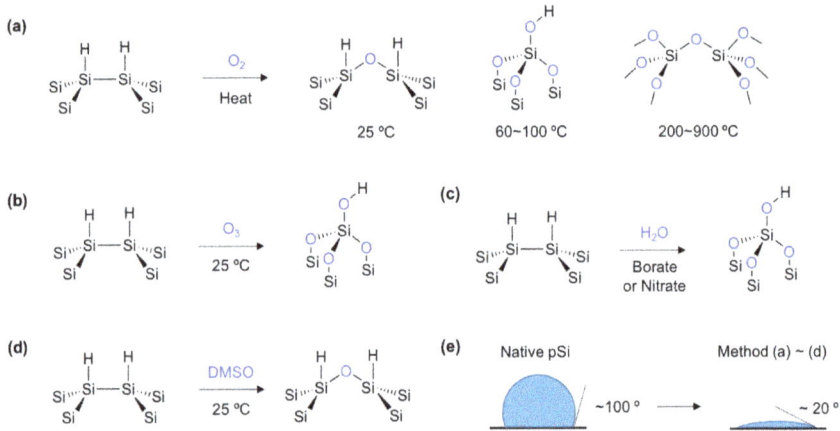

Figure 3. Surface oxidation of native porous silicon (pSi, Si–H). (**a**) Oxidation with a gas-phase oxidant, O_2, in different temperatures. (**b**) Ozone (O_3) oxidation. (**c**) Oxidation in aqueous media (H_2O) with an oxidant (borate or nitrate). (**d**) Oxidation by organic oxidant (DMSO). (**e**) Water contact angle of native pSi and its surface oxidized product.

Solution based oxidation of pSi is also widely used due to the facile operation and controllable oxidation states. In aqueous solutions, a silicon atom on the pSi surface can make a 5-coordinate intermediate with water molecules (H_2O) and its cascade condensation, giving a Si–O–Si and hydroxylated Si–OH oxide layer (Figure 3c) [41]. The oxidation rate and thickness of the oxide layer also can be boosted or regulated in the presence of an additional oxidant, such as borate [42] or nitrate [43]. Similarly, organic solvents can also be used. Dimethyl sulfoxide (DMSO) can also oxidize the surface of pSi in a mild way and generate the Si–O–Si layer on the surface [44]. The Deuterium tracer study revealed that DMSO tended to break the Si–Si rather than Si–H bond (Figure 3d).

In terms of bonding energy, the Si–Si bond is weaker than the Si–H bond, so the mild oxidant (DMSO, pyridine, etc.) [45] and thermal condition (25 °C) tends to break the Si–Si bond, and generate a Si–O–Si layer on the surface. Conversely, relatively reactive oxidants (O_3, borate, nitrate, etc.) and harsh conditions (>100 °C) tend to break both the Si–Si and Si–H and generate Si–O–Si and Si–OH. The generation of an oxide layer on the pSi surface can be monitored by the water contact angle measurement, and all oxidation products show hydrophilic properties below a 20° angle (Figure 3e).

The important point to consider about the oxidation of the native pSi surface can be summarized as: (i) enhancing hydrophilicity utilizing its bio-related work, (ii) improving functionalization as much as possible, (iii) enhancing quantum confinement effects on the surface, and (iv) enhancing the generation of the reactive Si–O intermediate to improve modification.

2.3. Hydrolytic Condensation

The oxidized pSi is a good platform for further surface modification. The hydrolytic condensation with organotrialkoxysilane reagents generates new Si–O–Si bonds on the surface in a protic solvent, such as ethanol at mild temperatures (25 °C) (Figure 4a) [46]. This simple silanol chemistry is well-known and widely used within bio-related works, such as biomolecule conjugation, PEGylation (polyethylene glycol attachment reaction), and controlled degradation of pSi materials in bio-fluid [17, 18,47–49]. In tests, there are two standard coupling reagents; (i) 3-aminopropyltriethoxy-silane (APTES, X = NH₂ in Figure 4b) [50,51], and (ii) 3-mercaptopropyltriethoxy-silane (MPTES, X = SH in Figure 4b). The hydrolytic condensation product with APTES gives primary amine terminals on the surface of pSi, and this amine moiety can be chemically conjugated using proteins, DNA, antibodies, drugs, and many other molecules, via amide coupling with carboxylic acid or *N*-hydroxysuccinimide esters (Method A) [52,53]. In a similar way, the hydrolytic condensation product with MPTES gives a thiol terminal on the surface of pSi, and it can be conjugated with substrates via a thiol-ene reaction (also known as alkene hydrothiolation) with maleimide (Method B) [54]. Both chemistries show sufficiently high yields of reaction for most of the oxidized pSi materials. Recently, further functionalization of oxidized pSi using metal-free bioorthogonal click chemistry was applied for the peptide conjugation [55,56].

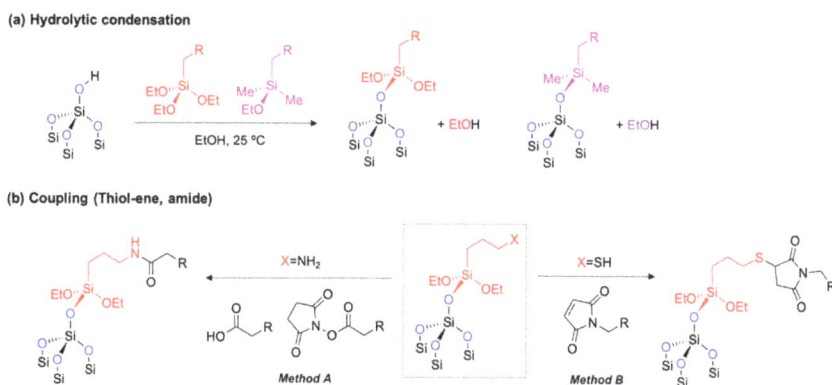

Figure 4. Surface modification of oxidized pSi. (**a**) Hydrolytic condensation with silanol reagents. (**b**) Further functionalization of hydrolytic condensation products via amide coupling (Method A) and thiol-ene addition (Method B). A box in Figure (**b**) indicates the surface functionality of hydrolytic condensation with APTES or MPTES on the oxidized pSi.

3. Recently Developed Surface Modification Methods

In the previous chapter, we summarized the existing surface modification methods for pSi materials. Although the known methods show superior surface modification abilities, the practical applications in industry and basic sciences have met difficulties for many reasons.

As an example, the thermally induced hydrosilylation of native pSi typically required high heat (>200 °C), a very specific light with sufficiently high power, an air-sensitive catalyst, such as rhodium complex (known as Wilkinson's catalyst), and special instruments. This kind of experiment could be conducted by an expert in an inert atmosphere like a nitrogen glovebox with home-built instruments. Occasionally, the resulting pSi material showed low efficiency of surface conversion and non-uniformed morphology. Surface modification by oxidation and hydrolytic condensation also came with problems, such as a loss of pSi during modification, long reaction times (> 12 h), undesired cross-linking of pSi, and low efficiency because of generation of the by-product. To overcome these issues, new surface modification methods have been developed recently, such as (i) thermally induced dehydrocoupling, (ii) ring-opening click chemistry, and (iii) calcium- or magnesium-silicate formation.

This chapter has summarized the key points of these recent advances in surface modification of pSi materials.

3.1. Thermally Induced Dehydrocoupling

The hydrosilylation reaction of the native pSi material, with alkene of alkyne moieties on the organic reagent, requires an exclusion of oxidants including air, moisture, and water in order to avoid production of an undesired silicon oxide. An alternative method is dehydrogenative coupling with a trihydridosilane reagent (H_3Si-R, R = aliphatic, aromatic) (Method A, Figure 5a) [57]. However, this dehydrogenative coupling method requires transition metal catalysts in order to effect the transformation, and the highly reactive catalysts can lead to an oxidative side reaction. Very recently, Kim and co-workers reported the non-catalytic dehydrocoupling reaction can proceed under mild thermal conditions on the pSi surface (Method B, Figure 5b) [58,59]. The first investigation was successfully carried out with the pSi film and an octadecysilane, $H_3Si(CH_2)_{17}CH_3$ (abbreviated as H_3Si-C_{18}). The pSi film was heated at 80 °C for 1–24 hours, in the presence of a neat silane reagent. Infrared spectrum analysis of native the pSi film and reaction product (pSi-Si(C_{18})) displayed bands associated with pSi-H and an organosilane reagent (Figure 6a). A strong Si lattice mode at 515 cm^{-1} was observed in the Raman spectrum of pSi-H and pSi-Si(C_{18}), which indicated a retained crystallinity of pSi after the reaction (Figure 6b). The preservation of an open pore structure was observed by scanning electron microscope (SEM) images (Figure 6c). The superhydrophobic surface property was observed in the water contact angle measurement (>150°) after rinsing with HF (Figure 6d), and it suggests that this reaction occurred between the Si element on the pSi and the Si element on the reagent, not through the oxidized pSi surface.

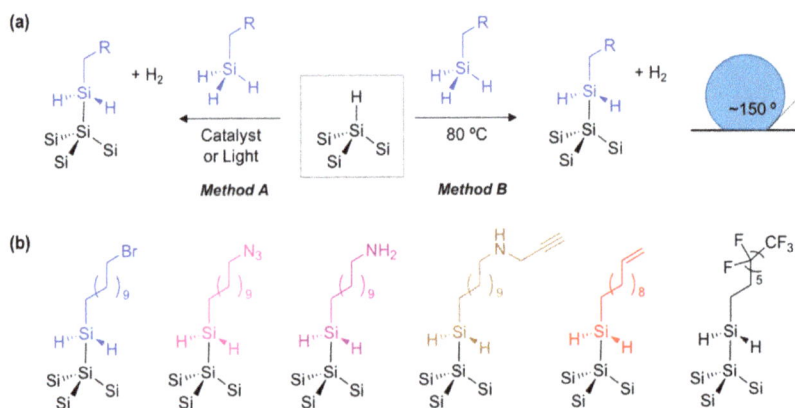

Figure 5. Thermally induced dehydrocoupling of the native pSi surface (Si–H) with trihydridosilane reagents (H_3Si-CH_2-R, marked in blue). (**a**) Dehydrocoupling by catalyst or light (Method A) and mild heat (Method B). Inset figure: water contact angle of reaction product. (**b**) Summary of reaction products of pSi with various trihydridosilane reagents.

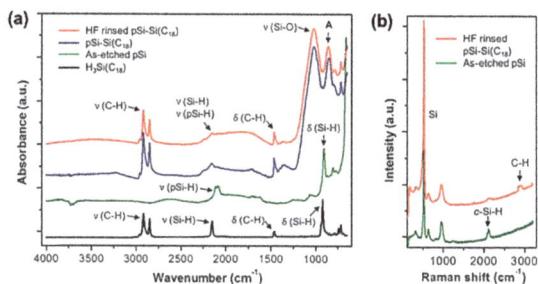

Figure 6. *Cont.*

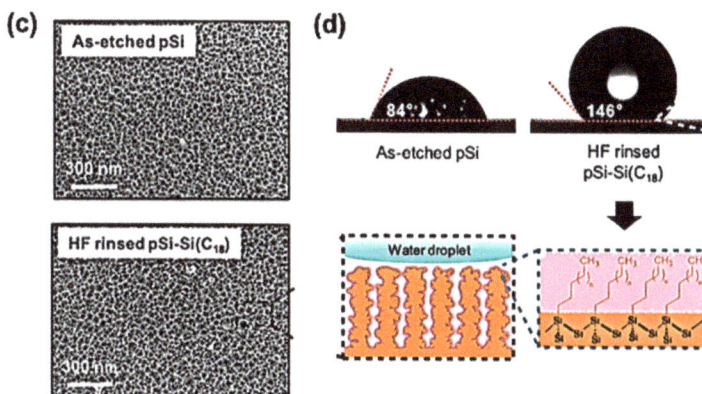

Figure 6. Characterization of a thermally induced dehydrocoupling reaction product of pSi with octadecysilane (H_3Si-C_{18}). (**a**) Attenuated total reflectance Fourier-transform infrared (ATR-FTIR) spectra of the reagent (H_3Si-C_{18}) and products (pSi-Si-C_{18}). (**b**) Raman spectra obtained before (As-etched pSi) and after reaction (HF rinsed pSi-Si-C_{18}). (**c**) Scanning electron microscope (SEM) images of pSi before (As-etched pSi) and after reaction (HF rinsed pSi-Si-C_{18}) (plan view). (**d**) Water contact angle image and illustration of the surface after reaction. Reproduced with permission from [58]. Copyright 2016 Wiley-VCH.

The thermally induced dehydrocoupling reaction with a trihydridosilane reagent shows remarkable tolerance to oxidants such as air and water. The reaction of pSi with a mixture of water and octadecysilane in open air conditions shows an adverse effect, and a grafting reaction. They also followed the effect of surface modification on the intrinsic photoluminescence quantum-confined pSi material. For this study, n-type pSi silicon film was prepared by electrochemical etching under UV light irradiation. Interestingly, the preserved photoluminescence was monitored after the reaction, and the product maintained their emissions for a long time (> 9 days) within the aerated water whilst immersed. They also demonstrated dehydrocoupling reactions with various trihydridosilane reagents and tested their suitability for further organic functionalization reaction. Many functional groups such as bromide, azide, primary amine, propargyl, and perfluorocarbon were able to be introduced onto the surface of the resulting productions. Further conjugation chemistry and electrostatic drug loading demonstrations were successfully carried out with a primary amine-terminal production toward fluorescein isothiocyanate and negatively charged drugs (ciprofloxacin in this study).

3.2. Ring-opening Click Chemistry

Driven by the desire for functionalization and stabilization of the pSi surface, the reactive Si-H surface often converted to Si–O moiety. As we described in the previous chapter, oxidized pSi surfaces can be functionalized by the grafting of organotrialkoxysilanes via a hydrolytic condensation reaction. Despite their utility, the critical limitations, such as undesired cross-linking, which results in an overly thick coating or clogging of the pore of pSi, long reaction times, and low coupling efficiency give many drawbacks to the practical applications of pSi [1,60].

In 2016, Kim and co-workers developed a facile surface modification method of oxidized pSi via ring-opening click chemistry, using 5-membered heterocyclic compounds containing Si–S or Si–N bonds within the ring (Figure 7) [61].

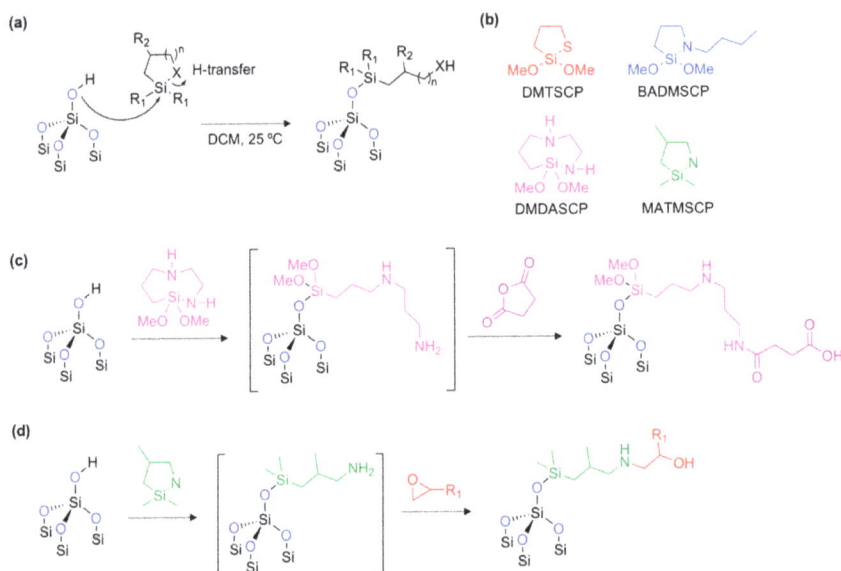

Figure 7. Ring-opening click chemistry of oxidized pSi surface (Si–OH) with cyclic-silane reagent. (a) Reaction mechanism. DCM: dicholoromethane. R_1 = OMe, Me. R_2 = H, Me. X = S, N. (b) Cyclic-silane reagents. (c,d) Further functionalization of ring-opening click chemistry products (pink and olive) via amide coupling with succinic anhydride (purple) and tandem coupling with epoxy containing reagent (red). DMTSCP: 2,2-dimethoxy-1-thia-2-silacyclopentane. BADMSCP: N-*n*-butyl-aza-2,2-dimethoxy-silacyclopentane. DMDASCP: 2,2-dimethoxy-1,6-diaza-2-silacyclooctane. MATMSCP: N-methyl-aza-2,2,4-trimethyl-silacyclopentane.

The name of the reaction originated from simple click chemistry that showed a lack of by-product with a high yield and wide scope of applicability. The hydroxy moiety on the oxidized pSi surface attacked the silicon element on the cyclic-silane reagent and the Si–X (X = S or N) bond was broken (ring-opening) with a proton transfer (H-transfer) within a mild condition (Figure 7a). Originally, this kind of click chemistry was proposed by Arkles and co-workers in Gelest Inc. USA in 2015 [62]. They found an unusually fast and efficient surface grafting ability of the cyclic azasilane reagent (Si–N bond in the ring). With promising results based on the amorphous silica materials, Kim and co-workers expanded the research field to oxidized pSi films and pSi micro- and nano-particles using other heterocyclic silanes (Figure 7b). The thermogravimetric analysis (TGA) of the resulting production, prepared through hydrolytic condensation and ring-opening click chemistry, gave ~2–4% and ~8% mass changes, representing a high efficiency for this new type of method for surface grafting.

The maintenance of the crystallinity of the silicon skeleton was then verified using a powder X-ray diffraction (XRD) measurement.

Within these promising surface grafting results, a one-pot tandem synthesis was demonstrated, using an amine-containing cyclic azasilane reagent (Figure 7c,d) [61,63]. The ring-opening click reaction generated a primary amine at the surface, which then reacted to the succinic anhydride or epoxy containing reagent. Through these tandem coupling reactions, new functional groups, such as carboxylic acid (–COOH) or hydroxyl (–OH) moiety were generated, and, therefore, showed a good case for application using bio-conjugation.

In this study, the authors also tested the compatibility of the surface functionalization with protein loaded porous silicon nano-particles (pSiNPs) (Figure 8). The nano-particle formulation of pSi maintained its pore structure, its protective sensitive payloads from denaturing in vitro or in vivo, and has proven to be a promising delivery vessel. A model protein lysozyme was loaded into the pSiNPs, and then the surface of resulting pSiNP was functionalized with thia-silane reagent in either DCM or *n*-Hex (Figure 8a). The generation of the surface thiol species was within 72 h and retained its activity; 98% (reaction in DCM) and 72% (reaction in *n*-Hex) (Figure 8b). By contrast, lysozyme loaded pSiNPs with the hydrolytic condensation reaction using MPTES gave only 68% activity of lysozyme after release.

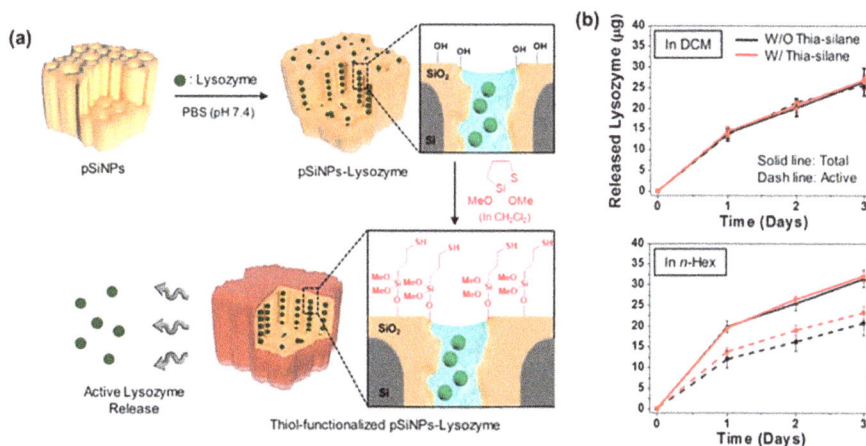

Figure 8. Application of ring-opening click chemistry for substrate loaded pSi nano-particles (pSiNPs). (a) procedure used to load lysozyme into pSiNPs and modify the resulting particles with the cyclic-silane reagent (thia-silane in this study). DCM: dichloromethane. (b) Release profile of lysozyme from unmodified pSiNPs and thia-silane functionalized pSiNPs that were prepared in different solvents, DCM or *n*-Hex. Reproduced with permission from [61]. Copyright 2016 American Chemical Society.

3.3. Calcium- or Magnesium-Silicate Formation

A formulation of a pSi core with an oxide layer shell is an attractive platform for further chemical conjugation, offering increased stability, enhanced photoluminescence intensity, and a controlled drug delivery release system. In some cases, the oxide layer formation of pSi has been used to trap the substrates within the nanostructure [43,49,64]. In 2016, Kang and co-workers reported a facile pSi surface oxidation method through the formation of a calcium silicate (Ca-silicate) insoluble shell (Figure 9) [48]. It was shown in this study that the high concentration calcium (II) ion can react with a hydroxylated silicon surface and hydrolyzed orthosilicic acid (Si(OH)$_4$) and form the Ca$_2$SiO$_4$ precipitation (calcium silicate shell) on the surface, trapping the substrate, siRNA. The calcium silicate shell could also be degraded in the aqueous media and release the substrate, but the release rate was slower than the known physical trapping method. In addition, the oxide shell formation process dramatically increased the quantum yield of pSiNPs from 0.1% to 21%. The calcium silicate formation with pore sealing was verified by transmission electron microscope (TEM) images, energy dispersive X-ray (EDC) analysis, Fourier transform infrared (FTIR) spectrum, and nitrogen adsorption–desorption isotherm analysis.

After basic characterizations, the authors went on to demonstrate a potential use for the disease specific gene delivery approach, by using a modification of the calcium silicate coated pSiNPs (Ca-pSiNPs) surface. First, the surface of the siRNA loaded Ca-pSiNPs (Ca-pSiNP-siRNA, ~20 wt % siRNA) was functionalized with 3-aminopropyl-dimethylethoxysilane (APDMES) via hydrolytic condensation. The Ca-pSiNP-siRNA was then conjugated with a bi-functional polyethyleneglycol (PEG), maleimide-PEG-succinimidyl carboxy methyl ester (MAL-PEG-SCM), via amide coupling between amine and succinimidyl carboxy methyl ester. The remaining distal end of the PEG was additionally conjugated, targeting RVG (rabies virus glycoprotein) and cell penetrating peptides (mTP; myristolated transportan). A mice study was conducted using mice with brain injuries, and it concluded that the formulation was successfully delivered to the injured site and released the active siRNA payload.

As a follow-up study, Wang and co-workers reported that a similar demonstration based on the magnesium silicate (Mg-silicate) trapping of bovine serum albumin (BSA) into the pSi micro-particles with their photoluminescence intensity analysis [20]. They found that the Mg-silicate can suppress burst releases of the payload, and this kinetic release can be tracked using the photoluminescence of pSi micro-particles.

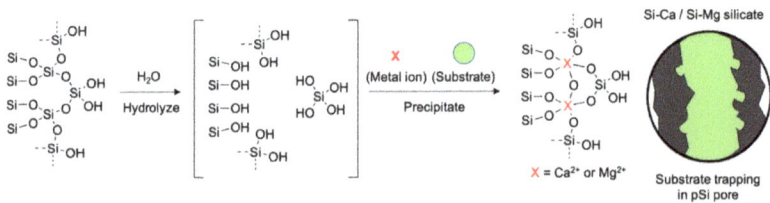

Figure 9. Schematic illustration of the calcium (Ca^{2+})- or magnesium (Mg^{2+})-silicate formation of pSiNPs surface. A thin oxide layer and orthosilicic acid ($Si(OH)_4$) is generated from the pSiNPs in aqueous media and forms a precipitate that traps the substrate (olive) within the nanostructure.

In this chapter, we cover more recently developed surface modification methods for pSi materials with their detailed mechanisms, key benefits, and applications. We also summarize the key advantages and disadvantages compared to existing surface modification methods (Table 1).

Table 1. Summary of advantages, disadvantages, and applications for each surface modification methods of pSi materials.

Methods	Advantages	Disadvantages
Carbonization (Si–H to Si–C)	- Single-step - Enhance hydrophobicity - Enhance stability	- Requires harsh reaction condition - Requires special instruments - Need practiced hands
Oxidation (Si–H to Si–OH, Si–O–Si)	- Facile method - Enhance hydrophilicity - Bio friendly	- Loss of pSi during modification - Difficult to control oxidation state - Undesired pore clogging
Hydrolytic condensation (Si–OH to Si–O–Si–R)	- Facile method - Functional group diversification - Bio friendly	- Loss of pSi during modification - Time consuming process - Undesired cross-linking of pSi
Thermal dehydrocoupling (Si–H to Si–Si–R)	- Single-step, mild condition - Enhanced hydrophobicity - Functional group diversification - High yield	- Undesired Si-O formation - Large amount of reagent (neat)
Ring opening click chemistry (Si–OH to Si–O–Si–R)	- Single-step, facile method - Functional group diversification - Inert to payload	- Reaction only in aprotic solvent - Only for hydroxylated pSi
Ca-/Mg-silicate formation (Si–OH to Si–O–Ca/Mg–O–Si)	- Single-step, facile method - High loading yield - Photoluminescence generation - Further surface modification	- Exothermic reaction - Require further modification

4. Summary and Outlook

Surface modification can provide physical- or chemical-stability, functionality, and chemically reactive sites, which can be used to interact or conjugate substrates. For this reason, there has been a desperate need for the development of a more advanced surface modification method of the materials. The existing methods have been confronted by many challenges, such as hydrosilylation, oxidation, and hydrolytic condensation. In this focused review, we introduced three recently developed and differing surface modification methods of porous silicon; hydrolytic condensation, ring-opening click chemistry, and calcium- or magnesium-silicate formation. These new methods have shown remarkable advantages compared to more traditional existing methods, in an attempt to develop this research to improve the properties of porous silicon. As scientists undertaking research within the field of material science, especially based on porous silicon, we believe that the next-generation of surface modification methods and reagents for porous silicon will have several focuses: (i) ease-of-use, (ii) high reproducibility, (iii) high bio-applicability (low toxicity), (iv) controllable surface modification ratio, (v) minimizing or maximizing the pore collapse during the modification, and (vi) multi-functional surface modified moiety. Advances in surface modification methods and reagents can expand their application throughout many different fields. In particular, those that are currently attracting attention, such as polymer fabrication, food preservation coating, cell growth regulation, tissue engineering, and antifungal coating for medical devices. In addition, this new fabrication technique is expected to be used in certain applications (i.e., water-repellent coating, photolithography, etc.). We hope this review offers some basic information about porous silicon and surface modification methods, and that it can inspire other scientists to develop more advanced methods in order to improve these materials.

Author Contributions: S.H.L., J.S.K., and D.K. designed the content and wrote this review.

Funding: This research was supported by the Bio & Medical Technology Development Program of the National Research Foundation (NRF) of Korea funded by the Ministry of Science & ICT (NRF-2018-M3A9H3021707). D. Kim gives thanks to the financial support from the Basic Science Research Program through the National Research Foundation (NRF) of Korea funded by the ministry of Education (NRF-2018-R1A6A1A03025124, NRF-2018-R1D1A1B07043383). J.S. Kang gives thanks to the financial support from the Basic Science Research Program through the National Research Foundation (NRF) of Korea funded by the Ministry of Education (2017R1A2B2010948).

Acknowledgments: Thanks to Neil P. George and Sujin Jung for the linguistic editing.

Conflicts of Interest: The authors declare no conflict of interest.

References

1. Sailor, M.J. *Porous Silicon in Practice: Preparation, Characterization and Applications*; John Wiley & Sons: Hoboken, NJ, USA, 2012.

2. Dubey, R.S.; Gautam, D.K. Porous silicon layers prepared by electrochemical etching for application in silicon thin film solar cells. *Superlattices Microst.* **2011**, *50*, 269–276. [CrossRef]

3. Korotcenkov, G.; Cho, B.K. Silicon Porosification: State of the Art. *Crit. Rev. Solid State Mater. Sci.* **2010**, *35*, 153–260. [CrossRef]

4. Kolasinski, K.W.; Gimbar, N.J.; Yu, H.; Aindow, M.; Mäkilä, E.; Salonen, J. Regenerative Electroless Etching of Silicon. *Angew. Chem. Int. Ed.* **2017**, *56*, 624–627. [CrossRef] [PubMed]

5. Lockwood, D.J.; Wang, A.G. Quantum confinement induced photoluminescence in porous silicon. *Solid State Commun.* **1995**, *94*, 905–909. [CrossRef]

6. Canham, L.T. Silicon quantum wire array fabrication by electrochemical and chemical dissolution of wafers. *Appl. Phys. Lett.* **1990**, *57*, 1046–1048. [CrossRef]

7. Lehmann, V.; Gösele, U. Porous silicon formation: A quantum wire effect. *Appl. Phys. Lett.* **1991**, *58*, 856–858. [CrossRef]

8. Estevez, J.O.; Agarwal, V. Porous Silicon Photonic Crystals. In *Handbook of Porous Silicon*; Canham, L., Ed.; Springer International Publishing: Cham, Switzerland, 2014; pp. 805–814.

9. Joo, J.; Defforge, T.; Loni, A.; Kim, D.; Li, Z.Y.; Sailor, M.J.; Gautier, G.; Canham, L.T. Enhanced quantum yield of photoluminescent porous silicon prepared by supercritical drying. *Appl. Phys. Lett.* **2016**, *108*, 153111. [CrossRef]

10. Hérino, R. Porous silicon for microelectronics and optoelectronics. *Mater. Sci. Technol.* **1997**, *13*, 965–970. [CrossRef]

11. Canham, L.T. Nanostructured Silicon as an Active Optoelectronic Material. In *Frontiers of Nano-Optoelectronic Systems*; Pavesi, L., Buzaneva, E., Eds.; Springer: Dordrecht, The Netherlands, 2000; pp. 85–97.

12. Kleps, I.; Nicolaescu, D.; Lungu, C.; Musa, G.; Bostan, C.; Caccavale, F. Porous silicon field emitters for display applications. *Appl. Surf. Sci.* **1997**, *111*, 228–232. [CrossRef]

13. Fauchet, P.M.; Tsybeskov, L.; Peng, C.; Duttagupta, S.P.; Behren, J.V.; Kostoulas, Y.; Vandyshev, J.M.V.; Hirschman, K.D. Light-emitting porous silicon: Materials science, properties, and device applications. *IEEE J. Sel. Top. Quantum Electron.* **1995**, *1*, 1126–1139. [CrossRef]

14. Harraz, F.A. Porous silicon chemical sensors and biosensors: A review. *Sens. Actuator B-Chem.* **2014**, *202*, 897–912. [CrossRef]

15. Tsamis, C.; Nassiopoulou, A.G. *Porous Silicon for Chemical Sensors*; Springer: Dordrecht, The Netherlands, 2005; pp. 399–408.

16. Song, J.H.; Sailor, M.J. Quenching of Photoluminescence from Porous Silicon by Aromatic Molecules. *J. Am. Chem. Soc.* **1997**, *119*, 7381–7385. [CrossRef]

17. Kim, D.; Kang, J.; Wang, T.; Ryu, H.G.; Zuidema, J.M.; Joo, J.; Kim, M.; Huh, Y.; Jung, J.; Ahn, K.H.; et al. Two-Photon In Vivo Imaging with Porous Silicon Nanoparticles. *Adv. Mater.* **2017**, *29*, 1703309. [CrossRef] [PubMed]

18. Park, J.-H.; Gu, L.; von Maltzahn, G.; Ruoslahti, E.; Bhatia, S.N.; Sailor, M.J. Biodegradable luminescent porous silicon nanoparticles for in vivo applications. *Nat. Mater.* **2009**, *8*, 331. [CrossRef] [PubMed]

19. Kumeria, T.; McInnes, S.J.P.; Maher, S.; Santos, A. Porous silicon for drug delivery applications and theranostics: Recent advances, critical review and perspectives. *Expert Opin. Drug Deliv.* **2017**, *14*, 1407–1422. [CrossRef] [PubMed]

20. Wang, J.; Kumeria, T.; Bezem, M.T.; Wang, J.; Sailor, M.J. Self-Reporting Photoluminescent Porous Silicon Microparticles for Drug Delivery. *ACS Appl. Mater. Interfaces* **2018**, *10*, 3200–3209. [CrossRef] [PubMed]

21. Nieto, A.; Hou, H.; Moon, S.W.; Sailor, M.J.; Freeman, W.R.; Cheng, L. Surface Engineering of Porous Silicon Microparticles for Intravitreal Sustained Delivery of Rapamycin. *Investig. Ophthalmol. Vis. Sci.* **2015**, *56*, 1070–1080. [CrossRef] [PubMed]

22. Gu, L.; Hall, D.J.; Qin, Z.; Anglin, E.; Joo, J.; Mooney, D.J.; Howell, S.B.; Sailor, M.J. In vivo time-gated fluorescence imaging with biodegradable luminescent porous silicon nanoparticles. *Nat. Commun.* **2013**, *4*, 2326. [CrossRef]

23. Li, X.; Gu, M.; Hu, S.; Kennard, R.; Yan, P.; Chen, X.; Wang, C.; Sailor, M.J.; Zhang, J.-G.; Liu, J. Mesoporous silicon sponge as an anti-pulverization structure for high-performance lithium-ion battery anodes. *Nat. Commun.* **2014**, *5*, 4105. [CrossRef]

24. Ge, M.; Fang, X.; Rong, J.; Zhou, C. Review of porous silicon preparation and its application for lithium-ion battery anodes. *Nanotechnology* **2013**, *24*, 422001. [CrossRef]

25. Ogata, Y.; Niki, H.; Sakka, T.; Iwasaki, M. Oxidation of Porous Silicon under Water Vapor Environment. *J. Electrochem. Soc.* **1995**, *142*, 1595–1601. [CrossRef]

26. Mawhinney, D.B.; Glass, J.A.; Yates, J.T. FTIR Study of the Oxidation of Porous Silicon. *J. Phys. Chem. B* **1997**, *101*, 1202–1206. [CrossRef]

27. Peng, W.; Rupich, S.M.; Shafiq, N.; Gartstein, Y.N.; Malko, A.V.; Chabal, Y.J. Silicon Surface Modification and Characterization for Emergent Photovoltaic Applications Based on Energy Transfer. *Chem. Rev.* **2015**, *115*, 12764–12796. [CrossRef] [PubMed]

28. Thissen, P.; Seitz, O.; Chabal, Y.J. Wet chemical surface functionalization of oxide-free silicon. *Prog. Surf. Sci.* **2012**, *87*, 272–290. [CrossRef]

29. Wong, K.T.; Lewis, N.S. What a Difference a Bond Makes: The Structural, Chemical, and Physical Properties of Methyl-Terminated Si(111) Surfaces. *Acc. Chem. Res.* **2014**, *47*, 3037–3044. [CrossRef] [PubMed]

30. Bhairamadgi, N.S.; Pujari, S.P.; Trovela, F.G.; Debrassi, A.; Khamis, A.A.; Alonso, J.M.; Al Zahrani, A.A.; Wennekes, T.; Al-Turaif, H.A.; van Rijn, C.; et al. Hydrolytic and Thermal Stability of Organic Monolayers on Various Inorganic Substrates. *Langmuir* **2014**, *30*, 5829–5839. [CrossRef] [PubMed]

31. Li, Y.; Calder, S.; Yaffe, O.; Cahen, D.; Haick, H.; Kronik, L.; Zuilhof, H. Hybrids of Organic Molecules and Flat, Oxide-Free Silicon: High-Density Monolayers, Electronic Properties, and Functionalization. *Langmuir* **2012**, *28*, 9920–9929. [CrossRef]

32. Buriak, J.M. Organometallic chemistry on silicon and germanium surfaces. *Chem. Rev.* **2002**, *102*, 1271–1308. [CrossRef]

33. Ciampi, S.; Harper, J.B.; Gooding, J.J. Wet chemical routes to the assembly of organic monolayers on silicon surfaces via the formation of Si–C bonds: Surface preparation, passivation and functionalization. *Chem. Soc. Rev.* **2010**, *39*, 2158–2183. [CrossRef]

34. Boukherroub, R.; Wojtyk, J.T.C.; Wayner, D.D.M.; Lockwood, D.J. Thermal Hydrosilylation of Undecylenic Acid with Porous Silicon. *J. Electrochem. Soc.* **2002**, *149*, H59–H63. [CrossRef]

35. Buriak, J.M. Illuminating Silicon Surface Hydrosilylation: An Unexpected Plurality of Mechanisms. *Chem. Mater.* **2014**, *26*, 763–772. [CrossRef]

36. Gurtner, C.; Wun, A.W.; Sailor, M.J. Surface Modification of Porous Silicon by Electrochemical Reduction of Organo Halides. *Angew. Chem. Int. Ed.* **1999**, *38*, 1966–1968. [CrossRef]

37. Lees, I.N.; Lin, H.; Canaria, C.A.; Gurtner, C.; Sailor, M.J.; Miskelly, G.M. Chemical Stability of Porous Silicon Surfaces Electrochemically Modified with Functional Alkyl Species. *Langmuir* **2003**, *19*, 9812–9817. [CrossRef]

38. Salonen, J.; Lehto, V.-P.; Björkqvist, M.; Laine, E.; Niinistö, L. Studies of Thermally-Carbonized Porous Silicon Surfaces. *Phys. Stat. Solidi (a)* **2000**, *182*, 123–126. [CrossRef]

39. Canaria, C.A.; Huang, M.; Cho, Y.; Heinrich, J.L.; Lee, L.I.; Shane, M.J.; Smith, R.C.; Sailor, M.J.; Miskelly, G.M. The Effect of Surfactants on the Reactivity and Photophysics of Luminescent Nanocrystalline Porous Silicon. *Adv. Funct. Mater.* **2002**, *12*, 495–500. [CrossRef]

40. Frotscher, U.; Rossow, U.; Ebert, M.; Pietryga, C.; Richter, W.; Berger, M.G.; Arens-Fischer, R.; Münder, H. Investigation of different oxidation processes for porous silicon studied by spectroscopic ellipsometry. *Thin Solid Films* **1996**, *276*, 36–39. [CrossRef]

41. Gupta, P.; Dillon, A.C.; Bracker, A.S.; George, S.M. FTIR studies of H_2O and D_2O decomposition on porous silicon surfaces. *Surf. Sci.* **1991**, *245*, 360–372. [CrossRef]

42. Joo, J.; Cruz, J.F.; Vijayakumar, S.; Grondek, J.; Sailor, M.J. Photoluminescent porous Si/SiO_2 core/shell nanoparticles prepared by borate oxidation. *Adv. Funct. Mater.* **2014**, *24*, 5688–5694. [CrossRef]

43. Fry, N.L.; Boss, G.R.; Sailor, M.J. Oxidation-induced trapping of drugs in porous silicon microparticles. *Chem. Mater.* **2014**, *26*, 2758–2764. [CrossRef]

44. Song, J.H.; Sailor, M.J. Dimethyl sulfoxide as a mild oxidizing agent for porous silicon and its effect on photoluminescence. *Inorg. Chem.* **1998**, *37*, 3355–3360. [CrossRef]
45. Mattei, G.; Alieva, E.V.; Petrov, J.E.; Yakovlev, V.A. Quick Oxidation of Porous Silicon in Presence of Pyridine Vapor. *Phys. Status Solidi A-Appl. Res.* **2000**, *182*, 139–143. [CrossRef]
46. Schwartz, M.P.; Cunin, F.; Cheung, R.W.; Sailor, M.J. Chemical modification of silicon surfaces for biological applications. *Phys. Status Solidi A-Appl. Res.* **2005**, *202*, 1380–1384. [CrossRef]
47. Kwon, E.J.; Skalak, M.; Bertucci, A.; Braun, G.; Ricci, F.; Ruoslahti, E.; Sailor, M.J.; Bhatia, S.N. Silicon Nanoparticles: Porous Silicon Nanoparticle Delivery of Tandem Peptide Anti-Infectives for the Treatment of *Pseudomonas aeruginosa* Lung Infections (Adv. Mater. 35/2017). *Adv. Mater.* **2017**, *29*, 1701527. [CrossRef]
48. Kang, J.; Joo, J.; Kwon, E.J.; Skalak, M.; Hussain, S.; She, Z.G.; Ruoslahti, E.; Bhatia, S.N.; Sailor, M.J. Self-Sealing Porous Silicon-Calcium Silicate Core–Shell Nanoparticles for Targeted siRNA Delivery to the Injured Brain. *Adv. Mater.* **2016**, *28*, 7962–7969. [CrossRef] [PubMed]
49. Hussain, S.; Joo, J.; Kang, J.; Kim, B.; Braun, G.B.; She, Z.-G.; Kim, D.; Mann, A.P.; Mölder, T.; Teesalu, T.; et al. Antibiotic-loaded nanoparticles targeted to the site of infection enhance antibacterial efficacy. *Nat. Biomed. Eng.* **2018**, *2*, 95. [CrossRef] [PubMed]
50. Anderson, A.S.; Dattelbaum, A.M.; Montaño, G.A.; Price, D.N.; Schmidt, J.G.; Martinez, J.S.; Grace, W.K.; Grace, K.M.; Swanson, B.I. Functional PEG-Modified Thin Films for Biological Detection. *Langmuir* **2008**, *24*, 2240–2247. [CrossRef] [PubMed]
51. Nijdam, A.J.; Cheng, M.M.C.; Geho, D.H.; Fedele, R.; Herrmann, P.; Killian, K.; Espina, V.; Petricoin, E.F.; Liotta, L.A.; Ferrari, M. Physicochemically modified silicon as a substrate for protein microarrays. *Biomaterials* **2007**, *28*, 550–558. [CrossRef]
52. Anderson, G.W.; Zimmerman, J.E.; Callahan, F.M. N-Hydroxysuccinimide Esters in Peptide Synthesis. *J. Am. Chem. Soc.* **1963**, *85*, 3039. [CrossRef]
53. Valeur, E.; Bradley, M. Amide bond formation: Beyond the myth of coupling reagents. *Chem. Soc. Rev.* **2009**, *38*, 606–631. [CrossRef]
54. Nair, D.P.; Podgórski, M.; Chatani, S.; Gong, T.; Xi, W.; Fenoli, C.R.; Bowman, C.N. The Thiol-Michael Addition Click Reaction: A Powerful and Widely Used Tool in Materials Chemistry. *Chem. Mater.* **2014**, *26*, 724–744. [CrossRef]
55. Wang, C.-F.; Sarparanta, M.P.; Mäkilä, E.M.; Hyvönen, M.L.K.; Laakkonen, P.M.; Salonen, J.J.; Hirvonen, J.T.; Airaksinen, A.J.; Santos, H.A. Multifunctional porous silicon nanoparticles for cancer theranostics. *Biomaterials* **2015**, *48*, 108–118. [CrossRef] [PubMed]
56. Liu, D.; Zhang, H.; Mäkilä, E.; Fan, J.; Herranz-Blanco, B.; Wang, C.-F.; Rosa, R.; Ribeiro, A.J.; Salonen, J.; Hirvonen, J.; et al. Microfluidic assisted one-step fabrication of porous silicon@acetalated dextran nanocomposites for precisely controlled combination chemotherapy. *Biomaterials* **2015**, *39*, 249–259. [CrossRef] [PubMed]
57. Li, Y.-H.; Buriak, J.M. Dehydrogenative Silane Coupling on Silicon Surfaces via Early Transition Metal Catalysis. *Inorg. Chem.* **2006**, *45*, 1096–1102. [CrossRef]
58. Kim, D.; Joo, J.; Pan, Y.; Boarino, A.; Jun, Y.W.; Ahn, K.H.; Arkles, B.; Sailor, M.J. Thermally induced silane dehydrocoupling on silicon nanostructures. *Angew. Chem. Int. Ed.* **2016**, *128*, 6533–6537. [CrossRef]
59. Mao, Y.; Kim, D.; Hopson, R.; Sailor, M.J.; Wang, L.-Q. Investigation of grafted mesoporous silicon sponge using hyperpolarized 129 Xe NMR spectroscopy. *J. Mater. Res.* **2018**, *33*, 2637–2645. [CrossRef]
60. Acres, R.G.; Ellis, A.V.; Alvino, J.; Lenahan, C.E.; Khodakov, D.A.; Metha, G.F.; Andersson, G.G. Molecular Structure of 3-Aminopropyltriethoxysilane Layers Formed on Silanol-Terminated Silicon Surfaces. *J. Phys. Chem. C* **2012**, *116*, 6289–6297. [CrossRef]
61. Kim, D.; Zuidema, J.M.; Kang, J.; Pan, Y.; Wu, L.; Warther, D.; Arkles, B.; Sailor, M.J. Facile surface modification of hydroxylated silicon nanostructures using heterocyclic silanes. *J. Am. Chem. Soc.* **2016**, *138*, 15106–15109. [CrossRef]
62. Maddox, A.F.; Matisons, J.G.; Singh, M.; Zazyczny, J.; Arkles, B. Single Molecular Layer Adaption of Interfacial Surfaces by Cyclic Azasilane "Click-Chemistry". *MRS Online Proc. Libr. Arch.* **2015**, *1793*, 35–40. [CrossRef]

63. Pan, Y.; Maddox, A.; Min, T.; Gonzaga, F.; Goff, J.; Arkles, B. Surface-Triggered Tandem Coupling Reactions of Cyclic Azasilanes. *Chem. Asian J.* **2017**, *12*, 1198–1203. [CrossRef]

64. Kang, J.; Kim, D.; Wang, J.; Han, Y.; Zuidema, J.M.; Hariri, A.; Park, J.H.; Jokerst, J.V.; Sailor, M.J. Enhanced Performance of a Molecular Photoacoustic Imaging Agent by Encapsulation in Mesoporous Silicon Nanoparticles. *Adv. Mater.* **2018**, 1800512. [CrossRef]

materials

MDPI

Review

Superhydrophobic Natural and Artificial Surfaces—A Structural Approach

Roxana-Elena Avrămescu, Mihaela Violeta Ghica *, Cristina Dinu-Pîrvu, Răzvan Prisada and Lăcrămioara Popa

Department of Physical and Colloidal Chemistry, Faculty of Pharmacy, University of Medicine and Pharmacy "Carol Davila", Bucharest 020956, Romania; roxana.avramescu@drd.umfcd.ro (R.-E.A.);
ecristinaparvu@yahoo.com (C.D.-P.); razvan.prisada@gmail.com (R.P.); lacramioara.popa@umfcd.ro (L.P.)
* Correspondence: mihaelaghica@umfcd.ro

Received: 27 April 2018; Accepted: 18 May 2018; Published: 22 May 2018

Abstract: Since ancient times humans observed animal and plants features and tried to adapt them according to their own needs. Biomimetics represents the foundation of many inventions from various fields: From transportation devices (helicopter, airplane, submarine) and flying techniques, to sports' wear industry (swimming suits, scuba diving gear, Velcro closure system), bullet proof vests made from Kevlar etc. It is true that nature provides numerous noteworthy models (shark skin, spider web, lotus leaves), referring both to the plant and animal kingdom. This review paper summarizes a few of "nature's interventions" in human evolution, regarding understanding of surface wettability and development of innovative special surfaces. Empirical models are described in order to reveal the science behind special wettable surfaces (superhydrophobic /superhydrophilic). Materials and methods used in order to artificially obtain special wettable surfaces are described in correlation with plants' and animals' unique features. Emphasis is placed on joining superhydrophobic and superhydrophilic surfaces, with important applications in cell culturing, microorganism isolation/separation and molecule screening techniques. Bio-inspired wettability is presented as a constitutive part of traditional devices/systems, intended to improve their characteristics and extend performances.

Keywords: special surfaces; wettability; superhydrophobic; cell cultures; anti-bio adhesion; self-cleaning fabrics

---

1. Introduction

Over the last decades, surface and interfacial phenomena gained interest among researchers, especially through applications which ease medical or industrial procedures, giving the latter the attribute of being "environmental friendly". The present paper focuses on the progress made in understanding and discovering new superficial properties of certain surfaces. The journey into the micronical size world begins with a short introduction into basic surface chemistry elements, in relation with the chemical structure of solids. Understanding surface phenomena makes it easier to unravel some yet unexplained superficial behavior and represents the starting point in developing useful applications for human kind. A detailed description of surfaces encountered in the natural (vegetal/animal) environment, is followed by a summary of methods used to obtain solid supports exhibiting special surface properties. Fast development in wettable surface engineering led to the discovery of novel applications in medical fields (biomolecule monitoring, cancer cell isolation), transportation, cleaning and other industries.

2. Superficial Properties

2.1. Special Wettable Surfaces

Surface wettability characterizes the interfacial phenomena between a liquid and a solid support. The liquid's behavior is in fact the wettability indicator. This superficial property is studied in order to establish hydrophilicity/hydrophobicity of a solid, offering an open gate to numerous every-day life applications (self-cleaning/anti-bacterial fabrics, anti-reflection, transparent coatings) and also industrial ones (anti-icing surfaces, special surface patterns). Superhydrophobic surfaces which display a contact angle higher than 150°, or 180°, (according to other authors) and a sliding angle smaller than 10° attracted attention of researchers. They were first observed in the natural environment (Lotus leave, butterfly wings, fish scales etc.). Following the principles of biomimetics, special superficial surface properties were adapted to human necessities and used as a model in many industrial areas, including nanotechnologies. In time, many applications related to superhydrophobicity were unraveled: Development of self-cleaning and low friction surfaces, satellite antennas, solar and photovoltaic panels, exterior glass, swimming suits etc. Studies show that these superhydrophobic surfaces have many other attributes: They prevent bacteria adhesion, metal corrosion, improve blood type compatibility, lower surface icing in humid atmosphere and low temperature conditions, are constitutive parts of water storage systems and of microreactors, in which new reaction compounds are produced [1].

Apart from superhydrophobic surfaces, superoleophobic ones gain researchers' attention. The suffix "oleo-" refers to liquids of low-surface tension (oils) and other organic liquids. Superoleophobicity represents the wetting phenomenon characterized through oil droplets displaying low surface tension on solid supports, along with contact angle values greater than 150°. Similar to superhydrophobic surfaces, superoleophobic ones find their applicability in self-cleaning, oil–water separation, controllable oil adhesion, oil caption etc. Superhydrophobic surfaces properties and applications will be accompanied by brief correlations with the superoleophobicity fast developing filed [2].

2.2. Superhydrophobic Surfaces' Structure

A complete understanding of surface properties was mandatory before discovering numerous applications of superhydrophobic surfaces. The first experiments developed in this direction involved studying the structure of certain substances which confer special wettability. Thus, it has come to the hypothesis that the chemical structure of a solid support is responsible for surface heterogeneity and roughness. By manipulating these properties, one depending on the other, diverse surfaces with different properties are obtained.

In order for a surface to be called "superhydrophobic", three conditions should be fulfilled: High apparent contact angle, low contact angle hysteresis and a high stability of the Cassie wetting state. A very interesting structure related to superhydrophobic surfaces was described by Mahadevan and Pomeau [3]. They observed that liquid droplets (water) rolled onto a hydrophobic powder bed (*Lycopodium*), result in formations called "liquid marbles", exhibiting a superhydrophobic-like behavior.

Contrariwise, in some cases, a drop maintained stable on a hydrophobic support can be mistaken with Leidenfrost droplets which slide off a heated support, due to the so called "vapor film levitation". It happens only as long as the support is heated over a certain temperature (the Leidenfrost temperature) and the film disappears as the stand cools. Vakarelski et al. (2012) [4] prove that superhydrophobic surface topography is important when stabilizing the vapor layer, implicitly in liquid-gas transitions on heated surfaces. The explanation lies in the fact that rugosity and porosities sustain the vapor layer, as the drop only makes contact with the rugosities' tips. This hypothesis was adjusted, as the following demonstration was made: The contact angle attains 180° and the levitating film regime is possible at superheat temperatures. Thus, special coatings (superhydrophobic, superamphiphobic, anti-frost etc.) were developed, allowing optimal heat exchange and aqueous drag reduction.

A transition from the surface phenomena analysis, to the study of molecular interactions reveals that the strength of the hydrophobic interactions between molecules is influenced by the ionic charge. Structural modification of hydrophobic surfaces follows an optimization towards molecular recognition processes, i.e., the ability to manipulate interactions between proteins. Negative ions inserted on to hydrophobic binding sites at the antibody's surface, generate special antibodies witch link to beta-amyloid fibrillar fragments. Thus, fibers cannot link to each other anymore, and the senile plate which induces Alzheimer's disease no longer forms [5]. In direct correlation with cell membrane formation and protein folding, is the hydrophobic hydration phenomenon. It is also responsible for improving the hydrogen-bonding network between water molecules surrounding hydrophobic radicals. Davis et al. (2012) [6] state that the structure formed by hydrogen bonds around hydrophobic groups disappears along with the increase of temperature. The tendency of hydrophobic compounds to dispose of as "clusters" in an aqueous media is a key phenomenon, paving the way into understanding biomolecules' dynamics.

Natural superhydrophobic surfaces exhibit hierarchical roughness at two scale ranges: Micro-and nano-roughness, as earlier presented. Hierarchical structures are unique in conferring quality superhydrophobic attributes due to nano-scaled asperities imbedded on top of micro-scaled rugosities. These rugosities stabilize the Cassie state and lower contact angle hysteresis. Experiments regarding artificial obtaining of superhydrophobic surfaces reveal that by decreasing surface roughness, the contact area between the drop and the support increases. Thus, any damage done to the hydrophobic surface affects the hydrophobicity and leads to unstable Cassie states or unwanted increasing contact angle hysteresis [7,8]. Experimental studies by Bhushan et al. [9] show that hierarchically structured plant surfaces exhibit both adhesive and non-adhesive properties. Water droplets penetrate into the micro-rugosities. Thus, they strongly adhere to the surface. Nano-rugosities are responsible for the high contact angle values. High contact angles coexist with strong adhesion to the same surface. The well-known wetting regimes: Wenzel, Cassie, Lotus and Petal may exhibit both nano- and micro-filled structures, resulting in nine wetting scenarios: Lotus, Rose petal, Rose filled microstructure, Cassie, Wenzel, Cassie filled nano- and microstructure, Wenzel filled nano- and microstructure.

Experiments by Zimmermann et al. report that superhydrophobic surfaces' properties (performance and durability) are improved by annealing. Other investigations show how hierarchical surfaces are fabricated on silicon by etching with KOH and catalyzed etching HF/H_2O_2. Rugosities stand in nanostructures build on micro-pyramids. If the surface undergoes abrasion, its self-cleaning properties are reduced and hysteresis increases [7,8].

2.3. Superficial Energy: Empirical Models Describing Surface Phenomena

Even if superhydrophobic surfaces are ubiquitous in the environment, advanced techniques were needed to fully understand them. In order to elucidate the structure of the hierarchical surface at a micro- and nano-metric level, the research went into detail. Techniques such as: Scanning electron microscopy (SEM), transmission electron microscopy (TEM) and atomic force microscopy (AFM) were used. Thus, the obtained data along with already known surface properties, concluded that a surface is superhydrophobic if it has low surface energy and a hierarchical nano-metric structure, conferring a water contact angle greater than 150°. Young's equation (Equation (1)) describes the equilibrium between the forces acting on a droplet placed on a solid support [1]:

$$\cos \theta = \frac{\gamma_{SV} - \gamma_{SL}}{\gamma_{LV}} \tag{1}$$

where θ is the contact angle, γ_{SV} is the solid-vapor superficial energy, γ_{SL} is the solid–liquid superficial energy and γ_{LV} is the liquid-vapor superficial energy.

Taking into account the fact that a minimal solid-gas superficial energy leads to a maximum contact angle, a list of surface energies for some chemical groups was established: $-CH_2- > -CH_3 > -CF_2-CF_2H > -CF_3$. Nishino et al. (1999) [10], measures a minimum surface energy and a $120°$ contact angle corresponding to regularly arranged, close and packed groups of $-CF_3$. However, Young's equation can only be applied to smooth, homogeneous surfaces and inert to the fluid they come in contact with, as showed in Figure 1a. Following this principle, the surfaces encountered in nature do not follow Young's wetting regime.

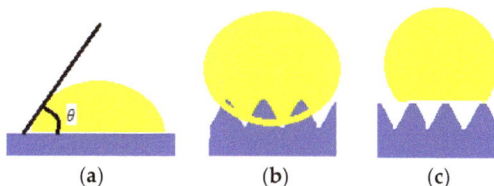

| (a) | (b) | (c) |

Figure 1. (a) Young wetting regime; (b) Wenzel wetting regime; (c) Cassie wetting regime.

In 1936, Wenzel [11] proposed an adaptation after Young's equation (Equation (1)). The wetting regime is presented in Figure 1b. It considers a roughness factor r, defined as ratio of the actual area of the rough surface to the geometric area projected on a relatively smooth surface, and the adapted (apparent) contact angle θ' is given by (Equation (2)):

$$\cos\theta' = \frac{r(\gamma_{SV} - \gamma_{SL})}{\gamma_{LV}} = r\cos\theta \tag{2}$$

The Wenzel model applies both to hydrophobicity and to hydrophilicity, where r is a roughness measure favoring both states.

Another model describing the behavior of the liquid droplet in contact with a solid support is the Cassie–Baxter model (1944) [12]. In this case, the surface displays rugosities, and air "pockets" between them, which a liquid drop cannot penetrate, as displayed in Figure 1c.

Calculation of the adapted contact angle θ' considers the surface in direct contact with the liquid. The fraction f is calculated as follows (Equation (3)):

$$f = \frac{\sum a}{\sum(a+b)} \tag{3}$$

where a and b are the contact areas with the drop (a) and respectively air (b). $(1-f)$ stands for the drop-air contact area. Considering a contact angle of $180°$, the calculation expression is (Equation (4)):

$$\cos\theta' = f\cos\theta + (1-f)\cos 180° = f\cos\theta + f - 1 \tag{4}$$

Werner's model (2005) [13] takes into account the possibility of the drop's penetration between the rugosities of the support. It shows a continuous increase of the contact angle, due to heavily hydrophobic "pockets" of air, which promote hydrophobicity, but only in case of rugged supports.

According to some authors, Quéré et al. (2003, 2004) [14,15] there is a critical value of the fraction f, (respectively of the critical contact angle θ_c), under which the Cassie model can exist, and the Wenzel model is thermodynamically more stable. This state is in fact evidence that the Wenzel regime is the state of equilibrium of the Cassie model. The corresponding critical contact angle θ_c is determined by the following equation (Equation (5)):

$$\cos\theta_c = \frac{1-f}{r-f} \tag{5}$$

In 2006, Yang et al. [16] develop a study with droplets placed in contact with surfaces displaying a fractal structure. It provides evidence that the contact angle depends on the average square root of surface roughness and is independent of the fractal dimension of surface D_f at a nano-metric level.

Researches' opinions are divided when discussing wetting states equilibrium or transitions.

More recent studies (2017) show that both Cassie and Wenzel regimes are meta-stable and co-occur at the same surface. A "bi-stable" wetting regime is assigned to hydrophobic textured (linear, pillar) surfaces. Experiments show the following: A Cassie levitating state corresponds to drops placed onto a hydrophobic support, whilst a Wenzel "impaled" (pinned) state refers to drops after an impact with the same surface. Wetting transitions between these states were reported as being responsible for spontaneous/external stimuli (pressure, vibration) triggered changes of a drop's contact angle. This barrier is a result of increasing liquid-air interface, as the liquid penetrates through the support. The value of the energy barrier separating Cassie and Wenzel states is attributed to a hierarchical organization i.e., surface roughness of the support. Revealing the wetting transition mechanism represents the answer in engineering highly stable superhydrophobic materials. Thus, experiments carried out by same authors reveal how wetting transitions are irreversible, due to asymmetry of the energy barrier (low from the side of the metastable state and high from the side of the stable state). Trend on future investigations are proposed [17,18].

Experiments by Yanshen et al. [18] were carried out to suppress the energy barrier, as a starter-guideline in optimizing future design of super-repellent materials. In this case, transitions between the Cassie and Wenzel state proved to be, in fact, spontaneously reversible. The method proposed by authors to probe Wenzel to Cassie (W2C) and Cassie to Wenzel (C2W) transitions implied analyzing a drop's behavior while squeezing and releasing between a textured surface and a non-adhesive plate. C2W transitions triggered by pressure, impact, underwater submerging proved to be reversible. In addition, it was demonstrated that it is possible for the two Cassie and Wenzel wetting regimes to co-exist on a double-scaled textured surface, similar to those found in the natural environment. Thus, the Wenzel state corresponds to the larger texture and Cassie to the smaller one (nano-Cassie state). The smaller rugosities do not allow irreversible trapping of water drops. The surface in the nano-Cassie state preserves its hydrophobicity and ability to induce reversible penetration of drops through larger rugosities (slippery character). The spectacular dynamics of water drops meeting such hydrophobic/superhydrophobic textured materials remains a challenge for future investigations concerning the development of robust super-repellent materials.

Investigations concerning wetting transitions of different rough surfaces (natural and synthetic) reveal how transitions may also occur as follows: Mixed Cassie air trapping/Wenzel state to Cassie impregnating state, mixed Cassie air trapping/Wenzel state to Wenzel state, Wenzel state to Cassie impregnating state and Cassie air trapping state to Cassie impregnating state. The Cassie impregnating state is characterized by the lowest energy [19,20].

The above mentioned four wetting states (Young, Wenzel, transitions and Cassie states) also apply for an oil droplet on a flat or rough solid substrate. In this case, the corresponding liquid in the equations refers to corresponding oils [2].

In order to achieve superoleophobicity, the formation of the Cassie wetting state is crucial. Since the liquid drops exhibit low surface tension, not all rough microstructures display a Cassie wetting state. Thus, a third parameter is introduced: Re-entrant surface curvature. Along with surface microstructure and low-surface energy, it is essential in obtaining superoleophobic surfaces [2]. The importance of re-entrant surface curvature in obtaining superoleophobicity was demonstrated using POSS (polyhedral oligomeric silsequioxane) covering fiber mats. Fluorodecyl POSS displayed with oil drops contact angle smaller than 90°, whilst the covered mats showed re-entrant surface curvature and proved to be superoleophobic [21].

3. Special Surfaces

3.1. Natural Special Surfaces

Throughout history, people were captivated by the special survival skills of animals and even plants. Their close observation and study, helped human kind understand and adapt those properties into a useful approach to their evolution. Thus, following the principle of biomimetics, man built the plane after observing bird's flight, the helicopter inspired by the body of the dragonfly, the submarine resembling a whale, and the Velcro closure system according to the way burdock (*Arctium* sp.) spreads its seeds.

Since ancient times, humans noticed the ability of plants to keep clean in marshy environments, to provide their water needs in arid areas. However, these skills remained a mystery until the 1960's when the development of SEM analysis techniques allowed a detailed investigation of surface properties. The plants and insects that raised questions on their survival skills were studied. In 2007, it was concluded that there are two types of microstructures conferring superhydrophobicity to the leave. The first model corresponds to hierarchical micro-/nano-metric structures (Lotus, rice, taro), and the second consists of a unitary structure, an ordered fiber network, with diameters of 1–2 μm (Chinese watermelon, Ramee leaves). The idea that the surface is superhydrophobic only if it has a hierarchical structure with micronic roughness was demystified [22]. Literature brings to light leaves whose capacity to reject the water resides in the presence of vertical hairs (*Alchemilla vulgaris*) [23] or horizontal hairs (*Populus* sp.) [24].

An illustrative model for superhydrophobicity is the lotus leaf (*Nelumbo nucifera*), displaying a surface network which allows dust particles to be removed by rain drops [25], as illustrated in Figure 2. It's considered to be derived from the Cassie-Baxter wetting model. At a micro-metric level, convex papillae are distinguished. At a nano-metric level, wax needles appear to be responsible for superhydrophobicity (contact angles greater than 150°). The veins placed on the top of the Tropaeolum plant leaves, secrete a wax-like substance, similar to the one on lotus leaves, providing cleansing through rolling water droplets. Curiously, the lower side of the lotus leaf has a different chemical structure and architecture, thus, inverse wettability. No waxy crystals are present, but tabular nano-grooved convex lumps cover the leaf's lower side surface. Superoleophobicity and low oil-adhesion in water were demonstrated by a measured contact angle of 155° and sliding angle of 12.1° for a 1,2-dichloroetahne [2].

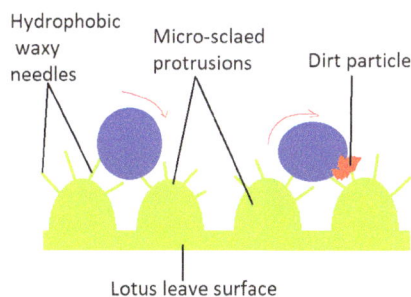

Figure 2. Dirt particle removed by rain drops from the Lotus leave's surface.

Superficial features of plants allow them to survive while floating on water, or while submerged. *Salvinia molesta* (water fern) is equipped with hydrophobic hairs which end in hydrophilic peaks [26]. They retain a layer of air, stabilize the air-water interface, while submerged under water and allow "respiration" (The Salvinia Paradox) [27]. Photosynthesis also continues in submerged *Oryza sativa* (rice), through the air film retained at the superhydrophobic leaves surface [28]. Leaf gas films enhance gas exchange, conferring plants the ability to survive during floods. They also delay salt entry in

Melilotus siculus (melilot) by diminishing Na$^+$ and Cl$^-$ intrusion through the submerged leaves, allowing short-term survival in salty water [29].

Out of the plant kingdom, the carnivorous plant *Nepenthes alata* stands out as an example of a surface with special superficial properties. Experiments show that its peristome has unique structural characteristics, which allow directed water transport through a microcavity system, in the absence of any chemical gradient. The plant feeds on insects. A continuous fluid film rejects the oils from insect's feet, sending them from the surface of the peristome into the "jug" type structure, where they are digested [30]. Many plants, such as taro (*Colocasia esculenta*), India canna (*Canna generalis baley*), rice (*Oryza sativa*), have leaves with contact angles higher than 150° and sliding angles of less than 10°, depending on the papillae arrangement [31]. Unique structures are found in *Strelitzia reginae*, which leaves are furrowed by parallel grooves. They show anisotropic superhydrophobicity, so that a drop of water remains anchored as the leaf is inclined in the grooves direction, or slides off, as the leaf leans perpendicularly to the direction of the grooves [32].

In most cases encountered in plants, a primordial importance of superhydrophobicity in leaf cleaning is conferred by the micro-modeled surfaces. There are special cases, like the *Cladonia chlorophea* lichen, in which cup shaped structures placed on hydrophobic strains, limit water storage only to the pores at the base of the stem [33]. The lichen retains only the required amount of water, banning the accumulation of any excess water, which would prevent spore spreading during reproduction.

Recent research divides superhydrophobic surfaces into opposite categories, judging by interactions with a solid. Thus, there are "slippery" surfaces (Lotus leaves), which present minimum water resistance, and "adherent" surfaces (gecko lizard) [34]. It has been thought that the gecko lizard's ability to climb, is due to structures on its fingers, which secrete adhesive substances, similar to those which allow the rise and hanging of ivy (*Hedera helix*) [35]. The lizard can climb even vertical surfaces, due to microscopic, aligned hairs, divided into nanometric formations called *setae* [36], as illustrated in Figure 3. A drop placed on this surface (a highly adhesive superhydrophobic surface) retains its shape, even in an anti-gravity position [37]. Following adhesive force's model, climbing a glass building using Kevlar special gloves and polyurethane was accomplished (2009) [35]. The category of superhydrophobic and super-adhesive surfaces includes rose petals. They possess a network of micro- and nano-structures, similar to the lotus, but of larger size.

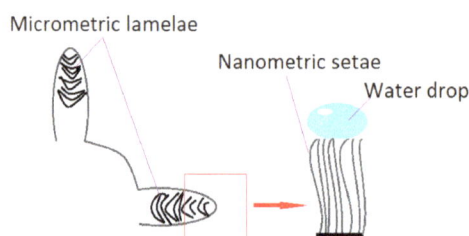

Figure 3. Gecko feet structure. Water drops on *setae*.

The work of Jiang et al. reveal the micro-architecture of the red rose petal. Superhydrophobicity (contact angle of 152° and high hysteresis) is given by micro papillae covered in nano-folds. These rugosities make a water droplet adhere to its surface and maintain a spherical shape. Even if turned upside-down water does not fall/roll off the surface. The phenomenon is called "petal effect", assumed to correspond to the Cassie impregnating wetting model. The Cassie impregnating wetting state is characterized by a liquid film impregnating the grooves but leaving some plateaus dry, as presented in Figure 4a. Moreover, it was demonstrated that the drop's volume is the one conditioning this special behavior, with a different dynamic compared to the lotus effect. Thus, a small droplet sticks to the surface because its volume is smaller than the surface tension. When receding 10 μL in volume, a volume-surface tension balance is reached, and overcoming it triggers the droplet's fall.

This explanation stands up when discussing why smaller drops maintain stable spherical shape on petals and why rain drops do not and roll of [38].

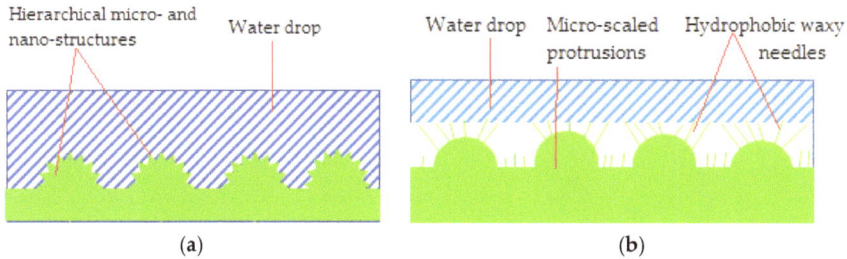

Figure 4. Petal surface structure (Cassie impregnating wetting state) (**a**) and Lotus leaf surface structure (Cassie state) (**b**).

When comparing the lotus leave's hierarchical structure with the rose petal one, differences stand in microstructures and also chemical composition. Thus, a drop on a lotus leaf has a small contact angle hysteresis, waxy protrusions prevent water from entering the micrometrical structure and so the droplet is free to roll of (advancing and receding at different contact points) as the leaf is tilted. Micro and nano-structures covering the petal are both bigger than those of the lotus. That makes the drops adhere to the surface, due to water being sealed in micro papillae, while exhibiting a high hysteresis when the petal is tilted/turned upside-down [38].

Experiments were carried out in order to artificially recreate the rose petal hierarchical structure. By mimicking the rose petal effect, superhydrophobic and also adhesive polymer films were developed. The natural petal is used as a template during the fabrication process, making the method an environmental friendly one compared to other techniques [38].

Following the path of herbal substances unusual applications, Zhang (2009) [36] proposes the use of ivy nano-particles in sunscreen lotions. The additional advantage over conventional creams is that of uniform topical application, very good skin surface adhesion, and the ability to successfully block a wide range of UV radiation wavelengths. An unprecedented breakthrough, with a high impact in the medical field, is the use of tree moss to obtain adhesives. In 2014, moss was found to be non-toxic and stronger than super-glue. It proves to work successfully in wet environments, joining both soft tissues and bones [35].

Researchers interest regarding the functionality of insect wings lead to the discovery of the following: Superhydrophobicity of insect wings derives either from tiny, cape-shaped structures, miniscule hairs or from scaly formations [39]. Numerous investigated insect species reveal wings with nanometric, hierarchically disposed scale structures, which enable flight and do not allow dust contamination, as illustrated by Choi et al. [40]. Study of the cicadas, highlighted the possibility of encompassing both transparency and superhydrophobicity, by alternation of nanostructure dimensions, conferring homogeneity. Their wings are able to selectively kill Gram-negative bacteria, without attacking Gram-positive ones [41].

Nature has its own ways when it comes to special wettable surfaces. The desert beetle (*Stenocara gracilipes*) is a special exponent of superhydrophillic surfaces joined with superhydrophobic ones. The usefulness of the association is to ensure the beetle's need of water, in high temperature conditions. Analysis highlight that the back of the beetle is full of hydrophilic non-waxy peaks, which capture water from the fog, in form of droplets (Figure 5), as shown by Ueda and Levkin (2013) [33]. As the droplets become too large, they slip on wrinkled hydrophobic waxy edges.

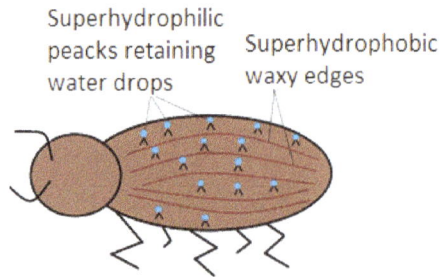

Figure 5. Namib desert beetle displaying superhydrophobic edges and superhydrophilic back protrusions.

Many authors [33], emphasize the importance of joining superhydrophilic surfaces (large surface energies, low contact angles ~0°) and superhydrophobic ones (low surface energies, high contact angles > 150°). Wettability variations function as advantages. Among them, the following are listed: Easy drop positioning as close as possible to each other, since the hydrophobic support does not allow interactions; simple surface geometry elaboration, including superhydrophilic water filled micro-channels. If the pattern follows the Cassie–Baxter wetting model, then the droplet's bio-adhesion does not occur and subsequent sampling is done, keeping the contents intact. The proposed model may be a source of inspiration in creating devices capable of collecting water from fog, in arid areas.

Although the behavior of water droplets placed on Lotus leaves was studied in air, it was unknown what happens if the superhydrophobic side is turned face to water. Thus, in 2009, it was shown that droplets break as the leaf turns, showing superaerophilicity. In connection with this phenomenon, the diving spider's (*Argyroneta aquatica*) survival underwater was explained. In order for it to breathe while submerged, it creates an artificial lung, an oxygen bubble that remains trapped between its feet and abdomen. The novelty lies in the bubble's silk outer shell, which is hydrophobic and gas permeable, allowing underwater breathing. This approach represents a model in developing methane transport systems, in preventing undesirable underwater discharges and global warming [42].

It is important to notice chemical structure in correlation with surface architecture of the seaweed (*Saccharina japonica*). The porous structures on its surface combined with the effect of salt-insensitive polysaccharides translate into durable underwater superoleophobicity, even in high-salinity and high-ionic water. In the same wettability regime, the clam's two region-divided shell architecture, proves to have anti-oil properties. The hydrophilic $CaCO_3$ composition along with the rough hierarchical microstructure makes the shell oleophobic under water (region 2) keeping it clean all the time, whereas the other region (region 1) is polluted by oil [2].

3.2. Superhydrophobic Surfaces: Learning from Nature

Following the principles of biomimetics, superhydrophobic surfaces were artificially obtained. Once the surface properties (contact angle, superficial energy) of natural solids are established, they can be varied one depending on the other and applied to synthetic materials. Thus, innovative raw materials are born.

The generic method used to artificially obtain superhydrophobic surfaces follows the lotus leaf model i.e., its self-cleaning capacity conferred by wax epicuticular crystals. Its hydrophobicity arises from the –C–H and –C–O– groups [43].

Fabrication of superhydrophobic surfaces following the principle of biomimetics began in 1990. In 1992, a submicrometer-roughed glass plate was hydrophobized using fluoroalkyltrichlorosilane, with contact angles approaching 155° [44]. Among early synthetized super water repellent surfaces, the one prepared by Shibuichi et al. (1996) [45,46] included n-alkyl ketene, with contact angles of 174°, due to the fractal nature of the surface. Alumina coatings were obtained with fluorialkyltrimethoxysilane on porous alumina gel (1997). Ion-plated PTFE coatings with nano-metrical rugosities were reported in the same year [47]. Experiments were carried out by McCarthy [48] to determine the effects of topography on wettability. Patterned silicone surfaces prepared by photolithography and salinization were obtained. Their wettability was also investigated in correlation with square posts dimensions. Contact angles appeared to be independent of square posts heights and of surface chemistry, when their dimensions ranged 20–140 µm. Surfaces with square spots of 64–128 µm dimensions were not ultra-hydrophobic as expected. Water drops penetrated between the square posts and were pinned to the surface. The phenomena intensified as the distance between posts increased or as the shape of the posts was changed to rhombus, star. Thus, the maximum length scale that confers surface hydrophobicity was established at 32 µm for surfaces covered in square posts.

Ever since the first artificially obtained special surfaces were reported, various methods were improved in order to confer surface roughness, transparency, possibility of color change, reversibility, permeability, anisotropy [49–51]. Some methods involved the use of fluorocarbon derivatives. Nature has shown that the presence of such compounds is not mandatory in order to obtain low surface energy [43]. Techniques that produce replicas of natural surfaces were developed. Among methods used to artificially create special surfaces, depositing hierarchical micro- and nanostructures on a hydrophobic substrate or chemically modifying a low superficial energy surface are proposed by some authors [52–54].

The artificial fabricated surfaces should display a hierarchical/unitary structure. Among known procedures, the most popular are: Chemical reactions in a humid atmosphere [55,56], thermic reactions [57,58], electrochemical deposition [59], individual/layer-by-layer assembling [60,61], etching [62], chemical vapor deposition [63], polymerization reactions [64]. These techniques are applied to silicone, copper, zinc, titanium, aluminum, cotton or glass substrates, depending on the procedure. Surfaces obtained after these modifications, display contact angles greater than 150°. Among the easiest and fastest to apply techniques are hydrothermal and chemical reactions in humid air. They can be adjusted to obtain objects of any shape or size [22]. Another versatile method used to deposit a salt solution on to a metal, is electrochemical deposition, which also confers the support a furrowed structure with numerous micro-grooves [59]. Depositing carbon nanotubes onto a cotton support, creates an artificial structure similar to the lotus leaf [65]. Chemical vapor deposition proved to be efficient in "constructing" micro-pyramid like patterns with contact angles greater than 170° [66]. Sol-gel techniques which confer hexagonal ordered superficial structures may be applied on to many substrates like glass, metal, silicone, textile materials [67].

By applying the techniques previously mentioned it is possible to mimic natural special surfaces. For instance, the duplication of petals' surfaces was achieved. Polymer films were obtained by using the red rose petal as a duplicated template. Imprinting of the nano-metric roughness is made through solvent-vapor techniques. Practically, a PVA 10% and PS 15% solutions were poured separately on different petals and evaporated. What is left behind is a PVA, respectively PS film imprinted with the micro structured pattern of the petal. The obtained films exhibit exactly the same wettability as the original petal i.e., high contact angles, hysteresis which does not allow rolling of droplets even if the plane is tilted [38].

The lotus leaf double-scaled surface pattern was also achieved. Experimental studies reveal the possibility to obtain polymeric superhydrophobic surfaces through a solvent-free ultrafine powder coating technique. Contact angles attain values of 160° and sliding angles do not exceed 5°. This method proved efficiency in mimicking the lotus leaf surface micro-and nano-scaled pattern without the use of any solvents that prove to emit toxic compounds into the environment. The method represents a breakthrough in the coating industry [68]. Large scale fabrication of special superhydrophobic surfaces is a continuously developing domain and "natural templates" still serve as models in designing innovative coatings.

3.3. Innovative Superhydrophobic Materials and Coatings

The previously submitted production techniques applied to obtain superhydrophobic surfaces/coatings, along with physical and chemical adsorption adjustments are applicable to non-reusable substrates (polymers, minerals).

Among low surface energy materials, fluorocarbon and silicon derivatives, some organic and inorganic compounds have been preferred for many years. Hsieh et al. (2006) [69] demonstrate that the fluorine/carbon ratio is the one influencing the superhydrophobic degree of the surface. Thus, the more fluorine atoms are included in the structure, the higher becomes the hydrophobicity. Genzer and Efimenko (2000) [70] assert that another determinant of stability and hydrophobicity is the density and layout of certain chemical groups. Experiments prove that $F(CF_2)(CH_2)xSiCl_3$ molecules applied on a polydimethylsiloxane (PDMS) stretched substrate, immediately form an organized layer, whose contact angle increases by $30°$, as the stretching force is removed.

When it comes to roughness, statistical parameters are indicative and cannot be extrapolated at a micro-/nano-metric level, for any type of surface. Geometrical roughness is considered and a direct correlation between roughness data corresponding to different sample sizes is attempted.

Polymers and minerals gained importance due to their potential as supports, with small to very small surface energies. However, these hierarchically organized structures have an inconvenient: Possible unwanted hydrophilic components. Thereby, they do not lend to the rigorous requirements of a hydrophilic barrier imposed in many fields (printing, packaging, perishable storage). In attempt to preserve the environment, these non-renewable supports were replaced with bio-based materials, derived from wood, plant fibers, agricultural residues. Lignocellulose is one example, thanks to its ease of purchase and transport, low mass and abundance. Progress has been made in the cardboard and cotton industry through producing superhydrophilic materials using cellulose as a base [1].

Following the direction of environmental-friendly materials, cellulose nano-crystals and composites receive attention. Cellulose undergoes roughening processes in order to lower its free surface energy. The structure and surface properties of cellulosic fibers are adapted through sol-gel processes or nano-particle deposition (metal oxides, minerals, polymers). Thus, the modified cellulose has both static and dynamic contact angles (hysteresis). The durability of the applied rugose layer allows any required improvements [71]. Recent experiments focused on durability and robustness of lignin-coated cellulose nano-crystal (L-CNC) particles. Commercial biodegradable L-CNC particles were used after hydrophobization (used to confer roughness). Two adhesives were used to support sticking between L-CNC particles and substrates. The resulted coatings exhibit astonishing attributes: Self-cleaning properties, water repellency, durability against sandpaper abrasion, finger-wipe, knife scratches, water jet, UV radiation, high temperature exposure, acid and alkali solution [72].

Remaining in the same field of textile materials, superhydrophobic flame-retardant cotton was developed by a researcher group using layer-by-layer assembly of branched poly(ethylamine) (bPEI), ammonium polyphosphate (APP) and F-POSS. Experiments reveal that exposed to fire, the cotton fabric and bPEI dehydrate, catalyzed by APP. Thus, a heat insulating char layer with porosities is generated. Pores are formed due to decomposition of bPEI and so the formation of flames is delayed. This is a very useful discovery for the field of flame-proof materials [73]. An effective method to use lignocellulose (LC) as base-support to fabricate superhydrophobic surfaces with flame retardant properties was recently discovered (2018). PDMS-stearic acid-modified kaolin-coated LC attains contact angles of $156°$. Due to kaolin particles, it also displays good flame retardant properties [74].

Since the UV-radiations were proved to be the cause of many human health affections, UV blocking products gained popularity. Besides the well-known cosmetic products, it seems that UV blocking textile products are receiving well-deserved attention. Some studies developed UV blocking textiles using white pigments. The expansion of the green technologies reflects on the development of multifunctional textiles (self-cleaning, antibacterial). A research group demonstrated UV radiation is absorbed by PET fabric covered in ZnO/SiO_2 pencil-shaped rods. The explanation lies in the nano-scaled rugosities and superhydrophobicity conferred by ZnO/SiO_2, exhibiting a receding contact angle (from 160° to 90°) when exposed to UV light, due to hydrophilic groups [73]. Novel methods (2018) to produce superhydrophobic superoleophobic-covered silk textiles are reported by Aslanidou et al. Alkoxy silanes, organic fluoropolymers, silane quaternary ammonium salt and silica nanoparticles are included in an aqueous solution which is sprayed onto silk. Thus, it gains both superhydrophobicity, superoleophobicity (contact angles greater than 150°) and also antibacterial properties. The coating confers a double roughed surface architecture which also acts as an antimicrobial agent, hindering microbial growth [75].

Another way of exploiting superhydrophobicity as an advantageous property is the use of waterborne resins made of aqueous silanes and siloxanes solutions containing silica nanoparticles. Once deposited onto marble, sandstone, mortar, wood, cotton, ceramic artifacts, the composite superhydrophobic protective film turned out to be a simple, cost effective and most of all environmental friendly (no organic solvents are used) technique to protect cultural heritage [76].

Superhydrophobic self-contained surfaces, which retain their properties over time have been obtained. They represent a complete and independent unit in and of itself and are autonomous. Even so, it is difficult to mimic the self-healing properties of natural surfaces and even harder to confer wear resistance, without altering the area's characteristics [26]. A novelty in this field is the SLIPS (slippery liquid-infused porous surfaces) technology, inspired by the lotus leaf. Basically, it refers to smooth and slippery coatings applied on sports shoes, or used in the construction field, military uniforms, medical gowns, in order to avoid biological fluid contamination. SLIPS surfaces were described as having the ability to clean themselves, due to surface fluids incorporated into the micro-/nano-porous substrate, forming a smooth surface. They are able to remove impurities with complex structures (oils, blood). No synthetic surface meets all the qualities of SLIPS surfaces: Large contact angle, lack of hysteresis, low slip angle, optical transparency, instant repair due to capillary action given by surface energy. Surfaces that serve as omniphobic materials, with possibilities to apply in fluid/biological materials, fuel transportation, glass surfaces that do not freeze and clean individually are still developing [77]. Another area in which SLIPS is being popularized is the sports footwear industry. Tests on sports shoes show that the SLIPS technology provides the best protection against moisture, but as a compromise, it does not allow the skin to be aerated. In 2009, a material whose structure mimics pine cones, with structures that open or close according to humidity was placed on the market. A single fiber includes two distinct polymers: A hydrophilic and a hydrophobic one which react according to the environmental conditions. Since 2016, another type of hybrid material has been used for yachting equipment. It is particularly useful in special activities, allowing opening and closing of temperature-dependent fibers, with the release of hot and humid air [78].

4. Applications of Superhydrophobic Surfaces in the Medical Field

As previously described, superficial properties determine superhydrophobic surfaces applications. An evolution was noticed in the use of artificial special surfaces in various fields. Continuous efforts are made in order to discover innovative means of use, in both industrial and laboratory fields, as well as in the medical field. [79].

4.1. Anti-Bio Adhesion

Biochemical phenomena such as protein adsorption, bacterial adhesion, and development of cell cultures are influenced by the support's wettability, which can be manipulated in the desired direction [79]. Thus, Lampin et al. (1997) [80] experimentally demonstrate that protein adhesion is favored by hydrophobicity conferred through a PMMA (polymethyl methacrylate) coating. More recent studies by Zelzer et al. (2008) [81], show that fibroblasts adhere differentially to a surface exhibiting a chemical gradient: From hydrophobic (polymerized hexane plasma) to hydrophilic (allylamine-polymerized plasma). In order to minimize activation and adhesion of blood platelets on to an implant/prosthesis surface, in vitro experiments have been carried out. It was demonstrated that platelets do not adhere or propagate further, on supports covered in TiO_2 nanotubes matrices, which are compatible with blood. In 2009 [82], four types of PDMS (polydimethylsiloxane) surfaces with rugosities of various sizes were developed: Superposed scale plates, sub-micron structures, nano-structured and smooth surfaces. The superposed scale plate surface proves to be the most effective against adhesion of blood platelets, under biological blood flow conditions. Statistical results also show a low surface adhesion, whilst the highest degree of adhesion corresponds to the smooth surface.

4.2. Anti-Bacterial Fabrics

Surface roughness combined with a low superficial energy, led to the artificial production of surfaces exhibiting special wettabilities, i.e., superhydrophobic surfaces. Considering liquids with a superficial energy lower than water (decan, octane [21], oils) and taking into account an additional parameter, surface curvature, superoleophobic surfaces with contact angles greater than 150° were developed [2]. Superoleophobic cellulose fibers have been obtained by modifying siloxane with silver nanoparticles, with the help of an active organic/inorganic binder. The inherent property of these fibers resides in antibacterial activity against *Escherichia coli* (100% suppression rate) and *Staphylococcus aureus*. Additional treatments applied to cotton fabric improved antibacterial activity, conferring durability and ease of washing [83]. Through special treatments with fluoroalkyl silanes, glass acquires its antibacterial action against small concentrations of *Pseudomonas aeruginosa* and *Staphylococcus aureus*. Aizenberg et al. [84] show that PEG films exhibit much lower antibacterial activity compared to fluorinated silicone oil infused slippery surfaces capable of removing 96–99% of the bacteria. Special coated (Siloxane + antimicrobial agent (am) + SiO_2) superhydrophobic silk shows a decrease in spore adhesion and growth compared to uncoated or Siloxane + am agent coated samples, kept in the same conditions [76].

4.3. Cancer Cell Isolation

All cell types (including cancerous cells) display certain protrusions on the membrane's surface, through which they interact with the biological medium: Bind to target cells/tissues or reject possible harmful agents (macrophages, antibodies). Thus, any medical material/implant/device that comes in contact with the human body, should be carefully designed. Attached cytophilic/cytophobic moieties (superhydrophilic/superhydrophobic), influence detection of circulating cancer cells [85,86]. Nanometrical printed roughness confers superhydrophobicity, captures only cancerous cells and does not allow adhesion of healthy blood cells [42]. Improvements have been made so that nanostructures–cell interactions successfully remove ill cells from the patient's blood. This method opens new gates in early diagnosis of rare cells which cannot be done using existing technologies (biopsy) [87]. Galactozylated carbon nanotubes are also cytotoxic structures, and can be used to capture viruses or bacteria [88].

5. Special Patterns. Joining both Superhydrophobic and Superhydrophilic Surfaces

In order to find more effective methods to study biomolecules' (peptides, oligonucleotides, enzymes) activity, innovative techniques have been developed at a small scale i.e., micro-pattern supports, made from superhydrophobic and superhydrophilic adjacent areas, as illustrated in Figure 6. Basically, molecules of interest, in the form of aqueous solutions, are deposited on hydrophilic areas. They remain separated from each other, due to hydrophobic zones which surround them, which do not allow migration or mixing. A single support may comprise hundreds of droplets, with volumes ranging from pico to microliters. Another proposed model implies that pre-impregnated spots carry cells of interest, selected from a library of molecules. Biomolecules are included both in aqueous solutions and isolated hydrogels, allowing 3D screening and complete immersion in a common environment. Comparative analysis of cell behavior and creation of a supportive culture medium is carried out. Microscopic analysis can be achieved due to the transparency of hydrophilic spots, separated by opaque superhydrophobic barriers [34,89,90].

Superhydrophobic delimitations

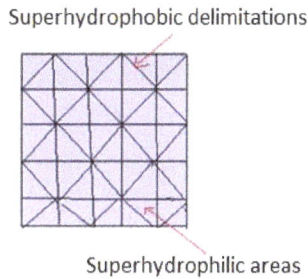

Superhydrophilic areas

Figure 6. Schematic representation of patterns joining superhydrophobic with superhydrophilic areas.

5.1. Peptides Separation

Wettability can also be set up as a grading criterion. Thus, 2D thin layer chromatography techniques are applied to separate peptides with different hydrophilicities and isoelectric points. Separation happens due to a superhydrophobic porous polymeric support, engraved with superhydrophilic channels. In fact, there are two processes that lead to separation: One inside the microchannels, as the aqueous mobile phase migrates to the hydrophilic areas, guided by the superhydrophobic pattern on support; the second process is based on the use of a mobile phase with acetonitrile (the separation itself). The separation occurs according to the hydrophobicity of the peptides. Detection is made through desorption or electrospray ionization. This method gives rise to the use of micron-sized diagnostic systems, by joining superhydrophobic patterns with superhydrophilic ones [91].

5.2. Molecules Screening

Hydrophobic/hydrophilic merged areas were developed at a micrometric level, in order to synthesize new inhibitors of serine protease NS3/4A, a promoter of hepatitis C virus. The pattern consists of areas with hydrophilic points, which ensure the stability of nano-droplets, placed on a hydrophobic support. NS3/4A inhibitors were synthesized into the drops [92]. Microdroplets placed on these special surfaces, allowed analysis of auto-fluorescent molecules, with possible applications in non-invasive diagnosis and real-time imaging [93].

Techniques based on merging different wettability surfaces, proved to be useful in assessing molecular and enzymatic kinetics, completing studies on drugs' mechanisms. The advantages of these techniques include: The possibility to control pattern's geometry, droplets position and volumes, safety by means of droplets stability which cannot migrate to another adjacent formation, ability to easily handle small volume drops, space and reagent saving, as well as fast by means of preparation and analysis methods [89].

6. Applications Derived from Water's Behavior

6.1. Anti-Icing Properties

The optimal functioning of airplanes, boats, telecommunication routes and highways is influenced by ice formation. Over the years, procedures were developed to avoid the frost of these surfaces. In recent years, the use of superhydrophobic materials that prevent/reduce condensation and ice formation became popular [42].

The anti-freeze property of superhydrophobic surfaces is already well known. Liangliang Cao et al. (2009) [94] discover a correlation between particle size and anti-freeze properties. An important aspect is brought to light: There is a difference between the particle size which confers superhydrophobicity and the one conferring anti-freeze characteristics. Obtaining a surface with both attributes has been a challenge in terms of factors influencing water's frost on a surface: The adhesion of ice on superhydrophobic surfaces and environmental conditions. The study is based on including a polymeric binder (silicone resin, acrylic polymer, silica particles) in the composition of nano-particulate polymeric surfaces. Further research is needed to establish a superhydrophobic surface design that satisfies the rigors of anti-freeze surfaces [94]. It is desirable to develop coatings which maintain dryness of the support, determining the drops to bounce off the surface. This way, metal rusting and airplane wings frosting can be avoided [26]. Attempts were made to link surface frost to the "ascending/descending" contact angle observed in ice adhesion on steel surfaces. Thus, by increasing the "withdrawal" contact angle, ice adhesion intensifies [95]. Two-leveled highly hydrophobic surfaces which do not allow formation of ice crystals, not even at extreme temperatures (−30 °C) were developed [96].

In 2013, experiments were conducted to analyze the behavior of a droplet falling on a fluorosilane-coated surface. Within milliseconds, the following events take place: Surface scattering, kickback, lifting. The events start with the drop's impact, which then stimulates the center's lift and the entire drop rebound [97]. In 2015, the droplet's jump off the superhydrophobic support was elucidated through an experiment developed in a dry-aired room and under low pressure. The drop rests on the rugosities of the support. Beneath it, are air voids including water vapors. Here the pressure is higher compared to the environmental one, resulting in the droplet's lifting. Another explanation is based on the sudden freezing of the already cooled droplet, which causes a rise in pressure and jump of the droplet. Of course, methods involving lowering ambient pressure cannot be applied to avoid frost in open spaces, but can be used at laboratory level [98].

6.2. Oil–Water Separation

Industrial accidents and massive spills resulted in enormous quantities of ecosystem-damaging oils and mixtures being discharged in seas and oceans. In order to support water cleaning, systems based on superhydrophobic/superoleophilic materials are developed [42].

Meshes of porous materials superhydrophobic–superoleophilicity or superoleophobicity–superhydrophilicity were a success as oil–water separation devices. An oil removing mesh removes oil form a water-oil mixture. Water did not wet the mesh due to the superhydrophobicity, but the oil fully wetted and permeated due to superoleophilicity. Separation occurs successfully. The only issue stands in the fact that oil blocks the mesh, decreasing the efficiency of the separating material. Other meshes, able to remove most oils (with lower density than water) were developed [2]. Diesel can be separated from water using Teflon-coated stainless-steel mesh systems. Organic solvents can be separated by absorption through nano-fiber membrane systems, which combine superhydrophobicity/superoleophobicity with capillarity [99]. In addition, by combining the two surface properties and adapting them within an aqueous medium, a net covered in a hydrogel was obtained. It separates water from crude oil, gasoline, and diesel oil. It is easy to clean and can be reused, putting an end to the waste water pollution process [100].

7. Applications of Superhydrophobicity in Other Domains

Nowadays, as our planet becomes more polluted, self-cleaning and anti-fouling materials are of much need. By means of self-cleaning, many superhydrophobic applications are included, such as anti-bio adhesion, as earlier presented. In addition, anti-reflective, anti-icing/fogging materials, water purification systems are of high demand when it comes to extreme situations (industrial oil spills, foggy airplane windows etc.) [36].

7.1. Self-Cleaning Textiles

Most self-cleaning surfaces exhibit a contact angle greater than 160° and an architecture similar to the lotus leaf. Multi scale roughness and low energy waxes are responsible for the superhydrophobic and self-cleaning property of artificially made surfaces which mimic the lotus leaf pattern. Recent progress in developing such surfaces rely on two appropriate yet different techniques: Constructing hierarchical rugosities on a hydrophobic surface and coating a rough surface with a low energy material. Windshields, windows, building or ship paintings, solar panels benefit upon the possibility of drag reduction, lower sliding angles [101].

As a drop rolls off a hydrophobic support, dirt particles are displaced to the drop's sides and re-deposited as the liquid slides off. Superhydrophobic surfaces allow the drop to easily slide off their surface and also gather solid unwanted particles. A low adhesion degree corresponds to the so called "self-cleaning" property [35]. The most illustrative examples are the previously presented lotus leaves and rose petals.

Textiles and materials with such self-cleaning properties are desirable. Thus, emerging procedures are applied to textiles/surfaces in order to confer water, oil, dirt-repellency. Techniques are still being developed in terms of cost-effectiveness, durability by means of extreme temperature exposure, wash resistance. Among popular methods, sol-gel processes, using fluoroalkyl water born siloxane (FAC), silver nanoparticles and inorganic binders prove effectiveness in preventing adhesion and growth of bacteria. The "plasma" technique, compared to classical chemical methods, offers the simultaneous advantage of roughness and low surface energy. After applying this technique, the material undergoes structural changes, gains nano-scale roughness and preserves its color and texture. Experiments by Vasilievic et al. regarding modifications of cellulose surface in order to induce superhydrophobicity, oleophobicity confirmed that a cotton fabric surface can behave similar to the lotus leave due to low-pressure water vapor plasma pre- treatment followed by the addition of a sol-gel coating using FAS precursor. FAS coating proved to offer superhydrophobicity to the cotton fiber and the plasma pre-treatment prior to that coating provided water and oil repellent properties. The plasma pre-treating process does not ensure durability of the lotus effect during repeated washing processes but enhances the formation of a FAS concentrated coating network [102]. Other studies by the same authors were made to establish the influence of oxygen plasma treatment on the water repellency of cotton fibers coated with an inorganic-organic hybrid sol-gel perfluoroalkil-functionaized polyselsequioxane (SiF). Regardless of the applied time and opperating current, the oxygen plasma treated fabric experienced an increase in contact angle value from 135° to 150°. It became obvious that plasma treatment influenced the water repellency induced by the SiF coating. The resulted cotton surface gained rugosities. It can be concluded that the plasma pre-treated and SiF-coated cotton fabric attaines micro- and nano-asperities which strongly determine hydrophobicity [103].

Xi Yao et al. developed a technique that produces superamphiphobic cashmere textiles (superhydrophobic and superoleophobic). Even more, the above presented techniques are economical and weakly polluting processes, that can be applied at an industrial level [79].

Some authors attain the self-cleaning property by modifying textile surfaces loaded with hydrophilic TiO_2. The surface also exhibits an antibacterial effect (due to Ag deposited on the activated cotton). The interest in using such bactericide/antifungal/antiviral textiles resides in the necessity to use topical treatments of skin diseases [104].

Self-cleaning properties of silk were achieved also by covering it in siloxane enriched with 2% SiO_2 nanoparticles. Any liquid is absorbed by the untreated silk. Superhydrophobicity and superoleophobicity are obtained. The wax is easily removed mechanically from the treated surface leaving behind no residue. On the untreated silk, a stain remains after wax removal. Durability of the coatings over a wide range of pH is demonstrated: Surfaces maintained their properties except for pH basic conditions. Additionally, a recent "green cleaning method" is presented using supercritical CO_2 at bar and 40 °C. Dirt is removed without affecting natural dyes. An equally important aspect is the fact that siloxane and SiO_2 coatings do not affect appearance of silk. Since no organic solvents are used, the coating method is friendly to the user and the environment. Moreover, the coating technique is reversible, allowing removal from the silk substrate using compressed CO_2 mixed with methanol [105].

Unlike superhydrophobic surfaces, whose cleaning properties rely on the Lotus effect i.e., the dust is collected through water drops rolling off the leaf function, superamphiphobic surfaces can be kept clean by droplets, which during the rolling process, adhere to the surface and thus remove dirt particles. Underwater superoleophobic materials with ultralow oil-adhesion also exhibit self-cleaning properties.

The oil adheres to the surface. After immersion it is completely removed from the surface. Self-cleaning effects underwater for superoleophobic surfaces is due to the intrinsic superhydrophilicity. Through immersion, water is injected into the microstructures and oil is pushed out [2].

7.2. Anti-Reflective Transparent Coatings

Glass used to fabricate mirrors, lenses, optical devices, exterior windows and solar panels should display self-cleaning abilities. Apart from self-cleaning, superhydrophobic surfaces can be exploited by means of anti-reflective properties. A special interest is also attributed to transparency and the ability not to reflect light beams. These characteristics can be conferred to glass by intervention upon roughness, a property closely related to transparency. Rugosities of nanometric dimensions are conferred by silicon derivatives (fluoroalkyl silanes), aluminum acetyl acetonate [106–108]. A surface film is formed, the thickness of which can be adjusted according to the desired degree of transparency. The surface film may be multilayered, each layer having its own adjustable characteristics. Polymeric films are the most flexible and resistant compared to inorganic ones [109].

7.3. Corrosion-Resistant Metals

Metal corrosion is one of the contemporary problems faced by humanity, which leads to large losses of money because of damaged areas, which require later replacement. Coating processes with chromium derivatives protect metallic corrosion surfaces, but they are environmentally harmful methods, which tend to disappear. New corrosion protection techniques are being sought, one of them being: Metal coating with superhydrophobic protective layers [110]. Between the protective layers and the support, an air cushion is created, thus preventing penetration of the corrosive agents. Differential coating techniques (microwave chemical vapor deposition, followed by immersion) with fluorochloride-silanes of magnesium alloys are used. Stable corrosion resistant coatings have been obtained, which reveal a color change from silver to green [111,112]. Subsequent studies refer to superhydrophobic aluminum substrates modified with hydroxides, zinc immersed in superhydrophobic solutions, proving resistance against acids, alkaline or saline solutions [113].

7.4. Microreactors

When studying chemical reactions, rationalization of reagents and limitation of secondary explosive/toxic compounds is primordial. Minimization of the entire chemical reaction scale starts with modifications applied to the reaction medium, reactants and by-products [114]. Microemulsions and microfluidic systems gained attention through their ability to miniaturize the chemical reaction, at levels corresponding to nano- or micro-liters [115].

While handled with superhydrophobic tweezers, liquid droplets placed on a non-adhesive support, constitute a reaction medium/reagent. The drops retain their shape, do not lose components, and can even be transported using a nano-particle composite wrapped tweezers. By forcing the drops together, their coalescence takes place. The resulting droplet functions as a "reaction plant". Two components from different droplets react, giving rise to a new reaction product. The advantages of the method are: The possibility to obtain new, fully collectable compounds in small amounts and by using minimal quantities of reactants, in a controlled environment. The resulted new products can be detected due to color change (i.e., yellow becomes colorless as the coalescence between a drop of tetrachloromethane bromine and styrene tetrachloromethane occurs) and sampled using tweezers [42]. Decomposition, etherification, combining and controlled temperature reactions can occur. A drop of water in oil can accommodate a reaction produced by heating over time, without the aqueous phase's evaporation. Restriction of experimental chemical reactions on a micrometric scale finds its applications in DNA analysis, synthesis of new molecules, new active substances, and may constitute basics of innovative analysis methods [116,117]. It is noteworthy that microdroplet manipulation was achieved through an oil-based microreactor, which relies on the controllable oil-adhesive superoleophobic surfaces. The droplet-based microreactors are important in enzymatic kinetics, protein crystallization processes as alternative controlled transporters to expensive micropumps, microvalves, microchannels [2].

7.5. Friction Reduction

The field of aeronautics and water transportation (ships, submarines) are governed by the unwanted phenomenon of water and air friction [118]. Removal strategies for this undesirable factor include development of surfaces with special structures inspired by the shark skin or Lotus leaf pattern. Experiments show that superhydrophobic surfaces have the ability to reduce fluid friction, both in laminar and turbulent flow. Air bubbles embedded between the rugosities (responsible for superhydrophobicity), result in a continuous surface film, thus reducing friction. The problem arises in the fact that the film disappears under high pressure. Recently, a hydrophobic pattern support, whose gaseous film is stable in extreme conditions was created [42].

7.6. Novel Transportation Devices

Transportation means invented by humans are inspired by nature (birds, insects, whales). The same applies in case of the water strider's needle-like rugosities, which confer superhidrophobicity to the spider's feet, allowing it to walk on the water surface. These insect's special features represent a starting point in developing prototypes of miniature robots similar in appearance and structure. Having the ability to move in a straight line and jump at the water's surface, while collecting information on its composition, these robots are used to monitor environmental water. Other models of robots have feet made of nanoparticles of organic semiconductors. They display high transport capacities in relation to their own dimensions. These suggestions materialize in models for developing innovative aquatic transport devices [119,120].

Micrometrical droplet transportation devices rely on superoleophobic surfaces exhibiting low oil-adhesion, which act as anti-oil "agents". If the oil-adhesion on a such surface is high, then the oil sticks to the surface, even if tilted. This may constitute a micro-oil-droplet transportation device [2].

Passing from transportation means on water, to water transportation itself, it comes to the idea that liquids can be transported by simply modifying surface's wettability. The roughness of the surface determines wettability, directing the movements of the water droplet. In addition, the intervention of external stimuli (light, electric current, magnetic field) can intervene and guide the fluid on the support surface [121,122]. By adjusting rugosities according to needs, the adhesion degree of the support can be controlled. An illustrative example of these assumptions is the lossless transport of micro-particulate water on superhydrophobic surfaces with PS nano-tubes [79].

7.7. Water Storage

Hydrophilic prominences alternating with superhydrophobic channels, make up the Namib beetle's chitin pattern. It represents an inspirational model for researchers in developing water collecting devices [123]. The alternation between different wettability areas, determines water collection. Various arrangements (circular, vertical, horizontal), can be adopted, depending on the case. Experiments show that droplet volume does not influence the force keeping the droplet anchored to the surface. Studies are being carried out to optimize these surfaces, through the contrast between surface wettability [42]. Promising applications spring from these patterns and refer to the ability of capturing water from the fog in desert areas, to save fresh water during drought periods.

7.8. Electronic Components

Surface characteristics have been explored in order to obtain small electronic components such as electrodes, inductors, transistors. They have textured surfaces with varying degrees of wettability. Obtaining methods include inkjet printing of surfaces with hydrophobic regions, which reject ink, channeling it to hydrophilic regions, thereby producing self-aligned, printed patterns [124].

8. Conclusions

This paper concentrates interdisciplinary researches on surface wettability, bringing together studies on superhydrophobic special surfaces. It focuses on structure, means of obtaining, respectively on practical applications of superhydrophobic surfaces, starting from models existing in the natural environment (lotus leaf, butterfly wings, etc.). Following the principle of biomimetics, researchers developed systems that allow 3D screening of biomolecules, isolation of cancer cells, self-cleaning textiles to prevent biological fluid adhesion. In addition to high-impact medical discoveries, systems being able to analyze the chemical composition of water were also fabricated. Superficial properties of superhydrophobic surfaces allowed the micrometric study of chemical reactions (microreactors) using ecological and economical techniques. All these researches have a common direction: Elaboration of efficient, fast and economical methods, applicable at an industrial level, in order to obtain special wettable surfaces or protective coatings. Exploration of the advantageous features of surfaces with special superficial properties continues.

Funding: This research received no external funding.

Conflicts of Interest: The authors declare no conflict of interest.

References

1. Song, J.; Rojas, O.J. Approaching super-hydrophobicity from cellulosic materials: A review. *Nord. Pulp Pap. Res. J.* **2013**, *28*, 216–238. [CrossRef]
2. Yong, J.; Chen, F.; Yang, Q.; Huo, J.; Hou, X. Superoleophobic surfaces. *Chem. Soc. Rev.* **2017**, *46*, 4168–4217. [CrossRef] [PubMed]
3. Mahadevan, L.; Pomeau, Y. Rolling droplets. *Phys. Fluids* **1999**, *11*, 2449–2453. [CrossRef]
4. Vakarelski, I.U.; Patankar, N.A.; Marston, J.O.; Chan, D.Y.C.; Thoroddsen, S.T. Stabilization of Leidenfrost vapour layer by textured superhydrophobic surfaces. *Nature* **2012**, *489*, 274–277. [CrossRef] [PubMed]
5. Garde, S. Physical chemistry Hydrophobic interactions in context. *Nature* **2015**, *517*, 277–279. [CrossRef] [PubMed]
6. Davis, J.G.; Gierszal, K.P.; Wang, P.; Ben-Amotz, D. Water structural transformation at molecular hydrophobic interfaces. *Nature* **2012**, *491*, 582–585. [CrossRef] [PubMed]
7. Herminghaus, S. Roughness-induced non-wetting. *EPL* **2000**, *52*. [CrossRef]
8. Mortazavi, V.; Khonsari, M.M. On the degradation of superhydrophobic surfaces: A review. *Wear* **2017**, *372–373*, 145–157. [CrossRef]
9. Bhushan, B.; Nosonovsky, M. The rose petal effect and the modes of superhydrophobicity. *Philos. Trans. R. Soc. A Math. Phys. Eng. Sci.* **2010**, *368*, 4713–4728. [CrossRef] [PubMed]
10. Nishino, T.; Meguro, M.; Nakamae, K.; Matsushita, M.; Ueda, Y. The lowest surface free energy based on −CF3Alignment. *Langmuir* **1999**, *15*, 4321–4323. [CrossRef]
11. Wenzel, R.N. Resistance of solid surfaces to wetting by water. *Ind. Eng. Chem.* **1936**, *28*, 988–994. [CrossRef]
12. Cassie, A.B.D.; Baxter, S. Wettability of porous surfaces. *Trans. Faraday Soc.* **1944**, *40*, 546. [CrossRef]
13. Werner, O.; Wågberg, L.; Lindström, T. Wetting of structured hydrophobic surfaces by water droplets. *Langmuir* **2005**, *21*, 12235–12243. [CrossRef] [PubMed]
14. Lafuma, A.; Quéré, D. Superhydrophobic states. *Nat. Mater.* **2003**, *2*, 457. [CrossRef] [PubMed]
15. Quéré, D. Model droplets. *Nat. Mater.* **2004**, *3*, 79. [CrossRef] [PubMed]
16. Yang, C.; Tartaglino, U.; Persson, B.N. Influence of surface roughness on superhydrophobicity. *Phys. Rev. Lett.* **2006**, *97*, 116103. [CrossRef] [PubMed]
17. Bormashenko, E. Progress in understanding wetting transitions on rough surfaces. *Adv. Colloid Interface Sci.* **2015**, *222*, 92–103. [CrossRef] [PubMed]
18. Li, Y.; Quéré, D.; Lv, C.; Zheng, Q. Monostable superrepellent materials. *Proc. Natl. Acad. Sci. USA* **2017**, *114*, 3387–3392. [CrossRef] [PubMed]
19. Bormashenko, E.; Pogreb, R.; Stein, T.; Whyman, G.; Erlich, M.; Musin, A.; Machavariani, V.; Aurbach, D. Characterization of rough surfaces with vibrated drops. *Phys. Chem. Chem. Phys.* **2008**, *10*, 4056–4061. [CrossRef] [PubMed]

20. Bico, J.; Thiele, U.; Quéré, D. Wetting of textured surfaces. *Colloids Surf. A Physicochem. Eng. Asp.* **2002**, *206*, 41–46. [CrossRef]

21. Tuteja, A.; Choi, W.; Ma, M.; Mabry, J.M.; Mazzella, S.A.; Rutledge, G.C.; McKinley, G.H.; Cohen, R.E. Designing Superoleophobic Surfaces. *Science* **2007**, *318*, 1618–1622. [CrossRef] [PubMed]

22. Guo, Z.; Liu, W.; Su, B.L. Superhydrophobic surfaces: From natural to biomimetic to functional. *J. Colloid Interface Sci.* **2011**, *353*, 335–355. [CrossRef] [PubMed]

23. Brewer, S.A.; Willis, C.R. Structure and oil repellency: Textiles with liquid repellency to hexane. *Appl. Surf. Sci.* **2008**, *254*, 6450–6454. [CrossRef]

24. Gu, Z.-Z.; Wei, H.-M.; Zhang, R.-Q.; Han, G.-Z.; Pan, C.; Zhang, H.; Tian, X.-J.; Chen, Z.-M. Artificial silver ragwort surface. *Appl. Phys. Lett.* **2005**, *86*, 201915. [CrossRef]

25. Brinker, C.J. Superhydrophobic Coating. US Patent US7485343 B1, 2008.

26. Barthlott, W.; Schimmel, T.; Wiersch, S.; Koch, K.; Brede, M.; Barczewski, M.; Walheim, S.; Weis, A.; Kaltenmaier, A.; Leder, A.; et al. The salvmia paradox: Superhydrophobic surfaces with hydrophilic pins for air retention under water. *Adv. Mater.* **2010**, *22*, 2325–2328. [CrossRef] [PubMed]

27. Mayser, M.; Bohn, H.; Reker, M.; Barthlott, W. Measuring air layer volumes retained by submerged floating-ferns salvinia and biomimetic superhydrophobic surfaces. *Beilstein J. Nanotechnol.* **2014**, *5*, 812–821. [CrossRef] [PubMed]

28. Verboven, P.; Pedersen, O.; Ho, Q.T.; Nicolai, B.M.; Colmer, T.D. The mechanism of improved aeration due to gas films on leaves of submerged rice. *Plant Cell Environ.* **2014**, *37*, 2433–2452. [CrossRef] [PubMed]

29. Winkel, A.; Visser, E.J.W.; Colmer, T.D.; Brodersen, K.P.; Voesenek, L.A.C.J.; Sand-Jensen, K.; Pedersen, O. Leaf gas films, underwater photosynthesis and plant species distributions in a flood gradient. *Plant Cell Environ.* **2016**, *39*, 1537–1548. [CrossRef] [PubMed]

30. Chen, H.; Zhang, P.; Zhang, L.; Liu, H.; Jiang, Y.; Zhang, D.; Han, Z.; Jiang, L. Continuous directional water transport on the peristome surface of Nepenthes alata. *Nature* **2016**, *532*, 85–89. [CrossRef] [PubMed]

31. Yoon, Y.; Kim, D.; Lee, J.-B. Hierarchical micro/nano structures for super-hydrophobic surfaces and super-lyophobic surface against liquid metal. *Micro Nano Syst. Lett.* **2014**, *2*. [CrossRef]

32. Bixler, G.D.; Bhushan, B. Bioinspired rice leaf and butterfly wing surface structures combining shark skin and lotus effects. *Soft Matter* **2012**, *8*, 11271. [CrossRef]

33. Ueda, E.; Levkin, P.A. Emerging applications of superhydrophilic-superhydrophobic micropatterns. *Adv. Mater.* **2013**, *25*, 1234–1247. [CrossRef] [PubMed]

34. Lai, Y.; Chen, Z.; Lin, C. Recent progress on the superhydrophobic surfaces with special adhesion: From natural to biomimetic to functional. *J. Nanoeng. Nanomanuf.* **2011**, *1*, 18–34. [CrossRef]

35. Scott, A.R. Polymers: Secrets from the deep sea. *Nature* **2015**, *519*, S12–S13. [CrossRef] [PubMed]

36. Savage, N. Synthetic coatings: Super surfaces. *Nature* **2015**, *519*, S7–S9. [CrossRef] [PubMed]

37. Liu, K.; Du, J.; Wu, J.; Jiang, L. Superhydrophobic gecko feet with high adhesive forces towards water and their bio-inspired materials. *Nanoscale* **2012**, *4*, 768–772. [CrossRef] [PubMed]

38. Feng, L.; Zhang, Y.; Xi, J.; Zhu, Y.; Wang, N.; Xia, F.; Jiang, L. Petal Effect: A Superhydrophobic state with high adhesive force. *Langmuir* **2008**, *24*, 4114–4119. [CrossRef] [PubMed]

39. Wagner, T.; Neinhuis, C.; Barthlott, W. Wettability and contaminability of insect wings as a function of their surface sculptures. *Acta Zool.* **1996**, *77*, 213–225. [CrossRef]

40. Choi, W.; Tuteja, A.; Mabry, J.M.; Cohen, R.E.; McKinley, G.H. A modified Cassie-Baxter relationship to explain contact angle hysteresis and anisotropy on non-wetting textured surfaces. *J. Colloid Interface Sci.* **2009**, *339*, 208–216. [CrossRef] [PubMed]

41. Ivanova, E.P.; Hasan, J.; Webb, H.K.; Truong, V.K.; Watson, G.S.; Watson, J.A.; Baulin, V.A.; Pogodin, S.; Wang, J.Y.; Tobin, M.J.; et al. Natural bactericidal surfaces: Mechanical rupture of pseudomonas aeruginosa cells by cicada wings. *Small* **2012**, *8*, 2489–2494. [CrossRef] [PubMed]

42. Tian, Y.; Su, B.; Jiang, L. Interfacial material system exhibiting superwettability. *Adv. Mater.* **2014**, *26*, 6872–6897. [CrossRef] [PubMed]

43. Barthlott, W.; Neinhuis, C. Purity of the sacred lotus, or esCape from contamination in biological surfaces. *Planta* **1997**, *202*, 1–8. [CrossRef]

44. Kazufumi, O.; Mamoru, S.; Yusuke, T.; Ichiro, N. Development of a transparent and ultrahydrophobic glass plate. *Jpn. J. Appl. Phys.* **1993**, *32*, L614.

45. Onda, T.; Shibuichi, S.; Satoh, N.; Tsujii, K. Super-water-repellent fractal surfaces. *Langmuir* **1996**, *12*, 2125–2127. [CrossRef]

46. Shibuichi, S.; Onda, T.; Satoh, N.; Tsujii, K. Super water-repellent surfaces resulting from fractal structure. *J. Phys. Chem.* **1996**, *100*, 19512–19517. [CrossRef]

47. Chen, W.; Fadeev, A.Y.; Hsieh, M.C.; Öner, D.; Youngblood, J.; McCarthy, T.J. Ultrahydrophobic and ultralyophobic surfaces: some comments and examples. *Langmuir* **1999**, *15*, 3395–3399. [CrossRef]

48. Öner, D.; McCarthy, T.J. Ultrahydrophobic surfaces. Effects of topography length scales on wettability. *Langmuir* **2000**, *16*, 7777–7782.

49. Li, N.; Xia, T.; Heng, L.; Liu, L. Superhydrophobic Zr-based metallic glass surface with high adhesive force. *Appl. Phys. Lett.* **2013**, *102*, 251603. [CrossRef]

50. Cheng, Y.-T.; Rodak, D.E.; Angelopoulos, A.; Gacek, T. Microscopic observations of condensation of water on lotus leaves. *Appl. Phys. Lett.* **2005**, *87*, 194112. [CrossRef]

51. Li, S.; Xie, H.; Zhang, S.; Wang, X. Facile transformation of hydrophilic cellulose into superhydrophobic cellulose. *Chem. Commun.* **2007**, 4857. [CrossRef]

52. Zhu, W.; Feng, X.; Feng, L.; Jiang, L. UV-Manipulated wettability between superhydrophobicity and superhydrophilicity on a transparent and conductive SnO_2 nanorod film. *Chem. Commun.* **2006**, 2753–2755. [CrossRef]

53. Lee, W.; Jin, M.K.; Yoo, W.C.; Lee, J.K. Nanostructuring of a polymeric substrate with well-defined nanometer-scale topography and tailored surface wettability. *Langmuir* **2004**, *20*, 7665–7669. [CrossRef] [PubMed]

54. Erbil, H.Y.; Demirel, A.L.; Avcı, Y.; Mert, O. Transformation of a simple plastic into a superhydrophobic surface. *Science* **2003**, *299*, 1377–1380. [CrossRef] [PubMed]

55. Gao, Y.; Feng, S.J.; Wang, Q.; Huang, Y.G.; Qing, F.L. Superhydrophobic and highly oleophobic cotton textile: achieved by silica particles and PFPE. *Adv. Mater. Res.* **2009**, *79–82*, 683–686. [CrossRef]

56. Guo, Z.; Zhou, F.; Hao, J.; Liu, W. Stable biomimetic super-hydrophobic engineering materials. *J. Am. Chem. Soc.* **2005**, *127*, 15670–15671. [CrossRef] [PubMed]

57. Guo, M.; Diao, P.; Wang, X.; Cai, S. The effect of hydrothermal growth temperature on preparation and photoelectrochemical performance of ZnO nanorod array films. *J. Solid State Chem.* **2005**, *178*, 3210–3215. [CrossRef]

58. Jiang, P.; Zhou, J.J.; Fang, H.F.; Wang, C.Y.; Wang, Z.L.; Xie, S.S. Hierarchical shelled ZnO structures made of bunched nanowire arrays. *Adv. Funct. Mater.* **2007**, *17*, 1303–1310. [CrossRef]

59. Safaee, A.; Sarkar, D.K.; Farzaneh, M. Superhydrophobic properties of silver-coated films on copper surface by galvanic exchange reaction. *Appl. Surf. Sci.* **2008**, *254*, 2493–2498. [CrossRef]

60. Bravo, J.; Zhai, L.; Wu, Z.; Cohen, R.E.; Rubner, M.F. Transparent superhydrophobic films based on silica nanoparticles. *Langmuir* **2007**, *23*, 7293–7298. [CrossRef] [PubMed]

61. Kinge, S.; Crego-Calama, M.; Reinhoudt, D.N. Self-assembling nanoparticles at surfaces and interfaces. *ChemPhysChem* **2008**, *9*, 20–42. [CrossRef] [PubMed]

62. Satyaprasad, A.; Jain, V.; Nema, S.K. Deposition of superhydrophobic nanostructured Teflon-like coating using expanding plasma arc. *Appl. Surf. Sci.* **2007**, *253*, 5462–5466. [CrossRef]

63. Ci, L.; Vajtai, R.; Ajayan, P.M. Vertically aligned large-diameter double-walled carbon nanotube arrays having ultralow density. *J. Phys. Chem. C* **2007**, *111*, 9077–9080. [CrossRef]

64. Cho, W.K.; Park, S.; Jon, S.; Choi, I.S. Water-repellent coating: Formation of polymeric self-assembled monolayers on nanostructured surfaces. *Nanotechnology* **2007**, *18*, 395602. [CrossRef] [PubMed]

65. Xue, C.-H.; Jia, S.-T.; Zhang, J.; Ma, J.-Z. Large-area fabrication of superhydrophobic surfaces for practical applications: an overview. *Sci. Technol. Adv. Mater.* **2010**, *11*, 033002. [CrossRef] [PubMed]

66. Gao, X.; Jiang, L. Biophysics: Water-repellent legs of water striders. *Nature* **2004**, *432*, 36. [CrossRef] [PubMed]

67. Wang, J.; Wen, Y.; Hu, J.; Song, Y.; Jiang, L. Fine control of the wettability transition temperature of colloidal-crystal films: From superhydrophilic to superhydrophobic. *Adv. Funct. Mater.* **2007**, *17*, 219–225. [CrossRef]

68. Mozumder, M.S.; Zhang, H.; Zhu, J. Mimicking Lotus Leaf: Development of micro-nanostructured biomimetic superhydrophobic polymeric surfaces by ultrafine powder coating technology. *Macromol. Mater. Eng.* **2011**, *296*, 929–936. [CrossRef]

69. Hsieh, C.-T.; Chen, J.-M.; Huang, Y.-H.; Kuo, R.-R.; Li, C.-T.; Shih, H.-C.; Lin, T.-S.; Wu, C.-F. Influence of fluorine/carbon atomic ratio on superhydrophobic behavior of carbon nanofiber arrays. *J. Vac. Sci. Technol. B* **2006**, *24*, 113–117. [CrossRef]

70. Genzer, J.; Efimenko, K. Creating long-lived superhydrophobic polymer surfaces through mechanically assembled monolayers. *Science* **2000**, *290*, 2130–2133. [CrossRef] [PubMed]

71. Verho, T.; Bower, C.; Andrew, P.; Franssila, S.; Ikkala, O.; Ras, R.H.A. Mechanically durable superhydrophobic surfaces. *Adv. Mater.* **2011**, *23*, 673–678. [CrossRef] [PubMed]

72. Huang, J.; Wang, S. Facile Preparation of a robust and durable superhydrophobic coating using biodegradable lignin-coated cellulose nanocrystal particles. *Materials* **2017**, *10*. [CrossRef] [PubMed]

73. Ahmad, I.; Kan, C.-W. A Review on development and applications of bio-inspired superhydrophobic textiles. *Materials* **2016**, *9*. [CrossRef] [PubMed]

74. Wang, Z.; Shen, X.; Qian, T.; Wang, J.; Sun, Q.; Jin, C. Facile fabrication of a PDMS@Stearic acid-kaolin coating on lignocellulose composites with superhydrophobicity and flame retardancy. *Materials* **2018**, *11*. [CrossRef] [PubMed]

75. Aslanidou, D.; Karapanagiotis, I. Superhydrophobic, superoleophobic and antimicrobial coatings for the protection of silk textiles. *Coatings* **2018**, *8*. [CrossRef]

76. Chatzigrigoriou, A.; Manoudis Panagiotis, N.; Karapanagiotis, I. Fabrication of water repellent coatings using waterborne resins for the protection of the cultural heritage. *Macromol. Symp.* **2013**, *331–332*, 158–165. [CrossRef]

77. Wong, T.-S.; Kang, S.H.; Tang, S.K.Y.; Smythe, E.J.; Hatton, B.D.; Grinthal, A.; Aizenberg, J. Bioinspired self-repairing slippery surfaces with pressure-stable omniphobicity. *Nature* **2011**, *477*, 443–447. [CrossRef] [PubMed]

78. Dolgin, E. Textiles: Fabrics of life. *Nature* **2015**, *519*, S10–S11. [CrossRef] [PubMed]

79. Yao, X.; Song, Y.; Jiang, L. Applications of bio-inspired special wettable surfaces. *Adv. Mater.* **2011**, *23*, 719–734. [CrossRef] [PubMed]

80. Lampin, M.; Warocquier-Clérout, R.; Legris, C.; Degrange, M.; Sigot-Luizard, M.F. Correlation between substratum roughness and wettability, cell adhesion, and cell migration. *J. Biomed. Mater. Res.* **1997**, *36*, 99–108. [CrossRef]

81. Zelzer, M.; Majani, R.; Bradley, J.W.; Rose, F.R.A.J.; Davies, M.C.; Alexander, M.R. Investigation of cell-surface interactions using chemical gradients formed from plasma polymers. *Biomaterials* **2008**, *29*, 172–184. [CrossRef] [PubMed]

82. Fan, H.; Chen, P.; Qi, R.; Zhai, J.; Wang, J.; Chen, L.; Chen, L.; Sun, Q.; Song, Y.; Han, D.; et al. Greatly improved blood compatibility by microscopic multiscale design of surface architectures. *Small* **2009**, *5*, 2144–2148. [CrossRef] [PubMed]

83. Tomsic, B.; Simoncic, B.; Orel, B.; Černe, L.; Tavčar, P.; Zorko, M.; Jerman, I.; Vilcnik, A.; Kovač, J. Sol-gel coating of cellulose fibres with antimicrobial and repellent properties. *J. Sol-Gel Sci. Technol.* **2008**, *47*, 44–57. [CrossRef]

84. Epstein, A.K.; Wong, T.S.; Belisle, R.A.; Boggs, E.M.; Aizenberg, J. Liquid-infused structured surfaces with exceptional anti-biofouling performance. *Proc. Natl. Acad. Sci. USA* **2012**, *109*, 13182–13187. [CrossRef] [PubMed]

85. Nagrath, S.; Sequist, L.V.; Maheswaran, S.; Bell, D.W.; Irimia, D.; Ulkus, L.; Smith, M.R.; Kwak, E.L.; Digumarthy, S.; Muzikansky, A.; et al. Isolation of rare circulating tumour cells in cancer patients by microchip technology. *Nature* **2007**, *450*, 1235–1239. [CrossRef] [PubMed]

86. Yu, X.; Liu, Z.; Janzen, J.; Chafeeva, I.; Horte, S.; Chen, W.; Kainthan, R.K.; Kizhakkedathu, J.N.; Brooks, D.E. Polyvalent choline phosphate as a universal biomembrane adhesive. *Nat. Mater.* **2012**, *11*, 468. [CrossRef] [PubMed]

87. Wang, S.; Liu, K.; Liu, J.; Yu, Z.T.F.; Xu, X.; Zhao, L.; Lee, T.; Lee, E.K.; Reiss, J.; Lee, Y.K.; et al. Highly efficient capture of circulating tumor cells by using nanostructured silicon substrates with integrated chaotic micromixers. *Angew. Chem. Int. Ed.* **2011**, *50*, 3084–3088. [CrossRef] [PubMed]

88. Vikesland, P.J.; Wigginton, K.R. Nanomaterial enabled biosensors for pathogen monitoring—A review. *Environ. Sci. Technol.* **2010**, *44*, 3656–3669. [CrossRef] [PubMed]

89. Geyer, F.L.; Ueda, E.; Liebel, U.; Grau, N.; Levkin, P.A. Superhydrophobic-superhydrophilic micropatterning: Towards genome-on-a-chip cell microarrays. *Angew. Chem. Int. Ed.* **2011**, *50*, 8424–8427. [CrossRef] [PubMed]

90. Lee, Y.Y.; Narayanan, K.; Gao, S.J.; Ying, J.Y. Elucidating drug resistance properties in scarce cancer stem cells using droplet microarray. *Nano Today* **2012**, *7*, 29–34. [CrossRef]

91. Yuan, Y.; Lee, T.R. *Surface Science Techniques*; Springer: Berlin/Heidelberg, Germany, 2013; Volume 51, pp. 1–33.

92. Burchak, O.N.; Mugherli, L.; Ostuni, M.; Lacapere, J.J.; Balakirev, M.Y. Combinatorial discovery of fluorescent pharmacophores by multicomponent reactions in droplet arrays. *J. Am. Chem. Soc.* **2011**, *133*, 10058–10061. [CrossRef] [PubMed]

93. Cao, L.; Jones, A.K.; Sikka, V.K.; Wu, J.; Gao, D. Anti-Icing superhydrophobic coatings. *Langmuir* **2009**, *25*, 12444–12448. [CrossRef] [PubMed]

94. Meuler, A.J.; Smith, J.D.; Varanasi, K.K.; Mabry, J.M.; McKinley, G.H.; Cohen, R.E. Relationships between water wettability and ice adhesion. *ACS Appl. Mater. Interfaces* **2010**, *2*, 3100–3110. [CrossRef] [PubMed]

95. Kim, P.; Wong, T.S.; Alvarenga, J.; Kreder, M.J.; Adorno-Martinez, W.E.; Aizenberg, J. Liquid-infused nanostructured surfaces with extreme anti-ice and anti-frost performance. *ACS Nano* **2012**, *6*, 6569–6577. [CrossRef] [PubMed]

96. Bird, J.C.; Dhiman, R.; Kwon, H.-M.; Varanasi, K.K. Reducing the contact time of a bouncing drop. *Nature* **2013**, *503*, 385–388. [CrossRef] [PubMed]

97. Schutzius, T.M.; Jung, S.; Maitra, T.; Graeber, G.; Köhme, M.; Poulikakos, D. Spontaneous droplet trampolining on rigid superhydrophobic surfaces. *Nature* **2015**, *527*, 82–85. [CrossRef] [PubMed]

98. Yuan, J.; Liu, X.; Akbulut, O.; Hu, J.; Suib, S.L.; Kong, J.; Stellacci, F. Superwetting nanowire membranes for selective absorption. *Nat. Nanotechnol.* **2008**, *3*, 332–336. [CrossRef] [PubMed]

99. Xue, Z.; Wang, S.; Lin, L.; Chen, L.; Liu, M.; Feng, L.; Jiang, L. A novel superhydrophilic and underwater superoleophobic hydrogel-coated mesh for oil/water separation. *Adv. Mater.* **2011**, *23*, 4270–4273. [CrossRef] [PubMed]

100. Zhang, M.; Feng, S.; Wang, L.; Zheng, Y. Lotus effect in wetting and self-cleaning. *Biotribology* **2016**, *5*, 31–43. [CrossRef]

101. Vasiljević, J.; Gorjanc, M.; Tomšič, B.; Orel, B.; Jerman, I.; Mozetič, M.; Vesel, A.; Simončič, B. The surface modification of cellulose fibres to create super-hydrophobic, oleophobic and self-cleaning properties. *Cellulose* **2013**, *20*, 277–289. [CrossRef]

102. Vasiljević, J.; Gorjanc, M.; Jerman, I.; Tomšič, B.; Modic, M.; Mozetič, M.; Orel, B.; Simončič, B. Influence of oxygen plasma pre-treatment on the water repellency of cotton fibers coated with perfluoroalkyl-functionalized polysilsesquioxane. *Fibers Polym.* **2016**, *17*, 695–704. [CrossRef]

103. Kiwi, J.; Pulgarin, C. Innovative self-cleaning and bactericide textiles. *Catal. Today* **2010**, *151*, 2–7. [CrossRef]

104. Aslanidou, D.; Karapanagiotis, I. Waterborne superhydrophobic and superoleophobic coatings for the protection of marble and sandstone. *Materials* **2018**, *11*. [CrossRef] [PubMed]

105. Prevo, B.G.; Hon, E.W.; Velev, O.D. Assembly and characterization of colloid-based antireflective coatings on multicrystalline silicon solar cells. *J. Mater. Chem.* **2007**, *17*, 791–799. [CrossRef]

106. Nakajima, A.; Abe, K.; Hashimoto, K.; Watanabe, T. Preparation of hard super-hydrophobic films with visible light transmission. *Thin Solid Films* **2000**, *376*, 140–143. [CrossRef]

107. Manca, M.; Cannavale, A.; De Marco, L.; Aricò, A.S.; Cingolani, R.; Gigli, G. Durable superhydrophobic and antireflective surfaces by trimethylsilanized silica nanoparticles-based sol-gel processing. *Langmuir* **2009**, *25*, 6357–6362. [CrossRef] [PubMed]

108. Yabu, H.; Shimomura, M. Single-step fabrication of transparent superhydrophobic porous polymer films. *Chem. Mater.* **2005**, *17*, 5231–5234. [CrossRef]

109. Liu, K.; Jiang, L. Metallic surfaces with special wettability. *Nanoscale* **2011**, *3*, 825–838. [CrossRef] [PubMed]

110. Ishizaki, T.; Sakamoto, M. Facile formation of biomimetic color-tuned superhydrophobic magnesium alloy with corrosion resistance. *Langmuir* **2011**, *27*, 2375–2381. [CrossRef] [PubMed]

111. Ishizaki, T.; Saito, N. Rapid formation of a superhydrophobic surface on a magnesium alloy coated with a cerium oxide film by a simple immersion process at room temperature and its chemical stability. *Langmuir* **2010**, *26*, 9749–9755. [CrossRef] [PubMed]

112. Zhang, F.; Zhao, L.; Chen, H.; Xu, S.; Evans, D.G.; Duan, X. Corrosion resistance of superhydrophobic layered double hydroxide films on aluminum. *Angew. Chem. Int. Ed.* **2008**, *47*, 2466–2469. [CrossRef] [PubMed]

113. Wen, L.; Tian, Y.; Jiang, L. Bioinspired super-wettability from fundamental research to practical applications. *Angew. Chem. Int. Ed.* **2015**, *54*, 3387–3399. [CrossRef] [PubMed]

114. López-Quintela, M.A.; Tojo, C.; Blanco, M.C.; García Rio, L.; Leis, J.R. Microemulsion dynamics and reactions in microemulsions. *Curr. Opin. Colloid Interface Sci.* **2004**, *9*, 264–278. [CrossRef]

115. Su, B.; Wang, S.; Song, Y.; Jiang, L. A heatable and evaporation-free miniature reactor upon superhydrophobic pedestals. *Soft Matter* **2012**, *8*, 631–635. [CrossRef]

116. Su, B.; Wang, S.; Song, Y.; Jiang, L. A miniature droplet reactor built on nanoparticle-derived superhydrophobic pedestals. *Nano Res.* **2010**, *4*, 266–273. [CrossRef]

117. McHale, G.; Newton, M.I.; Shirtcliffe, N.J. Immersed superhydrophobic surfaces: Gas exchange, slip and drag reduction properties. *Soft Matter* **2010**, *6*, 714–719. [CrossRef]

118. Bhushan, B. Bioinspired structured surfaces. *Langmuir* **2012**, *28*, 1698–1714. [CrossRef] [PubMed]

119. Koh, J.S.; Yang, E.; Jung, G.P.; Jung, S.P.; Son, J.H.; Lee, S.I.; Jablonski, P.G.; Wood, R.J.; Kim, H.Y.; Cho, K.J. Jumping on water: Surface tension-dominated jumping of water striders and robotic insects. *Science* **2015**, *349*, 517–521. [CrossRef] [PubMed]

120. Zhao, J.; Zhang, X.; Chen, N.; Pan, Q. Why superhydrophobicity is crucial for a water-jumping microrobot? Experimental and theoretical investigations. *ACS Appl. Mater. Interfaces* **2012**, *4*, 3706–3711. [CrossRef] [PubMed]

121. Gallardo, B.S. Electrochemical principles for active control of liquids on submillimeter scales. *Science* **1999**, *283*, 57–60. [CrossRef] [PubMed]

122. Kundu, P.K.; Samanta, D.; Leizrowice, R.; Margulis, B.; Zhao, H.; Börner, M.; Udayabhaskararao, T.; Manna, D.; Klajn, R. Light-controlled self-assembly of non-photoresponsive nanoparticles. *Nat. Chem.* **2015**, *7*, 646–652. [CrossRef] [PubMed]

123. Zhai, L.; Berg, M.C.; Cebeci, F.Ç.; Kim, Y.; Milwid, J.M.; Rubner, M.F.; Cohen, R.E. Patterned superhydrophobic surfaces: Toward a synthetic mimic of the namib desert beetle. *Nano Lett.* **2006**, *6*, 1213–1217. [CrossRef] [PubMed]

124. Lee, C.; Kang, B.J.; Oh, J.H. High-resolution conductive patterns fabricated by inkjet printing and spin coating on wettability-controlled surfaces. *Thin Solid Films* **2016**, *616*, 238–246. [CrossRef]

materials

MDPI

Review

Surface Texture-Based Surface Treatments on Ti6Al4V Titanium Alloys for Tribological and Biological Applications: A Mini Review

Naiming Lin [1,2,3,*], Dali Li [1], Jiaojuan Zou [1], Ruizhen Xie [4,*], Zhihua Wang [2] and Bin Tang [1]

[1] Research Institute of Surface Engineering, Taiyuan University of Technology, Taiyuan 030024, Shanxi, China; lidali0197@link.tyut.edu.cn (D.L.); zoujiaojuan@tyut.edu.cn (J.Z.); tangbin@tyut.edu.cn (B.T.)
[2] Shanxi Key Laboratory of Material Strength and Structure Impact, Taiyuan University of Technology, Taiyuan 030024, Shanxi, China; wangzhihua@tyut.edu.cn
[3] Department of Chemical and Materials Engineering, University of Alberta, Edmonton, AB T6G 1H9, Canada
[4] Department of Civil Engineering, Taiyuan University of Technology, Taiyuan, 030024, Shanxi, China
* Correspondence: lnmlz33@163.comcn (N.L.); xieruizhen0094@link.tyut.edu.cn (R.X.);
   Tel.: +86-351-601-0540 (N.L.)

Received: 26 January 2018; Accepted: 23 March 2018; Published: 24 March 2018

Abstract: Surface texture (ST) has been confirmed as an effective and economical surface treatment technique that can be applied to a great range of materials and presents growing interests in various engineering fields. Ti6Al4V which is the most frequently and successfully used titanium alloy has long been restricted in tribological-related operations due to the shortcomings of low surface hardness, high friction coefficient, and poor abrasive wear resistance. Ti6Al4V has benefited from surface texture-based surface treatments over the last decade. This review begins with a brief introduction, analysis approaches, and processing methods of surface texture. The specific applications of the surface texture-based surface treatments for improving surface performance of Ti6Al4V are thoroughly reviewed from the point of view of tribology and biology.

Keywords: surface texture; surface treatment; Ti6Al4V alloy; tribology; biology

1. Introduction

Titanium (Ti) was once considered a rare metal, however, Ti ranks as the ninth Clarke number, it is the fourth most abundant structural metal on the earth, and Ti presents a level of about 0.6% in the world [1]. Titanium and titanium alloys have been rapidly developed since the pure metal first became commercially available about sixty years ago [1]. Thanks to their extraordinary merits of high strength to weight ratio, relatively low modulus, high yield strength and toughness, excellent corrosion resistance as well as promising biocompatibility, titanium alloys have received extensive attention in various fields ranging from civilian products to military equipment for decades [2–5]. Series of titanium alloys have been designed and produced for different purposes with great success. Ti6Al4V alloy (referred to Ti6Al4V hereafter) which was developed and made its name in the 1950s, was the first practical titanium alloy in the world. Ti6Al4V got its reputation of ace titanium alloy from the fact that most other existing titanium alloys for various applications were obtained by optimizing and improvement based on it. Up to now, Ti6Al4V is still the most frequently and successfully used titanium alloy, it occupies about one half of the total world production of titanium alloys [6–8]. However, Ti6Al4V cannot meet all of the engineering demands, e.g., it is seldom operated in tribological-related engineering conditions due to its drawbacks of low surface hardness, high coefficient of friction, and poor abrasive wear resistance. These shortcomings have greatly limited or even prevented Ti6Al4V larger scale use for various applications [9–13]. It is well known that degradation/failure of materials in engineering, e.g., wear and/or corrosion are mainly determined by the surface performance of the material rather than by bulk properties. Therefore, endowing improved properties on the surfaces of materials by surface modification technologies are attractive and suitable approaches to overcome the aforementioned issues [14]. Appropriate surface treatment is able to improve the surface performances (hardness, chemical stability, friction-reduction or wear-resistant) and retain the desirable bulk attributes of the materials, and then further expand the applications of materials in different fields [15]. On the other hand, the surface modification can also make a promising compromise between the cost and the performance of engineering components [16]. A variety of surface modification technologies have been applied to enhance the tribological performance of Ti6Al4V on the surface by forming coatings/films/layers [17–34].

According to recent bionic studies, apart from conventional surface modification technologies, appropriate design on the surface topography/pattern is able to improve the tribological performance of materials [35]. From the earlier point of view, smoother surfaces should be more favorable to improve the tribological performance of engineering components. However, recent achievements of bionics have suggested that non-smooth surfaces with regular arranged topographies/patterns usually exhibit promising tribological behaviors. For examples, dung beetle with non-smooth epidermis is able to resist wear and extrusion; the non-smooth skin of dolphin can effectively reduce its swimming resistance; the micro-rhombus structures on the surface of shark skin contribute to noise reduction in the process of diving [36–39]. The non-smooth surfaces in the natural world have brought about many meaningful inspirations to material scientists and engineers [40–42]. It has been confirmed that artificial design on the surface topography/pattern of materials by imitating the non-smooth surfaces in the natural world can even realize some similar functions [35–39]. The way to obtain artificial surface patterns with typical distributing characteristics such as dimples, grooves, pimples and so on were collectively named as surface texture (ST) [43–45]. Surface texturing which has been confirmed as an effective method to improve the tribological behaviors of materials and tools in tribology-related fields, has been considered a hot issue in material science and mechanical engineering over the last decade [46–56]. In general, the active roles of ST in tribological performance lie in the main aspects as follows (see Figure 1) [57,58]: service as storage of the solid/grease lubricant to provide continuous lubrication and improve the elastohydrodynamic effect under liquid lubrication; trapping wear debris generated during service and minimizing abrasive wear, reduction in nominal contact area. ST has been extensively applied on various materials for different purposes [52,53,59–62].

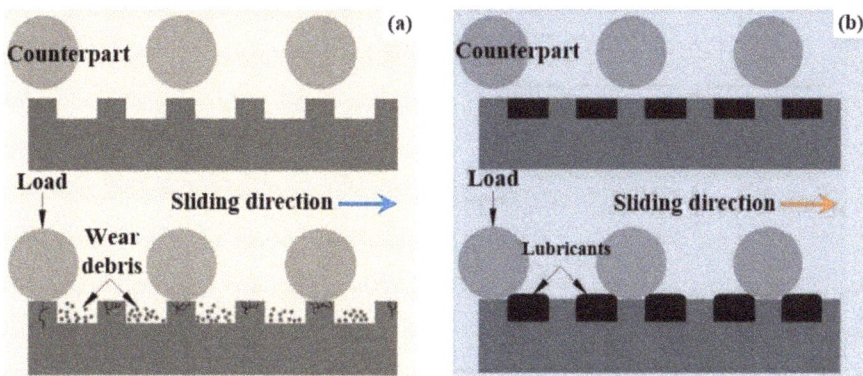

Figure 1. Schematic diagram of active roles of surface texturing in tribological performance: (**a**) continuous lubrication and (**b**) capturing friction debris.

With the advantages of surface modification technologies and surface texturing in improving the tribological performance of materials, some duplex treatments of surface modification-surface texturing have been developed. One kind is the "surface texturing + surface modification", the other kind is the "surface modification + surface texturing", as shown in Figure 2 [63–76]. These works have created a database and provided reference information for practical applications of surface modification-surface texturing duplex treatments. Due to its promising merits above, Ti6Al4V has long been used as hard tissue replacements (e.g., dental implant and artificial joint) and services in the human body. Before an implanting operation, Ti6Al4V biomedical devices and components are usually surface modified to achieve the expected properties of better biocompatibility, excellent antibacterial property, promising osteogenesis capability, low biotoxicity, benefiting cell adhesion and proliferation, controllable ion release rate, and so on [77,78]. The surface modification–surface texturing duplex treatment is considered to make use of the inherent positive effects of the surface modification layer as well as take the advantages of surface texture. Meanwhile a higher specific surface area which is favorable to hydrophilicity or hydrophobicity, adhesion and proliferation of cell and antibacterial behavior, might be realized on the surfaces of Ti6Al4V biomedical devices and components after the mentioned duplex treatment (Figure 3) [79]. It seems that the combination of the surface modification and surface texturing can achieve a "1 + 1 > 2" effect on improving the surface performance of Ti6Al4V.

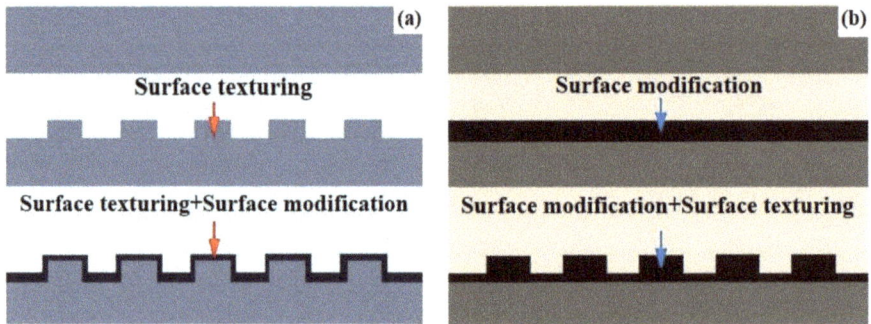

Figure 2. Schematic diagram of duplex treatment: (**a**) type one: surface texturing + surface modification; (**b**) type two: surface modification + surface texturing.

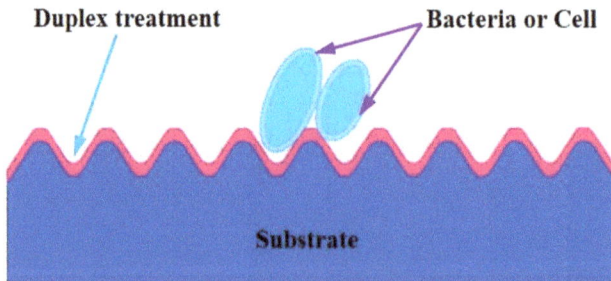

Figure 3. Schematic diagram of adhesion of bacteria or cell on a duplex treated surface.

Surfaces with promising super-hydrophobicity have received much attention from scientists due to their great potential in many applications, particularly in fields such as waterproofing, antifouling, self-cleaning, anti-corrosion, drag-reducing, anti-frosting, and anti-icing [80,81]. As a multi-purpose material, Ti6Al4V has also been endowed with super-hydrophobicity by surface texturing in some cases of the mentioned applications [80,81].

The relevant information with emphasis on the latest progress in the research on the surface texture-based surface treatments on Ti6Al4V for tribological and biological applications is summarized in this mini review. Although the surface texture-based surface treatments are not fire-new technologies, the demand for failure protection and to prolong service life drive the on-going interests of the scientific community making them still worthy of further studies. This work is expected to create a database and provide reference information, thereby broadening the practical applications of surface texture-based surface treatments on Ti6Al4V and other metallic materials.

2. Surface Texture-Based Surface Treatments with Improved Performance

Based on the above discussed background, it has been confirmed that ST possesses outstanding advantages. Therefore, many researchers have dedicated themselves to the studies of surface texture, including the characterizations, fabrications, and applications of surface texture. Surface patterns on textured surfaces usually have regular distributing characteristics at nano- or micro-scales, therefore microscopic analysis is required. Scanning electron microscope (SEM), atomic force microscope (AFM) and laser scanning confocal microscope (LSCM) are often used to observe the geometrical characteristics of surface texture [82–84]. Additionally, some researchers have paid attention to the residual stresses in surface layers on the textured surfaces [85,86].

With respect to the research on fabrications of surface texture, researchers mainly focus on the shapes of the ST unit, the processing methods of surface texturing, and related parameters [87–93]. Dimple and pimple with geometric configurations of ellipse, circle, triangle, square, hexagons, and grooves in the straight and zigzag lines have been adopted to form the shape of ST [40–45,87]. Up to now, there have been several ways to obtain ST on the surfaces of materials [41,94], including laser surface texturing (LST), laser shock peening (LSP), electro spark surface texturing (ESST)/electrical discharge machining (EDM), chemical reactive ion etching (RIE), lithography and anisotropic etching (LIAE), abrasive jet machining (AJM), lithography, galvanoformung, and abformung (LIGA), vibrorolling, undulated surfaces and so on. Actually, amongst all the practical surface texturing methods, it seems that LST is the most promising concept. This is because the laser is extremely fast and allows short processing times, it is clean to the environment, and provides excellent control of the shape and size of the micropatterns, which allows realization of optimum designs [40–42,90]. Generally, the fabricating parameters of ST are case-by-case. Taking LST as an example, the width and depth of the textures are related to scanning velocity, power, frequency, and pulse width of the laser [93]. On the other hand, the distribution characteristic (area density: the ratio of the ST area to the whole surface area) of ST play an important role in the surface performance of a textured surface. Depending on the specific situation, parameter optimization of surface texturing is usually conducted through trial and error tests [87–94]. The existing applications in tribology and biology of surface texture-based surface treated Ti6Al4V were reviewed, and the relevant research on fabrications of ST are also presented.

In the following, surface texture-based surface treatments, including the single surface texturing and surface texturing/surface modification duplex treatment on Ti6Al4V alloy for tribological and biological applications are suggested in Sections 2.1 and 2.2.

2.1. Tribological Applications

Under different service conditions, a friction phenomenon usually exists between relative movement interfaces of solid–solid, solid–liquid and solid–gas [95–97]. Friction normally can lead to damage of wear and/or fatigue, lowering work efficiency and producing noise pollution when the engineering components are used under different service environments where the above-mentioned friction interfaces are involved [22,98,99]. While under the negative effects of friction in normal applications, the service performance of Ti6Al4V engineering components would be decreased and then the lifecycle of the whole equipment might also be reduced [100,101]. Surface texture-based surface treatments developed by different scientists and engineers have provided promising approaches to overcome the relevant issues [42–94,102,103].

Guo and Caslaru [90,101] conducted a novel micro laser shock peening (LSP) on Ti6Al4V to fabricate micro-circle dent arrays with different densities based on a surface patterning technique at various power levels, 1 W, 2 W, and 3 W. It was found that higher laser power could produce deeper dent. The highest depth of approximately 1 μm was achieved using a 3 W laser power, and the dent diameter also increased as the laser power was increased. A pile-up region appeared on the outer edge of each dent. It was believed that tensile stress developed in the pile-up region. Meanwhile the LST treatment could create a hard surface, the hardness in the central zone of the dent was enhanced by about 15%. Strain hardening, strain-rate hardening, and compressive residual stress (an increase in compressive residual stress on the LSP treated surface was found after removing the pile-up effect via mild polishing) contributed to the hardening effect in the dented area. Two surfaces with dent densities of 10% and 20% were selected to investigate the tribological behaviors under flooded and boundary lubricated conditions, countered with chrome steel balls. The 20% dent density produced a higher pile-up zone density which indicated a negative influence on tribological behavior. However, the beneficial effects of surfaces with 10% dent density were more pronounced at reducing the coefficient of friction (CoF) and wear rate under both lubricated conditions. The results confirmed that LSP is a reliable process for fabricating micro-dent arrays with different densities, micro-dents with low density were suitable for Ti6Al4V in sliding contact applications.

In order to improve the tribological performance of Ti6Al4V, Hu et al. [104] applied laser surface texturing (LST) to form regular circle dimple ST with three different diameters of 45 μm, 160 μm and 300 μm, as well as with a spacing of 100 μm and a depth of 25 μm on Ti6Al4V. Two types of poly-alpha-olefin (PAO) lubricants with different kinetic viscosities were evaluated as lubricants on raw Ti6Al4V and LST treated Ti6Al4V samples using a pin-disc tribometer and steel pins. It was observed that the textured surfaces exhibited lower friction coefficients and wear compared with the un-textured surface. Furthermore, the textured surfaces exhibited better tribological properties when the tests were conducted under higher speeds and loads and with higher viscosity oil. However, when the load increased, the effect of friction reduction for all the textured surfaces decreased for the lubricating oil with lower viscosity. It was found that the lubricant film thickened due to the enhancement of hydrodynamic pressure near the dimples, and higher viscosity oil was prone to provide a secondary lubrication effect. The formed micro-dimples on Ti6Al4V surfaces shifted the transition from boundary to mixed lubrication as the tests were operated under much higher loads when higher viscosity oil was used. The effect of dimple sizes on the tribological properties the LST surfaces under the same testing conditions was also analyzed. The results showed that the dimple with a diameter size of 160 μm was more favorable to reduce friction when the interval of dimples was fixed.

By using LST technique, Lian et al. [105] fabricated three kinds of ST on Ti6Al4V surfaces: groove, crosshatch (included angle 90°), and dimple with interval spacing of 100, 200, and 300 μm, respectively. All of the LST treated surfaces showed higher surface hardness values than that of the raw Ti6Al4V. Actually, the hardness values could be arranged as follows: raw Ti6Al4V < groove-Ti6Al4V < crosshatch-Ti6Al4V < dimple-Ti6Al4V. It was seen that the dimple textured samples with interval spacing of 200 and 300 μm had remarkably reduced the coefficient of friction, while groove and crosshatch textures were helpful for friction reduction. The textured surfaces presented shallower wear traces than that of the raw Ti6Al4V, the distribution characteristics of the three kinds of ST received much slighter damage compared with the untreated Ti6Al4V. In this work, a conclusion could be drawn that groove, crosshatch (included angle 90°), and dimple obtained by LST managed to improve the wear resistance of Ti6Al4V.

Xu et al. [106] manufactured crosshatch STs on Ti6Al4V surfaces by employment of electrical discharge machining (EDM). The orthogonal design was applied to investigate the effects of geometric parameters (width, depth, interval/width ratio, and angles) of the crosshatches on the tribological characteristics of Ti6Al4V alloy in water lubrication against Si_3N_4. The width, depth, interval/width ratio, and angles of the crosshatch were factored in, and each of the above factors had four levels: width values (0.15, 0.20, 0.25, 0.30 mm), depth values (0.05, 0.075, 0.1, 0.125 mm), interval/width ratio

(2, 6, 10, 14) and angles (15°, 30°, 45°, 60°). The results suggested that the crosshatch ST with suitable geometric parameters could effectively reduce the friction coefficient and wear rate of tribopairs in water lubrication. It was found that width (0.2 mm), depth (0.125 mm), interval/width ratio (10) and angle (45°) were the optimum factors and levels for realizing a lower and stable friction. While the combinations of factors and levels with width (0.2 mm), depth (0.05 mm), interval/width ratio (2), and angle (45°) revealed the lowest wear loss. When the angle of the crosshatch was 45°, both the friction coefficient and the wear of the tribopairs simultaneously decreased. The extent of the geometric parameters impact on friction coefficient was arranged in the following sequence: width > depth > angles > interval/width ratio. The depth and angle of the crosshatch were the main factors which influenced the wear loss of the tribopairs.

Bonse et al. [107] obtained homogeneous ripple ST with a spatial period of about 600 nm on Ti6Al4V surface after multiple femtosecond laser pulse irradiation. Compared to the blank sample, the ripple textured Ti6Al4V demonstrated lower and more stable friction coefficient in reciprocal sliding against a hardened steel ball under engine oil lubrication. The treated Ti6Al4V received a shallow and narrow wear trace, which was also hardly visible by microscope. Slight surface damage was generated after the tribological test. Meanwhile, it was found that the additives in the engine oil efficiently covered the ST, where a complete gliding intermediate layer was formed. This intermediate layer which played an important role in reducing friction and wear during the reciprocal sliding motion, prevented direct intermetallic contact of the metallic tribopairs.

Lian et al. [108] concentrated on enhancing the tribological performance of Ti6Al4V in seawater and built crosshatch and dimple textures on Ti6Al4V surfaces via LST. Tribological performance was evaluated by reciprocating friction tests against Si_3N_4 balls in artificial seawater and distilled water, respectively. The results showed that the friction coefficients and wear losses in volume of crosshatch and dimple textured Ti6Al4V surfaces were far smaller than those of the raw Ti6Al4V substrates. The friction coefficients belonging to the surfaces with crosshatch and dimple were decreased by 11.7% and 17.8%, and wear losses in volume were reduced by 57.5% and 36.8% in artificial seawater. Both textured Ti6Al4V surfaces revealed lower friction coefficients in artificial seawater than those in distilled water, while the wear losses in volume led to the opposite results. Ti6Al4V was prone to form a passive film on the surface when it was put in artificial seawater, the passive film could act as a solid lubricant due to its lower shear strength. Meanwhile artificial seawater was more benefit for the passivation process than distilled water, therefore all the Ti6Al4V samples showed lower friction coefficients in artificial seawater compared with the samples which were measured in distilled water. However, passivation–depassivation which continuously took place on the surfaces of Ti6Al4V during the reciprocating friction tests, could significantly lead to material removal on the surfaces. That was why the Ti6Al4V samples presented lower friction coefficients but higher wear losses in volume in artificial seawater than those in distilled water. Fortunately, the tribological performances of Ti6Al4V in artificial seawater were dramatically improved as the crosshatch and dimple STs were prepared, as expected.

Further, Lian et al. [109] first fabricated crosshatch and dimple ST (with a spacing of 100 μm) on Ti6Al4V by LST, and then both of the textured surfaces were coated with nano SiO_2 particles via the sol-gel method. The ultimately obtained surfaces (SiO_2-crosshatch and SiO_2-dimple) indicated excellent super-hydrophobic property. The super-hydrophobic property of SiO_2-crosshatch textured surface was a bit better than the SiO_2-dimple texture surface. Meanwhile the subsequent measurements of tribological performance under dry sliding demonstrated that the wear rates of the SiO_2-crosshatch textured surface and SiO_2-dimple textured surface were decreased by 53.8% and 32.3%, respectively. While both of the fluctuation and friction coefficient values decreased on SiO_2-crosshatch textured and SiO_2-dimple textured surfaces. Ti6Al4V samples were well endowed with promising tribological property on surfaces using the adopted duplex treatments.

By employment of the LST method, Hu et al. [110] obtained three dimple STs on Ti6Al4V. The geometric parameters were: diameter ~150 μm, average depth ~40 μm and three dimple intervals were designed by controlling the dimple area densities (13%, 23%, and 44%). Meanwhile the formed STs were burnished using commercially available MoS$_2$ solid lubricant. Under dry sliding against bearing steel, it was found that the textured Ti6Al4V with higher dimple density showed lower friction and wear compared with blank Ti6Al4V only if the tribological tests were conducted under low load and speed. ST with higher dimple density led to a lower friction coefficient and also contributed to a more promising wear debris capturing effect. As the LST surfaces were burnished MoS$_2$ film, all of them exhibited excellent friction-reducing and wear resistance under any applied load. It was considered that by transferring of MoS$_2$ reserved in dimples, into the space between the dimples, a continuous solid lubricant film was maintained on the surface. The textured surfaces with 23% dimple density revealed the lowest friction coefficients under different testing conditions. However, a longer wear lifetime was realized by increasing the dimple density, which can be ascribed to the higher remaining amount of MoS$_2$ and higher transferring efficiency of MoS$_2$ from the dimples to the friction interface.

Ripoll et al. [67] used a Nd:YAG nanosecond pulsed laser to form hexagonally arranged dimples STs (with diameter of 40 μm, depth of 16–20 μm, spacing of 40, 50, 60, and 70 μm) on Ti6Al4V. The blank Ti6Al4V and textured Ti6Al4V were coated with MoS$_2$ films with a thickness of 2 μm by sputtering. Dry reciprocating sliding tests were performed on MoS$_2$-coated textured Ti6Al4V and MoS$_2$-coated Ti6Al4V using a ball on flat configuration (against 100Cr6 steel balls) at two different oscillation amplitudes. The results displayed that under certain conditions, surface texturing was able to reduce friction, prolonged the lifetime of the film, and gave progressive film degradation until failure. It was seen that the friction coefficients significantly decreased, especially for higher dimple densities owing to the better wear debris capturing function. With respect to low amplitudes, ST could effectively act as wear debris traps and then increase the lifetime of MoS$_2$ film. However, in respect of high amplitudes, ST had a negative impact on tribological behaviors. Additionally, aiming to avoid yielding excessive contact pressure and low amount of MoS$_2$, the normal distance between dimples was preferred to be not smaller than 50 μm. For the testing amplitude of 0.5 mm, an optimum dimple density was found to be between 40% and 67%.

Auezhan Amanov [111] produced dimple ST with diameter (100 μm) and depth (25 μm) on Ti6Al4V by LST, and then a Cr-doped diamond-like carbon (DLC) film was successively fabricated on the dimple textured Ti6Al4V via unbalanced magnetron sputtering (UBMS). The tribological characteristics of the un-textured (polished), dimple textured and coated dimple textured Ti6Al4V samples were investigated against Cr-plated bear steel SAE 52100 pin at 50 °C under poly-alpha-olefin (PAO) oil lubricating conditions. It was seen that a noticeable enhancement in hardness and H/E ratio of the Ti6Al4V after LST process was found, which meant the dimple textured Ti6Al4V had higher mechanical properties and resistance to plastic deformation. This was ascribed to the microstructural modification due to the high energy action of the pulsating laser beam. At normal loads of 10 N and 20 N, the friction coefficient of the coated dimple textured Ti6Al4V was reduced by about 67% and 50%, and 75% and 65% in comparison to those of the un-textured and dimple textured samples, respectively. It was found that the coated dimple textured Ti6Al4V specimen showed significantly lower wear rate compared with those of the un-textured and dimple textured samples for the applied normal loads. The wear rate of the un-textured specimen was slightly higher than that of the dimple textured specimen and the coated dimple textured Ti6Al4V showed far lower wear rate than those of the un-textured and dimple textured samples. After LST + deposition of Cr-doped DLC film, Ti6Al4V benefited a lot from the functions of ST: storage of oil and wear debris capturing, as well as the merits of DLC: high hardness and H/E ratio.

Micro arc oxidation (MAO) has long been applied to improve the tribological performance of Ti6Al4V with great success by receiving series of ceramic coatings on the surface [15,24]. Wang et al. [70] employed a fine particle shot-peening (FPSP) process to form half-ball dimple ST on Ti6Al4V using ball-shape γ-Al$_2$O$_3$ particles with an average diameter of 28 μm to avoid introducing iron pollutants. The half-ball dimples distributed on the FPSP treated Ti6Al4V substrate surface resulted in an obvious increase in surface roughness from Ra 0.26 μm to Ra 1.65 μm. The pretreated process of FPSP was followed by micro arc oxidation (MAO) treatment. The related samples in that work were original Ti6Al4V (polished), FPSP-Ti6Al4V, MAO coatings (with thickness of 5 μm and 10 μm: MAO5 and MAO10) and FPSP-MAO coatings (with thickness of 5 μm and 10 μm: FPSP-MAO5 and FPSP-MAO10). The tribological behaviors of the mentioned samples were investigated on a pin-on-disk tester against SAE 52 100 steel ball (with a diameter of 6.0 mm, a surface roughness Ra 0.05 μm and a hardness HRC 61) with normal load of 1 N under dry sliding condition. There was no obvious difference in variation trends of friction coefficient between original Ti6Al4V FPSP-Ti6Al4V. However, both of the MAO coatings and FPSP-MAO coatings showed no obvious friction reduction effect on Ti6Al4V, which was in good agreement with the previous publications [32,112]. The MAO5 and MAO10 coatings were completely worn out after 900 and 1700 sliding cycles implying equal friction coefficient values of MAO coatings and original Ti6Al4V. Throughout the sliding cycle number, there was no evidence that could reveal complete worn damage of FPSP-MAO5 and FPSP-MAO10 coatings. The dimples on the surfaces of FPSP-MAO5 and FPSP-MAO10 coatings could act as sink for trapping wear debris, which was helpful to alleviate three body wear, while the restrained three body wear was responsible for the enhancement in wear resistance of FPSP-MAO5 and FPSP-MAO10. The resultant specific wear volume of each tested sample could be arranged as follows: original Ti6Al4V > FPSP–Ti6Al4V > MAO5 > MAO10 > FPSP–MAO5 > FPSP–MAO10. FPSP–MAO was considered as an effective means to improve the wear resistance of Ti6Al4V. Meanwhile the fatigue life of FPSP-Ti6Al4V sample was increased by 39% in comparison to the original Ti6Al4V, which was ascribed to the compressive residual stress caused by FPSP treatment. With respect to the simples of MAO5 and MAO10, the fatigue life decreased from 20, 618 cycles for MAO5, to 16, 282 cycles for MAO10. While in respect of the FPSP-MAO samples, the fatigue life decreased from 23, 223 cycles for FPSP-MAO5, to 17, 653 cycles for FPSP-MAO10. At the same thickness of 5 μm and 10 μm, FPSP-MAO5 enhanced fatigue life by 12.6% compared with simple MAO5, while FPSP-MAO10 suggested a slight improved fatigue life only by 8.4% compared with sample MAO10. In general, the FPSP-MAO coatings presented better fatigue resistance in cyclic loading conditions. Nevertheless, because of the fragile nature of the ceramic coatings, all the MAO coated samples exhibited deterioration in fatigue lives.

Using the LST technique, Prem Ananth et al. [113] attained dimple ST with geometric parameters of: dimple densities of 38%–42%, diameter of 100 μm, depth of 2.5 μm, and the interval between two consecutive dimples was 160 μm. Magnetron sputtered physical vapor deposition (MSPVD) technique was applied to prepare nano AlCrN composite coatings on raw and dimple textured Ti6Al4V surfaces. Due to the positive mechanical locking effect, the coating cover on the LST Ti6Al4V surface showed appreciable improved bonding and shear strength. Under different normal loads (9.8–29.7 N), the LST + MSPVD treated Ti6Al4V suggested more stable friction coefficients and longer wear lives than that of the AlCrN coating on raw Ti6Al4V. Up to now, LST has been usually followed by advanced surface modification technologies to improve the surface performance of Ti6Al4V. Prem Ananth et al. [114] used to produce dimple ST on Ti6Al4V surface with geometric parameters of: dimple density (38%–42%), diameter (100 μm), depth (2.5 μm), and dimple-dimple interval (160 μm). The blank and dimple textured Ti6Al4V samples were coated with titanium aluminum nitride (TiAlN) nanocomposite coatings with chromium interlayer via cathodic arc physical vapor deposition (CAPVD) system. The obtained coatings reached a total thickness of 4–5 μm. Using scratch tester and pin-on-disc tribometer, the original Ti6Al4V, dimple textured Ti6Al4V and TiAlN coated samples were subjected to bonding strength and tribological evaluations. It was found that the bonding strength of the coating on dimple textured Ti6Al4V with a chromium inter layer increased by 8% compared to that of the coating on the original Ti6Al4V surface. Under dry sliding contact conditions (variation in load: 9.8 N, 14.7 N, 19.6 N, and 29.4 N), increasing normal load led to an increasing friction coefficient, however the TiAlN coated dimple textured Ti6Al4V presented slightly lower increasing rates in comparison to the TiAlN coating on original Ti6Al4V. Meanwhile it was seen that samples with textured surface revealed comparatively lower steady friction coefficients compared to those of lapped surfaces. It was found that increasing normal load contributed to early coating failure and then resulted in metal-to-metal contact. However, it was confirmed that the LST + CAPVD treatment could significantly improve the tribological performance of Ti6Al4V alloy.

Qin et al. [115] investigated the tribological properties of LST treated and plasma electrolytic oxidation (PEO) duplex-treated Ti6Al4V deposited with MoS_2 film. First, the dimple ST formed a Ti6Al4V plate surface using a pulsed Nd:YAG laser with a wave length of 1064 nm and a pulse width of 450 ns. Then the LST-Ti6Al4V plate was ground by 800 mesh SiC emery paper to remove the raised ripples. The electrolyte then used for PEO contained 12 g/L sodium aluminate ($NaAlO_2$), 1.6 g/L trisodium phosphate ($Na_3PO_4 \cdot 12H_2O$) and a small amount of sodium hydroxide (NaOH). The PEO treatment was conducted for 1 h with an initial voltage of 440 V, adopting a duty cycle of 20% and a frequency of 400 Hz to form PEO-Ti6Al4V and LST + PEO-Ti6Al4V. The temperature of electrolyte was maintained at about 30 °C through a water cooling system. The lubricating MoS_2 film was bonded onto the tested samples using E44 epoxy and acetone as the adhesion agent and the solvent, and the acquired samples were MoS_2-Ti6Al4V, LST + MoS_2-Ti6Al4V PEO + MoS_2-Ti6Al4V and LST + PEO + MoS_2-Ti6Al4V. It was seen that micro dimples with a diameter (D) of 260 μm, an interval between the centers of two dimples of 500 μm, and an area density (s) of 21.2% were regularly distributed on the polished Ti6Al4V surface. The PEO coatings on both the un-textured and LST-Ti6Al4V surfaces reached a similar thickness of 30 ± 5 μm.

Friction and wear tests were conducted using a ball-on-disk tribometer under dry sliding conditions at the applied loads of 0.3–5 N and a constant velocity of 0.1 m/s against a GCr15 steel ball with a diameter of 4 mm. Under a load of 5 N, polished Ti6Al4V substrate suffered severe adhesive wear and plough, the dimples on the LST-Ti6Al4V were nearly worn out and some wear debris was trapped in the residual dimples. The predominant wearing modes of PEO-Ti6Al4V were deformation and abrading. LST + PEO - Ti6Al4V received the lightest damage, its worn trace, a mild polishing of the raised rims of dimples, occurred on the space between the micro-dimples in the sliding direction. The wear rate of polished Ti6Al4V (4.27×10^{-4} mm^3/Nm) was slightly above the LST-Ti6Al4V (3.93×10^{-4} mm^3/Nm). The PEO-Ti6Al4V showed a lower wear rate of 1.88×10^{-4} mm^3/Nm. It was observed that the LST + PEO - Ti6Al4V provided the most promising improvement in wear resistance of Ti6Al4V, indicated by the lowest wear rate of 8.45×10^{-5} mm^3/Nm as compared with other samples. Meanwhile all of the samples benefited from the bonded MoS_2 film in tribological performance varying in low friction duration. It just took 5 min to completely remove the surface of MoS_2-Ti6Al4V under a load of 1 N. The LST + MoS_2-Ti6Al4V survived for a longer period of 20 min with a lower friction coefficient than that of MoS_2-Ti6Al4V under the same load. However even as the load was raised up to 5 N, the PEO + MoS_2-Ti6Al4V still revealed a stable lower friction coefficient than MoS_2-Ti6Al4V and LST + MoS_2-Ti6Al4V for about 200 min. The LST + PEO + MoS_2-Ti6Al4V presented surpassed friction behavior in comparison to PEO + MoS_2-Ti6Al4V, as expected. The MoS_2 which was stored in the dimple-reservoirs was squeezed out to the contact region and formed a thin lubricating transfer film on the friction interface during dry sliding. The thin lubricating transfer film was beneficial to friction reduction and was helpful to sustain longer duration. Additionally, the partially open micro-dimples could act as sinks for trapping wear debris, which was also beneficial in reducing the damage and prolonging the effective life of the lubricating film.

Furthermore, the effects of textured dimple area densities (referred to S, S = dimple area/total surface area) and the surface roughness values of oxide ceramic underlay which influenced the lifetime of MoS_2 films were thoroughly studied by Qin et al. [116]. Similar results showed that the LST + PEO + MoS_2-Ti6Al4V which was superior to the MoS_2-Ti6Al4V, LST + MoS_2-Ti6Al4V and PEO + MoS_2-Ti6Al4V, exhibited much longer low friction life. It was obvious that the low friction life of the LST + PEO + MoS_2-Ti6Al4V was prolonged with increasing the S values from 8% to 55%, due to a higher area density which could preserve a larger amount of MoS_2 in the dimples. The lubricants were continuously and effectively replenished at a shorter interval between the dimples by adequate supply of fresh MoS_2 both from the large textured dimples and the small discharged dimples during sliding and resulted in a longer sliding life with low friction. In addition, the hard and polished LST-PEO surfaces could also provide a high load support for the soft MoS_2 film. The received LST + PEO + MoS_2-Ti6Al4V (with S = 55%) with a Ra = 2.9 μm showed disappointing friction behavior. However, when the surface roughness values of LST + PEO + MoS_2-Ti6Al4V were polished down to Ra = 1.6 μm and Ra = 1.0 μm, the obtained sample revealed the best friction behaviors. In this work, a much longer low friction life LST + PEO + MoS_2-Ti6Al4V was realized by choosing an LST surface with S = 55% and polishing the LST + PEO + MoS_2 to a surface roughness of Ra = 1.0 μm.

In He et al.'s work [64], STs of micro-dimples with various textured dimple area densities and diamond-like carbon (DLC) films were fabricated on the surfaces of Ti6Al4V by LST and close-field magnetron sputtering (CFMS), respectively. To be specific, the LST was conducted using a Nd:YAG laser with a wave length of 1064 nm, a frequency of 10 kHz and a 90% overlapping rate of laser spot. An average power of 10 W at a 5 mm/s traverse speed was applied to treat the samples. After LST, all the received samples (with textured dimple area densities of 13%, 24%, and 44%) were subjected to two step polishing processes, aimed to eliminate bulges or burrs around the edge of the dimples and obtaining a roughness of Ra ≤0.02 μm. High-energy ion implantation of nitrogen was performed on the textured and smooth Ti6Al4V samples to improve their surface hardness (pulse voltage −60 kV; frequency 60 Hz; pulse width 30 μs; working gas pressure 3×10^{-2} Pa). Subsequently, thin chromium films (220 nm thick) were deposited on the as-nitrogen-implanted samples as interlayers to improve bonding strength, and DLC films with a total thickness of 3.3 μm were prepared by CFMS using high-purity graphite targets at a DC power supply of 2500 W. Each DLC deposition process began at an original pressure of about 3×10^{-3} Pa, and the deposition was performed at a substrate pulsed bias of −70 V, a frequency of 250 kHz, and a process pressure of 0.2 Pa under Ar flow. The obtained specimens were marked as DLC-smooth, DLC-T13%, DLC-T24%, and DLC-T44%. Tribological tests were performed on a ball-on-plate reciprocating mode with a reciprocatory displacement of 5 mm, a normal load of 5N, a sliding frequency of 5 Hz, and the sliding direction parallel to the LST patterns. The Ø10 mm AISI 52100 steel balls, with hardness of HV 725, were used as counterparts. Dry friction and liquid lubrication (1-butyl-3-methylimidazolium hexafluorophosphate ionic liquid) conditions were conducted for 50,000 cycles with the same test parameters in ambient air (temperature of 20 °C and relative humidity of 35%).

The effects of dimple area densities and DLC phase transformation on the properties of Ti6Al4V under dry friction and liquid lubrication conditions were thoroughly investigated. The DLC-smooth suggested a critical load of 55 N, while the DLC-T13%, DLC-T24%, and DLC-T44% samples indicated almost the same critical load of about 37 N. Damage to the DLC films, particularly to the DLC-T44%, was more severe in comparison to DLC-T13% and DLC-T24% samples. The stress concentration in the films around the rims of the dimples resulted in low load carrying capacities. Due to the combined action of dimple-induced graphitizing transformation and the function of the fluid/wear debris reservoirs of the dimples, DLC-T44% always revealed remarkably lower friction coefficients than those of DLC-smooth and DLC-textured samples with lower dimple area densities regardless of under dry friction or under liquid lubrication conditions. However, the most outstanding wear resistance (with the lowest wear rate) under dry friction was observed on DLC-T24%, which could be explained in that an appropriate number of dimples on the DLC surface were not only trapped the most wear debris but also maintained film hardness (a low level of graphitization transformation) during friction. It was concluded that combining appropriate surface texturing with DLC films was able to effectively reduce the friction and wear of Ti6Al4V and thus could be beneficial for its wider applications.

Thermal oxidation (TO) is an ideal surface technology which has been extensively used to strengthen Ti-based materials. Its popularity is mainly assigned to its cost effectiveness, simplicity, and rapidity. Furthermore, the TO treatment has no special requirements for substrate geometric shape [117]. Taking the advantages of LST and TO processes, Sun et al. [118] formed a series of TO films of various thicknesses and composition on regular dimple textured Ti6Al4V surfaces. The STs on Ti6Al4V were realized by LST. The TO processes were performed at 500 °C, 650 °C, and 800 °C for 5–50 h. The LST treated Ti6Al4V were well covered with uniform and continuous TO coatings. It was also found that ST could decrease the internal stress and effectively improve bonding strength of the oxide film to the Ti6Al4V substrate. However, despite higher TO temperature and longer TO duration being favorable for increasing the thickness of the TO coatings, the obtained TO coatings under these conditions did not necessarily show excellent properties. Under dry friction conditions, TO-LST Ti6Al4V revealed much lower wear rates compared to the original Ti6Al4V under the same applied loads against GCr15 steel and two ZrO_2 balls. The TO-LST Ti6Al4V surface received at 650 °C for 25 h suggested that the most excellent tribological properties were attributable to good comprehensive properties: a strongly bonded rutile coating with high hardness (7.73 GPa), increased hardness to elastic modulus ratio (0.096), and improved load-bearing capacity. The TO-LST Ti6Al4V obtained at optimal parameters with promising surface performance could withstand or weakened the wearing damage, on the other hand, the ST was able to capture the wear debris and then decreased the abrasive wear. The TO + LST duplex treatment indicated satisfactory complementation on improving tribological behavior of Ti6Al4V.

Martinez et al. [119] simultaneously realized thermochemical oxidation of UNS R56400 (Ti6Al4V) alloy through LST treatment in an open-air atmosphere with varied parameters. The treated UNS R56400 alloy indicated increased surface hardness and varied color tonality. The formation of titanium oxides and a rapid cooling rate contributed to the increase of hardness. The color tonality was especially affected by the laser treatment parameters of pulse rate (F (kHz)) and scan speed of the beam (Vs (mm/s)). Pin on disc tribological tests showed that design and development of topographies on the surfaces of UNS R56400 alloy were favorable for obtaining high sliding-friction-wear resistance. The most advantageous surface reduced the friction coefficient values by approximately 20%, which was obtained using a Vs of 150 mm/s.

As a special wearing mode, cavitation erosion which is fairly complicated, is usually generated under mechanical, chemical, and electrochemical interactions. Actually, cavitation erosion is a type of dynamic damage, and the fatigue damage on the material surface is caused by the impact of cavitation bubbles [120–122]. It has been confirmed that surface texture-based surface treatment certainly has positive effect on reducing cavitation erosion damage of Ti6Al4V. LST was conducted on Ti6Al4V by Pang et al. [123] to enhance its cavitation erosion resistance by forming groove, crosshatch (included angle 90°) and dimple STs with interval spacing of 50 μm and 100 μm. LST resulted in increasing surface hardness values of the textured Ti6Al4V samples by quenching effect (martensitic transformation and fine grain strengthening). The tested surfaces varied in surface hardness which could be arranged as: dimple (interval spacing of 100 μm) > crosshatch (interval spacing of 50 μm) > crosshatch (interval spacing of 100 μm) > groove (interval spacing of 50 μm) > groove (interval spacing of 100 μm) > original Ti6Al4V. By detecting the surface morphologies of the cavitation eroded samples, it was found that the dimple textured Ti6Al4V showed the best cavitation erosion resistance, followed by crosshatch textured Ti6Al4V, groove textured Ti6Al4V, and original Ti6Al4V. The arrangement in cavitation erosion resistance of the related Ti6Al4V samples was in good agreement with the results of surface hardness tests. Cavitation erosion resistance of Ti6Al4V specimens was significantly enhanced by LST.

On the basis of Pang et al.' s work, Lian et al. [124] first fabricated groove and crosshatch (included angle 90°) STs on Ti6Al4V surfaces with interval spacing of 50, 100, 150 μm, and then prepared 1H, 1H, 2H, 2H-Perfluorooctyltrichlorosilane (FOTS) self-assembled monolayers (SAMs) on the blank and textured Ti6Al4V samples. The variations in surface hardness values of the textured Ti6Al4V were consistent with Pang et al.'s work: crosshatch textured Ti6Al4V showed higher surface hardness than that of groove textured Ti6Al4V. Furthermore, as the two textured surfaces were covered with FOTS-SAMs, both of them revealed better hydrophobicity than that of the original Ti6Al4V. By employment of microscopy observation, the original Ti6Al4V indicated a cavitation area of 75% on the total surface area. The crosshatch textured Ti6Al4V with different interval spacing revealed cavitation area values of about 18% (50 μm), 21% (100 μm), 25% (150 μm); the groove textured Ti6Al4V with different interval spacing presented cavitation area values of about 24% (50 μm), 29% (100 μm), 45% (150 μm). After FOTS-SAMs covering, both of the crosshatch textured and the groove textured Ti6Al4V with different interval spacing suggested far lower cavitation area, the largest one was only about 15%. FOTS-SAMs with hydrophobicity could reduce the flow resistance of water and then weakened the shock effect on the surface by bubble collapse. FOTS-SAMs on a textured surface were able to bring out more excellent cavitation erosion resistance of Ti6Al4V on the surface in comparison to a single FOTS-SAMs covering or single surface texturing treatment.

2.2. Biological Applications

Specially designed surface patterns (surface roughening) by surface texturing on Ti6Al4V are good for it when used as hard tissue biomaterial. For example, a textured surface is able to stabilize the bone-implant interface (bonding strength), minimize micro-motion during arthrodesis, promote osseo-integration, carry drug or biomaterial particles, improve antibacterial properties, and enhance the related bio-properties of bioactive coating/film/layer, and so on [69,77,125–128]. Meanwhile a surface distributed with ordered TiO$_2$ nanotubes (NTs) can also be considered as a kind of surface texture. The nanostructure topography, TiO$_2$ NTs diameter and length, the spacing in between the nanotubes and the protein physical properties as electric charge and size are all crucial to the interactions of cells, proteins, molecules, and bacteria with textured titanium [129].

In Wang et al.'s work [69] a subsequent treatment of fine particle shot-peening (FPSP) process was performed on micro arc oxidation (MAO) coatings on Ti6Al4V, the obtained rougher dimple surfaces interspersed by fine pore structure was beneficial to induce the deposition of biomimetic apatite. Kumari et al. [77] found that there was an increased bioactivity on LST treated Ti6Al4V surface according to the calcium phosphate deposition rate in Hank's solution. As expected, LST did not produce any cytotoxic substances reflected by XTT ($C_{22}H_{19}N_7Na_2O_{14}S_2$) assay test. It was shown that cell adherence preferred ridges and corners and was less in the dimple textured surface, in respect of linear textured surface, cells were preferentially attached along the direction of texturing in the textured zone. Olivares-Navarrete et al. [125] prepared macro/micro/nano-textures on Ti6Al4V surfaces by employment of sand blasting and acid etching. It was seen that average surface roughness played an important role in determining the response between cells in the osteoblast lineage and Ti6Al4V implants. Additionally, macroscale textured surfaces were favorable for mechanical stability during arthrodesis, however, these features generated no positive effect on the healing response at a cellular level. Surface features formed with micro-texture and nano-texture were able to be discriminated by committed osteoblasts and multipotent MSCs, which is to say they were sensitive to surfaces with microscale and nanoscale features. Meng et al. [126] first fabricated blind micro-hole array on Ti6Al4V via LST by trial and error, the formed textured surface obtained from optimal laser parameters was successively coated with hydroxyapatite (HA) coating through electrophoretic deposition technology. It was found that the textured surface was convenient for trapping nano HA particles and depositing a coating on Ti6Al4V. LST + electrophoretic deposition HA coating was considered to improve the carrying capacity of Ti6Al4V used as biomaterial. Inspired by dragonfly wings, Bhadra et al. [127] prepared nano-patterned surface arrays on titanium samples via a facile one-step hydrothermal etching process. The fabricated titanium surfaces revealed similar surface architecture to dragonfly wings, and the received surfaces possessed selective bactericidal activity indicating by reduction of almost 50% of *Pseudomonas aeruginosa* cells and about 20% of the *Staphylococcus aureus* cells, respectively. Kurella et al. [128] conducted laser processing to coat zirconia (ZrO$_2$) and texture simultaneously on Ti6Al4V to produce surfaces that were hierarchically integrated and organized at multiple scales. Cataphracted surfaces with high specific surface area were finally obtained on the ZrO$_2$ coating. Such chemical and physical transformations were expected in a Ti6Al4V bio-implant, which was good for effective contact with protein, cells, and tissues at various scales and was also helpful to enhance its chemical and mechanical (tribological) performance in the bio-environment.

By Chen et al. [130], the effects of ST on the interactions between human osteosarcoma (HOS) cells and different Ti6Al4V coupons were studied. The Ti6Al4V coupons differed in surface topography: polished Ti6Al4V (control); roughened Ti6Al4V (Al$_2$O$_3$ blasted), and LST grooved Ti6Al4V samples with controlled interval spacing (20, 30, 40, 50, and 60 μm). Immuno-fluorescence staining of adhesion proteins (actin and vinculin) was applied to investigate the spreading and adhesion of HOS cells in 48-h culture experiments. Quantitative measures of adhesion were also realized by employment of an enzymatic detachment assay. The results revealed that the HOS cells were strongly affected by variations in the ST at micron-scale. Cell spreading on polished and roughened surfaces presented irregular orientations. It was also found that cell spreading reduced with increased surface roughness. After a 2-day culture duration, the stress fibers of the actin cytoskeleton were observed co-localizing with the focal adhesions at the ends of the stress fibers on all the investigated surfaces. On the micro-grooved textured surfaces, actin microfilament alignment reflected the orientation as a whole and focal adhesion concentration was found to scale with the level of contact guidance. Enhanced orientation and attachment were observed on the Ti6Al4V micro-grooved textured surfaces with groove interval spacing of 20 μm and with a micro-roughness characterized by higher rms surface roughness. Contact guidance was revealed to increase as grooved spacing decreased. The lower enzymatic detachment rates obtained for the LST treated Ti6Al4V showed that LST provided improved adhesion between HOS cells and laser textured surfaces. It was also found that ST had a strong effect on cell detachment rates. For the range of micro-grooved geometries studied, micro-grooves with depth of ~10 μm, width of ~11 μm, and interval spacing of 20 μm indicated the most promising combination of cell orientation and adhesion of HOS cells to LST grooved Ti6Al4V surface. Furthermore, Chen et al. [131] conducted the initial cell spreading and adhesion on longitudinally- and transversally-oriented micro-grooved Ti6Al4V surfaces formed by LST. The results showed that cell-spreading and adhesion were both enhanced by longitudinally-oriented and transversally-oriented micro-grooves. Contact guidance was found to promote cell adhesion due to the increasing interactions between the focal adhesions and the patterned extra-cellular matrix (ECM) proteins on the micro-grooved surfaces.

In Fasasi et al.'s work [132] the diode pumped solid-state (DPSS) 355 nm (UV) laser operating with a pulse repetition frequency (PRF) of 50 kHz, a focal length of 100 nm, and scan speeds of ~200 mm/s to 300 mm/s produced micro-groove geometries that were close to the 'optimal' groove depth and width of 8 to 12 μm. Such groove dimensions were in the range that could promote cell integration and contact guidance. The results from this study suggest that nano-second DPSS UV lasers can be used to introduce the desired micro-groove geometries without micro-cracks in the heat-affected zones. The desired 8~12 μm groove depths and widths can be achieved by control of pulse frequency, scan speed, and lens focal length that controls spot size. The appearance of the physical surface features (resolidification packets, ripples, and wall deformations) obtained using DPSS UV lasers, warrants further studies. This may lead to further optimized groove geometries that promote increased cell adhesion.

Wettability on the implant material surface which could modulate the protein adsorption and thereby affect cell attachment and tissue integration at the interface, usually plays an important role in the success of an implanting operation. In order to improve the wettability, both the surface topography and the surface chemistry, Dahotre and Paital et al. [133,134] first sprayed Ca-P (calcium phosphate tribasic, Ca$_5$(OH)(PO$_4$)$_3$) slurry onto Ti6Al4V substrate, and then conducted direct laser writing on the Ca-P coating. Various phases such as, CaTiO$_3$, Ca$_3$(PO$_4$)$_2$, TiO$_2$ (Anatase and Rutile) were detected in the coated regions. The received STs obtained using direct laser writing technique suggested a remarkable decrease in the apparent contact angle to simulated body fluid (SBF) and distilled water. Meanwhile the textured Ca-P coating surface was favorable to cell spreading, which was confirmed by comparative investigations with a representative Ca-P-coated Ti6Al4V in the spreading of the MC3T3-E1 osteoblast cells after culture for 24 h. The behavior of the cells on a surface is significantly influenced by the amount and direction of stress on the cytoskeleton, while the surface chemistry and surface topography also play key roles on cell behavior. Mukherjee et al. [135] found that the surface with narrower or sharper secondary texture features on a groove textured Ti6Al4V surface was able to promote cell attachment and differentiation. As the cell cytoskeleton held low stiffness, getting attached to a surface with sharp features was able to induce the type of stress that was beneficial for the cell activities. A surface with such feature dimensions could be effective in influencing the cellular activities on the surface and thus enhancing the biocompatibility of groove textured Ti6Al4V on the surface. Mirhosseini et al. [136] found that small holes formed by LST on Ti6Al4V samples could not only change the surface roughness in comparison to the shot blasting treated Ti6Al4V, but also created better cell integration and increased 2T3 osteoblast cell growth. LST increased the surface energy of Ti6Al4V and resulted in a more active surface to attach cells.

Anodizing titanium to fabrication nanotubes (NTs) is a simple strategy for providing inexpensive and well-ordered nanotopographies on implant surfaces to enhance their biological behavior. It has been suggested that using anodizing to create NTs on Ti surfaces presents great promise in vitro; it is possible to improve stem cell differentiation, increase bone growth, improve bladder stent urothelialization, increase vascular stent endothelialization, decrease bacteria function and inflammation. Meanwhile the Ti surfaces distributed with NTs produced by anodizing, exhibited continued promise in various orthopedic applications after in vivo estimations [137]. Kummer et al. [138] conducted anodizing on Ti to form TiO_2 NTs with 20, and 80 nm tube diameters under applied voltages changed to 5, and 20 V, respectively, in a 0.5% hydrofluoric acid electrolyte solution for 30, and 15 min, respectively. The anodizing treated Ti samples were also subjected to heat treatment to remove fluorin. It was revealed that after ultraviolet (UV) light, ethanol soaking, and autoclaving, the treated Ti samples which possessed 20 nm NTs presented the greatest promise as an antibacterial implant material.

Fibronectin and vitronectin are two major proteins which play important roles in osteoblast adhesion. It was reported that TiO_2 NTs significantly increased fibronectin (15%) and vitronectin (18%) adsorption on anodized titanium in comparison to the raw titanium samples, due to promoting adherence of cells. The fibronectin and vitronectin adsorption step was increased on anodized titanium substrates with TiO_2 NTs, which benefited from biomechanical interlocking (high specific surface area) and biological interactions (enhance bone cell function) [139].

Balasundaram et al. [140] found that TiO_2 NTs obtained by anodizing titanium could promote osteoblast adhesion through BMP-2 knuckle peptide functionalization. The received TiO_2 NTs could also act as drug molecules or bone building agents (such as RGD, KRSR, etc.) carrier for new bone formation. Furthermore, Zile et al. [141] performed functionalization of TiO_2 NTs with fibroblast growth factor-2 (FGF-2). It was seen that the FGF-2 functionalized TiO_2 NTs were able to increase keratinocyte density, reduce bacteria adhesion and promote bone tissue formation, as expected.

Liu et al. [142] conducted temperature-controlled atomic layer deposition (ALD) to fabricate unique nano-TiO_2 coatings on Ti substrates. Increased surface nano-roughness and surface energy contributed to these antibacterial properties. The prepared nano-TiO_2 coatings showed promising antimicrobial effects against gram-positive bacteria (*S. aureus*), gram-negative bacteria (*E. coli*), and antibiotic-resistant bacteria (MRSA) all without resorting to the application of antibiotics. Meanwhile in vitro results revealed that TiO_2 coating stimulated osteoblast adhesion and proliferation while suppressing fibroblast adhesion and proliferation, as compared with the original materials.

Bhardwaj et al. [143] obtained a nanophase titanium dioxide surface texture on Ti6Al4V alloy using electrophoretic deposition (EPD). Two distinct nanotopographies (Ti-160 and Ti-120) both presented a certain reduction in *Staphylococcus aureus, Pseudomonas aeruginosa,* and *Escherichia coli* compared to the untreated controls. There were 95.6%, 90.2%, and 81.1% reductions for Ti-160 samples, respectively. Similarly, Ti-120 sample respectively displayed reductions of 86.8%, 82.1%, and 48.6%. In addition, osteoblast proliferation on Ti-120 at day 3 and day 5 was increased by 120.7% and 168.7% over the controls.

Hosseini et al. [144] prepared four different photoanodes by sol-gel spin coating onto a glassy substrate of fluorine-doped tin oxide. The photocatalytic activities of TiO_2, $TiO_2/C/TiO_2$, $TiO_2/C/C/TiO_2$, and $TiO_2/C/TiO_2/C/TiO_2$ photoanodes were evaluated under UV-Vis light irradiation. A higher photocurrent density was detected with double layers of mesoporous carbon between TiO_2 as compared with a single layer of mesoporous carbon. A double layer of mesoporous carbon between TiO_2 also indicated a higher degree of surface roughness in comparison to a single layer of mesoporous carbon. A remarkable improvement in photocurrent was observed by adding additional layers as shown for the two photoanodes: $TiO_2/C/TiO_2$ and $TiO_2/C/TiO_2/C/TiO_2$. The addition of two carbon layers enhanced the graphite sheets between the TiO_2 layers. The graphite sheets might facilitate rapid transport of charges or contribute to generation of charged carriers owing to the functional groups of mesoporous carbon.

Hanson et al. [145] used modified poly(L-lactic acid) (PLLA) scaffolds using oxygen plasma treatment to increase surface hydroxyl groups and thereby improve substrate hydrophilicity. The contact angle of water on PLLA decreased from 75.6° to 58.2° after oxygen plasma treatment, which meant the surface hydrophilicity of PLLA was increased. The DNA analysis results suggested that there was an increased number of human mesenchymal stem cells (hMSCs) on oxygen plasma treated scaffolds. Oxygen plasma treatment also promoted a more even distribution of hMSCs throughout the scaffold and enhanced cell spreading at earlier time points without affecting cell viability.

TiO_2-based nanotubes with high specific surface area and ion-changeable ability have been considered for extensive biomedical applications (osseointegration, antibacterial activity, and drug delivery) [146,147]. Sterilization is usually the final surface treatment procedure of all implantable devices and it also must be considered before implementation. Different sterilization procedures for all implantable devices can influence mechanical properties and biological responses [146–148]. Junkar et al. [148] investigated the effect of different sterilization techniques (sterilization with autoclave, sterilization with ultra-violet (UV) light radiations, commercial hydrogen peroxide (H_2O_2) plasma, and oxygen plasma treatment) on titanium dioxide nanotubes (TiO_2 NTs). TiO_2 NTs with three different diameters of 15 nm (NT15), 50 nm (NT50), and 100 nm (NT100) were prepared using anodization. It was seen that different sterilization procedures did not influence the wettability of TiO_2 NTs. However, steam autoclaving destroyed the nanostructure of TiO_2 NTs, while UV-light, commercial H_2O_2 plasma sterilization, and oxygen plasma treatment techniques showed no nagetive effects on TiO_2 NTs surface features.

Kulkarni et al. [149] conducted investigation of protein interactions with layers of TiO_2 nanotubes (NTs) and nanopores (NPs). The proteins presence on the nanostructures was evaluated by XPS and ToF-SIMS. It was found that there was significant difference in surface charge density between the inner NTs/NPs. NTs adsorbed 31% more histone and albumin than those of NPs. The differences were due to the distribution of NTs: the albumin/histone could also bind to the inner and partially to the outer surface of NTs due to steric and charge restrictions, as compared with the one edge NPs. The size, net charge, and internal charge distribution of proteins had obviously affected their binding ability to the negatively charged TiO_2 surface. Meanwhile longer NTs which had a higher total surface area, absorbed more protein. NTs with small diameter were able to bind more small-sized positively charged proteins per surface area, e.g., histone. According to theoretical modelling, small diameter TiO_2 NTs could lead to an increased magnitude of the surface charge density (negative) at the wall edge, which was favorable to more histone adhesion. In addition, protein adhesion on the top surface revealed a higher protein amount on the NTs tops for histone in comparison to albumin. All in all, the importance of TiO_2 nanostructures' topography was suggested for biomedical applications such as drug delivery or implant materials, where interactions with small size proteins or molecules were primordial.

Interactions between the implant surface and the surrounding bone tissue are essential for the successful integration of a bone implant. In respect of titanium (Ti) implant, it has been confirmed that the contact between the cell membrane of osteoblasts and the Ti oxide surface is established in two steps: first, the osteoblast's cell membrane might set a non-specific contact due to electrostatics (originating in the Coulomb interaction between the negatively charged surface and positively charge proteins), followed by a second step, where the specific binding was made [150]. Kabaso et al. [150] found that adhesion of osteoblast-like cells to a Ti surface implant was a dynamic process driven by interaction with the extracellular matrix and intracellular mechanisms after the Monte Carlo (MC) simulations. The free energy of the system was decreased as the osteoblasts bound to the Ti surface. On the other hand, the strong interactions at contact regions could capture the membrane and increased the local lateral membrane tension leading to an increase in free energy of the cell membrane. It was shown that membrane-bound protein complexes (PCs) increased the membrane protrusion growth between the osteoblast and the groove-textured on the titanium (Ti) surface and thereby promoted the adhesion of osteoblasts to the Ti surface.

Hamlekhan et al. [151] optimized the anodizing and annealing conditions to achieve non-aging hydrophilic surfaces (TiO$_2$ nanotubular surface) on Ti6Al4V alloy. It was found that the nanotubes obtained by anodizing at 60 V and followed by annealing at 600 °C maintained their hydrophilicity significantly longer. Due to presence of nanotubes with larger dimensions and higher surface roughness, 60 V anodized samples revealed a lower water contact angle (WCA). On the other hand, the annealing temperature was the main factor that could affect the maintenance and stability of the obtained hydrophilic TiO$_2$ nanotubular surfaces. Anodizing at high voltages partially promoted the formation of a crystalline structure, and then enhanced surface hydrophilicity. It was suggested that as the surface hydroxylation/dehydroxylation equilibrium was reached, the aged surfaces lost their hydrophilicity. Transformation of the amorphous structure to an anatase crystalline structure was able to slow down the hydroxylation/dehydroxylation equilibrium progress on the TiO$_2$ nanotubular surface while the slowest equilibrium process occurred as the anatase was transformed to rutile.

Cunha et al. [152] formed textured surfaces on Ti6Al4V by a femtosecond laser treatment. Four types of STs were obtained on Ti6Al4V samples: (ST-1) nanoscale laser-induced periodic surface structures (LIPSS); (ST-2) nanopillars; (ST-3) a bimodal roughness distribution texture formed of LIPSS overlapping microcolumns; (ST-4) a complex texture formed of LIPSS overlapping microcolumns with a periodic variation of the columns size in the laser scanning direction. The roughness values of the textured surfaces were characterized by the arithmetic mean surface roughness (Ra) and the mean peak to valley height (Rz), calculated from the surface profiles. Distilled-deionized (DD) water and Hank's balanced salt solution (HBSS) were used to evaluate surface wettability of polished and surface textured Ti6Al4V by the sessile drop method. It was found that the surface roughness values of the four STs were as follows: ST-1 Ra = 290 ± 20 nm and Rz = 2.4 ± 0.3 µm; ST-2 Ra = 260 ± 10 nm and Rz = 2.1 ± 0.1 µm; ST-3 Ra = 1.1 ± 0.1 µm and Rz = 8.5 ± 1.0 µm; ST-4 Ra = 4.7 ± 0.6 µm and Rz = 26.8 ± 2.3 µm, respectively. The polished Ti6Al4V was wetted by both liquids and the contact angles were very similar for the two liquids (68.0° and 63.4° for DD water and HBSS, respectively after 600 s of contact). However, the surface textured surfaces revealed a time-dependent wetting behavior. At t = 0 s, besides ST-3, all the surfaces were wetted by water (θ <90°). HBSS could wet all the textured surfaces. ST-1 and ST-4 showed the best wetting for both liquids, with equilibrium contact angles of 43.4° and 24.1° for water and 21.9° and 8.4° for HBSS, respectively. The reduction of the contact angle with HBSS was maximum for surfaces with ST-4 and then reached a very low contact angle of about 8.0° the droplets spread over all the textured surface. ST-2 and ST-3 suggested similar wetting behavior to the control polished one, with equilibrium contact angles of 56.2° and 76.2° for water and 61.8° and 47.6° for HBSS, respectively. It could be concluded that the HBSS spread much faster than water on the textured surfaces, reflecting an average value of the spreading coefficient 60% higher than that of water. The anisotropy of the surfaces was a key factor in controlling the wetting behavior. Surface texturing on Ti6Al4V by femtosecond laser certainly was an effective route to refine its surface wettability, and also held potential application in improving mesenchymal stem cells adhesion when Ti6Al4V was used as biomaterial.

Super-hydrophobicity (water contact angle $\geq 150°$ and sliding angle $\leq 10°$) on the surface of Ti6Al4V can also be achieved by surface texture-based surface treatments to meet the required service behaviors in some fields outside of biomaterials [80,81,153–160]. Shen et al. [80] obtained hierarchical surfaces on Ti6Al4V with super-hydrophobicity via a multi-step process: Step-1 traditional sand blasting of the polished Ti6Al4V samples was carried out with aluminum oxide (60 mesh, 150 mesh, and 300 mesh) at 0.5 MPa for 10 s to form uneven rough structured surfaces like microhills (textured surfaces); Step-2 hydrothermal treatment of the sand blasted samples was conducted in an autoclave with 30 mL of 1 M NaOH solution in a 220 °C oven and reacted for different times (1 h, 2 h, 4 h, 6 h, 8 h, and 12 h) and cooled down to ambient temperature in the oven. Thereafter the samples were immersed in 1 M HCl solution for 30 min. The samples were rinsed with deionized water and put into a muffle furnace (heating rate was 2 °C s^{-1}) and heated at 500 °C for 3 h leading to the growth of one-dimensional (1D) TiO$_2$ nanowires on the surfaces with microscale rough structures. This process significantly increased the specific surface area of the samples, resulting in an increase of grafted area with the fluorine-containing low-surface-energy groups. Step-3 all of the hydrothermally treated samples were immersed in 1 wt % FAS-17 ethanol solution for 24 h and then dried in a 120 °C oven for 2 h to obtain the final samples.

Sand blasting with 60 mesh aluminum oxide produced a large-size concave-convex structure (~60 μm). While sand blasting with 150 mesh aluminum oxide received a relatively even concave-convex structure (~35 μm). Sand blasting by employment of 300 mesh aluminum oxide resulted in an even finer structure with a smooth overall topography. As fluorination modifications were conducted on the sand blasted surfaces with FAS-17, hydrophobic surfaces were obtained in comparison to fluorination modification on the polished surface. The sand blasted sample with 150 mesh aluminum oxide followed by fluorination modification showed the largest contact angle of liquid droplets (~135°) among the measured surfaces. The reaction duration of hydrothermal treatment had certain effects on the sizes of the formed nanowires, and furthermore influenced the contact angles of the fluorination modified sand blasted surfaces with 150 mesh aluminum oxide. It was found that with a short hydrothermal reaction time (1–2 h), the formed nanowires were relatively short and small, which led to the surface wetting state in the transition state between the Wenzel wetting state and the Cassie wetting state; the apparent contact angle of the droplets on the surfaces slightly increased from about 135° to 138° (sliding angle 8° to 7 °). As the hydrothermal reaction time was prolonged to 4 h, the length of nanowire increased with relatively even distribution, and the endings gradually gathered, the spacing distance between the nanowires with each other was far less than 100 nm, forming a larger continuous air layer, the wetting regime of the liquid droplets on the surface successfully changed from the Wenzel wetting state to the Cassie wetting sate (apparent contact angle 155° and sliding angle 6°). When the hydrothermal reaction time was kept to 8 h, the nanowire length was further extended, and a more dense distribution could be found on the surfaces of the microscale structures. The generated secondary nanowires and microscale structure could capture a large quantity of air owing to the higher specific surface area, hence the apparent contact angle of droplets on the surfaces presented a noticeable increasing to 161° (sliding angle 3°). It was also found that there was no obvious change in the apparent contact angle and sliding angle as the hydrothermal reaction time was even longer.

In addition, Shen et al. [153] conducted anodic oxidation on sand blasted Ti6Al4V, followed by fluorination modification with FAS-17. The ordered nanotube arrays were formed on sand blasted Ti6Al4V after anodic oxidation, and the formed surface could be considered as some kind of textured surface. As the ordered nanotube arrays were modified by FAS-17, the final surface of Ti6Al4V revealed an apparent contact angle of about 151° and a sliding angle of about 8°. It was seen that sand blasting + anodic oxidation + fluorination modification with FAS-17 was able to endow Ti6Al4V with a certain super-hydrophobicity on the surface.

On the basis of the above results, Shen et al. [80] expanded the research of super-hydrophobicity to icephobicity (anti-icing property) on the surface of Ti6Al4V. The multi-step treatment of sand blasting (with 150 grit alumina at 0.5 MPa for 10 s) + hydrothermal treatment (in 30 mL 1 M NaOH aqueous solution at 220 °C for 8 h) + fluorination modification with FAS-17 (in 1 wt % FAS-17 ethanol solution for 24 h and dried at 120 °C for 2 h) was conducted on Ti6Al4V. Icephobicity was evaluated by comparing how long the water droplets were maintained before completely freezing on these surfaces under different temperatures (−10 °C, −20 °C, and −30 °C). A self-made ice adhesion strength measurement device including a cooling plate with a temperature in the range from 0 °C to −40 °C was applied to measure the ice adhesion strength. There were four types of samples tested: the polished Ti6Al4V (S1), the sand blasted+ fluorination modification with FAS-17 treated Ti6Al4V (S2), the hydrothermal treatment + fluorination modification with FAS-17 treated Ti6Al4V (S3), and the entire multi-step treated Ti6Al4V (S4). The results showed that when the icing process of water droplets was performed at −10 °C, water droplets were quickly frozen on S1 and S2 (11.3 s and 12.1 s), while water droplets took a much longer time to completely freeze on the S3 (623.9 s) and S4 (750.4 s). The ice adhesion strength values of the tested samples could be arranged as follows: S1 (~730 kPa) > S2 (~320 kPa) > S3 (~160 kPa) > S4 (~80 kPa). As water droplets were conducted at −20 °C and −30 °C, the icing-delay durations of water droplets on all surfaces significantly decreased. However, the icing-delay performance of S4 is obviously superior to the S1, S2, and S3 as expected. Meanwhile it was seen that the ice adhesion strength values of the tested samples under −20 °C and −30 °C slightly increased compared with those at −10 °C, but the order was not changed. The S4 sample obtained from multi-step treatment of sand blasting + hydrothermal treatment + fluorination modification with FAS-17 suggested the most promising icephobicity.

Additionally, Shen et al. [154] first formed a microscale array textured surface on Ti6Al4V via chemical micromachining, then the textured surface was successively treated by hydrothermal treatment and fluorination modification with FAS-17 using similar processing parameters to the previous research. There were three types of samples tested: the polished + fluorination modification with FAS-17 treat Ti6Al4V (S1), the hydrothermal treatment + fluorination modification with FAS-17 treated Ti6Al4V (S2), and the surface texturing + hydrothermal treatment + fluorination modification with FAS-17 treated Ti6Al4V (S3). It was found that when the icing process of water droplets was measured at −10 °C, S1 presented a icing-delay time of 13.2 s, which meant the water droplet was quickly frozen on S1, while water droplets took a far longer time to completely freeze on the S2 (599 s) and S3 (765 s). The ice adhesion strength values of the tested samples were ranked in the following sequence: S3 (~70 kPa) < S2 (~190 kPa) < S1 (~700 kPa). Combining with the measurement results of icing process and ice adhesion strength, it was possible to draw a conclusion that Ti6Al4V obtained extraordinary anti-icing property on the surface after surface texturing + hydrothermal treatment + fluorination modification with FAS-17.

By employment of LST technique, Lian et al. [155,156] prepared three kinds of ST on Ti6Al4V surfaces: groove, crosshatch (included angle 90°), and space-lattice (in the shape of frustum or cylinder), respectively. All of the LST treated surfaces and the raw Ti6Al4V were hydroxylation modified on the surfaces by ultraviolet (UV) radiation for 1 h. Then the hydroxylated samples were immersed in a solution of 1 mL isooctane 15 μL 1H, 1H, 2H, 2H-Perfluorooctyltrichlorosilane (FOTS) for 12 h. Self-assembled monolayers (SAMs) were obtained on the blank and textured Ti6Al4V samples. In this work the polished Ti6Al4V showed a close contact angle to the textured Ti6Al4V of 50–70°, which meant they were hydrophilic surfaces. On the contrary, all of the SAMs-coated samples presented hydrophobicity. The contact angles of the water droplet on the tested samples could be arranged as follows: SAMs-space-lattice (151.6°) > SAMs-crosshatch (126.1°) > SAMs-groove (124.8°) > SAMs-Ti6Al4V (117°). SAMs-space-lattice revealed obvious super-hydrophobicity, meanwhile it was seen that the measured angles were more aligned with the Cassie model.

Based on the above results, Lian et al. [157] examined the effects of surface film on super-hydrophobic characteristics of Ti6Al4V samples with a dimple surface texture. Four different self-assembled monolayers (SAMs): 1H, 1H, 2H, 2H-Perfluorodecyltrichlorosilane (FDTS), 1H, 1H, 2H, 2H-Perfluorooctyltrichlorosilane (FOTS), Octadecyltrichlorosilane (OTS) and

3-Mercaptopropy-trimethoxysilan (MPS) were selected. The influence of spacing values (50 μm, 60 μm, 70 μm, 80 μm, 90 μm, 100 μm) on the super-hydrophobicity of the SAMs coated-textured Ti6Al4V was also investigated. The results showed that when the spacing of the dimple surface texture was 50 μm, all the SAMs coated-textured Ti6Al4V samples held super-hydrophobicity indicated by the contact angles which were higher than 150°. The contact angles of the tested samples were in the following sequence: FDTS coated-textured Ti6Al4V > FOTS coated-textured Ti6Al4V > OTS coated-textured Ti6Al4V > MPS coated-textured Ti6Al4V. The maximum contact angle of 164.5° was achieved on the surface of FDTS coated-textured Ti6Al4V. The contact angles of the tested samples decreased with increasing dimple spacing. Even so, the contact angles of the FDTS, FOTS, and OTS coated Ti6Al4V with surface texture were still larger than 150° and kept super-hydrophobicity on these surfaces. However a noticeable decrease in contact angle was found on the surfaces of MPS coated-textured Ti6Al4V samples, when MPS were coated on textured Ti6Al4V samples with dimple spacing values of 60 μm, 70 μm, 80 μm, 90 μm, 100 μm; the observed contact angles were in the range of 145–130°. The MPS coated-textured Ti6Al4V samples with higher dimple spacing values showed certain hydrophobicity on these surfaces.

Further taking full advantage of the existing results above, Lian et al. [158,159] turned to the specific application of Ti6Al4V with super-hydrophobicity in the marine environment. At first, dimple surface texture and surface pattern inspired from shell surface were produced on Ti6Al4V by LST. Then both of the textured surfaces were coated with nano SiO₂ particles. A low surface energy solution was prepared with 0.05 mL 1H, 1H, 2H, 2H-perfluoroalkyltriethoxysilanes (PFO) and 0.1 mL ethanol. Droplets of the mentioned solution were dropped on the nano SiO₂ coated samples. The final received Ti6Al4V samples with two kinds of surface texture not only demonstrated excellent super-hydrophobicity, but also exhibited promising performance of antifouling of halobios in comparison to the polished Ti6Al4V after a 45 d exposure test in neritic region. According to the results above, it was clear that surface textured Ti6Al4V could realize that the functions of icephobicity and antifouling were mainly premised on promising super-hydrophobicity [160].

Wael Att et al. [161] discovered that both UVA (ultraviolet light treatment, peak wavelength of 365 nm) and UVC (ultraviolet light treatment, peak wavelength of 365 nm) treatment could convert the 4-week-old titanium surfaces from hydrophobic to superhydrophilic. However, the UVC phototreatment surmounted the innate bioactivity of new surfaces, and the aged surface increased its rat bone marrow-derived osteoblastic cell attachment capacity to a level 50% higher than that of the new surfaces. In addition, proliferation, alkaline phosphatase activity, and mineralization of cells were higher on the UVC-treated 4-week-old surfaces compared with the new surfaces.

3. Summary and Outlook

Mmaterial scientists and engineers have long devoted themselves to the design and production of new materials with better properties to meet the increasing challenges and demands over a wide range of applications under special, aggressive and harsh service conditions. Fabrication of coating onto the surfaces of existing materials by employment of surface modification technologies can obtain the expected properties and improve their surface performance. Titanium and its alloys have been rapidly developed as the pure metal first became commercially available after the 1950s. Its excellent advantages make Ti6Al4V titanium alloy valued in the titanium alloy family, and it is the most frequently and successfully used titanium alloy. However, the disadvantages of Ti6Al4V cannot be ignored, Ti6Al4V holds shortcomings of low surface hardness, high friction coefficient, and insufficient abrasive wear resistance. Surface modification technologies are attractive and suitable approaches to solve the above problems that occur on the surface of Ti6Al4V. A series of surface modification technologies have been used to improve the tribological performance of Ti6Al4V. In addition, surface texture which is inspired from the non-smooth surfaces in the natural world has been considered as an effective approach to refine the tribological behavior of materials and tools in tribology-related fields.

Meanwhile along with the expansion and deepening of research on surface texture, Ti6Al4V has also benefited a great deal from surface texture in biological-related applications.

In this mini review, the surface texture-based surface treatments on Ti6Al4V for tribological (with/without lubricated, cavitation erosion) and biological (where Ti6Al4V was used as biomaterials and for the applications of super-hydrophobicity, icephobicity and antifouling) applications were suggested and consolidated. The following conclusions and prospects were based on published literature.

(1) As Ti6Al4V was treated by single surface texturing, exceptional properties and fascinating functions were simultaneously imparted. Meanwhile the combination of surface texturing and surface treatment could take advantage of the mentioned techniques and realized a "1 + 1 > 2" effect. The surface performance of Ti6Al4V titanium alloy was further tuned by surface texturing-based surface treatments and resulted in more extensive applications as expected.

(2) The positive effects of the surface texture at macro/micro/nano scales on Ti6Al4V in tribological performance also lay in the following aspects: storing solid lubricant and grease to supply continuous lubrication or re-lubrication, improving the elastohydrodynamic effect under liquid lubrication, capturing wear debris, minimizing abrasive wear, and reducing nominal contact area. Meanwhile it was confirmed that surface treatment could provide a surface texture with a strongly helping hand.

(3) Ti6Al4V with different surface textures on the surface was able to achieve a higher specific surface area, which has had a significant effect on the adhesion and proliferation of cell and antibacterial behavior, hydrophilicity or hydrophobicity. Nano-scale texture seems to be more effective compared with the macro- and micro- scale textures. Textured surface covered by coating/film with certain functions has increasingly influenced and enhanced the performance of Ti6Al4V.

(4) The exceptional performance brought by surface texture has been greatly inspired by the non-smooth surfaces of some flora and fauna in nature. There are many species on the world, the specific property of each species with a non-smooth surface has an important significance for bionic research. Such investigations might make a contribution to enrich the studies and applications of surface texture.

(5) There is no one method that appears to work for all conditions. Therefore, the practical application of surface texture-based surface treatment on Ti6Al4V should be conducted case by case rather than having a direct and indiscriminate adoption. Meanwhile the essence and mechanism of physical and chemical reactions between the surface texture and lubricant, cell, bacteria, and liquid drops that occur on the textured surface are worthy of attention.

(6) Despite the surface texture-based surface treatments on Ti6Al4V which have demonstrated significant advances in the tribological and biological applications, the research in this area is still in an early stage. Establishing material systems with different conpositions and multiple functions on e textured surfaces is helpful to accelerate the practical applications of surface texture-based surface treatments. Besides the conventional way of trial and error, computer simulation and big data technologies are also capable of having potential benefits in the design and application of surface texture-based surface treatments.

Acknowledgments: This work was supported by the National Natural Science Foundation of China (No. 51501125), the China Postdoctoral Science Foundation (No. 2016M591415).

Author Contributions: Naiming Lin, Jiaojuan Zou, and Ruizhen Xie conceived and designed the content and structure of this review; Naiming Lin, Dali Li, Ruizhen Xie, and Jiaojuan Zou wrote this review; Dali Li and Jiaojuan Zou contributed to corrections of spelling and grammatical mistakes; Zhihua Wang and Bin Tang made useful comments during the writing process.

Conflicts of Interest: The authors declare no conflict of interest.

References

1. Liu, X.Y.; Chu, P.K.; Ding, C.X. Surface modification of titanium, titanium alloys, and related materials for biomedical applications. *Mater. Sci. Eng. R* **2004**, *47*, 49–121. [CrossRef]
2. Sidambe, A.T. Biocompatibility of advanced manufactured titanium implants-a review. *Materials* **2014**, *7*, 8168–8188. [CrossRef] [PubMed]
3. Attanasio, A.; Gelfi, M.; Pola, A.; Ceretti, E.; Giardini, C. Influence of material microstructures in micromilling of Ti6Al4V alloy. *Materials* **2013**, *6*, 4268–4283. [CrossRef] [PubMed]
4. Li, Y.H.; Yang, C.; Zhao, H.D.; Qu, S.G.; Li, X.Q.; Li, Y.Y. New developments of Ti-based alloys for biomedical applications. *Materials* **2014**, *7*, 1709–1800. [CrossRef] [PubMed]
5. Attar, H.; Ehtemam-Haghighi, S.; Kent, D.; Okulov, I.V.; Wendrock, H.; Bönisch, M.; Volegov, A.S.; Calin, M.; Eckert, J.; Dargusch, M.S. Nanoindentation and wear properties of Ti and Ti-TiB composite materials produced by selective laser melting. *Mater. Sci. Eng. A* **2017**, *688*, 20–26. [CrossRef]
6. Chien, C.-S.; Liu, C.-W.; Kuo, T.-Y. Effects of laser power level on microstructural properties and phase composition of laser-clad fluorapatite/zirconia composite coatings on Ti6Al4V substrates. *Materials* **2016**, *9*, 380. [CrossRef] [PubMed]
7. Barriobero-Vila, P.; Gussone, J.; Haubrich, J.; Sandlöbes, S.; Da Silva, J.C.; Cloetens, P.; Schell, N.; Requena, G. Inducing stable α + β microstructures during selective laser melting of Ti-6Al-4V using intensified intrinsic heat treatments. *Materials* **2017**, *10*, 268. [CrossRef] [PubMed]
8. Markhoff, J.; Krogull, M.; Schulze, C.; Rotsch, C.; Hunger, S.; Bader, R. Biocompatibility and inflammatory potential of titanium alloys cultivated with human osteoblasts, fibroblasts and macrophages. *Materials* **2017**, *10*, 52. [CrossRef] [PubMed]
9. Tang, J.; Liu, D.; Zhang, X.; Du, D.; Yu, S. Effects of plasma ZrN metallurgy and shot peening duplex treatment on fretting wear and fretting fatigue behavior of Ti6Al4V alloy. *Materials* **2016**, *9*, 217. [CrossRef] [PubMed]
10. Cadena, N.L.; Cue-Sampedro, R.; Siller, H.R.; Arizmendi-Morquecho, A.M.; Rivera-Solorio, C.I.; Di-Nardo, S. Study of PVD AlCrN coating for reducing carbide cutting tool deterioration in the machining of titanium alloys. *Materials* **2013**, *6*, 2143–2154. [CrossRef] [PubMed]
11. Campanelli, S.L.; Contuzzi, N.; Ludovico, A.D.; Caiazzo, F.; Cardaropoli, F.; Sergi, V. Manufacturing and characterization of Ti6Al4V lattice components manufactured by selective laser melting. *Materials* **2014**, *7*, 4803–4822. [CrossRef] [PubMed]
12. Ernesto, B.P.; Aldo, M.U.; Benjamín, V.S.; Cristina, V.; Monica, C.; Alan, E.; Ernesto, V.; Francisco, V. Improved osteoblast and chondrocyte adhesion and viability by surface-modified Ti6Al4V alloy with anodized TiO$_2$ nanotubes using a super-oxidative solution. *Materials* **2015**, *8*, 867–883.
13. Strantza, M.; Vafadari, R.; de Baere, D.; Vrancken, B.; van Paepegem, W.; Vandendael, I.; Terryn, H.; Guillaume, P.; van Hemelrijck, D. Fatigue of Ti6Al4V structural health monitoring systems produced by selective laser melting. *Materials* **2016**, *9*, 106. [CrossRef] [PubMed]
14. Bansal, D.G.; Eryilmaz, O.L.; Blau, P.J. Surface engineering to improve the durability and lubricity of Ti–6Al–4V alloy. *Wear* **2011**, *271*, 2006–2015. [CrossRef]
15. Fazel, M.; Salimijazi, H.R.; Golozar, M.A.; Garsivaz jazi, M.R. A comparison of corrosion, tribocorrosion and electrochemical impedance properties of pure Ti and Ti6Al4V alloy treated by micro-arc oxidation process. *Appl. Surf. Sci.* **2015**, *324*, 751–756. [CrossRef]
16. Montemor, M.F. Functional and smart coatings for corrosion protection: A review of recent advances. *Surf. Coat. Technol.* **2014**, *258*, 17–37. [CrossRef]
17. Lin, N.M.; Zhang, H.Y.; Zou, J.J.; Tang, B. Recent developments in improving tribological performance of TC4 titanium alloy via double glow plasma surface alloying in China: A literature review. *Rev. Adv. Mater. Sci.* **2014**, *38*, 61–74.
18. Liu, Y.J.; Luo, J.; Liu, B.; Zhang, J.Y. The cytocompatibility investigation of Ti6Al4V modified with a fluorine-contained copolymer thin film. *Appl. Surf. Sci.* **2011**, *257*, 6429–6434. [CrossRef]
19. Li, S.M.; Zhu, M.Q.; Liu, J.H.; Yu, M.; Wu, L.; Zhang, J.D.; Liang, H.X. Enhanced tribological behavior of anodic films containing SiC and PTFE nanoparticles on Ti6Al4V alloy. *Appl. Surf. Sci.* **2014**, *316*, 28–35. [CrossRef]

20. Cai, J.B.; Wang, X.L.; Bai, W.Q.; Zhao, X.Y.; Wang, T.Q.; Tu, J.P. Bias-graded deposition and tribological properties of Ti-contained a-C gradient composite film on Ti6Al4V alloy. *Appl. Surf. Sci.* **2013**, *279*, 450–457. [CrossRef]

21. Du, D.X.; Liu, D.X.; Ye, Z.Y.; Zhang, X.H.; Li, F.Q.; Zhou, Z.Q.; Yu, L. Fretting wear and fretting fatigue behaviors of diamond-like carbon and graphite-like carbon films deposited on Ti-6Al-4V alloy. *Appl. Surf. Sci.* **2014**, *313*, 462–469. [CrossRef]

22. Çelik, O.N. Microstructure and wear properties of WC particle reinforced composite coating on Ti6Al4V alloy produced by the plasma transferred arc method. *Appl. Surf. Sci.* **2013**, *274*, 334–340. [CrossRef]

23. Straffelini, G.; Molinari, A. Dry sliding wear of Ti-6Al-4V alloy as influenced by the counterface and sliding conditions. *Wear* **1999**, *236*, 328–338. [CrossRef]

24. Wang, C.; Hao, J.M.; Xing, Y.Z.; Guo, C.F.; Chen, H. High temperature oxidation behavior of $TiO_2 + ZrO_2$ composite ceramic coatings prepared by microarc oxidation on Ti6Al4V alloy. *Surf. Coat. Technol.* **2015**, *261*, 201–207. [CrossRef]

25. Xu, J.; Liu, L.; Li, Z.; Munroe, P.; Xie, Z.H. Niobium addition enhancing the corrosion resistance of nanocrystalline Ti_5Si_3 coating in H_2SO_4 solution. *Acta Mater.* **2014**, *63*, 245–260. [CrossRef]

26. Wang, Z.X.; Wu, H.R.; Shan, X.L.; Lin, N.M.; He, Z.Y.; Liu, X.P. Microstructure and erosive wear behaviors of Ti6Al4V alloy treated by plasma Ni alloying. *Appl. Surf. Sci.* **2015**, 1–7. [CrossRef]

27. Liu, L.L.; Xu, J.; Munroe, P.; Xie, Z.H. Microstructure, mechanical and electrochemical properties of in situ synthesized TiC reinforced Ti5Si3 nanocomposite coatings on Ti-6Al-4V substrates. *Electrochim. Acta* **2014**, *115*, 86–95. [CrossRef]

28. Masmoudi, M.; Assoul, M.; Wery, M.; Abdelhedi, R.; El Halouani, F.; Monteil, G. Friction and wear behaviour of cp Ti and Ti6Al4V following nitric acid passivation. *Appl. Surf. Sci.* **2006**, *253*, 2237–2243. [CrossRef]

29. Obadele, B.A.; Andrews, A.; Mathew, M.T.; Olubambi, P.A.; Pityana, S. Improving the tribocorrosion resistance of Ti6Al4V surface by laser surface cladding with $TiNiZrO_2$ composite coating. *Appl. Surf. Sci.* **2015**, *345*, 99–108. [CrossRef]

30. Pawlak, W.; Kubiak, K.J.; Wendler, B.G.; Mathia, T.G. Wear resistant multilayer nanocomposite WC_{1-x}/C coating on Ti-6Al-4V titanium alloy. *Tribol. Int.* **2015**, *82*, 400–406. [CrossRef]

31. Xiang, Z.F.; Liu, X.B.; Ren, J.; Luo, J.; Shi, S.H.; Chen, Y.; Shi, G.L.; Wu, S.H. Investigation of laser cladding high temperature anti-wear composite coatings on Ti6Al4V alloy with the addition of self-lubricant CaF_2. *Appl. Surf. Sci.* **2014**, *313*, 243–250. [CrossRef]

32. Durdu, S.; Usta, M. The tribological properties of bioceramic coatings produced on Ti6Al4V alloy by plasma electrolytic oxidation. *Ceram. Int.* **2014**, *40*, 3627–3635. [CrossRef]

33. Pyka, G.; Kerckhofs, G.; Papantoniou, I.; Speirs, M.; Schrooten, J.; Wevers, M. Surface roughness and morphology customization of additive manufactured open porous Ti6Al4V structures. *Materials* **2013**, *6*, 4737–4757. [CrossRef] [PubMed]

34. Yu, H.P.; Tian, X.; Luo, H.; Ma, X.L. Hierarchically textured surfaces of versatile alloys for superamphiphobicity. *Mater. Lett.* **2015**, *138*, 184–187. [CrossRef]

35. Wang, Z.Z.; Han, Z.W.; Ren, L.Q. Research on wear resistance of unsmoothed surface with regular burrs. *J. Jilin Univ. (Eng. Technol. Ed.)* **2002**, *32*, 45–48.

36. Yu, H.W.; Huang, W.; Wang, X.L. Dimple patterns design for different circumstances. *Lubr. Sci.* **2013**, *25*, 67–78. [CrossRef]

37. Bixler, G.D.; Bhushan, B. Bioinspired rice leaf and butterfly wing surface structures combining shark skin and lotus effects. *Soft Matter* **2012**, *8*, 11271–11284. [CrossRef]

38. Wen, L.; Weaver, J.C.; Lauder, G.V. Biomimetic shark skin: Design; fabrication and hydrodynamic function. *J. Exp. Biol.* **2014**, *217*, 1656–1666. [CrossRef] [PubMed]

39. Autumn, K.; Gravish, N. Gecko adhesion: Evolutionary nanotechnology. *Philos. Trans. R. Soc. A* **2008**, *366*, 1575–1590. [CrossRef] [PubMed]

40. Etsion, I. Modeling of surface texturing in hydrodynamic lubrication. *Friction* **2013**, *1*, 195–209. [CrossRef]

41. Etsion, I. State of the art in laser surface texturing. *J. Tribol.* **2005**, *127*, 248–253. [CrossRef]

42. Ibatan, T.; Uddin, M.S.; Chowdhury, M.A.K. Recent development on surface texturing in enhancing tribological performance of bearing sliders. *Surf. Coat. Technol.* **2015**, *272*, 102–120. [CrossRef]

43. Cho, M.H. Effect of contact configuration on the tribological performance of micro-textured AISI 1045 steel under oscillating conditions. *Mater. Trans.* **2014**, *55*, 363–370. [CrossRef]

44. Wu, Z.; Deng, J.X.; Xing, Y.Q.; Cheng, H.W.; Zhao, J. Effect of surface texturing on friction properties of WC/Co cemented carbide. *Mater. Des.* **2012**, *41*, 142–149. [CrossRef]

45. Huang, W.; Wang, X.L. Biomimetic design of elastomer surface pattern for friction control under wet conditions. *Bioinspir. Biomim.* **2013**, *8*, 046001. [CrossRef] [PubMed]

46. Hu, T.C.; Hu, L.T. Tribological properties of lubricating films on the Al-Si alloy surface via laser surface texturing. *Tribol. Trans.* **2011**, *54*, 800–805. [CrossRef]

47. Ye, Y.W.; Wang, C.T.; Chen, H.; Wang, Y.X.; Zhao, W.J.; Mu, Y.T. Micro/Nanotexture design for improving tribological properties of Cr/GLC films in seawater. *Trobol. Trans.* **2017**, *60*, 95–105. [CrossRef]

48. Yu, H.W. Optimal Design of Surface Texture Based on Hydrodynamic Lubrication. Ph.D. Thesis, Nanjing University of Aeronautics and Astronautics, Nanjing, China, 2011.

49. Sun, Z.A. Theoretical and Experimental Study on Surface Texturing Piston Ring. Master's Thesis, Nanjing University of Aeronautics and Astronautics, Nanjing, China, 2011.

50. Liu, W. Research on Surface Texture of Piston Skirt by Model Test and FEM Simulation. Master's Thesis, Nanjing University of Aeronautics and Astronautics, Nanjing, China, 2009.

51. Yang, Z.W. The Manufacture of Array Micro-protrudes and Micro-Pits in Non-traditional Machining. Master's Thesis, Nanjing University of Aeronautics and Astronautics, Nanjing, China, 2007.

52. Zhang, Y.H. Improving the Tribological Performance of UHMWPE with Surface Texture. Master's Thesis, Nanjing University of Aeronautics and Astronautics, Nanjing, China, 2009.

53. Xu, Y. A Study on Tribological Properties of Ceramic Class Composite Materials in Water. Master's Thesis, Nanjing University of Aeronautics and Astronautics, Nanjing, China, 2010.

54. Yuan, S.H. Research on Friction Properties of Micro-Grooves Surface Textures. Master's Thesis, Nanjing University of Aeronautics and Astronaoutics, Nanjing, China, 2011.

55. Yan, D.S. Fundamental Research on Tribological Performance of Textured Piston Ring. Master's Thesis, Nanjing University of Aeronautics and Astronautics, Nanjing, China, 2009.

56. Mu, Q. Research on Tribological Properties of Bionic Hexagonal Pillar-Textured Surfaces. Master's Thesis, Nanjing University of Aeronautics and Astronautics, Nanjing, China, 2013.

57. Tang, W.; Zhou, Y.K.; Zhu, H.; Yang, H.F. The effect of surface texturing on reducing the friction and wear of steel under lubricated sliding contact. *Appl. Surf. Sci.* **2013**, *273*, 199–204. [CrossRef]

58. Li, J.L.; Xiong, D.S.; Wu, H.Y.; Zhang, Y.K.; Qin, Y.K. Tribological properties of laser surface texturing and molybdenizing duplex-treated stainless steel at elevated temperatures. *Surf. Coat. Technol.* **2012**, *228*, S219–S223. [CrossRef]

59. Reinert, L.; Lasserre, F.; Gachot, C.; Grützmacher, P.; MacLucas, T.; Souza, N.; Mücklich, F.; Suarez, S. Long-lasting solid lubrication by CNT-coated patterned surfaces. *Sci. Rep.* **2017**, *7*, 42873. [CrossRef] [PubMed]

60. Fan, H.Z.; Hu, T.C.; Zhang, Y.S.; Fang, Y.; Song, J.J.; Hu, L.T. Tribological properties of micro-textured surfaces of ZTA ceramic nanocomposites under the combined effect of test conditions and environments. *Tribol. Int.* **2014**, *78*, 134–141. [CrossRef]

61. Dong, Y.C.; Svoboda, P.; Vrbka, M.; David, K.; Urban, F.; Cizek, J.R.P.; Dong, H.S.; Krupka, I.; Hartl, M. Towards near-permanent CoCrMo prosthesis surface by combining micro-texturing and low temperature plasma carburising. *J. Mech. Behav. Biomed. Mater.* **2015**, *55*, 215–227. [CrossRef] [PubMed]

62. Dobrzański, L.A.; Drygała, A. Surface texturing of multicrystalline silicon solar cells. *J. Achiev. Mater. Manuf. Eng.* **2008**, *31*, 77–82.

63. Higuera Garrido, A.; González, R.; Cadenas, M.; Hernández Battez, A. Tribological behavior of laser-textured NiCrBSi coatings. *Wear* **2011**, *271*, 925–933. [CrossRef]

64. He, D.Q.; Zheng, S.X.; Pu, J.B.; Zhang, G.A.; Hu, L.T. Improving tribological properties of titanium alloys by combining laser surface texturing and diamond-like carbon film. *Tribol. Int.* **2015**, *82*, 20–27. [CrossRef]

65. Hu, T.C.; Zhang, Y.S.; Hu, L.T. Tribological investigation of MoS_2 coatings deposited on the laser textured surface. *Wear* **2012**, *278–279*, 77–82. [CrossRef]

66. Zhang, X.H.; Tan, J.; Zhang, Q.; Wang, M.; Meng, L.D. Effect of laser surface texturing depth on the adhesion of electroless plated nickel coating on alumina. *Surf. Coat. Technol.* **2017**, *311*, 151–156. [CrossRef]

67. Ripoll, M.R.; Simič, R.; Brenner, J.; Podgornik, B. Friction and lifetime of laser surface-textured and MoS_2-coated Ti6Al4V under dry reciprocating sliding. *Tribol. Lett.* **2013**, *51*, 261–271.

68. Rapoport, L.; Moshkovich, A.; Perfilyev, V.; Gedanken, A.; Koltypin, Y.; Sominski, E.; Halperin, G.; Etsion, I. Wear life and adhesion of solid lubricant films on laser-textured steel surfaces. *Wear* **2009**, *267*, 1203–1207. [CrossRef]

69. Li, J.L.; Xiong, D.S.; Zhang, Y.K.; Zhu, H.G.; Qin, Y.K.; Kong, J. Friction and wear properties of MoS$_2$-overcoated laser surface-textured silver-containing nickel-based alloy at elevated temperatures. *Tribol. Lett.* **2011**, *43*, 221–228. [CrossRef]

70. Wang, Y.M.; Guo, J.W.; Zhuang, J.P.; Jing, Y.B.; Shao, Z.K.; Jin, M.S.; Zhang, J.; Wei, D.Q.; Zhou, Y. Development and characterization of MAO bioactive ceramic coatinggrown on micro-patterned Ti6Al4V alloy surface. *Appl. Surf. Sci.* **2014**, *299*, 58–65. [CrossRef]

71. Wan, Y.; Xiong, D.S.; Wang, J. Tribological properties of dimpled surface alloying layer on carbon steel. *J. Wuhan Univ. Technol. Mater. Sci. Ed.* **2009**, *24*, 218–222. [CrossRef]

72. Lamraoui, A.; Costil, S.; Langlade, C.; Coddet, C. Laser surface texturing (LST) treatment before thermal spraying: A new process to improve the substrate-coating adherence. *Surf. Coat. Technol.* **2010**, *205*, S164–S167. [CrossRef]

73. Li, J.L.; Xiong, D.S.; Wu, H.Y.; Huang, Z.J.; Dai, J.H.; Tyagi, R. Tribological properties of laser surface texturing and molybdenizing duplex-treated Ni-base alloy. *Tribol. Trans.* **2010**, *53*, 195–202. [CrossRef]

74. Tripathi, K.; Gyawali, G.; Amanov, A.; Lee, S.W. Synergy effect of ultrasonic nanocrystalline surface modification and laser surface texturing on friction and wear behavior of graphite cast iron. *Tribol. Trans.* **2017**, *2*, 226–237. [CrossRef]

75. Shum, P.W.; Zhou, Z.F.; Li, K.Y. To increase the hydrophobicity; non-stickiness and wear resistance of DLC surface by surface texturing using a laser ablation process. *Tribol. Int.* **2014**, *78*, 1–6. [CrossRef]

76. Mello, D.J.D.B.; Gonçalves, J.J.L.; Costa, H.L. Influence of surface texturing and hard chromium coating on the wear of steels used in cold rolling mill rolls. *Wear* **2013**, *302*, 1295–1309. [CrossRef]

77. Guo, L.T.; Tian, J.L.; Wu, J.; Li, B.; Zhu, Y.B.; Xu, C.; Qiang, Y.H. Effect of surface texturing on the bonding strength of titanium-porcelain. *Mater. Lett.* **2014**, *131*, 321–323. [CrossRef]

78. Kumari, R.; Scharnweber, T.; Pfleging, W.; Besser, H.; Majumdar, J.D. Laser surface textured titanium alloy (Ti-6Al-4V)-Part II-Studies on bio-compatibility. *Appl. Surf. Sci.* **2015**, *357*, 750–758. [CrossRef]

79. Martínez-Calderon, M.; Manso-Silván, M.; Rodríguez, A.; Gómez-Aranzadi, M.; García-Ruiz, J.P.; Olaizola, S.M.; Martín-Palma, R.J. Surface micro- and nano-texturing of stainless steel by femtosecond laser for the control of cell migration. *Sci. Rep.* **2016**, *6*, 36296. [CrossRef] [PubMed]

80. Shen, Y.Z.; Tao, J.; Tao, H.J.; Chen, S.L.; Pan, L.; Wang, T. Nanostructures in superhydrophobic Ti6Al4V hierarchical surfaces control wetting state transitions. *Soft Matter* **2015**, *11*, 3806–3811. [CrossRef] [PubMed]

81. Shen, Y.Z.; Tao, H.J.; Chen, S.L.; Zhu, L.M.; Wang, T.; Tao, J. Icephobic/anti-icing potential of superhydrophobic Ti6Al4V surfaces with hierarchical textures. *RSC Adv.* **2015**, *5*, 1666–1672. [CrossRef]

82. Pfleging, W.; Kumari, R.; Besser, H.; Scharnweber, T.; Majumdar, J.D. Laser surface textured titanium alloy (Ti-6Al-4V)-Part I-Surface characterization. *Appl. Surf. Sci.* **2015**, *355*, 104–111. [CrossRef]

83. Zhang, X.L.; Jia, J.H. Frictional behavior of micro/nanotextured surfaces investigated by atomic force microscope: A review. *Surf. Rev. Lett.* **2015**, *22*, 1530001. [CrossRef]

84. Lin, N.M.; Liu, Q.; Zou, J.J.; Guo, J.W.; Li, D.L.; Yuan, S.; Ma, Y.; Wang, Z.X.; Wang, Z.H.; Tang, B. Surface texturing-plasma nitriding duplex treatment for improving tribological performance of AISI 316 stainless steel. *Materials* **2016**, *9*, 875. [CrossRef] [PubMed]

85. Han, L.; Wu, Y.X.; Gong, H.; Shi, W.Z. Effect of surface texturing on stresses during rapid changes in temperature. *Metals* **2016**, *6*, 290. [CrossRef]

86. Kolobov, Y.R.; Golosov, E.V.; Vershinina, T.N.; Zhidkov, M.V.; Ionin, A.A.; Kudryashov, S.I.; Makarov, S.V.; Seleznev, L.V.; Sinitsyn, D.V.; Ligachev, A.E. Structural transformation and residual stresses in surface layers of α + β titanium alloys nanotextured by femtosecond laser pulses. *Appl. Phys. A Mater. Sci. Process.* **2015**, *119*, 241–247. [CrossRef]

87. Zenebe, S.D.; Hwang, P. Friction control by multi-shape textured surface under pin-on-disc test. *Tribol. Int.* **2015**, *91*, 111–117. [CrossRef]

88. Mukherjee, S. Laser Surface Modification of Ti6Al4V Implants. In Proceedings of the 1st International Electronic Conference on Materials, 30 May 2014; pp. 1–6.

89. Soveja, A.; Cicală, E.; Grevey, D.; Jouvard, J.M. Optimisation of TA6V alloy surface laser texturing using an experimental design approach. *Opt. Lasers Eng.* **2008**, *46*, 671–678. [CrossRef]

90. Guo, Y.B.; Caslaru, R. Fabrication and characterization of micro dent arrays produced by laser shock peening on titanium Ti-6Al-4V surfaces. *J. Mater. Process. Technol.* **2011**, *211*, 729–736. [CrossRef]

91. Wang, M.L.; Zhang, C.T.; Wang, X.L. The wear behavior of textured steel sliding against polymers. *Materials* **2017**, *10*, 330. [CrossRef] [PubMed]

92. Men, B.; Wan, Y.; Zhang, R.; Zhang, D.; Liu, C. Fabrication of micro-featured array with laser and parameter optimization. *Tool Eng.* **2015**, *49*, 17–20.

93. Anand, P.; Bajpai, V.; Singh, R.K. Experimental Characterization of Fiber Laser Based Surface Texturing. In Proceedings of the 7th International Conference on Micro Manufacturing, Evanston, IL, USA, 17–19 March 2012; pp. 60–65.

94. Arslan, A.; Masjuki, H.H.; Kalam, M.A.; Varman, M.; Mufti, R.A.; Mosarof, M.H.; Khuong, L.S.; Quazi, M.M. Surface Texture Manufacturing Techniques and Tribological Effect of Surface Texturing on Cutting Tool Performance: A Review. *Crit. Rev. Solid State Mater. Sci.* **2016**, *41*, 447–481. [CrossRef]

95. Faure, L.; Bolle, B.; Philippon, S.; Schuman, C.; Chevrier, P.; Tidu, A. Friction experiments for titanium alloy tribopairs sliding in dry conditions: Sub-surface and surface analysis. *Tribol. Int.* **2012**, *54*, 17–25. [CrossRef]

96. Saranadhi, D.; Chen, D.Y.; Kleingartner, J.A.; Srinivasan, S.; Cohen, R.E.; McKinley, G.H. Sustained drag reduction in a turbulent flow using a low-temperature Leidenfrost surface. *Sci. Adv.* **2016**, *2*, e1600686. [CrossRef] [PubMed]

97. Mo, J.L.; Wang, Z.G.; Chen, G.X.; Shao, T.M.; Zhu, M.H.; Zhou, Z.R. The effect of groove-textured surface on friction and wear and friction-induced vibration and noise. *Wear* **2013**, *301*, 671–681. [CrossRef]

98. Obadelea, B.A.; Lepule, M.L.; Andrews, A.; Olubambi, P.A. Tribocorrosion characteristics of laser deposited Ti-Ni-ZrO$_2$ composite coatings on AISI 316 stainless steel. *Tribol. Int.* **2014**, *78*, 160–167. [CrossRef]

99. Obrosov, A.; Sutygina, A.N.; Volinsky, A.A.; Manakhov, A.; Weiß, S.; Kashkarov, E.B. Effect of hydrogen exposure on mechanical and tribological behavior of Cr$_x$N coatings deposited at different pressures on IN718. *Materials* **2017**, *10*, 563. [CrossRef] [PubMed]

100. Kumar, D.; Akhtar, S.N.; Anup, K.P.; Ramkumar, J.; Balani, K. Tribological performance of laser peened Ti-6Al-4V. *Wear* **2015**, *322–323*, 203–217. [CrossRef]

101. Caslaru, R. *Frbrication, Characterization, and Tribological Performance of Micro Dent Arrays Produced by Laser Shock Peening on Ti-6Al-4V Alloy*; The University of Alabama: Tuscaloosa, AL, USA, 2010.

102. Tripathi, K.; Joshi, B.; Gyawali, G.; Amanov, A.; Lee, S.W. A study on the effect of laser surface texturing on friction and wear behavior of graphite cast iron. *J. Tribol.* **2016**, *138*, 011601. [CrossRef]

103. Amanov, A.; Watabe, T.; Tsuboi, R.; Sasaki, S. Improvement in the tribological characteristics of Si-DLC coating by laser surface texturing under oil-lubricated point contacts at various temperatures. *Surf. Coat. Technol.* **2013**, *2632*, 549–560. [CrossRef]

104. Hu, T.C.; Hu, L.T.; Ding, Q. The effect of laser surface texturing on the tribological behavior of Ti-6Al-4V. *Proc. Inst. Mech. Eng. Part J J. Eng. Tribol.* **2012**, *226*, 854–863. [CrossRef]

105. Lian, F.; Zhang, H.C.; Pang, L.Y. Laser texture manufacturing on Ti6A14V surface and its dry tribological characteristics. *Lubr. Eng.* **2011**, *36*, 1–5.

106. Xu, P.F.; Zhou, F.; Wang, Q.Z.; Peng, Y.J.; Chen, J.N.; Yun, N.Z. Influence of meshwork pattern grooves on the tribological characteristics of Ti-6Al-4V alloy in water lubrication. *J. Tribol.* **2012**, *32*, 377–383.

107. Bonse, J.; Koter, R.; Hartelt, M.; Spaltmann, D.; Pentzien, S.; Höhm, S.; Rosenfeld, A.; Krüger, J. Femtosecond laser-induced periodic surface structures on steel and titanium alloy for tribological applications. *Appl. Phys. A Mater. Sci. Process.* **2014**, *117*, 103–110. [CrossRef]

108. Lian, F.; Zang, L.P.; Xiang, Q.K.; Zhang, H.C. Tribological performance of surper hydrophobic titanium alloy surface in artificial seawater. *Acta Metall. Sin.* **2016**, *52*, 592–598.

109. Lian, F.; Ren, H.M.; Guan, S.K.; Zhang, H.C. Preparation of super hydrophobic titanium alloy surface and its tribological performance. *Chin. J. Nonferrous Met.* **2015**, *25*, 2421–2427.

110. Hu, T.C.; Hu, L.T.; Ding, Q. Effective solution for the tribological problems of Ti-6Al-4V: Combination of laser surface texturing and solid lubricant film. *Surf. Coat. Technol.* **2012**, *206*, 5060–5066. [CrossRef]

111. Amanov, A.; Sasaki, S. A study on the tribological characteristics of duplex-treated Ti-6Al-4V alloy under oil-lubricated sliding conditions. *Tribol. Int.* **2013**, *64*, 155–163. [CrossRef]

112. Arslan, E.; Totik, Y.; Demirci, E.E.; Efeoglu, I. Wear and adhesion resistance of duplex coatings deposited on Ti6Al4V alloy using MAO and CFUBMS. *Surf. Coat. Technol.* **2013**, *214*, 1–7. [CrossRef]

113. Prem Ananth, M.; Ramesh, R. Influence of surface texture on tribological performance of AlCrN nanocomposite coated titanium alloy surfaces. *Proc. Inst. Mech. Eng. Part J J. Eng. Tribol.* **2013**, *227*, 1157–1164.

114. Prem Ananth, M.; Ramesh, R. Tribological improvement of titanium alloy surfaces through texturing and TiAlN coating. *Surf. Eng.* **2014**, *30*, 758–762. [CrossRef]

115. Qin, Y.K.; Xiong, D.S.; Li, J.L. Tribological properties of laser surface textured and plasma electrolytic oxidation duplex-treated Ti6Al4V alloy deposited with MoS_2 film. *Surf. Coat. Technol.* **2015**, *269*, 266–272. [CrossRef]

116. Qin, Y.K.; Xiong, D.S.; Li, J.L. Characterization and friction behavior of LST/PEO duplex-treated Ti6Al4V alloy with burnished MoS_2 film. *Appl. Surf. Sci.* **2015**, *347*, 475–484. [CrossRef]

117. Lin, N.M.; Liu, Q.; Zou, J.J.; Li, D.L.; Yuan, S.; Wang, Z.H.; Tang, B. Surface damage mitigation of Ti6Al4V alloy via thermal oxidation for oil and gas exploitation application: Characterization of the microstructure and evaluation of the surface performance. *RSC Adv.* **2017**, *7*, 13517–13535. [CrossRef]

118. Sun, Q.C.; Hu, T.C.; Fan, H.Z.; Zhang, Y.S.; Hu, L.T. Thermal oxidation behavior and tribological properties of textured TC4 surface: Influence of thermal oxidation temperature and time. *Tribol. Int.* **2016**, *94*, 479–489. [CrossRef]

119. Martinez, J.M.V.; Pedemonte, F.J.B.; Galvin, M.B.; Gomez, J.S.; Barcena, M.M. Sliding wear behavior of UNS R56400 titanium alloy samples thermally oxidized by laser. *Materials* **2017**, *10*, 830. [CrossRef] [PubMed]

120. Sreedhar, B.K.; Albert, S.K.; Pandit, A.B. Cavitation damage: Theory and measurements—A review. *Wear* **2017**, *372–373*, 177–196. [CrossRef]

121. Pola, A.; Montesano, L.; Tocci, M.; La Vecchia, G.M. Influence of ultrasound treatment on cavitation erosion resistance of AlSi7 alloy. *Materials* **2017**, *10*, 256. [CrossRef] [PubMed]

122. Zhang, L.; Lu, J.-Z.; Zhang, Y.-K.; Ma, H.-L.; Luo, K.-Y.; Dai, F.-Z. Effects of laser shock processing on morphologies and mechanical properties of ANSI 304 stainless steel weldments subjected to cavitation erosion. *Materials* **2017**, *10*, 292. [CrossRef] [PubMed]

123. Pang, L.Y.; Lian, F.; Gao, Y.Z.; Zhang, H.C. Topographical characteristics of cavitation erosion on Ti6A14V alloy with surface texture manufactured by laser. *J. Dalian Marit. Univ.* **2010**, *36*, 101–103.

124. Lian, F.; Zhang, H.C.; Gao, Y.Z.; Pang, L.Y. Influence of surface texture and surface film on cavitation erosion characteristics of Ti6A14V alloy. *Rare Met. Mater. Eng.* **2011**, *40*, 793–796.

125. Olivares-Navarrete, R.; Hyzy, S.L.; Berg, M.E.; Schneider, J.M.; Hotchkiss, K.; Schwartz, Z.; Boyano, B.D. Osteoblast lineage cells can discriminate microscale topographic features on titanium-aluminum-vanadium surfaces. *Ann. Biomed. Eng.* **2014**, *42*, 2551–2561. [CrossRef] [PubMed]

126. Meng, L.N.; Wang, A.H.; Wu, Y.; Wang, X.; Xia, H.B.; Wang, Y.N. Blind micro-hole array Ti6Al4V templates for carrying biomaterialsfabricated by fiber laser drilling. *J. Mater. Process. Technol.* **2015**, *222*, 335–343. [CrossRef]

127. Bhadra, C.M.; Truong, V.K.; Pham, V.T.H.; Kobaisi, M.A.; Seniutinas, G.; Wang, J.Y.; Juodkazis, S.; Crawford, R.J.; Ivanova, E.P. Antibacterial titanium nanopatterned arrays inspired by dragonfly wings. *Sci. Rep.* **2015**, *5*, 16817. [CrossRef] [PubMed]

128. Kurella, A.; Dahotre, N.B. Laser induced multi-scale textured zirconia coating on Ti-6Al-4V. *J. Mater. Sci. Mater. Med.* **2006**, *17*, 565–572. [CrossRef] [PubMed]

129. Zhang, Y.N.; Zhang, L.; Li, B.; Han, Y. Enhancement in sustained release of antimicrobial peptide from dual-diameter-structured TiO_2 nanotubes for long-lasting antibacterial activity and cytocompatibility. *ACS Appl. Mater. Interf.* **2017**, *9*, 9449–9461. [CrossRef] [PubMed]

130. Chen, J.; Mwenifumbo, S.; Langhammer, C.; McGovern, J.-P.; Li, M.; Beye, A.; Soboyejo, W.O. Cell/Surface interactions and adhesion on Ti-6Al-4V: Effects of surface texture. *J. Biomed. Mater. Res. Part B Appl. Biomater.* **2007**, *82*, 360–373. [CrossRef] [PubMed]

131. Chen, J.; Ulerich, J.P.; Abelev, E.; Fasasi, A.; Arnold, C.B.; Soboyejo, W.O. An investigation of the initial attachment and orientation of osteoblast-like cells on laser grooved Ti-6Al-4V surfaces. *Mater. Sci. Eng. C* **2009**, *29*, 1442–1452. [CrossRef]

132. Fasasi, A.Y.; Mwenifumbo, S.; Rahbar, N.; Chen, J.; Li, M.; Beye, A.C.; Arnold, C.B.; Soboyejo, W.O. Nano-second UV laser processed micro-grooves on Ti6Al4V for biomedical applications. *Mater. Sci. Eng. C-Bio S* **2009**, *29*, 5–13. [CrossRef]

133. Dahotre, N.B.; Paital, S.R.; Samant, A.N.; Daniel, C. Wetting behaviour of laser synthetic surface microtextures on Ti-6Al-4V for bioapplication. *Phil. Trans. R. Soc. A* **2010**, *368*, 1863–1889. [CrossRef] [PubMed]

134. Paital, S.R.; He, W.; Dahotre, N.B. Laser pulse dependent micro textured calcium phosphate coatings for improved wettability and cell compatibility. *J. Mater. Sci. Mater. Med.* **2010**, *21*, 2187–2200. [CrossRef] [PubMed]

135. Mukherjee, S.; Dhara, S.; Saha, P. Enhancing the biocompatibility of Ti6Al4V implants by laser surface microtexturing: An in vitro study. *Int. J. Adv. Manuf. Technol.* **2013**, *76*, 5–15. [CrossRef]

136. Mirhosseini, N.; Crouse, P.L.; Schmidth, M.J.J.; Li, L.; Garrod, D. Laser surface micro-texturing of Ti-6Al-4V substrates for improved cell integration. *Appl. Surf. Sci.* **2007**, *253*, 7738–7743. [CrossRef]

137. Kummer, K.M.; Taylor, E.; Webster, T.J. Biological applications of anodized TiO_2 nanostructures: A review from orthopedic to stent applications. *Nanosci. Nanotechnol. Lett.* **2012**, *4*, 483–493. [CrossRef]

138. Kummer, K.M.; Taylor, E.N.; Durmas, N.G.; Tarquinio, K.M.; Ercan, B.; Webster, T.J. Effects of different sterilization techniques and varying anodized TiO_2 nanotube dimensions on bacteria growth. *J. Biomed. Mater. Res. Part B Appl. Biomater.* **2013**, *101*, 677–688. [CrossRef] [PubMed]

139. Yao, C.; Webster, T.J. Anodization: A promising nano-modification technique of titanium implants for orthopedic applications. *J. Nanosci. Nanotechnol.* **2006**, *6*, 2682–2692. [CrossRef] [PubMed]

140. Balasundaram, G.; Yao, C.; Webster, T.J. TiO_2 nanotubes functionalized with regions of bone morphogenetic protein-2 increases osteoblast adhesion. *J. Biomed. Mater. Res. Part A* **2008**, *84*, 447–453. [CrossRef] [PubMed]

141. Zile, M.A.; Puckett, S.; Webster, T.J. Nanostructured titanium promotes keratinocyte density. *J. Biomed. Mater. Res. Part B Appl. Biomater.* **2011**, *97*, 59–65. [CrossRef] [PubMed]

142. Liu, L.; Bhatia, R.; Webster, T.J. Atomic layer deposition of nano-TiO_2 thin films with enhanced biocompatibility and antimicrobial activity for orthopedic implants. *Int. J. Nanomed.* **2017**, *12*, 8711–8723. [CrossRef] [PubMed]

143. Bhardwaj, G.; Webster, T.J. Reduced bacterial growth and increased osteoblast proliferation on titanium with a nanophase TiO_2 surface treatment. *Int. J. Nanomed.* **2017**, *12*, 363–369. [CrossRef] [PubMed]

144. Hosseini, S.; Jahangirian, H.; Webster, T.J.; Soltani, S.M.; Aroua, M.K. Synthesis, characterization, and performance evaluation of multilayered photoanodes by introducing mesoporous carbon and TiO_2 for humic acid adsorption. *Int. J. Nanomed.* **2016**, *11*, 3969–3978.

145. Hanson, A.D.; Wall, M.E.; Pourdeyhimi, B.; Loboa, E.G. Effects of oxygen plasma treatment on adipose-derived human mesenchymal stem cell adherence to poly(L-lactic acid) scaffolds. *J. Biomater. Sci. Polym. Ed.* **2007**, *18*, 1387–1400. [CrossRef] [PubMed]

146. Kulkarni, M.; Mazare, A.; Gongadze, E.; Perutková, Š.; Iglič, V.; Milosev, I.; Schmuki, P.; Iglič, A.; Mozetic, M. Titanium nanostructures for biomedical applications. *Nanotechnol.* **2015**, *26*, 062002. [CrossRef] [PubMed]

147. Shin, D.H.; Shokuhfar, T.; Choi, C.K.; Lee, S.H.; Friedrich, C. Wettability changes of TiO_2 nanotube surfaces. *Nanotechnology* **2011**, *22*, 315704. [CrossRef] [PubMed]

148. Junkar, I.; Kulkarni, M.; Drasler, B.; Rugelj, N.; Mazare, A.; Flasker, A.; Drobne, D.; Humpolicek, P.; Resnik, M.; Schmuki, P.; et al. Influence of various sterilization procedures on TiO_2 nanotubes used for biomedical devices. *Bioelectrochemistry* **2016**, *109*, 79–86. [CrossRef] [PubMed]

149. Kulkarni, M.; Mazare, A.; Park, J.; Gongadze, E.; Killian, M.S.; Kralj, S.; Mark, K.; Iglič, A.; Schmuki, P. Protein interactions with layers of TiO_2 nanotube and nanopore arrays: Morphology and surface charge influence. *Acta Biomater.* **2016**, *45*, 357–366. [CrossRef] [PubMed]

150. Kabaso, D.; Gongadze, E.; Perutková, Š.; Matschegewski, C.; Kralj-Iglič, V.; Beck, U.; van Rienen, U.; Iglič, A. Mechanics and electrostatics of the interactions between osteoblasts and titanium surface. *Comput. Methods Biomech. Biomed. Eng.* **2011**, *14*, 469–482. [CrossRef] [PubMed]

151. Hamlekhan, A.; Butt, A.; Patel, S.; Royhman, D.; Takoudis, C.; Sukotjo, C.; Yuan, J.; Jursich, G.; Mathew, M.T.; Hendrickson, W.; et al. Fabrication of anti-aging TiO_2 nanotubes on biomedical Ti alloys. *PLoS ONE* **2014**, *9*, e96213. [CrossRef] [PubMed]

152. Cunha, A.; Serro, A.P.; Oliveira, V.; Almeida, A.; Vilar, R.; Durrieu, M.C. Wetting behaviour of femtosecond laser textured Ti–6Al–4V surfaces. *Appl. Surf. Sci.* **2013**, *265*, 688–696. [CrossRef]

153. Shen, Y.Z.; Tao, J.; Tao, H.J.; Chen, S.L.; Pan, L.; Wang, T. Superhydrophobic Ti_6Al_4V surfaces with regular array patterns for anti-icing applications. *RSC Adv.* **2015**, *5*, 32813–32818. [CrossRef]

154. Shen, Y.Z.; Tao, H.J.; Chen, S.L.; Xie, Y.J.; Zhou, T.; Wang, T.; Tao, J. Water repellency of hierarchical superhydrophobic Ti6Al4V surfaces improved by secondary nanostructures. *Appl. Surf. Sci.* **2014**, *321*, 469–474. [CrossRef]

155. Lian, F.; Tan, J.Z.; Zhang, H.C. The impacts of the surface pattern on its wettability and antifouling performance. *Funct. Mater.* **2014**, *45*, 2105–2109.

156. Lian, F.; Tan, J.Z.; Zhang, H.C. Preparation of superhydrophobic titanium alloy surface and its antifouling of halobios. *Rare Met. Mater. Eng.* **2014**, *43*, 2267–2271.

157. Lian, F.; Zhang, H.C.; Pang, L.Y.; Li, J. Fabrication of superhydrophobic surfaces on Ti6Al4V alloy and its wettability. *Nanotechnol. Precis. Eng.* **2011**, *9*, 6–10.

158. Lian, F.; Zhang, H.C.; Pang, L.Y.; Zhu, H.B. Effects of surface film on superhydrophobic characteristics of Ti6Al4V with dotted matrix structure. *Rare Met. Mater. Eng.* **2012**, *41*, 612–616.

159. Lian, F.; Zhang, H.C.; Pang, L.Y. Fabrication of surface texture on Ti6Al4V alloy and its wettability. *Funct. Mater.* **2011**, *42*, 464–467.

160. Drelich, J.; Chibowski, E.; Meng, D.D.S.; Terpilowski, K. Hydrophilic and superhydrophilic surfa ces and materials. *Soft Matter* **2012**, *7*, 9804–9828. [CrossRef]

161. Att, W.; Hori, N.; Iwasa, F.; Yamada, M.; Ueno, T.; Ogawa, T. The effect of UV-photofunctionalization on the time-related bioactivity of titanium and chromium-cobalt alloys. *Biomaterials* **2009**, *30*, 4268–4276. [CrossRef] [PubMed]

MDPI

St. Alban-Anlage 66

4052 Basel

Switzerland

Tel. +41 61 683 77 34

Fax +41 61 302 89 18

www.mdpi.com

Materials Editorial Office

E-mail: materials@mdpi.com

www.mdpi.com/journal/materials

www.ingramcontent.com/pod-product-compliance
Lightning Source LLC
Chambersburg PA
CBHW051711210326
41597CB00032B/5438